INTRODUCTION TO MATHEMATICAL PROBABILITY THEORY

PRENTICE-HALL INTERNATIONAL, INC., *London*
PRENTICE-HALL OF AUSTRALIA, PTY. LTD., *Sydney*
PRENTICE-HALL OF CANADA, LTD., *Toronto*
PRENTICE-HALL OF INDIA PRIVATE LTD., *New Delhi*
PRENTICE-HALL OF JAPAN, INC., *Tokyo*

MARTIN EISEN
Department of Mathematics
Temple University

INTRODUCTION TO MATHEMATICAL PROBABILITY THEORY

PRENTICE-HALL, INC.
Englewood Cliffs, New Jersey

 Library of Congress Catalog Card Number 69—10324. Printed in the United States of America.

13-487488-9

Current Printing (last digit):

10 9 8 7 6 5 4 3 2

TO
CAROLE, RUTH, DEBORAH, AND NAOMI

PREFACE

This book was written as an introduction to elementary mathematical probability theory. In the first part of the book, probability theory in discrete spaces is used to introduce measure theory. Then measure theory and integration are studied and applied in general probability spaces. The last part of the book is devoted to laws of large numbers and classical limit theorems.

Some of the material covered in this book could be treated without measure theory by the judicious use of characteristic and distribution functions. However, this procedure gives a false simplification of these problems and conceals their intuitive basis. At present it is impossible to study topics in advanced probability theory such as stochastic processes without continually making use of measure theory.[1]

The following courses can be based on this book:

1. *A first course in probability theory.*

This is a one-semester undergraduate course with no prerequisites other than mathematical ability. The course can consist of:
(a) an introduction to permutations and combinations with a large number of probability problems based on these concepts.[2]

[1]J. L. Doob, *Stochastic Processes*, New York, Wiley, 1953; M. Loeve, *Probability Theory*, 3rd ed. Princeton: Van Nostrand, 1963; J. Neveu, *Mathematical Foundations of the Calculus of Probability*. San Francisco: Holden-Day, 1965.

[2]This material can be found in M. Eisen, *Elementary Combinatorial Analysis*, Chapters 1 and 5, to be published by Gordon and Breach; W. Feller, *An Introduction To Probability Theory and Its Applications*, Vol. 1, John Wiley and Sons (1960), Chapter 2; W. A. Whitworth, *Choice and Chance* 5th ed., London, 1901.

(b) the material on basic operations of set theory described below in the section on notation and set-theoretical preliminaries.

(c) Chapters 1–3 (1–5 if calculus is a prerequisite).

The student completing this course is prepared and motivated for a measure-theoretical approach to probability theory as in the remainder of the book. He has been introduced to such concepts as measure, field, measurable function, and product measure by simple concrete examples in discrete probability spaces.

2. *An introductory graduate course in probability.*

The prerequisites are (a), (b), and a course in advanced calculus. Chapters 1–13 can be covered in three semesters. The first semester's work would consist of the first five chapters and can be taken by qualified seniors. Chapters 6–10 could be covered in the second semester while the remainder would be covered in the third semester. The book can be covered in two semesters by omitting topics in Chapters 9–11.

3. *A one semester course in probability theory for students who have had a course in measure theory.*

The student can review Chapters 1–7 and use them for reference and motivation while the course would consist of Chapters 8–13.

References and credits for theorems have been assigned for historical interest and not for priority. Frequently, results were discovered independently by several people at approximately the same time and often special cases were known earlier by others.

The author is guilty of peccadillos common in mathematics. Sometimes a function is denoted by $f(x)$ instead of by f and many definitions are phrased using "if" instead of "iff" (if and only if).

METHOD OF ENUMERATION OF THEOREMS AND REFERENCES

The text is divided into chapters and sections. Theorems, equations, and definitions are numbered consecutively within each section. Theorem i. j. k stands for the kth theorem in Section j of Chapter i. Theorem j of Section i appearing in the same chapter in which the reference appears is denoted by Theorem i. j. Finally, Theorem i refers to the ith theorem of the section in which it appears. A similar convention is used for lemmas, definitions, and equations. Equation numbers are written in parenthesis and the word "equation" is often omitted. For example, (1.2.3) stands for equation (3) in Section 2 of Chapter 1.

References are given at the end of each chapter and are denoted by numbers enclosed in brackets, [ ].

ACKNOWLEDGEMENTS

This book was begun at the T. J. Watson Research Center. The author is grateful to the I.B.M. Corp.: in particular, to his former manager, Dr. B. J. Flehinger, for fostering an atmosphere conducive to mathematical scholarship. He also wishes to thank his former colleagues Doctors M. Leibowitz, J. Miller, and M. Tainiter for their encouragement and helpful suggestions.

Various portions of the manuscript were read by J. L. Doob, E. S. Elyash, J. E. Freund, H. Gindler, and J. L. Snell. Their advice and criticism were beneficial.

The author's greatest debt of gratitude is due his wife not only for her encouragement but for the active part that she played in shaping the book. She read the entire manuscript and found many grammatical and mathematical errors which she corrected by frequently rewriting whole sections.

MARTIN EISEN

CONTENTS

Chapter 7

INTEGRATION

Chapter 8

PROBABILITY AND MEASURE

Chapter 9

DISTRIBUTIONS AND MOMENTS

Chapter 10

CHARACTERISTIC FUNCTIONS

Chapter **11**

INDEPENDENCE

Chapter **12**

SERIES OF INDEPENDENT RANDOM VARIABLES

Chapter **13**

LIMIT THEOREMS FOR SUMS OF INDEPENDENT RANDOM VARIABLES

NOTATION AND SET-THEORETICAL PRELIMINARIES

Sets will be denoted by upper case Latin or Greek letters. The *members* (*points*, *elements*) of a set will be denoted by lower case Latin or Greek letters. The symbol $\in$ will indicate membership in a set; thus $x \in A$ means that x *is a member* (or *element*) of the set A. The negation of "$x \in A$" is written "$x \notin A$."

A *class* is a set whose members are all sets. Classes will be denoted by upper case script Latin letters.

If $P(x)$ is a proposition concerning x, the symbol $(x : P(x))$ (or the symbol $\{x|P(x)\}$) denotes the set of all x satisfying the proposition $P(x)$.

Finite sets are often displayed by enclosing their members in braces and separating them by commas (when there is more than one element). The symbol $\{x_1, x_2, \ldots, x_n\}$ is used for the set whose elements are $x_1, x_2, \ldots, x_n$. The three dots indicate that there are other elements in the set which have been omitted but are assumed known.

The *empty* (*void* or *null*) set $\emptyset$ is the set without members. It may be specified as $\{x : x \neq x\}$.

Two sets A and B are the *same* (or *equal*) if they have the same elements and we write $A = B$; otherwise $A \neq B$.

We write $A \subset B$ (read as "A is *contained* in B" or "A is a *subset* of B" or "B *contains* A") if each member of A is a member of B. In this situation we also write $B \supset A$. It is always true that $\emptyset \subset A$. If $A \subset B$ and $B \subset C$, then $A \subset C$. In addition, $A \subset B$ and $B \subset A$ implies that $A = B$.

In mathematics, all sets under discussion are usually subsets of a given fixed set denoted by X (or Ω) and called the *space* (or *universe*).

$A \cup B$, called the *union* of A and B, is the set $\{x : x \in A$ or $x \in B$ (or both)$\}$. For all A and B it is true that if $B \supset A$, then $B \cup A = B$; in particular, $\emptyset \cup A = A$ and $X \cup A = X$. Moreover, $A \subset A \cup C$ for all A and C.

AB (or $A \cap B$), called the *intersection* of A and B, is the set $\{x : x \in A$ and $x \in B\}$. For all A and B it is true that if $B \supset A$, then $B \cap A = A$; in particular, $\emptyset \cap A = \emptyset$ and $X \cap A = A$. In addition, $A \supset A \cap C$ for all A and C.

The *relative complement* of A in B, written $B\text{-}A$, is the set $\{x : x \in B, x \notin A\}$. In particular, if $B = X$, $X\text{-}A$ is denoted by A' and is called the *complement* of A. Note that $B\text{-}A = BA'$. If $A \subset B$ then $A' \supset B'$. $(A')' = A$ for any set A.

It is easily shown that $\cup$ and $\cap$ are commutative and associative. Moreover the distributive laws

$$A \cap (B \cup C) = (A \cap B) \cup (A \cap C),$$

$$A \cup (B \cap C) = (A \cup B) \cap (A \cup C),$$

are valid. The following results

$$(A \cup B)' = A' \cap B' \text{ and } (A \cap B)' = A' \cup B'$$

are called de Morgan's laws.

One often needs to form unions and intersections of families of sets in which a set A_i is assigned to each member i of some index set I. Then $\bigcup_{i \in I} A_i$ (or $\bigcup A_i$) is by definition the set $\{x : x \in A_i$, for some $i \in I\}$. Similarly one defines $\bigcap_{i \in I} A_i$ to be $\{x : x \in A_i$ for every $i \in I\}$. In particular, if $I = \{1, 2, 3, \ldots n\}$ then $\bigcup A_i$ is denoted by $\bigcup_{i=1}^{n} A_i$ or $A_1 \cup A_2 \cup \ldots \cup A_n$ and if $I = \{1, 2, 3, \ldots\}$ then $\bigcup A_i$ is denoted by $\bigcup_{i=1}^{\infty} A_i$. In other words, the same shorthand symbolism is used for unions as for summation. A similar notation is employed for intersections.

The de Morgan formulae become

$$(\bigcup A_i)' = \bigcap A'_i \qquad (\bigcap A_i)' = \bigcup A'_i.$$

The distributive laws also generalize — for example,

$$A \cap (\cup B_i) = \cup (A \cap B_i).$$

A and B are termed *disjoint* if $A \cap B = \emptyset$. A class of sets $\mathcal{C}$ is said to be *disjoint* if whenever A and B are members of $\mathcal{C}$ one has either $A = B$ or $A \cap B = \emptyset$. The symbol ΣA_i (or $A_1 + A_2 + \cdots + A_n$ for a finite number of sets) is used instead of $\cup A_i$ if $\mathcal{C} = \{A_i : i \in I\}$ is a disjoint class. Any countable union of sets $\bigcup_{i=1}^{\infty} A_i$ can be written as a disjoint sum of sets $\sum_{i=1}^{\infty} B_i$ (i.e. $\bigcup_{i=1}^{\infty} A_i = \sum_{i=1}^{\infty} B_i$) by setting $B_1 = A_1$ and $B_i = A_i A'_{i-1} A'_{i-2} \cdots A'_3 A'_2 A'_1$ for $i = 2, 3, 4, \ldots$.

The (*Cartesian*) *product* $A \times B$ of two sets A and B is defined to be $\{(x,y) : x \in A,\ y \in B\}$. The *product* of an arbitrary family (A_i) of sets, denoted by $\underset{i \in I}{\times} A_i$ (or $\times A_i$) is defined to be the set of all points $x = (x_i)$ for $i \in I$ such that $x_i \in A_i$ for each i.

A *partition* of a set A is a disjoint class $\mathcal{P}$ of subsets of A whose union is A. A partition $\mathcal{P}_1$ is *finer* than $\mathcal{P}_2$ if each element of $\mathcal{P}_1$ is contained in some element of $\mathcal{P}_2$ or equivalently if every set in $\mathcal{P}_2$ is the union of sets in $\mathcal{P}_1$.

Let $\{A_n\}$ be a sequence of subsets of X. The set of all points of X which belong to A_n for all but a finite number of values of n is called the *inferior limit* of the sequence and is denoted by lim inf A_n (or $\underline{\lim}\, A_n$). The *superior limit* of the sequence, denoted by lim sup A_n (or $\overline{\lim}\, A_n$), is the set of all points of X which belong to infinitely many of the subsets A_n. Note that

$$\liminf A_n = \bigcup_{m=1}^{\infty} \bigcap_{n=m}^{\infty} A_n \subset \bigcap_{m=1}^{\infty} \bigcup_{n=m}^{\infty} A_n = \limsup A_n.$$

A *function* f is a subset of the product $X \times Y$ of two sets having the property that $(x,y) \in f$ and $(x,y_1) \in f$ implies that $y = y_1$. The terms *mapping*, *transformation* and *correspondence* are each synonymous with function. The *domain* of the function f is $\{x : \text{for some } y, (x,y) \in f\}$ and the *range* of f is $\{y : \text{for some } x, (x,y) \in f\}$. By abuse of notation, whenever $(x,y) \in f$ we write $y = f(x)$.

The symbol $f : A \rightarrow B$ will mean that f is a function with domain A and range contained in B. If $C \subset A$, then $f(C) = \{y : y = f(x) \text{ for some } x \in C\}$ is termed the *image* of C under (or by) f. If $D \subset B$, then $f^{-1}(D) = \{x : f(x) \in D,\ x \in A\}$ is the *inverse image* of D under (or by) f. One has the following formulae:

$$f(C_1 \cup C_2) = f(C_1) \cup f(C_2),$$

$$f(C_1 \cap C_2) \subset f(C_1) \cap f(C_2),$$

$$f^{-1}(D_1 \cup D_2) = f^{-1}(D_1) \cup f^{-1}(D_2),$$

$$f^{-1}(D_1 \cap D_2) = f^{-1}(D_1) \cap f^{-1}(D_2),$$

$$f^{-1}(D_1 - D_2) = f^{-1}(D_1) - f^{-1}(D_2),$$

C_i and D_i being arbitrary sets in A and B respectively for $i = 1, 2$.

If $f : A \to B$ and for each $b \in f(A)$ there is only one $a \in A$ with $f(a) = b$, f is said to be *one-to-one* or to have an *inverse*. If $f(a) = b$ the inverse function, denoted by f^{-1}, is defined by the equation $f^{-1}(b) = a$. The inverse function has domain $f(A)$ and range A. The function is said to map A *onto* B if $f(A) = B$ and *into* B if $f(A) \subset B$.

If $f : A \to B$ and $g : B \to C$ then the mapping $gf : A \to C$, called the *composition* (or *product*) of f and g, is defined by the equation $gf(a) = g(f(a))$ for $a \in A$.

The function g is said to be an *extension* of the function f and f a *restriction* of g if the domain of g contains the domain of f and $f(x) = g(x)$ for all x in the domain of f.

The *indicator* (or *characteristic*) function I_A of a set A is defined by $I_A(x) = 1$ for $x \in A$ and $I_A(x) = 0$ for $x \notin A$.

By the *extended real number system* we mean the real numbers with the symbols ∞ and $-\infty$ adjoined. These symbols and any real number x satisfy the following relations:

$$(\pm\infty) + (\pm\infty) = x + (\pm\infty) = (\pm\infty) + x = \pm\infty.$$

$$(\pm\infty)(\pm\infty) = +\infty.$$

$$(\pm\infty)(\mp\infty) = -\infty.$$

$$x/(\pm\infty) = 0.$$

$$x(\pm\infty) = (\pm\infty)x = \begin{cases} \pm\infty & (\text{if } x > 0) \\ 0 & (\text{if } x = 0) \\ \mp\infty & (\text{if } x < 0) \end{cases}$$

The *extended complex number system* is the set of complex numbers with the single symbol ∞ adjoined.

If A is a set of real numbers, then the *supremum* of A, denoted by sup A (or the *least upper bound* of A, denoted by lub A) is the smallest real number s, if it exists, such that $s \geq a$ for all $a \in A$. Otherwise, it is ∞. The inf A, the *infimum* of A (or glb A, the *greatest lower bound* of A) is the largest real number i, if it exists, such that $i \leq a$ for all $a \in A$. Otherwise it is $-\infty$.

If A is an infinite set of real numbers, then the symbol lim sup A (or $\overline{\lim}\ A$) denotes the infimum of all numbers s with the property that only a finite set of members of A exceed s; the definition of the symbol lim inf A (or $\underline{\lim}\ A$) is similar. In particular, if A is a sequence $\{x_n\}$ then $\limsup A = \lim_{m\to\infty} \sup_{n\geq m} x_n$ which is denoted by $\limsup x_n$ (or $\overline{\lim}\ x_n$) and $\liminf A = \lim_{m\to\infty} \inf_{n\geq m} x_n$ which is denoted by $\liminf x_n$ (or $\underline{\lim}\ x_n$).

R, R^2, R^n stand for the line, the plane, and the n-dimensional Euclidean space. A point $(x_1, x_2, \ldots, x_n) \in R^n$ will be denoted by x. If $a \in R$ and $b \in R$ then the symbol (a, b) denotes the *open interval* defined by $\{x : a < x < b\}$; the symbol $[a, b]$ denotes the *closed interval* defined by $\{x : a \leq x \leq b\}$ and the *semi-open* intervals $(a, b]$ and $[a, b)$ are given by $\{x : a < x \leq b\}$ and $\{x : a \leq x < b\}$ respectively. Similar definitions hold in R^n — for example,

$$(a, b] = \{x : a_1 < x_1 \leq b_1, a_2 < x_2 \leq b_2, \ldots, a_n < x_n \leq b_n\}$$

If $z = x + iy$ is a complex number (x and y are real) then x and y are called the *real part* and the *imaginary part* of z and are denoted by $\Re(z)$ and $\Im(z)$ respectively.

$f(x) = O(g(x))$ denotes that $|f(x)| < A\, g(x)$ if x is sufficiently near some given limit. The symbols $f(x) = o(g(x))$ and $f(x) \sim g(x)$ denote respectively that $f(x)/g(x) \to 0$ and $f(x)/g(x) \to 1$ as x tends to a given limit.

INTRODUCTION TO MATHEMATICAL PROBABILITY THEORY

CHAPTER 1

FINITE PROBABILITY SPACES

1.1 INTRODUCTION

A mathematical theory usually evolves from some concrete physical problem. The first step is the translation of the physical phenomena into mathematical terms. The crude mathematical models are gradually refined until a formal axiomatic description is obtained. In this abstract formulation there is a complete listing of the primitive terms together with a set of axioms, which are the only facts that can be used in the derivation of theorems. Of course, theorems that have already been derived can also be used. The primitive terms remain undefined, and great care is taken so that the intended meanings of these terms (i.e., the objects they might represent in the physical world) do not enter into a proof, unless explicitly expressed by the axioms. This last fact results in great economy of thought, since the same model may represent many different physical situations; for instance, we shall see that a probability space is just a measure space. In addition, the axiomatic method enables us to study subjects that have no physical realizations—for example, certain non-Euclidean geometries.

At first glance, it may seem strange to the reader that modern mathematics is only concerned with relations between undefined things. However, without realizing it, the reader has often indulged in this pastime. In elementary arithmetic, rules are given for operating with integers without a philosophical discussion on the nature of an integer. The game of checkers is defined by giving a set of rules for moving the pieces. The checkers and the board are unimportant; checkers can even be played mentally.

We will show how this process is carried out for probability theory. With the help of set theory, we will free ourselves from physical experimentation and the consideration of such particular problems as tossing coins, rolling dice, or placing balls in cells, and will obtain a set of axioms for a *finite* probability space. General probability spaces will be treated in Chapter 8, after the necessary mathematical tools have been mastered.

1.2 HISTORICAL BACKGROUND

Man has probably been speculating on games of chance since antiquity. The earliest indication that there are different probabilities for the various throws that can be made with three dice appears in 1477 in a commentary on Dante's *Divine Comedy. Liber de Ludo Aleae*, a treatise on gambling by Geronimo Cardan (1501–1576)[1] that was published posthumously in 1663, gives the number of favorable cases for each point when two dice are used. It also contains some obscure results for a certain game played with three dice.

The next investigation appears to be the one made in the early seventeenth century by Galileo (1564–1642). A friend was puzzled by observing that when three dice are thrown the number 10 appears more often than the number 9; he reasoned that there should be six different ways in which the three dice could add up to 9 and six ways in which they could total 10. Upon analyzing the problem carefully, Galileo showed that out of the 216 possible cases, 27 are favorable to the number 10 appearing and 25 to the number 9 appearing.

These early contributions are considered to be negligible; it is generally acknowledged that the founder of probability theory was Blaise Pascal (1623–1662). It appears that the Chevalier de Méré, a gambler, proposed certain questions about games of chance to Pascal. In a series of letters written around 1654, Pascal and Pierre de Fermat (1601–1665) discussed these problems. Their interest centered on variants of the following question, known as the Problem of Points: If two players of equal skill each need a certain number of points to win a game, how should the stake be divided between them if they stop playing before the game is completed?

Pascal solved the problem in an algebraic manner, while Fermat used the method of combinations. The problem is equivalent to finding the probability of winning that each player has at each stage of the game. An example illustrating Fermat's method is given below, as it is the basis of the elementary definition of probability. Suppose that there are two players of

[1]See *Cardano: The Gambling Scholar*, by Oystein Ore (Princeton, N. J.: Princeton University Press, 1953).

equal skill, say A and B, who respectively need two and three points to win. Let the letter a denote the fact that A has won and b the fact that B has won. There are at most four possible games; the 16 possible outcomes can be written as follows:

$$\begin{array}{cccc} (a, a, a, a) & (a, b, a, a) & (b, a, a, a) & (b, b, a, a) \\ (a, a, a, b) & (a, b, a, b) & (b, a, a, b) & (b, b, a, b) \\ (a, a, b, a) & (a, b, b, a) & (b, a, b, a) & (b, b, b, a) \\ (a, a, b, b) & (a, b, b, b) & (b, a, b, b) & (b, b, b, b) \end{array}$$

where, for example, (b, a, b, b) denotes the fact that B won games number 1, 3, and 4 while A won game 2. In this case B has three points and so is the winner. Assuming that these cases are all equally likely, there are 11 cases favorable to A and 5 favorable to B; therefore, A's chances of winning the game to B's chances are as 11 is to 5.

The use of the ratio of the number of favorable cases to the total number of cases as a definition of probability was first clearly stated by Pierre Simon de Laplace (1749–1827), although it had been used earlier in an intuitive manner. In fact, the following statement is found in *De Ratiociniis in Ludo Aleae* (*c.* 1657) by the Dutch physicist Christian Huygens (1629–1695), which summarized known results and proposed new problems: If a player has p chances of getting a and q chances of getting b, his expectation is $(pa + qb)/(p + q)$.

Fermat probably justified his results intuitively and connected them with reality in the following manner. Suppose that the above game was really played a large number of times, say N. Since each of the outcomes is equally likely, we would expect to obtain $N/16$ occurrences of any one of them. However, we are concerned only with the ratio of A's chances to B's chances; therefore, it is sufficient to merely list the 16 different outcomes. Another point in favor of the aforementioned convention is that it agrees with observation. For example, if we repeat the experiment of tossing a well-balanced coin a large number of times, say N, we will observe about $N/2$ heads in the majority of cases, but not always. It is possible to always get tails. It remained for Jacques Bernoulli (1654–1705) to formulate this fact mathematically, thus putting probability theory on a firm footing. We will discuss the Bernoulli law of large numbers in Section 3.6. This famous theorem, along with new results, and the solution of some problems posed by Huygens, are contained in Bernoulli's classic work on probability theory, *Ars Conjectandi*, which was printed posthumously in 1713.[2]

[2]The contributions of Thomas Bayes (1702–1761), G. L. Buffon (1707–1788), Abraham De Moivre (1667–1754), Siméon Poisson (1781–1840), K. F. Gauss (1777–1855), and others are described in *History of the Mathematical Theory of Probability*, by I. Todhunter (New York: Chelsea Publishing Company, 1949).

EXERCISES

1. List the possible outcomes that are obtained when two and three dice are thrown. Verify Galileo's statement.

2. An objection to Fermat's definition is that symbols such as (a, a, a, a) are meaningless, since the game really ends after two steps—i.e., (a, a). How can this criticism be answered?

3. Write down the outcomes of throwing a coin once, twice. Use your results to calculate the ratio of the chances of getting heads to the chances of getting tails in one throw of a coin. Why are these results the same?

4. Write down the outcomes if two players of equal skill play two games. Do the same for three and four games. What are the probabilities of A winning in exactly two games, exacly three games, and exactly four games? Add the results and compare the answer with that obtained by Fermat's method.

5. Throw a coin 1000 times and record the number of heads. (Or get ten people each to throw a coin 100 times.)

6. An urn contains two white balls and one black one. A ball is drawn and replaced, then another ball is chosen. By making a complete list of all outcomes, find the probability of getting two white balls, two black ones and one black and one white one. Answer the same question if the first ball is not replaced.

1.3 ABSTRACT MODEL

The reader will have undoubtedly noticed that the outcomes of an experiment can be considered as a set. The set of outcomes is called a *probability* or *sample space.* For example, if we toss three coins, denoting heads and tails by h and t respectively, the space Ω_1 of outcomes is

$$\Omega_1 = \{(h, h, h), (h, h, t), (h, t, h), (h, t, t), (t, h, h), (t, h, t), (t, t, h), (t, t, t)\}.$$

In probability theory, the outcomes of an experiment or the results of an observation are usually called *events.* The event A that at least two heads will occur can be represented by

$$A = \{(h, h, h), (h, h, t), (h, t, h), (t, h, h)\},$$

while the event B that exactly three heads will occur can be represented by

$$B = \{(h, h, h)\}.$$

Thus events correspond to subsets of the space Ω_1. To further illustrate this point, let us write down the possible outcomes that arise when two dice are thrown. This space, Ω_2, can be represented by ordered pairs of integers that represent the spots appearing on the dice. Let E_i be the event that the points on both dice total i. Then

$$\Omega_2 = E_2 \cup E_3 \cup \cdots \cup E_{12},$$

where

$$E_2 = \{(1, 1)\}$$
$$E_3 = \{(1, 2), (2, 1)\}$$
$$E_4 = \{(2, 2), (3, 1), (1, 3)\}$$
$$E_5 = \{(2, 3), (3, 2), (4, 1), (1, 4)\}$$
$$E_6 = \{(3, 3), (4, 2), (2, 4), (5, 1), (1, 5)\}$$
$$E_7 = \{(4, 3), (3, 4), (5, 2), (2, 5), (6, 1), (1, 6)\}$$
$$E_8 = \{(4, 4), (5, 3), (3, 5), (6, 2), (2, 6)\}$$
$$E_9 = \{(3, 6), (6, 3), (5, 4), (4, 5)\}$$
$$E_{10} = \{(5, 5), (6, 4), (4, 6)\}$$
$$E_{11} = \{(6, 5), (5, 6)\}$$
$$E_{12} = \{(6, 6)\}.$$

The event that a doublet is obtained is the subset

$$C = \{(1, 1), (2, 2), (3, 3), (4, 4), (5, 5), (6, 6)\},$$

while the event D that we obtain an even number is

$$D = E_2 \cup E_4 \cup E_6 \cup E_8 \cup E_{10} \cup E_{12}.$$

The reader who is familiar with set theory should have no difficulty in translating D into words; it merely states that in order to get an even number we must get a 2 or 4 or$\cdots$or 12. The events B, E_2, and E_{12} are called *indecomposable* or *simple events*, since they are sets which contain a single point of their spaces. The remaining sets given above are called *decomposable* or *compound events*, since they can be decomposed into simple events.

The elementary definition of probability—that is, the ratio of the number of favorable cases to the total number of cases—can be rephrased to define probability in our abstract model as follows. Give each point of our space Ω a weight of unity; i.e., count it once. Then the probability of an event A is

$$P(A) = \frac{W(A)}{W(\Omega)} = \frac{\text{sum of weights of points in } A}{\text{sum of weights of points in } \Omega}.$$

One reason for using the term "weight" is to emphasize that $P(A)$ is a ratio and that therefore it will have the same value even if we assign a weight of w to each point. If we assign the same weight to each point, we say the *outcomes* are *equally likely*. The reader can easily verify that $P(A) = \frac{1}{2}$, $P(B) = \frac{1}{8}$, $P(E_2) = \frac{1}{36}$, $P(C) = \frac{1}{6}$, and $P(D) = \frac{1}{2}$. Note that for any set A of a space Ω we have $0 \leq P(A) \leq 1$; also $P(\Omega) = 1$. The empty set ϕ contains no points, and so $P(\phi) = 0$.

The whole space Ω is called the *sure event* and ϕ is called the *impossible event*; we sometimes say that Ω occurs with certainty and ϕ is impossible. We also call disjoint events *mutually exclusive*.

Since events are, by definition, subsets of our space Ω, it follows that the union and intersection of a finite number of events and the complement of an event also are events. For our model to mirror reality the above result must hold, since $A \cup B$, $A \cap B$, and A' correspond to A or B occur, A and B occur, and A does not occur, respectively. Exercise 8 is designed to convince the reader that all statements about events can be written in terms of sets, using only $\cup$, $\cap$, and $'$.

Another reason for defining probability in terms of weights is that we can assign different weights (positive numbers) to each point to allow for the fact that some events are more likely to occur than others. The weight of a set is just the sum of the weights associated with each point of the set. Thus for a coin with two heads we assign the weight w to heads and the weight 0 to tails. As another example, consider a cube with three red faces, two white faces, and one blue face. Here the possible outcomes are red, white, and blue, to which we attach the weights 3, 2, and 1, respectively.

From our definition of probability and the fact that $W(E + F + G) = W(E) + W(F) + W(G)$,[3] we can prove the following theorem for any probability space Ω.

THEOREM 1

If A and B are any two events, the probability that either A or B or both occur is given by

$$P(A \cup B) = P(A) + P(B) - P(AB).$$

Proof

We need only show that both sides have equal weight. Noting that $B = AB + (B - A)$, $A = AB + (A - B)$, and $A \cup B = AB + (B - A) + (A - B)$, the result follows upon taking weights of the three expressions and eliminating $W(A - B)$ and $W(B - A)$.

COROLLARY

If A and B are disjoint events, then $P(A + B) = P(A) + P(B)$.

[3]The notation $E + F$ is used for $E \cup F$ if and only if $EF = \phi$. More generally, $\sum E_i$ denotes $\cup E_i$ if and only if $E_i E_j = \phi$ for $i \neq j$.

EXERCISES

1. Show that $W(E+F+G) = W(E) + W(F) + W(G)$.

2. If the probability of getting heads is $\frac{3}{4}$, what are the weights attached to heads and tails?

3. What is the probability of getting either a 3 or an odd number with a pair of dice?

4. Show that the sample space for a coin that is tossed n times can be written $\overset{n}{\underset{i=1}{\times}} A_i$, where $A_i = \{H, T\}$. If all the points are given equal weights, what is the probability of getting k heads?

5. Write down all the three-digit numbers that can be formed from the digits 1, 2, and 3, allowing repetitions [e.g., (3, 3, 3)]. Show that this is equivalent to all the different ways of placing three different balls into three cells, when any number of balls can be placed in a cell. What is the probability of any indecomposable event, if all arrangements are considered equally likely?

6. Show that all the solutions of the equation $x_1 + x_2 + x_3 = 3$, where $x_i = 0, 1, 2, 3$, correspond to the number of different ways of placing three similar balls in three different cells, if any number of balls can be placed in a cell. What is the probability of any indecomposable event, if each point is given the same weight?

7. Calculate the required probabilities in Exercises 5 and 6 for k balls and n cells.

8. Let E_1, E_2, and E_3 be any events of an arbitrary sample space Ω. Write the following equations symbolically in terms of $'$, $\cup$, and $\cap$.

(a) One and only one occurs.
(b) Only E_3 occurs.
(c) Two or more occur.
(d) E_1, E_2, and not E_3 occur.
(e) None occur.
(f) Not more than two occur.
(g) All three occur.
(h) At least one occurs.
(i) At least two occur.

1.4 AXIOMS

The essential features of our abstract model must be captured in our axiom system. In particular, we have seen that we can think of all possible outcomes as points of an abstract space Ω and that events correspond to a class $\mathscr{E}$ of subsets of Ω. A finite union of events, a finite intersection of events, and the complement of an event are all events. By means of weights, we have assigned a probability to each event A such that $P(A) \geq 0$, $P(\phi) = 0$, $P(\Omega) = 1$, and $P(A_1 + \cdots + A_n) = P(A_1) + \cdots + P(A_n)$. (See the corollary to Theorem 3.1.)

To proceed formally, we first list all the primitive terms. Let Ω or the *sure event* be a space of points; the *impossible event* or empty set will be denoted by ϕ. Let $\mathscr{E}$ be a nonempty class of subsets of Ω, which will be called *events*. Let P or the *probability* be a real-valued function defined on $\mathscr{E}$; the value of P for an event E will be called the *probability* of E and will be denoted by $P(E)$.

DEFINITION 1

The pair $(\mathscr{E}, P)$ is called a *probability field* and the triplet $(\Omega, \mathscr{E}, P)$ is called a *probability space*.

EXAMPLE 1

To see why $\Omega, \mathscr{E}$, and P must all be specified, consider the following probability spaces.

1. A fair coin is tossed once. Here $\Omega = \{h, t\}$. The class of events $\mathscr{E} = \{\phi, \Omega, \{h\}, \{t\}\}$ corresponds to the following statements about the outcomes: "heads and tails," "heads or tails," "heads," and "tails." The probability function P is defined by $P\{h\} = P\{t\} = \frac{1}{2}$, $P\phi = 0$, and $P\Omega = 1$.

2. An unfair coin is tossed once. Here Ω and $\mathscr{E}$ are defined as in number 1, while $P\{h\} = \frac{3}{4}$, $P\{t\} = \frac{1}{4}$, $P\phi = 0$, and $P\Omega = 1$.

3. Now suppose that a two-headed coin is tossed; then $\Omega = \{h\}$, $\mathscr{E} = \{\Omega, \phi\}$, $P\Omega = 1$, and $P\phi = 0$.

4. Finally, suppose that a fair die is rolled; then $\Omega = \{1, 2, 3, 4, 5, 6\}$. $\mathscr{E}$ is the class of all subsets of Ω, while $PA = \frac{1}{6}$ (the number of points in A) for any $A \subset \Omega$.

It is easy to see that the following axioms have the properties listed in the first paragraph of this section.

AXIOMS FOR A FINITE PROBABILITY SPACE

1. If $E_i \in \mathscr{E}$ for $i = 1, 2, \ldots, n$, then $\bigcup_{i=1}^{n} E_i \in \mathscr{E}$ and $\bigcap_{i=1}^{n} E_i \in \mathscr{E}$.

2. If $E \in \mathscr{E}$, then $E' \in \mathscr{E}$.

3. If $E \in \mathscr{E}$, then $P(E) \geq 0$; also $P(\Omega) = 1$.

4. If E and F are any two arbitrary disjoint events, then $P(E + F) = P(E) + P(F)$.

An axiomatic treatment of probability theory was first proposed by G. Bohlmann at the International Congress of Mathematics in 1908. Further historical data appear in Section 5.2.

All of the rules of the algebra of sets and logic can be used to deduce results from these axioms. Some authors also include axioms for set algebra[4] along with Axioms 1–4 above. If we wanted to be even more formal we could include a set of axioms that would describe the rules of logic.

To show that the axioms are *consistent*, we construct a model for which they are satisfied, with $\mathscr{E} = \{\Omega, \phi\}$, $P(\Omega) = 1$, $P(\phi) = 0$.

A probability function P must only satisfy Axioms 3 and 4. Therefore, for a given event E it is possible to assign different values to $P(E)$. For example, in tossing a coin we can set $P(\{h\}) = P(\{t\}) = \frac{1}{2}$ or $P(\{h\}) = \frac{3}{4}$ and $P(\{t\}) = \frac{1}{4}$. This means that the system of axioms is incomplete;[5] however, this is beneficial in probability. Frequently we must consider the same class of random events, but with different probabilities. In the example above, the first assignment would be used if we were tossing a fair coin; the second, a weighted coin.

A class of sets with the properties described in Axioms 1 and 2 has acquired a special name.

DEFINITION 2

A class of sets $\mathscr{E}$ is called a *field*[6] (or *algebra*) if and only if the following conditions hold true:

1. If $E_i \in \mathscr{E}$, then $\bigcup_{i=1}^{n} E_i \in \mathscr{E}$.
2. If $E \in \mathscr{E}$, then $E' \in \mathscr{E}$.

From this it follows that $E_i \in \mathscr{E}$ implies $\bigcap_{i=1}^{n} E_i \in \mathscr{E}$.

[4]Actually we need only add that the events form a Boolean algebra—namely,

1. $A \cup B = B \cup A \qquad A \cap B = B \cap A.$
2. $(A \cup B) \cup C = A \cup (B \cup C) \qquad A \cap (B \cap C) = (A \cap B) \cap C.$
3. $(A \cap B) \cup B = B \qquad (A \cup B) \cap B = B.$
4. $B \cap (A \cup C) = (B \cap A) \cup (B \cap C) \qquad B \cup (A \cap C) = (B \cup A) \cap (B \cup C).$
5. $(A \cup A') \cap B = B \qquad (A \cap A') \cup B = B.$

[5]The student interested in exploring the concepts of completeness and consistency can consult books on logic. An elementary discussion can be found on page 134 of *Introduction to Logic and to the Methodology of Deductive Sciences*, by A. Tarski (New York: Oxford University Press, Inc., 1951).

[6]This concept is not the same as the concept of "field" in algebra. However, if we define $A + B = A \Delta B$ and $A \times B = AB$, then $\mathscr{E}, \times, +$ is a commutative ring with an identity.

EXAMPLE 2

1. $\mathscr{E} = \{\Omega, \phi\}$ is a field.
2. $\mathscr{E} = \{A, A', \Omega, \phi\}$ is a field.
3. $\mathscr{E} = \{A, \Omega, \phi\}$ is not a field, since $A' \notin \mathscr{E}$.
4. $\mathscr{E} = \{A, B, A', B', A' \cup B, A \cup B', A' \cup B', A \cup B, A'B, AB', A'B', AB, \Omega, \phi\}$ is a field.
5. The class of all subsets of a given set Ω is a field.

Once the axioms are accepted we must forget about their physical interpretation in proving theorems. We cannot say that since the empty set ϕ has no elements, then obviously $P\phi = 0$. This fact must be rigorously proved from the axioms by using logical reasoning. At present, we do not even know that ϕ is an event, so $P\phi$ may be meaningless. As an analogy, consider the game of checkers. The object of this game is to win all your opponent's checkers. The obvious way of winning is simply to seize all his pieces. However, you cannot do this; the rules must be obeyed.

We illustrate the proper method of reasoning in the following examples.

EXAMPLE 3

1. Show that if $A \in \mathscr{E}$ and $B \in \mathscr{E}$, then $AB' \in \mathscr{E}$.

Since $B \in \mathscr{E}$, $B' \in \mathscr{E}$ using Axiom 2. Using Axiom 1, it follows that $AB' \in \mathscr{E}$.

2. If $A \in \mathscr{E}$, $B \in \mathscr{E}$, and $B \supset A$, then $PB \geq PA$.

The identity $B = (B - A) + A$ allows us to apply Axiom 4, since the events $B - A$ and A are disjoint. Hence,

$$PB = P(B - A) + PA. \tag{1}$$

Since $B - A = BA' \in \mathscr{E}$, then from Axiom 3, $P(BA') \geq 0$. Therefore, $PB \geq PA$.

3. For any event A, $0 \leq PA \leq 1$.

By Axiom 3, $PA \geq 0$. Since $\Omega \supset A$ and $P\Omega = 1$, then from number 2 of this example, $1 \geq PA$.

4. If B and A are events and $B \supset A$, then $P(B - A) = PB - PA$.

Since $PA < \infty$, PA may be subtracted from both sides of equation (1). This yields the desired result.

5. $P\phi = 0$.

Since $\mathscr{E}$ is not empty, then $A \in \mathscr{E}$ exists. Hence $A' \in \mathscr{E}$, from Axiom 2.

Therefore, $AA' = \phi \in \mathscr{E}$ from Axiom 1. Hence $P\phi$ is defined. Axiom 4 and the fact that $\phi + \phi = \phi$ imply that $P\phi + P\phi = P\phi$. Since $P\phi < \infty$, it follows that $P\phi = 0$.

Once we have accepted the axioms for a probability space we can construct probability spaces. Such spaces need not have any natural physical interpretation. The only prerequisite is that the axioms hold.

EXAMPLE 4

1. Let $\Omega = \{a, b, c, d\}$ and $\mathscr{E} = \{\{a\}, \{b, c, d\}, \Omega, \phi\}$, and let $P\{a\} = \frac{1}{1000}$, $P\{b, c, d\} = \frac{999}{1000}$, $P\Omega = 1$, and $P\phi = 0$. The triplet $(\Omega, \mathscr{E}, P)$ is a probability space, as can easily be verified by checking the axioms.
2. Let Ω and $\mathscr{E}$ be defined as in number 1 of this example and let $P\{a\} = \frac{999}{1000}$, $P\{b, c, d\} = \frac{1}{1000}$, $P\Omega = 1$, and $P\phi = 0$. Again, $(\Omega, \mathscr{E}, P)$ is a probability space.[7]
3. Retaining the same Ω, let $\mathscr{E} = \{\{a\}, \{b, c\}, \{d\}, \phi, \Omega\}$ and $P\{a\} = P\{d\} = \frac{1}{4}$, $P\{b, c\} = \frac{1}{2}$, $P\Omega = 1$, and $P\phi = 0$. Since the events $\mathscr{E}$ do not form a field, $(\Omega, \mathscr{E}, P)$ is not a probability space.

EXERCISES

1. Show by induction that $P(A_1 + A_2 + \cdots + A_n) = P(A_1) + P(A_2) + \cdots + P(A_n)$. P is said to be *finitely additive.*

2. By using the axioms, show that $PA' = 1 - PA$.

3. Prove the following theorem rigorously: $P(A \cup B) = P(A) + P(B) - P(AB)$.

4. Show that

$$P(A_1 \cup A_2 \cup \cdots \cup A_n) \leq P(A_1) + P(A_2) + \cdots + P(A_n).$$

5. Show that the following simple probability field satisfies the axioms. The events in $\mathscr{E}$ are ϕ and any sum of the disjoint events $E_1, E_2, \ldots, E_n$ that form a finite partition of Ω; i.e., $\Omega = \sum_{i=1}^{n} E_i$. Let $p_k = P(E_k)$ be such that $p_k \geq 0$ for $k = 1, 2, \ldots, n$, and $\sum_{k=1}^{n} p_k = 1$; then P is defined on every $E \in \mathscr{E}$ as follows: $P(E) = \sum_{E_k \subset E} p_k$.

[7]Note that the axioms do not specify how to attach probabilities to events. The only requirement is that the probability function be defined consistently as stated in the axioms. For example, the axioms will not tell us how to find the probability that heads will occur when a given coin is tossed. This problem will be considered later.

6. Show that the intersection of an arbitrary number of fields is a field.

1.5 REMARKS

We shall study algebras and fields in more detail in Chapter 6, since they appear in more general types of probability spaces, which turn out to be equivalent to measure spaces.

The abstract formulation does not refer to any physical experiments or tell us how to choose our space Ω in order to represent such experiments. Consequently there can be many different models for the same experiment. For example, consider the case of tossing a coin once. A simple space Ω consists of the points $+1$ and -1, which represent heads and tails respectively. The probability of a set is one-half the number of points it contains. A second model can be obtained as follows. Let $f(x) = 1$ on A and $f(x) = -1$ on B, where $A = (x : 0 \leq x \leq \frac{1}{2})$ and $B = (x : \frac{1}{2} \leq x \leq 1)$. The space Ω is the range of values of f and the probability of $\{1\}$ and $\{-1\}$ is the length of A and B. Another model of the same experiment is formed by letting Ω be the set of all real numbers. An event is any subset of Ω; its probability is one-half the number of the points of the set $\{-1, 1\}$ which lie in it.

Since this last model has an infinite number of events, it is an example of an infinite probability space. The reader might feel that using an infinite probability space complicates matters needlessly, especially when there are only a finite number of outcomes. However, if we are interested in the first time an event occurs (say the first time we get a head in repeated tosses of a coin), it is more convenient to use an infinite sample space, since the number of repetitions is not fixed in advance. In the coin-tossing example, each point of the space Ω is an infinite sequence of numbers $(x_1, x_2, x_3, \ldots)$, where x_i is $+1$ or -1. That is, each point represents all possible outcomes of an experiment in which the coin is tossed infinitely often. Moreover, by forming the set whose points consist of all sequences beginning with $i_1, i_2, \ldots, i_n$, by assigning a probability of $(\frac{1}{2})^n$ to this set, and by identifying this set with the event that the numbers $i_1, i_2, \ldots, i_n$ turn up on the first n tosses in that order, we achieve a model of a coin tossed n times. In particular, still another model for the coin-tossing experiment is obtained for $n = 1$.

CHAPTER 2

RANDOM VARIABLES AND COMBINATIONS OF EVENTS

2.1 INTRODUCTION

In games of chance we receive money or a prize in a favorable situation; this is one reason we may be interested in the numerical values that are assigned to each outcome. These functions, called *random variables*, can be described by linear combinations of indicators[1] of events in finite probability spaces. An *indicator* is a point function defined on a space Ω such that $I_A(\omega) = 1$ if $\omega \in A$, and $I_A(\omega) = 0$ if $\omega \notin A$. In other words, it indicates whether or not a point belongs to a set A.

The expectation of a random variable is the formal analog of our average gain when we play a game of chance many times. It turns out that the average value of an indicator of an event is the same as the probability of that event. This observation enables us to prove several theorems about combinations of events. In particular, given N events we can find the probabilities that exactly m and at least m out of the N events occur.

[1]Also called "characteristic functions."

2.2 RANDOM VARIABLES

Since the notion of random variables stems from early studies of gambling, it seems appropriate to introduce the subject by considering the following simple game of chance. A fair coin is tossed; if it falls heads we win one dollar and if it falls tails we lose one dollar. Roughly speaking, the random variable in this game consists of the values attached to the points.

A more precise definition can be given by using indicator functions. Our space is the set $\Omega = \{h, t\}$, where h and t represent heads and tails respectively. When we attach values of one dollar to the point h and minus one dollar to the point t, we are really mapping Ω onto the set $\{1, -1\}$. This function, which may be termed v, is usually called a random variable. By using indicator functions, we can represent v explicitly as $v = (1)I_h + (-1)I_t$, where we have abused our usual convention and denoted $\{h\}$ by h (h is not a set).

While we may be interested in the probability that heads will be tossed, the gambler is interested in the probability of winning one dollar, or the probability that $v = 1$. The problem therefore arises of attaching a probability to the statement that a random variable has a value x_j, or $v = x_j$. Now $v(\omega) = x_j$ if and only if ω belongs to the set $A = (\omega: v(\omega) = x_j)$ or $A = v^{-1}(x_j)$ in the space Ω under consideration.[2] Therefore, it is natural to say that the probability that $v = x_j$ is $P(A)$. Let $P(A) = p_j$. The set A is usually denoted by $\{v = x_j\}$ and the last statement is shortened to $P\{v = x_j\} = p_j$. For example, $P\{v = 1\} = \frac{1}{2}$, $P\{v = -1\} = \frac{1}{2}$, and $P\{v = 3\} = 0$, since $\{v = 3\} = \phi$.

As another illustration of these ideas, let us suppose that we have an urn containing four black balls and two white balls. We draw two balls from the urn. The number of black balls n in the sample is a random variable. Let the event A be represented by the pairs (B_i, W_j), where $i = 1, 2, 3, 4$ and $j = 1, 2$. B_i and W_j denote black and white balls, respectively; $B = \{(B_i, B_j)\}$; $i < j$, where $i, j = 1, 2, 3, 4$; and $C = \{(W_1, W_2)\}$. Then $n = 1I_A + 2I_B + 0I_C$ and $P\{n = 1\} = \frac{8}{15}$ (since there are eight possible cases in which we have one black ball and the 15 different outcomes are regarded as equally likely). Similarly, $P\{n = 2\} = \frac{2}{5}$, $P\{n = 0\} = \frac{1}{15}$, and $P\{n = 200\} = 0$.

Finally, let us consider the example of throwing a pair of dice given in 1.3 and the following random variables: d_1 = number of spots on the

[2]To every mapping f of a set Ω into X there corresponds the inverse mapping, denoted by f^{-1}, of the class of all subsets of X into the class of all subsets of Ω defined by $f^{-1}(A) = (\omega : f(\omega) \in A)$. This mapping has the following properties: $f^{-1}(\phi) = \phi, f^{-1}(X) = \Omega, f^{-1}(A') = (f^{-1}(A))'$, $f^{-1}(\cup A_i) = \cup f^{-1}(A_i)$, and $f^{-1}(\cap A_i) = \cap f^{-1}(A_i)$.

first die, d_2 = number of spots on the second die, s = total number of spots, and m = minimum of d_1 and d_2.

If we let $A_i = \{(i, 1), (i, 2), (i, 3), (i, 4), (i, 5), (i, 6)\}$ for $i = 1, 2, \ldots, 6$, then $d_1 = 1I_1 + 2I_2 + 3I_3 + \cdots + 6I_6$, where I_j is the indicator of A_j. The reader can verify that $P\{d_1 = i\} = \frac{1}{6}$. The random variable d_2 can be treated in an analogous fashion with $B_i = \{(1, i), \ldots, (6, i)\}$.

Using the notation of 1.3, we see that $s = \sum_{j=2}^{12} jI_j$, where I_j is the indicator of E_j and $P\{s = j\} = (j - 1)/36$ $(j = 2, 3, \ldots, 7)$, while $P\{s = j\} = (13 - j)/36$ $(j = 8, 9, 10, 11, 12)$.

We can represent m as $m = 1J_1 + 2J_2 + \cdots + 6J_6$, where J_k is the indicator of C_k and C_k consists of all pairs (k, j) and (j, k) with $1 \leq k \leq j \leq 6$. Since the outcomes are equally likely, we see that $P\{m = i\} = \{13 - 2i\}/36$, where $i = 1, 2, \ldots, 6$.

Let us examine the above examples to see what they have in common. In each case we had a finite number of possible outcomes and the sure event Ω was partitioned into a number of pairwise disjoint events $A_1, A_2, \ldots, A_n$. The events A_i were either simple as in the first example or decomposable as in the remainder. The random variable v is given by assigning to these outcomes the real numbers $x_1, \ldots, x_n$ respectively. The following definition summarizes these facts.

DEFINITION 1

A (*simple*) *random variable*[3] is a linear combination $v = \sum_{j=1}^{n} x_j I_j$ of indicators I_j of events A_j of a finite partition[4] of Ω, where the x_j are real numbers. Random variables will be denoted by small letters such as $r, s, v, \ldots$.

At this point the reader might wonder why we did not simply say that a random variable is a real-valued function on a finite probability space. The most important reason is that, in general, such a function could not serve as a random variable. To illustrate this point, let $\Omega = \{1, 2, 3, \ldots, 6\}$ and let the events be $\mathscr{E} = \{\phi, \Omega, A, B\}$, where $A = \{1\}$ and $B = \{2, 3, \ldots, 6\}$. We define a probability function P by $P(A) = \frac{2}{6}$, $P(B) = \frac{4}{6}$, $P(\Omega) = 1$, and $P(\phi) = 0$. The reader can easily verify that we have a finite probability space $(\Omega, \mathscr{E}, P)$ by showing that the axioms given in Section 1.4 are satisfied. If we define a function on Ω by $f(i) = i$, with $i = 1, 2, 3, \ldots, 6$, then f should not be considered a random variable.

[3]In this chapter only simple random variables are considered.

[4]This means that $\Omega = A_1 + A_2 + \cdots + A_n$ with $A_i A_j = \phi$ for $i \neq j$.

For as we have seen, it is expedient to attach a probability to statements such as $f = i$. However, if $i = 2, f^{-1}(2) = \{2\}$; this set is not a member of $\mathscr{E}$ and therefore it has no probability attached to it. Thus, only those functions f for which $f^{-1}(x) \in \mathscr{E}$ for any real number x should be called random variables. For example, if we define $g(1) = 1$ and $g(i) = 2$, when $i = 2, 3, \ldots, 6$, then g is a random variable. These are precisely the functions described in the definition above; in fact, $g = 1I_A + 2I_B$. Such functions are particular examples of measurable functions, described in Chapter 7, which are necessary for a rigorous treatment of noncountable spaces. The following definition of a random variable is equivalent to Definition 1 and is based upon the above example.

DEFINITION 1′

A real-valued function f defined on a finite probability space $(\Omega, \mathscr{E}, P)$ is called a (*simple*) *random variable* if $f^{-1}\{x\} \in \mathscr{E}$ for any real number x.

It is not difficult to see that the two definitions of a random variable are equivalent. Let f be a random variable according to Definition 1. Then $f = \sum_{i=1}^{n} x_i I_i$ (assume that the x_i are distinct), and so $f^{-1}\{x\} = A_i$ if $x = x_i$ and $f^{-1}\{x\} = \phi$ if $x \neq x_i$. Hence $f^{-1}\{x\} \in \mathscr{E}$ for all x, and so f is a random variable according to Definition 1′.

Now let f be a random variable according to Definition 1′. Then $\Omega = \sum_x f^{-1}\{x\}$ and f assumes only a finite number of distinct values, say $x_1, \ldots, x_n$. Let $A_i = f^{-1}\{x_i\}$; then $A_i \in \mathscr{E}$ and $\Omega = \sum_{i=1}^{n} A_i$. If I_j is the indicator function of A_j, then $f = \sum_{j=1}^{n} x_j I_j$ and so f is a random variable according to Definition 1. The proof for nondistinct x_i is left to the reader.

In elementary probability theory, random variables are defined as real-valued functions on finite probability spaces where every subset is an event. In this case the definition is correct since $f^{-1}\{x\} \in \mathscr{E}$.

DEFINITION 2

The function f defined for every real number x by $P(v = x) = f(x)$ is called the *probability density function*[5] of the random variable v.

The function f is well defined, since $\{v = x\}$ is an event. If v assumes n distinct values x_j, then for all real x, $f(x) \geq 0$ and $\sum_{j=1}^{n} f(x_j) = 1$.

[5]This is also known as a probability distribution function.

EXAMPLE 1

Find the density of the number of spades in a five-card hand drawn from an ordinary deck.

Let r be a random variable denoting the number of spades. Then x spades can be drawn in $\binom{13}{x}$ ways, when $0 \leq x \leq 5$, and $5 - x$ non-spades can be drawn in $\binom{39}{5-x}$ ways. Therefore, $f(x) = P(r = x) = \binom{13}{x}\binom{39}{5-x}\Big/\binom{52}{5}$ when $x = 0, 1, \ldots, 5$, and 0 otherwise.

A real-valued function F of a simple random variable v is a random variable. By definition, $P\{Fv = x_j\}$ is the probability of the set consisting of the points of Ω mapped on x_j by Fv. However, this set is an event since it is the inverse image under a random variable v of the set of real numbers mapped by F on x_j; succinctly stated, $(Fv)^{-1}(x_j) = v^{-1}F^{-1}(x) \in \mathscr{E}$. As an illustration, consider the square of the random variable v defined on page 14. The event $\{v^2 = 1\}$ is the union of the events $\{v = 1\}$ and $\{v = -1\}$.

There can be many different random variables defined on the same sample space. In the dice-throwing example, we discussed four random variables: d_1, d_2, s, and m. Frequently we are interested in the joint outcomes of these random variables. For dice, we might want to know the number of spots appearing on the first die and the number on the second die, which we can briefly denote by (d_1, d_2). Other illustrations appear in Example 2.

DEFINITION 3

A random vector $r = (r_1, r_2, \ldots, r_n)$ is a vector whose components r_i are random variables defined on the same probability space.

EXAMPLE 2

1. Let two coins be tossed. Let r_1 be the number of heads on the first toss and r_2 the number of heads on the second toss. Then the two-dimensional random vector (r_1, r_2) is defined by Table 1.

Table 1

Points in Ω	(h, h)	(h, t)	(t, h)	(t, t)
Value of (r_1, r_2)	$(1, 1)$	$(1, 0)$	$(0, 1)$	$(0, 0)$

2. A card drawn from an ordinary deck may be characterized by its suit and face value. Let r_1 assume the values 1, 2, 3, and 4 if the card drawn is a spade, heart, diamond, and club respectively, and let r_2 be a random variable that can assume the values $1, 2, \ldots, 13$, which correspond to the face values ace, $2, \ldots, 10$, J, Q, K. Then (r_1, r_2) is a random vector describing the drawn card.
3. Suppose that eight cards are drawn without replacement from an ordinary deck. Let r_1 be the number of aces, r_2 the number of deuces, r_3 the number of threes, and r_4 the number of fours. Then (r_1, r_2, r_3, r_4) is a random vector.

Just as in the case of one random variable, we wish to assign a probability to the statement $r_1 = x_1, r_2 = x_2, \ldots, r_n = x_n$, written more concisely as $r = x$.[6] Now the set $r_i^{-1}(x_i) = \{r_i = x_i\}$ is an event A_i (since r_i is a random variable); therefore, $r = x$ if and only if

$$r^{-1}(x) = A_1 \cap A_2 \cap \cdots \cap A_n.$$

Since the class of events $\mathscr{E}$ forms a field (see Section 1.5), $A_1 \cap A_2 \cap \cdots \cap A_n$ is an event and so has a probability attached to it.

DEFINITION 4

The function f defined for every real n-tuple $(x_1, \ldots, x_n)$ by

$$P\{r_1 = x_1, r_2 = x_2, \ldots, r_n = x_n\} = f(x_1, \ldots, x_n) \tag{1}$$

is called the *joint probability density function* of the r_i.

EXAMPLE 3

The density function f of the random vector described in number 3 of Example 2 is

$$f(x_1, x_2, x_3, x_4) = \binom{4}{x_1}\binom{4}{x_2}\binom{4}{x_3}\binom{4}{x_4}\binom{36}{8 - x_1 - x_2 - x_3 - x_4}\Big/\binom{52}{8} \qquad (0 \leq x_i \leq 4)$$

if $0 \leq x_1 + x_2 + x_3 + x_4 \leq 8$, and 0 otherwise.

Clearly, $\sum_{x_1, x_2, \ldots, x_n} f(x_1, \ldots, x_n) = 1$ [where the summation is over all possible values of the x_i and $f(x_1, \ldots, x_n) \geq 0$]. The random variables r_i, where $i = 1, 2, \ldots, n$, are said to be jointly distributed. Suppose that $r_i = \sum_{j=1}^{n_i} x_{ij} I_{ij}$, where I_{ij} is the indicator of A_{ij}, has the probability distribution $f_i(x)$, or

$$P\{r_i = x\} = f_i(x). \tag{2}$$

[6] $(r = x) = (r_1 = x_1, r_2 = x_2, \ldots, r_n = x_n) = (\omega : r_1(\omega) = x_1, \ldots, r_n(\omega) = x_n)$.

The distributions $f_i(x)$ are called *marginal distributions* and

(3) $$\sum_{x_1, x_2, \ldots, x_n}^{i} f(x_1, x_2, \ldots, x_n) = f_i(x_i),$$

where the summation is over all possible values of the $x_j, j \neq i$, and the superscript i indicates that there is no summation over x_i. It suffices to verify this result for $i = 1$ because of the symmetry. We begin with the simple observation that the set $r_1 = x_1, r_2 = x_2, \ldots, r_n = x_n$, where $x_1 = x_{1j_1}, x_2 = x_{2j_2}, x_3 = x_{3j_3}, \ldots, x_n = x_{nj_n}$, is the set $A_{1j_1} A_{2j_2} A_{3j_3} \cdots A_{nj_n}$. From Definition 4, Equation (3) can be rewritten as follows:

(4) $$P(A_{1j_1}) = \sum_{j_2, j_3, \ldots, j_n} P(A_{1j_1} A_{2j_2} \cdots A_{nj_n}),$$

where the summation is now over all possible values of $j_2, j_3, \ldots, j_n$. Hence the proof of Equation (3), when $i = 1$, reduces to showing that

(5) $$A_{1j_1} = \sum_{j_2, j_3, \ldots, j_n} A_{1j_1} A_{2j_2} \cdots A_{nj_n}.$$

However, this follows from the observation that $A_{2j_2} A_{3j_3} \cdots A_{nj_n}$ is just an element of the product[7] of the partitions $\mathscr{P}_2 \mathscr{P}_3 \cdots \mathscr{P}_n$, where $\mathscr{P}_i = \{A_{i1}, A_{i2}, \ldots, A_{in_i}\}$ and so $\Omega = \sum_{j_2, j_3, \ldots, j_n} A_{2j_2} A_{3j_3} \cdots A_{nj_n}$. Upon taking intersections with A_{1j}, Equation (5) is obtained.

For example, (d_1, d_2) is defined on the space consisting of the sets $A_i B_j = \{(i, j)\}$ and $P\{d_1 = i, d_2 = j\} = f(i, j) = \frac{1}{36}$. Therefore, $A_i = \sum_{j=1}^{6} A_i B_j$ and $P(A_i) = \sum_{j=1}^{6} P(A_i B_j)$ or $\sum_{j=1}^{6} f(i, j) = f_1(i) = \frac{1}{6}$.

In many elementary probability books random variables are not defined. First a density function f is defined as any nonnegative real-valued function such that $\sum_x f(x) = 1$, where the summation is over all real x such that $f(x) \neq 0$. The mysterious objects, random variables, are introduced by stating that $P(\bar{r} = x)$ is a density function $f(x)$, where x is a real number and $\bar{r}$ is a random variable. Let us try to clarify this last statement using our definition of a random variable. Let $r = \sum_{j=1}^{n} x_j I_j$ be a random variable, defined on a probability space $(\Omega, \mathscr{E}, P)$. Let $\bar{\Omega} = \{x_1, x_2, \ldots, x_n\}$ be a set of real numbers. We can construct a probability space on $\bar{\Omega}$ as follows. Define a transformation T between Ω and $\bar{\Omega}$ by setting $T(\omega) = x_i$ if and only if $\omega \in A_i = (\omega\colon I_i(\omega) = 1)$; in other words, $T(\omega) = r(\omega)$. Let $\bar{\mathscr{E}}$ be the class of all subsets of $\bar{\Omega}$ and define

[7]Recall that if $\mathscr{P}_1 = (A_i, i = 1, \ldots, m)$ and $\mathscr{P}_2 = (B_j, j = 1, \ldots, n)$, then the product of the two partitions is defined to be the partition

$$\mathscr{P} = (A_i B_j, i = 1, \ldots, m; j = 1, \ldots, n).$$

$$\bar{P}(\bar{E}) = P(T^{-1}(\bar{E}))$$

for $\bar{E} \in \bar{\mathscr{E}}$. It is not difficult to verify that $(\bar{\Omega}, \bar{\mathscr{E}}, \bar{P})$ is a probability space. Define a coordinate random variable $\bar{r}$ on $\bar{\Omega}$ by setting $\bar{r}(\bar{\omega}) = \bar{\omega}$. Then $\bar{r}$ and r are equivalent; that is, they both have the same density function. This follows directly from the definition of a density function when it is noted that

$$P(r = x_i) = PA_i$$

and

$$\bar{P}(\bar{r} = x_i) = P(T^{-1}\{x_i\}) = PA_i.$$

EXAMPLE 4

Suppose that an urn contains balls $b_1, b_2, \ldots, b_6$, numbered as indicated by the subscripts. We choose a ball from the urn, receiving twice the number on the ball (in dollars) if the number is even and paying twice the number on the ball if the number is odd.

Here $\Omega = \{b_1, b_2, \ldots, b_6\}$; $\mathscr{E}$ = class of all subsets of Ω; and $PE = \frac{1}{6}$ (number of points in E). The gain is a random variable r defined by $r = 4I_2 + 8I_4 + 12I_6 - 2I_1 - 6I_3 - 10I_5$, where I_j is the indicator function of the event $\{b_j\}$.

Using the procedure described above, $\bar{\Omega} = \{4, 8, 12, -2, -6, -10\}$; $\bar{\mathscr{E}}$ = class of all subsets of $\bar{\Omega}$; and $\bar{P}(\bar{E}) = P(T^{-1}(\bar{E}))$, where $T(\omega) = r(\omega)$. Note that r and $\bar{r}$ have the same density function.

As a generalization, suppose that we have n random variables $r_1, r_2, \ldots, r_n$ defined on Ω. We can now form a new space which consists of the set of points $(x_1, x_2, \ldots, x_n)$ in n-dimensional Euclidean space to which probabilities are assigned by the joint distribution of the r_i. Thus probability theory can be reduced to the study of spaces whose elements are n-tuples of real numbers. This is the usual approach in elementary probability theory. A probability distribution is attached to a set of n-tuples of real numbers and the random variables are just the projection functions on the space; that is, $r_i(x_1, x_2, \ldots, x_n) = x_i$.

EXERCISES

In Exercises 1-5, describe the probability space, write down an explicit expression for the random variable, and find its probability distribution, where the random variable is:

1. The number of times a fair coin is tossed until tails appear for the first time, or 0 if no tails appear after five tosses.

2. The number of letters in the wrong envelope if four letters are randomly placed in four envelopes (any of 24 combinations are equally likely).

3. The maximum (minimum) of two numbers drawn without replacement (i.e., the first ticket is not replaced) from a hat that contains tickets marked with the numbers 1 to 6.

4. The number of balls in the second urn if five balls are distributed randomly among three urns (i.e., a ball is equally likely to be in any urn).

5. The number of kings in a hand of five cards selected at random without replacement from an ordinary deck of playing cards.

6. Show that if f is a density function, then $\sum_{x_1, x_2, \ldots, x_n} f(x_1, \ldots, x_n) = 1$, where the summation is over all possible values of x_i.

7. Show that if v is a random variable and A is any event, then $v = vI_A + vI_{A'}$. Also, if $v \geq c$, then $vI_A \geq cI_A$.

8. If u and v are two random variables on a space Ω, show that uv and $u + v$ are random variables by obtaining their explicit representation in terms of indicator functions.

9. Show that any real-valued function of a finite number of random variables that are defined on the same probability space is also a random variable.

10. The distribution function F of a random variable r is defined by

$$F(x) = P(r \leq x).$$

(a) Verify the following properties of F:

(i) F is a real-valued function defined on the entire line.
(ii) F is increasing; that is, $x_1 < x_2$ implies $F(x_1) \leq F(x_2)$.
(iii) $\lim_{x \to -\infty} F(x) = 0$, $\lim_{x \to \infty} F(x) = 1$.
(iv) F is continuous except at a finite number of points. At its points of discontinuity, F is right continuous.

(b) Let f be the density function of r. Show that

$$F(x) = \sum_{y \leq x} f(y).$$

Conversely, given F, express f in terms of F.
(c) Verify that $F(b) - F(a) = P(a < r \leq b)$.

2.3 MOMENTS

Let us now turn to the more interesting problem of determining our expected winnings if we play a game of chance a large number of times. To be specific, consider the following game, which is similar to roulette.

The rim of a wheel is divided into a finite number of arcs a_i, where $i = 1, 2, \ldots, n$. The event A_i occurs if the wheel comes to rest at a point of a_i and we receive an amount x_i (positive, zero, or negative). Since the length of a_i is arbitrary, we can attach any probability we desire to A_i by defining $P(A_i) = (\text{length of } a_i)/(\text{circumference of wheel})$. If we played this game k times and outcome A_i occurred k_i times, then our average gain would be

$$x_1\left(\frac{k_1}{k}\right) + x_2\left(\frac{k_2}{k}\right) + \cdots + x_n\left(\frac{k_n}{k}\right).$$

Physical observations reveal that k_i/k clusters about PE_i (see the end of 1.2); therefore this average clusters about $x_1 PE_1 + x_2 PE_2 + \cdots + x_n PE_n$. Since any random variable $r = x_1 I_1 + \cdots + x_n I_n$ can be interpreted in this manner, the following definition is suggested.

DEFINITION 1

Let $r = x_1 I_1 + x_2 I_2 + \cdots + x_n I_n$, where I_j is the indicator of the event A_j, be a random variable. The *expected value* of r (or *mean*) is denoted by Er and is defined as follows:

$$Er = x_1 PA_1 + x_2 PA_2 + \cdots + x_n PA_n.$$

The mean is usually denoted by the letter μ.[8]

We must show that this definition is independent of the representation used for r. Suppose that

$$r = y_1 J_1 + \cdots + y_m J_m,$$

where the $\{J_k\}$ are indicator functions of the events $\{B_k\}$ that form a partition of Ω. Let us assume for the moment that the y_k are all distinct; that is, $y_i \neq y_j$ if $i \neq j$. Let $\{x_{ik}\}$ be the set of all x_i that are equal to y_k and let $\{A_{ik}\}$ be the corresponding sequence of events—that is, the events that imply $\{r = y_k\}$. Then $B_k = \sum_i A_{ik}$ and $PB_k = \sum_i PA_{ik}$. Therefore,

$$Er = \sum_{i=1}^{n} x_i PA_i = \sum_{k=1}^{m} y_k \sum_i PA_{ik} = \sum_{k=1}^{m} y_k PB_k.$$

This proves the uniqueness of the expectation even if the y_k are not distinct. There is a representation of r using only distinct values of r. By the above result both of the expectations will be equal to the expectation defined by this new representation.

[8]Frequently we speak of the mean of a density function without mentioning the random variable.

As a corollary of the above result, assuming the x_i are distinct,

$$Er = \sum_{i=1}^{n} x_i f(x_i),$$

where $f(x_i)$ is the density function of the random variable r.

In tossing a fair die, the expected number of spots appearing on a face is $\frac{1}{6}1 + \frac{1}{6}2 + \cdots + \frac{1}{6}6 = 3.5$. The expected number of heads obtained with an unbalanced coin with a probability p for heads is p. This last example is a particular case of the following lemma, whose proof follows from the fact that $I_A = 1I_A + 0I_{A'}$.

LEMMA 1

If I is the indicator of an event A, then

$$EI = PA.$$

Noting that $Er = \sum_{j=1}^{n} x_j EI_j$ leads us to speculate that the expectation of a sum of random variables is the sum of their expectations. To prove this result we need the following property of indicators: $I_{AB} = I_A I_B$. The proof is obvious from the definition of the indicator function.

THEOREM 1

If $r_1, r_2, \ldots, r_n$ are simple random variables, then

$$E(r_1 + r_2 + \cdots + r_n) = Er_1 + Er_2 + \cdots + Er_n.$$

Proof

It suffices to prove the assertion for two random variables, since the general case follows by induction. Let $v_1 = x_1 I_1 + \cdots + x_m I_m$ where I_k is the indicator of A_k, and let $v_2 = y_1 J_1 + \cdots + y_n J_n$ where J_k is the indicator of B_k. Now the random variable $v_1 + v_2$ is defined by means of the indicators of the product of the two given partitions; hence it can be written

$$v_1 + v_2 = \sum_{r=1}^{m} \sum_{s=1}^{n} (x_r + y_s) I_r I_s.$$

Since $I_r I_s$ is the indicator of $A_r B_s$, it follows from the definition of expectation that

$$E(v_1 + v_2) = \sum_{r=1}^{m} \sum_{s=1}^{n} (x_r + y_s) P(A_r B_s)$$

$$= \sum_{r=1}^{m} x_r PA_r + \sum_{s=1}^{n} y_s \left[\sum_{r=1}^{m} P(A_r B_s) \right].$$

The last line follows from the fact that the B_i's form a partition Ω and so

$$A_r = A_r B_1 + A_r B_2 + \cdots + A_r B_n.$$

Similarly, $\sum_{r=1}^{m} P(A_r B_s) = P(B_s)$ and the desired conclusion follows from the definition of expectation, namely, $Ev_1 = \sum_{r=1}^{m} x_r PA_r$.

As an application of these results we will prove Theorem 2, which concerns combinations of events, by using the next lemma.

LEMMA 2

The following formulas hold for indicator functions:

1. $I_{A'} = 1 - I_A$.
2. $I_{A+B} = I_A + I_B$.
3. $I_{A \cup B} = I_A + I_B - I_{AB} = \sup (I_A, I_B)$.
4. $I_{A \cap B} = I_A I_B = \inf (I_A, I_B)$.

The proofs are obvious.

THEOREM 2

$$P(A \cup B \cup C) = PA + PB + PC - PAB - PBC - PCA + PABC.$$

Proof

We note that $P(A \cup B \cup C) = E(I_{A \cup B \cup C})$ and $A \cup B \cup C = A + BA' + CB'A'$; hence

$$I_{A \cup B \cup C} = I_A + I_B(1 - I_A) + I_C(1 - I_A)(1 - I_B).$$

The result follows upon expanding the right-hand side and taking expectations.

We leave as an exercise for the reader the proof of the fact that the product of two random variables is also a random variable.

The above properties of indicators and expectations also yield *Chebyshev's inequality*.[9]

THEOREM 3

If v is a simple random variable, then for every $\epsilon > 0$,

[9]After P. L. Chebyshev, a Russian mathematician (1821–1894).

$$P(|v| \geq \epsilon) \leq \frac{1}{\epsilon^2} Ev^2.$$

Proof

Write v^2 as $v^2 = v^2 I_A + v^2 I_{A'}$, where $A = \{|v| \geq \epsilon\}$ and $A' = \{|v| < \epsilon\}$. Take expectations, using the fact that the expectation of a nonnegative random variable is nonnegative; this yields

$$Ev^2 = E(v^2 I_A) + E(v^2 I_{A'}) \geq E(v^2 I_A).$$

The desired result follows from the fact that

$$E(v^2 I_A) \geq \epsilon^2 EI_A = \epsilon^2 P(|v| \geq \epsilon).$$

COROLLARY

$$P(|v - \mu| \geq \epsilon) \leq \frac{1}{\epsilon^2} E(v - \mu)^2.$$

The proof follows directly from the theorem, since $v - \mu$ is a random variable.

Theorem 3 states that if Ev^2 is small the spread of v about the origin will be small—that is, large deviations from zero are improbable. In practice we are often concerned with deviations from the mean that will be small if $E(v - \mu)^2$ is small. Consequently, the quantity $E(v - \mu)^2$ has acquired a special name.

DEFINITION 2

Let v be a random variable and let $\mu = Ev$ be its mean. The *variance* of v is defined by

$$\text{Var } v = E(v - \mu)^2$$

and is often denoted by σ^2 or $\sigma^2 v$ (the square of sigma). Its positive square root σ is called the *standard deviation* of v.

By expanding $(v - \mu)^2$ and taking expectations (cf., Exercise 1), it follows that

$$\text{Var } v = Ev^2 - \mu^2.$$

The notion of variance is quite useful in connection with limit theorems (see Chapter 13).

In order to find the variance, we must calculate the expectation of the square of a random variable. The required procedure is contained in Theorem 4. We first illustrate the proof of the theorem by calculating

$E\phi(v)$ for $v = I_{A_1} - I_{A_2} + 2I_{A_3} - 2I_{A_4}$. Let f be the density function of v and suppose that $PA_i = \frac{1}{4}$, where $i = 1, 2, 3, 4$. If $\phi(v) = v^2$, then ϕ only assumes the values $y_1 = 1$ and $y_2 = 4$. The x_k are divided into disjoint sets $\{1, -1\}$ and $\{2, -2\}$, defined by $(x: \phi(x) = 1)$ and $(x: \phi(x) = 4)$ respectively. Let $J_1 = I_{A_1} + I_{A_2}$ be the indicator function of $v^{-1}\{1, -1\}$ and let $J_2 = I_{A_3} + I_{A_4}$ be the indicator function of $v^{-1}\{2, -2\}$. Then $\phi(v) = J_1 + 4J_2$ and

$$E\phi(v) = EJ_1 + 4EJ_2 = PA_1 + PA_2 + 4(PA_3 + PA_4)$$
$$= \sum_{i=1}^{4} \phi(x_i)f(x_i).$$

THEOREM 4

Any real function $\phi(x)$ defines a new random variable $\phi(v)$ and

$$E(\phi(v)) = \sum_{k=1}^{n} \phi(x_k)f(x_k),$$

where $f(x)$ is the distribution function of v.

Proof

Let $v = \sum_{k=1}^{n} x_k I_k$. Note that ϕ can take the same value for different values of x_k. Suppose that ϕ takes m distinct values $y_1, \ldots, y_m$. Then the x_k are divided into m disjoint sets $\{x_{r1}, x_{r2}, \ldots, x_{rn_r}\}$, defined by $(x: \phi(x) = y_r)$. If I_{rs} is the indicator of the set on which v assumes the value x_{rs}, then v can also be written as $\sum_{r=1}^{m} \sum_{s=1}^{n_r} x_{rs} I_{rs}$.

Then $\phi(v) = \sum_{r=1}^{m} y_r J_r$, where $J_r = \sum_{s=1}^{n_r} I_{rs}$, since $\phi(x_{rs}) = y_r$. By applying Theorem 1,

$$E\phi(v) = \sum_{r=1}^{m} y_r EJ_r = \sum_{r=1}^{m} \phi(x_{rs})E\left(\sum_{s=1}^{n_r} I_{rs}\right) = \sum_{r=1}^{m} \sum_{s=1}^{n_r} \phi(x_{rs})E(I_{rs}),$$

from which the result follows as a consequence of Definitions 2.2 and 1.

If v is the number scored with a perfect die, then

$$E(v^2) = \frac{1}{6}(1^2 + 2^2 + 3^2 + 4^2 + 5^2 + 6^2) = \frac{91}{6}$$

and

$$\sigma^2 = \frac{91}{6} - \left(\frac{7}{2}\right)^2,$$

where $Ev = \frac{7}{2}$.

Now let us calculate the variance of $v_1 + v_2$, where v_1 and v_2 are random variables on the same space. Denoting the means of v_1 and v_2 by μ_1 and μ_2 respectively, we have

$$\begin{aligned}\text{Var}\,(v_1 + v_2) &= E(v_1 + v_2 - \mu_1 - \mu_2)^2 \\ &= E(v_1 - \mu_1)^2 + 2E(v_1 - \mu_1)(v_2 - \mu_2) + E(v_2 - \mu_2)^2.\end{aligned}$$

DEFINITION 3

The *covariance* of v_1 and v_2 is defined by

$$\text{Cov}\,(v_1, v_2) = E(v_1 - \mu_1)(v_2 - \mu_2),$$

where μ_1 and μ_2 are the means of v_1 and v_2 respectively. The *correlation coefficient* of v_1 and v_2, denoted by $\rho(v_1, v_2)$, is defined by

$$\rho(v_1, v_2) = \text{Cov}\,(v_1, v_2)/\sigma v_1 \sigma v_2\,(\sigma v_1 \neq 0,\ \sigma v_2 \neq 0).$$

By expanding the right-hand side in the equation for the covariance, we have

$$\text{Cov}\,(v_1, v_2) = E(v_1 v_2) - \mu_1 \mu_2.$$

Note that the correlation coefficient is independent of the origin and units of measurement; that is, for any constants a_1, a_2, b_1, and b_2 with $a_1 > 0$ and $a_2 > 0$, we have $\rho(a_1 v_1 + b_1, a_2 v_2 + b_2) = \rho(v_1, v_2)$. Further properties of the correlation coefficient appear at the end of Section 3.3.

The expectations of v and v^2 do not characterize the probability distribution completely; we need the expectations of higher powers.

DEFINITION 4

Let v be a random variable; then Ev^k is called the k*th moment of* v *about the origin.*

If v assumes n distinct values x_i, and f is the density function of v, then $Ev^r = \sum_{k=1}^{n} x_k^r f(x_k)$. In this case, if enough moments are known we can find all the $f(x_i)$.

THEOREM 5

If the random variable v assumes n distinct values, then the probability density of v is uniquely determined by the first $n - 1$ moments about the origin.

We leave the details of the proof to the reader. It is based upon the fact that the determinant of the coefficients of the unknowns $f(x_j)$ does not vanish (see Exercise 5).

EXERCISES

1. Prove the following statements, where v_1 and v_2 are random variables.

(a) If c is a constant, $Ec = c$ and $E(cv) = cEv$.
(b) $v_1 \geq v_2$ implies $Ev_1 \geq Ev_2$.
(c) If $v_1 = av + b$, where a and b are constants, then $Ev_1 = a(Ev) + b$ and $\text{Var } v_1 = a^2 \text{ Var } v$.

2. Calculate the variance of the random variables n, d_1, d_2, s, and m given in the examples in Section 2.2.

3. Suppose that 10,000 tickets are sold for a raffle in which the prizes are one car valued at \$5000 and ten radios valued at \$150. How much should you be willing to pay for a ticket?

4. Let the random variables v_1 and v_2 have the density functions f and g respectively. Assume that v_1 takes the distinct values x_i, where $i = 1, 2, \ldots, m$, while v_2 takes the values y_j, where $j = 1, 2, \ldots, n$. Show that if ϕ and ψ are real valued functions,

$$E(\phi(v_1) + \psi(v_2)) = \sum_{i=1}^{n} \phi(x_i)f(x_i) + \sum_{j=1}^{m} \psi(y_j)g(y_j).$$

5. Prove that the Vandermonde[10] determinate $|a_{ij}|$, where $a_{ij} = x_i^{j-1}$ and i, $j = 1, 2, \ldots, n$, does not vanish for distinct values of x_i. (*Hint:* Assume that the rows are linearly dependent and deduce that a polynomial of degree $n - 1$ has n roots and so vanishes identically.)

6. Generalize Theorem 2 by showing that for N events,

$$P(A_1 \cup A_2 \cup \cdots \cup A_N) = \sum P(A_i) - \sum_{i,j} P(A_i A_j) + \sum_{i,j,k} P(A_i A_j A_k) - \cdots + (-)^{N-1} P(A_1 A_2 \cdots A_N).$$

Here $1 \leq i < j < k \cdots \leq N$, so that in the sum each combination of events only occurs once.

7. If $v_1, v_2, \ldots, v_n$ are random variables with finite variances $\sigma_1^2, \sigma_2^2, \ldots, \sigma_n^2$, then

$$\text{Var}\,(v_1 + v_2 + \cdots + v_n) = \sum_{k=1}^{n} \sigma_k^2 + 2 \sum_{j<k} \text{Cov}\,(v_j, v_k).$$

[10] A. T. Vandermonde (1735–1796), a French mathematician who was one of the founders of the theory of determinants.

8. Let μ'_k denote the kth moment—i.e., $\mu'_k = E(v^k)$. Two other moments are frequently used. These are the *kth central moment* $\mu_k = E((v - \mu)^k)$ and the *kth factorial moment* $\mu^*_k = E(v(v-1)\cdots(v-k+1))$. Verify that the following relations hold between these moments:

(a) $\mu_2 = \mu'_2 - \mu^2$, $\mu_3 = \mu'_3 - 3\mu\mu'_2 + 2\mu^3$.
(b) $\mu'_3 = \mu_3 + 3\mu\mu_2 + \mu^3$.
(c) $\mu^*_1 = \mu$, $\mu^*_2 = \mu'_2 - \mu$, $\mu^*_3 = \mu'_3 - 3\mu'_2 + 2\mu$.

9. Prove that the Schwarz inequality,[11] $E^2(v_1 v_2) \leq Ev_1^2 Ev_2^2$, holds for any two random variables v_1 and v_2. [*Hint:* Use the fact that the polynomial $E(xv_1 + v_2)^2$ is nonnegative.]

10. Eight single men and seven single women happen randomly to have purchased single tickets in the same 15-seat row of a theater. On the average, how many pairs of adjacent seats will be occupied by marriageable couples?

2.4 COMBINATIONS OF EVENTS

In practice many problems occur which involve finding the probability that exactly m and at least m out of N events occur. For example, a postman delivers N differently addressed letters to an apartment building with N mailboxes. The boxes are very old and the plates are illegible. What is the probability that m people will get their own mail if the postman places the letters in the boxes at random, one letter to each box?

If we let A_i be the event that letter i ($i = 1, 2, \ldots, N$) is in the correct box, then our problem reduces to finding the probability that exactly m of the events A_i occur.

The reader can easily verify that the probability that any m letters $i_1, i_2, \ldots, i_m$ are in the correct boxes is

$$P(A_{i_1} A_{i_2} \cdots A_{i_m}) = \frac{(N-m)!}{N!}$$

if we assume that all outcomes are equally likely. However, the answer cannot be obtained by merely adding the probabilities of all possible combinations of m events, since they are not disjoint; for example, $A_1 A_2 \cdots A_m \cap A_2 A_3 \cdots A_{m+1}$ is the event that the first $m + 1$ letters are delivered correctly.

Fortunately, there is a systematic procedure, known as the *method of indicators*,[12] for calculating the probability of any event B which is a Boolean function $f(A_1, \ldots, A_n)$ of the events $A_1 \cdots A_n$. By a Boolean func-

[11] H. A. Schwarz, a German mathematician (1843–1921). Also known as the Cauchy inequality and the Buniakovski inequality.

[12] M. Loève, "Sur les systèmes d'événementts," *Ann. Univ. Lyon*, Sec. A(3) **5**, 55–74 (1942).

tion, we mean any combination of the events and their complements using the operations of unions and intersections—for example, let

$$B = (A_1 \cup A_2' \cup A_3') \cap (A_2 \cup A_3') \cap A_3. \tag{1}$$

To apply this method we replace B by its indicator and use Lemma 3.2 to express I_B as an ordinary arithmetic function of the indicators of the events A_i. We simplify this expression and then take expectations; hence by Lemma 3.1 we obtain $P(B)$ in terms of the sums of probabilities of the intersections of certain A_i.

In the above example, Equation (1) would be replaced by

$$\begin{aligned} I_B &= [I_1 + (1 - I_2)(1 - I_1) + (1 - I_3)I_2(1 - I_1)] \\ &\quad \cdot [I_2 + (1 - I_3)(1 - I_2)](I_3) \\ &= (1 - I_2 I_3 + I_1 I_2 I_3)(1 - I_3 + I_1 I_2)I_3 = I_1 I_2 I_3, \end{aligned} \tag{2}$$

where I_j is the indicator of A_j. Taking expectations in Equations (2), we obtain $PB = PA_1 A_2 A_3$. The method of indicators was also used in proving Theorem 3.2 and can be used tc solve Exercise 3.6.

Of course, we can try to simplify our function (by taking complements, replacing unions by sums, using the distributive laws, etc.) before replacing the events by their indicators. If we simplify Equation (1) by two applications of the distributive law (starting with A_3), we obtain $B = A_1 A_2 A_3$ and $PB = P(A_1 A_2 A_3)$ directly.

As a final illustration we will apply the method of indicators to prove Theorem 1, which solves the problem stated at the beginning of this section. In order to shorten the statement of the theorem, we introduce the following notation.

$$\begin{aligned} &S_0 = 1, S_1 = \sum_{k=1}^{N} P(A_k), \ldots, S_r = \sum_{(k_1, k_2, \ldots, k_r)} P(A_{k_1} A_{k_2} \ldots A_{k_r}), \ldots, \\ &S_N = P(A_1 A_2 \ldots A_N). \end{aligned} \tag{3}$$

The summation in Equation 3 is over the $\binom{N}{r}$ possible subsets $\{k_1, k_2, \cdots, k_r\}$ of size r from the set $\{1, 2, 3, \ldots, N\}$. Although the order of the subscripts is irrelevant, for definiteness we shall suppose $k_1 < k_2 < k_3 < \ldots < k_r$. Hence S_r can be written as

$$S_r = \sum_{k_1=1}^{m} \sum_{k_2=k_1+1}^{m} \cdots \sum_{k_r=k_{r-1}+1}^{m} P(A_{k_1} A_{k_2} \ldots A_{k_r}).$$

THEOREM 1

The probability $P_{[m]}$ that exactly m $(0 \leq m \leq N)$ among the N events $A_1, \ldots, A_N$ occur simultaneously is given by

(4) $$P_{[m]} = \sum_{k=m}^{N} (-1)^{k-m} \binom{k}{m} S_k.$$

Proof

The event $B_{[m]}$ that exactly m of the N events $A_1, A_2, \ldots, A_N$ occur is

$$B_{[m]} = \sum_{(k_1,k_2,\ldots,k_m)} A_{k_1}A_{k_2}\ldots A_{k_m}A'_{k_{m+1}}\ldots A'_{k_N},$$

where the summation is over the $\binom{N}{m}$ possible subsets $\{k_1, k_2, \ldots, k_m\}$ of size m from the set $X = \{1, 2, 3, \ldots, N\}$, while the remaining set of subscripts is the complement—that is, $X - \{k_1, k_2, \ldots, k_m\}$. (For example, $X = \{1, 2, \ldots, 7\}$, $\{k_1, k_2, k_3\} = \{1, 2, 3\}$, and $\{k_4, k_5, k_6, k_7\} = \{4, 5, 6, 7\}$.) Let $I_{[m]}$ be the indicator of $B_{[m]}$; then, using Lemma 3.2

(5) $$I_{[m]} = \sum_{(k_1,k_2,\ldots,k_m)} I_{k_1}I_{k_2}\cdots I_{k_m}(1 - I_{k_{m+1}})(1 - I_{k_{m+2}})\cdots(1 - I_{k_N}),$$

since the events are mutually exclusive. However, the last product can be written as

$$\begin{aligned}(1 - I_{k_{m+1}})\cdots(1 - I_{k_N}) &= 1 - \sum_{j_1} I_{j_1} + \sum_{(j_1 j_2)} I_{j_1}I_{j_2} \\ &\quad - \cdots (-1)^r \sum_{(j_1,j_2,\ldots,j_r)} I_{j_1}I_{j_2}\cdots I_{j_r} + \cdots \\ &\quad + (-1)^{N-m} I_{j_1}\cdots I_{j_N},\end{aligned}$$

where the summation in the $(r + 1)$th term is over all $\binom{N-m}{r}$ subsets $\{j_1, j_2, \ldots, j_r\}$ of size r from the set $\{k_{m+1}, k_{m+2}, \ldots, k_N\}$. Thus a typical term of Equation 5 when the right-hand side is expanded is

(6) $$(-1)^r \sum_{(k_1,k_2,\ldots,k_m)} \sum_{(j_1,j_2,\ldots,j_r)} I_{k_1}I_{k_2}\cdots I_{k_m}I_{j_1}I_{j_2}\cdots I_r.$$

The set $\{k_1, k_2, \ldots, k_m, j_1, j_2, \ldots, j_r\}$ is just a subset of size $m + r$ of X, since the sets $\{k_1, k_2, \ldots, k_m\}$ and $\{j_1, j_2, \ldots, j_r\}$ are disjoint. (If $r = 2$, then $\{j_1 j_2\}$ can be $\{4, 5\}$, $\{4, 6\}$, $\{4, 7\}$, $\{5, 6\}$, $\{5, 7\}$, $\{6, 7\}$. However, some of these sets will be the same since the j's and k's assume all values $1, 2, \ldots, N$. The number of repetitions of a fixed set, say $\{k_1, k_2, \ldots, k_m, j_1, j_2, \ldots, j_r\}$, is just the number of different ways $\binom{m+r}{r}$ of dividing this set into two sets containing m and r elements. (The set $\{1, 2, 3, 4, 5\}$ can arise from $\{1, 2, 3\} \cup \{4, 5\}$, $\{1, 2, 4\} \cup \{3, 5\}$, $\{1, 2, 5\} \cup \{3, 4\}$, etc., and $I_1I_2I_3I_4I_5 = I_1I_2I_4I_3I_5 = I_1I_2I_5I_3I_4$, etc.) Hence, the term (6) can be rewritten as

$$(-1)^r \binom{m+r}{m} \sum_{(j_1,j_2,\ldots,j_{r+m})} I_{j_1}I_{j_2}\cdots I_{j_{r+m}},$$

where the summation is over all subsets $\{j_1, j_2, \ldots, j_{r+m}\}$ of size $r + m$ of X. Using this result, Equation 5 becomes

$$
\begin{aligned}
I_{[m]} &= \sum_{(j_1 \cdots j_m)} I_{j_1} \cdots I_{j_m} - \binom{m+1}{m} \sum_{(j_1 \cdots j_{m+1})} I_{j_1} \cdots I_{j_{m+1}} + \cdots \\
&\quad + (-1)^{r-m} \binom{m+r}{m} \sum_{(j_1, \ldots, j_r)} I_{j_1} \cdots I_{j_r} + \cdots \qquad (7) \\
&\quad + (-1)^{N-m} I_{j_1} \cdots I_{j_N}.
\end{aligned}
$$

The theorem follows by taking expectations.

Let $J_0 = 1$ and

$$J_r = \sum_{(k_1 k_2 \cdots k_r)} I_{k_1} I_{k_2} \cdots I_{k_r} \qquad (8)$$

so that $EJ_r = S_r$.

In order to prove the corollary stated below, we will need the following lemma.

LEMMA 1

For any sequence a_{kr} we have

$$\sum_{r=0}^{N} \sum_{k=r}^{N} a_{kr} = \sum_{k=0}^{N} \sum_{r=0}^{k} a_{kr}. \qquad (9)$$

Proof

Both sums are over all lattice points (points whose coordinates are integers) inside and on the triangle whose sides are $r = k$, $k = N$, and $r = 0$ (see Figure 1). On the right-hand side of Equation (9) we first sum over all lattice points on each vertical line and then add all these sums, while on the left-hand side we first sum over the lattice points on a horizontal line and then add the resulting sums.

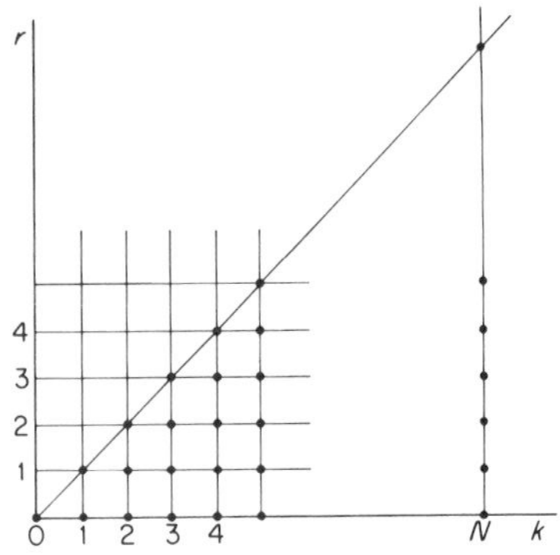

Figure 1

COROLLARY

(10) $$\sum_{r=0}^{N} I_{[r]} x^r = \sum_{s=0}^{N} J_s (x-1)^s. \qquad (x \text{ is an arbitrary variable})$$

Proof

Using Equation (8), Equation (7) can be rewritten in the following form:

(11) $$I_{[r]} = \sum_{k=r}^{N} (-1)^{k-r} \binom{k}{r} J_k.$$

Multiplying Equation (11) by x^r and summing with respect to r yields

$$\sum_{r=0}^{N} I_{[r]} x^r = \sum_{r=0}^{N} \left[\sum_{k=r}^{N} (-1)^{k-r} \binom{k}{r} x^r \right] J_k = \sum_{k=0}^{N} \left[\sum_{r=0}^{k} (-1)^{k-r} \binom{k}{r} x^r \right] J_k$$

by using Lemma 1. The result follows since

$$(x-1)^k = \sum_{r=0}^{k} (-1)^{k-r} \binom{k}{r} x^r.$$

Returning to our example, we see that the probability that exactly m people will get correctly addressed mail is

(12) $$\begin{aligned} P_{[m]} &= \frac{1}{m!} - \binom{m+1}{m} \frac{1}{(m+1)!} + \binom{m+2}{m} \frac{1}{(m+2)!} + \cdots \\ &\quad + (-1)^{N-m} \binom{N}{m} \frac{1}{N!} = \frac{1}{m!} \left[1 - \frac{1}{1!} + \frac{1}{2!} - \frac{1}{3!} + \cdots \right. \\ &\quad \left. + (-1)^{N-m} \frac{1}{(N-m)!} \right], \end{aligned}$$

since $S_k = \dfrac{1}{k!}$.

The brackets on the right-hand side of Equation (12) contain the initial terms of the expansion of e^{-1}; hence for large $N - m$ we have approximately

$$P_{[m]} \approx p_m = \frac{1}{m!} e^{-1}.$$

Since $\sum_{m=0}^{\infty} \dfrac{1}{m!} = e$, $\{p_m\}$ can be interpreted as a probability density function of a random variable v such that $P\{v = m\} = p_m$. This is an example of a countable probability space whose points are $\{1, 2, 3, \ldots\}$. Such spaces will be studied in Chapter 5.

We can use Theorem 1 to find the probability of the realization of at least m events.

THEOREM 2

The probability $P_{(m)}$ that at least m $(1 \leq m \leq N)$ of the N events $A_1, \ldots, A_N$ occur is given by

$$(13) \qquad P_{(m)} = S_m - \binom{m}{m-1} S_{m+1} + \cdots + (-1)^{r-m} \binom{r-1}{m-1} S_r + \cdots + (-1)^{N-m} \binom{N-1}{m-1} S_N.$$

Proof

Since $P_{(m)} = P_{[m]} + P_{[m+1]} + \cdots + P_{[N]}$, if we write out each of the expansions as in Equation (4) and add all the coefficients of S_r, the proof reduces to showing that

$$(14) \qquad (-1)^{r-m} \binom{r}{m} + (-1)^{r-m-1} \binom{r}{m+1} + (-1)^{r-m-2} \binom{r}{m+2} + \cdots + (-1)\binom{r}{r} = (-1)^{r-m} \binom{r-1}{m-1}.$$

However, if we replace $\binom{r}{k}$ $(k = m, m+1, \ldots, r)$ in the left-hand side of Equation (14) by

$$(15) \qquad \binom{r}{k} = \binom{r-1}{k-1} + \binom{r-1}{k}$$

and simplify, we obtain $(-1)^{r-m} \binom{r-1}{m-1}$.

EXERCISES

1. A psychologist is studying extrasensory perception by concentrating on a symbol marked on the face of a card, which the subject is supposed to guess. If there are five cards marked with the symbols star, square, circle, one line, and two wavy lines, what is the probability of getting i correct, where $i = 0, 1, 2, 3, 4, 5$?

2. Find the probability that no more than three dice will turn up with the same number of spots when four dice are thrown.

3. Four socks are selected at random from five pairs of socks in a drawer. What is the probability that there will be at least one pair selected? What is the minimum number of socks that must be drawn until the probability of getting a pair is greater than $\frac{1}{2}$?

4. Consider a modification of the problem described at the beginning of this section. Suppose that the postman has two letters for each person. Find the probability that exactly m people will receive their correct mail.

5. *The Matching Problem.* This problem was first investigated by the French mathematician P. R. Montmort (1678–1719). He considered the following game (known as Treize or Rencontre).

Suppose that we have 13 cards, numbered 1, 2, 3, . . . , 13, in a bag. The cards are drawn out singly. Find the probability that the number on a card will coincide with the number expressing the order in which it was drawn (i.e., match has occurred).

Generalize for N cards (this is the problem at the beginning of this section). If $P_{[m]}$ is the probability that exactly m matches occur, then $P_{[N-1]} = 0$. Show that the expected number of matches and the variance are both 1. [*Hint:* To find $E(r) = \sum_{r=0}^{N} rP_{[r]} = 1$, use the *generating function* $\sum_{r=0}^{N} P_{[r]}x^r$ (see Section 5.4).]

6. *The Card Collectors' Problem.* Suppose that N cards numbered $1 \cdots N$ are equally likely to be found in packages of cereal. If k packages are bought, find the probability that exactly m of the N cards will not be obtained. [*Hint:* Let A_i be the event that card i will not be found in the k packages. Then

$$P(A_{i_1}A_{i_2} \cdots A_{i_n}) = \left(1 - \frac{n}{N}\right)^k.]$$

7. Let R be the number of events A_j that occur. Show directly that

$$n!J_n = R(R-1) \cdots (R-n+1).$$

Verify the result by showing that $x^R = \sum_{r=0}^{N} x^r I_{[r]}$, differentiating the righthand side of Equation (10) n times, and setting $x = 1$.

8. Show that $S_r = \sum_{k=r}^{N} \binom{k}{r} P_{[k]} = \sum_{k=r}^{N} \binom{k-1}{r-1} P_{(k)}$. [*Hint:* Set $x = y + 1$ in Equation (10) and equate the coefficients of the powers of y^r. The second half follows from the fact that $P_{(m)} - P_{(m+1)} = P_{[m]}$ and from Equation (15).

9. (a) If $r \le m \le N$, then $I_{(r)} \ge I_{(m)} \ge I_{(N)}$.

(b) Under the above hypothesis, $P_{(r)} \ge P_{(m)} \ge P_{(N)}$.

(c) From Exercises 8 and 9(b), deduce that

$$S_r \ge \sum_{k=r}^{m} \binom{k-1}{r-1} P_{(m)} = \binom{m}{r} P_{(m)}.$$

[*Hint:* To deduce that $\sum_{k=r}^{m} \binom{k-1}{r-1} = \binom{m}{r}$, use Equation (15).]

(d) In a similar manner, deduce that

$$P_{(m)}\left[\sum_{k=m}^{N} \binom{k-1}{r-1}\right] + \sum_{k=r}^{m-1} \binom{k-1}{r-1} \ge S_r,$$

which simplifies to

$$P_{(m)} \ge -\frac{S_r - \binom{m-1}{r}}{\binom{N}{r} - \binom{m-1}{r}}.$$

(e) Prove Gumbel's inequality:

$$P_{(N)} \geq \frac{S_r - \binom{N-1}{r}}{\binom{N-1}{r-1}}.$$

(f) Prove Fréchet's inequality:

$$P_{(N)} \leq \frac{S_r}{\binom{N}{r}}.$$

10. Let $\Delta J(k) = J(k+1) - J(k)$, where $J(k) = 1 - J_k \Big/ \binom{m}{k}$. Prove the following facts:

(a) $$\frac{J_{k+1}}{\binom{m}{k+1}} = \sum_{s=k+1}^{m} \frac{\binom{m-k-1}{s-k-1}}{\binom{m}{s}} I_{[s]}.$$

(b) $$\Delta J(k) = \sum_{s=k}^{m-1} \frac{\binom{m-k-1}{s-k}}{\binom{m}{s}} I_{[s]}.$$

(c) $\Delta J(k) \geq I_{[r]} \binom{m-k-1}{r-k} \Big/ \binom{m}{r}$.

(d) $\Delta J(k) \geq 0$ implies $S_k \Big/ \binom{m}{k} \geq S_{k+1} \Big/ \binom{m}{k+1}$.

11. Prove the following:

(a) $$-m\Delta\left(\frac{J(k)}{k}\right) = \sum_{s=k}^{m} \frac{\binom{m-k-1}{s-k}}{\binom{m-1}{s-1}} (1 - I_{(s)}).$$

(b) $$1 - I_{(s)} \leq \frac{\binom{m-1}{k-1}}{\binom{m-k-1}{r-k}} \left[-m\Delta\left(\frac{J(k)}{k}\right)\right].$$

(c) $-m\Delta\left(\frac{J(k)}{k}\right) \geq 0$ implies $\frac{1}{k}\left(1 - S_k \Big/ \binom{m}{k}\right) \geq \frac{1}{k+1}\left(1 - S_{k+1} \Big/ \binom{m}{k+1}\right)$.

Further inequalities are given in Fréchet's monographs.[13]

[13] M. Fréchet, "Les Probabilités Associés à un Système d'Événements Compatibles et Dépendants," *Actualités Scientifiques et Industrielles,* Nos. 859, 942 (Paris: Hermann & Cie (Éditeurs) 1940, 1943).

CHAPTER 3

DEPENDENCE AND INDEPENDENCE

3.1 INTRODUCTION

Our basic method of finding the probability of an event was to list all possible outcomes and note the favorable ones. However, if there is a large number of possibilities this procedure is impractical.

Often such complicated problems can be simplified by using the notion of independence. Roughly speaking, two events A and B are independent if the occurrence of A has no influence on the occurrence of B and vice versa. This simplification arises from the fact that for independent events, $P(AB) = P(A)P(B)$. A similar result holds for n events. The notion of independence enables us to study repeated independent trials; this corresponds to the intuitive idea of repeating an experiment under identical conditions.

If the occurrence of B influences the occurrence of A, then we require the concept of conditional probability. This in turn leads to the theorem of total probability and Bayes' theorem,[1] which are quite useful for a large class of problems.

[1]After the Reverend Thomas Bayes, an English mathematician (1702–1761).

3.2 CONDITIONAL PROBABILITY

To see that the occurrence of an event B can alter the probability of an event A, consider the following example. Imagine three closed opaque boxes, one of which contains a coin. The probability of the coin being in one of the boxes is $\frac{1}{3}$. Suppose that we choose two boxes in an attempt to find the coin. Let B be the event that the first box is empty and let A be the event that the second box contains the coin. Then the probability of A given that B has occurred is $\frac{1}{2}$, while the probability of A given B' is zero.

As a generalization, let us consider a probability space consisting of n points with equal weights. If we want to find the probability of an event A given that an event B has occurred, we must take into account that the number of favorable cases for A has been reduced. Since B has occurred, the number of favorable cases is restricted to those common to A and B or the number of points $n(AB)$ in AB. The total number of outcomes, which is also restricted due to the occurrence of B, is just the number of points $n(B)$ in B. If we assume that $n(B) \neq 0$, then it follows from our elementary definition that the desired probability is

$$\text{(1)} \qquad \frac{n(AB)}{n(B)} = \frac{n(AB)/n}{n(B)/n} = \frac{P(AB)}{P(B)} .$$

As an illustration let us use Equation (1) to find the probability of getting at least two heads in three tosses of a coin (the event A), given that the first toss resulted in a head (the event B). The probability is $\frac{3}{4}$, since $n(AB) = 3$ and $n(B) = 4$. Using Equation (1) we obtain the same result, since $P(AB) = \frac{3}{8}$ while $P(B) = \frac{1}{2}$. Note that $P(A) = \frac{1}{2}$, since there are eight possible outcomes and four are favorable.

We can use Equation (1) to define conditional probability for a general finite probability space.

DEFINITION 1

Let $(\Omega, \mathscr{E}, P)$ be a probability space and let A be an event with positive probability. The *(conditional) probability of* B *given* A, denoted $P(B/A)$, is defined by

$$\text{(2)} \qquad P(B/A) = \frac{P(AB)}{P(A)} .$$

The parentheses in the above definition signify that the word "conditional" is sometimes omitted; also, $P(B/A)$ is often denoted by $P_A(B)$.

For events A with $P(A) > 0$ we can define a new probability space $(\Omega_1, \mathscr{E}_1, P_1)$, where $\Omega_1 = A$, $\mathscr{E}_1 = (AB : B \in \mathscr{E})$, and $P_1(BA) = P(B/A)$. We leave the proof of this statement to the reader as an exercise (cf., Exercise 9).

The so-called *theorem of compound probabilities* or *multiplication theorem* is just a restatement of Definition 1.

THEOREM 1

$$P(AB) = P(B/A)P(A). \tag{3}$$

COROLLARY

For n events $A_1, A_2, \ldots, A_n$, we have

$$P(A_1, A_2, \ldots, A_n) = P(A_1)P(A_2/A_1)P(A_3/A_1A_2)\cdots P(A_n/A_1A_2\cdots A_{n-1}). \tag{4}$$

Proof

The proof is by induction and follows from the theorem by observing that $P(A_1A_2\cdots A_n) = P(A_n/A_1A_2\cdots A_{n-1})P(A_1A_2\cdots A_{n-1})$.

The advantage of using the form given in this theorem is that we do not have to worry if $P(B) = 0$; in this situation we automatically have $P(AB) = 0$ and we set both sides equal to zero even though $P(A/B)$ is indeterminate.

EXAMPLE 1

Suppose that we have n cards numbered from 1 to n. If the cards are thoroughly shuffled and we draw them one at a time without replacing them (sampling without replacement), what is the probability of getting the sequence $i_1, i_2, \ldots, i_n$?

Let A_k be the event that on the kth drawing we get the card numbered i_k. Then $P(A_1) = 1/n$ and $P(A_2/A_1) = 1/n-1$, since all outcomes are equally likely. In general, $P(A_{r+1}/A_1A_2\cdots A_r) = 1/n-r$, $r = 2, \ldots, n-1$, since there are $n-r$ cards left after r drawings. The answer is found by applying Equation (4), which yields $P(A_1A_2\cdots A_n) = 1/n!$. This probability can also be calculated by noting that there are $n!$ different outcomes.

Another useful result known as the *rule of total probability* follows from Equation (3).

THEOREM 2

Let $\Omega = B_1 + B_2 + \cdots + B_n$; that is, we have a partition of the space Ω. For any event A we have

(5) $P(A) = P(A/B_1)P(B_1) + P(A/B_2)P(B_2) + \cdots + P(A/B_n)P(B_n).$

Proof

The events AB_i form a partition of A; since they are mutually exclusive, their probabilities add. Hence

$$P(A) = P(AB_1) + P(AB_2) + \cdots + P(AB_n).$$

The result follows by applying Equation (3) to every $P(AB_i)$.

Equation (4) is useful because it is sometimes easier to calculate the conditional probabilities $P(A|B_i)$ than to evaluate $P(A)$ directly.

EXAMPLE 2

A fair coin is tossed. If it lands heads we throw a pair of well-balanced dice; if it lands tails we pick a card from a well-shuffled deck of 11 cards numbered 2, 3, 4, . . . , 12. What is the probability of getting the number i?

Let H and T be the events that we get a head and a tail; let A_i be the event that we get the number i. Then $P(A_i|H)$ is just the probability of getting the number i with a pair of dice; it is given by

$$\begin{aligned} P(A_i|H) &= \frac{i-1}{36} \qquad (i = 2, \ldots, 7) \\ &= \frac{13-i}{36}. \qquad (i = 8, \ldots, 12) \end{aligned}$$

Since every outcome is equally likely, $P(A_i|T) = \frac{1}{11}$. From Equation (5) it follows that $P(A_i) = P(A_i|H)P(H) + P(A_i|T)P(T)$ and so

$$\begin{aligned} P(A_i) &= \frac{1}{2}\left(\frac{i-1}{36}\right) + \frac{1}{2}\frac{1}{11} \qquad (i = 2, \ldots, 7) \\ &= \frac{1}{2}\left(\frac{13-i}{36}\right) + \frac{1}{2}\frac{1}{11}. \qquad (i = 8, \ldots, 12) \end{aligned}$$

EXAMPLE 3

Suppose that we have a large water storage tank of uniform cross section filled to a height h. Accordingly, the volume is proportional to the height. The probability that a height k of the water will be required in any time interval of length t is

$$R_k(t) = \frac{(\mu t)^k}{k!} e^{-\mu t}. \qquad (6)$$

Assuming that the demand for the water is independent of its height, what is the probability of having k units of water left at time t? What is the mean height of the water at time t?

Let $A_i(t)$ be the event that the height of the water in the tank is i units at time t. Let $D_i(t)$ denote the event that i units are required in a time interval of length t.

Then,

$$A_0(t+\Delta) = A_0(t) + A_h(t)\sum_{i=h}^{\infty} D_i(\Delta) + A_{h-1}(t)\sum_{i=h-1}^{\infty} D_i(\Delta) + \cdots$$
$$+ A_1(t)\sum_{i=1}^{\infty} D_i(\Delta),$$

since the only way to have no water at time $t+\Delta$ is either to have no water at time t, or to have i units, where $1 \le i \le h$, at time t and i or more units required in a time Δ. Further,

$$A_k(t+\Delta) = A_k(t)D_0(\Delta) + A_{k+1}(t)D_1(\Delta) + \cdots + A_h(t)D_{h-k}(\Delta) \qquad (k = 1, \ldots, h-1) \tag{7}$$

$$A_h(t+\Delta) = A_h(t)D_0(\Delta). \tag{8}$$

Equation (7) follows from the fact that during a time Δ we may need $0, 1, 2, \ldots,$ or h units. If we required i units during Δ and if the height at time $t+\Delta$ is k, then the height at time t must have been $k+i$. Equation (8) states that the only way to have h units in the tank at time $t+\Delta$ is to have h units in the tank at time t and to have no demands during the time Δ.

Let $P_k(t)$ denote the probability of event $A_k(t)$ occurring. From Equation (6) (using the Taylor expansion for $e^{-\mu\Delta}$), $P(D_0(\Delta)) = 1 - \mu\Delta + o(\Delta)$, $P(D_1(\Delta)) = \mu\Delta + o(\Delta)$, and $P(D_i(t)) = o(\Delta)$, where $i = 2, 3, \ldots, h$.[2] Hence equations (7) and (8) can be rewritten as

$$P_0(t+\Delta) = P_0(t) + \mu\Delta\, P_1(t) + o(\Delta) \tag{9}$$

$$P_k(t+\Delta) = P_k(t)(1-\mu\Delta) + P_{k+1}(t)\,\mu\Delta + o(\Delta) \quad (k=1,\ldots,h-1) \tag{10}$$

$$P_h(t+\Delta) = P_h(t)(1-\mu\Delta) + o(\Delta). \tag{11}$$

Expanding Equations (9)–(11) in powers of Δ and letting $\Delta \to 0$ yields, after some simplifications,

$$\frac{dP_0}{dt} = \mu P_1 \tag{12}$$

$$\frac{dP_k}{dt} = -\mu P_k + \mu P_{k+1} \qquad (k = 1, 2, \ldots, h-1) \tag{13}$$

$$\frac{dP_h}{dt} = -\mu P_h. \tag{14}$$

[2]The notation $f(x) = 0(g(x))$ at $x = a$ means $\lim_{x\to a} f/g = 0$.

The solution of these equations is obtained by first solving Equation (14) and then substituting the answer in Equation (13), with $k = h - 1$. The resulting equation is solved and the answer substituted in Equation (13), with $k = h - 2$. A repetition of this procedure leads to

$$P_0(t) = 1 - e^{-\mu t}\left[\sum_{i=0}^{h-1} \frac{(\mu t)^i}{i!}\right] \tag{15}$$

$$P_i(t) = \frac{(\mu t)^{h-i}}{(h-i)!}\, e^{-\mu t}. \qquad (i = 1, 2, \ldots, h) \tag{16}$$

The mean height $h(t)$ at time t is $\sum_{i=0}^{h} iP_i(t)$, which reduces to

$$h(t) = (h - \mu t)e^{-\mu t}\left[\sum_{i=0}^{h-2} \frac{(\mu t)^i}{i!}\right] + he^{-\mu t}\,\frac{(\mu t)^{h-1}}{(h-1)!}\,. \tag{17}$$

For large h, Equation (17) reduces to $h(t) = h - \mu t$. It is interesting to compare this result with a deterministic situation in which μ units of water are required per unit of time; in this case $h(t) = h - \mu t$ also.

Frequently, the occurrence of an event A alters our estimation of the probabilities of events $B_1, B_2, \ldots, B_n$. A quantitative statement of our new estimates is given by Bayes' theorem or the *formula for probabilities of hypotheses*. Bayes' original essays[3] were published posthumously by R. Price.

THEOREM 3

Let $B_1, B_2, \ldots, B_n$ be events which form a partition of Ω, and let A be another event. Then the conditional probability of B_i, given A, is

$$P(B_i|A) = \frac{P(A|B_i)P(B_i)}{\sum_{k=1}^{n} P(A|B_k)P(B_k)}\,. \qquad (i = 1, 2, \ldots, n) \tag{18}$$

Proof

By using the multiplication theorem 1 twice,

$$P(B_i|A) = \frac{P(B_i A)}{P(A)} = \frac{P(A|B_i)P(B_i)}{P(A)}\,. \tag{19}$$

Expressing the denominator in Equation (19) in the form of the total probability equation (5), we obtain Equation (18).

[3]T. Bayes, "An Essay Towards Solving a Problem in the Doctrine of Chances," *Phil. Trans.*, **53** (1963), 376–398; **54** (1964), 298–310.

EXAMPLE 4

A ball is selected at random from two urns by first choosing an urn and then drawing the ball from the selected urn. The first urn contains three black balls and seven white balls, while the second contains nine white balls and one black ball. If the chosen ball is black, what is the probability that it came from the first urn? Suppose that the ball is replaced in the urn from which it came, the experiment is repeated, and we again obtain a black ball. Now what is the probability that the ball came from the first urn?

Let B_i be the event that the ball came from urn i, where $i = 1, 2$, and let A be the event that the ball is black. It is reasonable to assume that $P(B_1) = P(B_2) = \frac{1}{2}$, since either urn is equally likely to be chosen. The probability $P(B_i) = \frac{1}{2}$ is called the *a priori* (or *prior*) probability of the event B_i and is known before any observations are taken. Equation (18) yields

$$P(B_1|A) = \frac{(0.3)(0.5)}{(0.3)(0.5) + (0.1)(0.5)} = \frac{3}{4},$$

since $P(A|B_1) = \frac{3}{10}$ and $P(A|B_2) = \frac{1}{10}$. Similarly, $P(B_2|A) = \frac{1}{4}$, which can also be obtained by noting that $P(B_1|A) + P(B_2|A) = 1$. The conditional probability $P(B_i|A)$ is called the *a posteriori* (or *posterior*) probability; it is a reappraisal of $P(B_i)$ after the experiment has been performed—that is, after event A has occurred.

When the experiment is repeated a second time, all the conditions are similar to those described above. However, for our prior probabilities, we use the new estimates obtained above—$P(B_1) = \frac{3}{4}$ and $P(B_2) = \frac{1}{4}$. Hence Bayes' theorem yields $P(B_1|A) = \frac{9}{10}$ and $P(B_2|A) = \frac{1}{10}$. Alternatively, we could let A be the event that two black balls were drawn from the urn and could then find $P(B_i|A)$ in one step; the same result would be obtained.

Frequently Bayes' theorem is applied incorrectly because in some situations the quantities $P(A|B_i)$ and $P(B_i)$ are not known. The cardinal error is usually the mistaken belief that complete ignorance of $P(B_i)$ is equivalent to equal likelihood of all the B_i. Sometimes this assumption makes sense, as in the above example; at other times it does not. An oftenquoted consequence of Bayes' theorem, Laplace's rule of succession, has a long history of misuse. This law can easily be derived by reinterpreting the following example, which clearly illustrates the underlying assumptions.

EXAMPLE 5

Suppose that we have $N+1$ boxes; the ith box, where $i = 0, 1, 2, \ldots, N$, contains $N - i$ cards marked with an F and i cards marked with an S. First let us choose a box at random and then draw m cards successively from this box. We replace the drawn card each time before proceeding to the next draw (sampling with replacement). If we obtain m cards marked with an S, what is the probability of drawing another card marked with an S from the same box, assuming that all outcomes are equally likely?

Let A be the event that m cards bearing S are chosen on the first m draws, and let B be the event that a card marked S is obtained on the $(m + 1)$st draw. For $i = 0, 1, \ldots, N$, let H_i be the event that we choose box i. Since we are required to find $P(B|A)$, we begin by finding $P(AB)$, the probability of getting $m + 1$ cards marked S.

Since a box is chosen at random, $P(H_j) = 1/(N + 1)$; since there are N^{m+1} different outcomes, of which j^{m+1} are favorable under the hypothesis that the jth box was chosen, $P(AB|H_j) = (j/N)^{m+1}$.

Therefore,

$$P(AB) = \sum_{j=0}^{N} P(AB|H_j)P(H_j) = \sum_{j=0}^{N} \left(\frac{j}{N}\right)^{m+1} \frac{1}{N+1}.$$

Similarly,

$$P(A) = \sum_{j=0}^{N} P(A|H_j)PH_j = \sum_{j=0}^{N} \left(\frac{j}{N}\right)^{m} \frac{1}{N+1}.$$

From our definition of conditional probability, Equation (2),

$$P(B|A) = \frac{[1/(N+1)] \sum_{j=0}^{N} (j/N)^{m+1}}{[1/(N+1)] \sum_{j=0}^{N} (j/N)^{m}}.$$

An alternate method for obtaining this result is to calculate the *a posteriori* probabilities of the H_i, given that m S's were drawn, and to use these estimates for PH_i in calculating the probability that the $(m + 1)$st card is S. Bayes' theorem yields

$$P(H_i|A) = C\left(\frac{i}{N}\right)^{m} \frac{1}{N+1}, \qquad (i = 0, \ldots, N) \tag{20}$$

where C represents the denominator appearing in Bayes' result. The value of C is obtained from the condition $\sum_{i=0}^{N} P(H_i|A) = 1$. By the total probability theorem,

$$P(B) = \sum_{i=0}^{N} P(B|H_i)PH_i = \sum_{i=0}^{N} \left(\frac{i}{N}\right) PH_i,$$

where the estimate in Equation (20) is used for PH_i. By carrying out the details of the calculation, the reader can verify that this expression for PB agrees with $P(B|A)$ obtained above.

The same result is also obtained if we estimate PH_i in m steps. At each stage we must reappraise PH_i because we have obtained another card marked S. The procedure is outlined in Exercise 8.

For large N, the above expression can be simplified. We can regard the sums as approximating the Riemann integral;[4] therefore,

$$\frac{1}{N+1}\sum_{j=0}^{N}\left(\frac{j}{N}\right)^{m+1} \approx \int_0^1 x^{m+1}\,dx = \frac{1}{m+2}$$

$$\frac{1}{N+1}\sum_{j=0}^{N}\left(\frac{j}{N}\right)^{m} \approx \int_0^1 x^{m}\,dx = \frac{1}{m+1}.$$

Hence

$$P(B|A) \approx \frac{m+1}{m+2}.$$

This result is often loosely stated as follows: If an event has occurred successively m times, the probability that it will happen again the next time is $(m+1)/(m+2)$.

If we interpret success as a card marked S and failure as a card marked F, then we see that the previous statement is true only under the assumptions we have made in the above example: that (0 successes, N failures), . . . , (N successes, 0 failures) are all equally likely.

Laplace illustrated his rule of succession by finding the probability that the sun will rise the next day, given that it has risen daily for 5000 years, or 1,826,213 days. Laplace was willing to bet 1,826,214 to 1 that it would rise the next day.

If we accept this reasoning, we must also believe that it was just as likely for the sun not to have risen during those 5000 years. It is difficult to see what Laplace really meant, since the statement appears in *Essai Philosophique sur les Probabilités* (1814), which was originally a popular translation of his classic technical work *Théorie Analytique des Probabilités*, written in 1812.

It is easy to manufacture "applications" that are even more absurd. A man is contemplating suicide by jumping from a 100-story building. Being a firm believer in Laplace's law, he calculates the probability that he will come to another floor before he hits the ground. Since he would have passed 100 floors, this probability is $\frac{101}{102}$. Accordingly he abandons his plan, since he is practically certain that he will never hit the ground.

Bayes' theorem is often used as a basis for statistical inference. To illustrate this procedure, suppose that we are given a box containing N cards marked S or F. The problem is to determine the number of cards marked S and the number marked F by sampling M cards (with replacement). The

[4]After G. F. B. Riemann, a German mathematician (1826–1866).

method of solution is to assume that instead of one box we have $N+1$ boxes of the composition described in Example 5. An *a priori* distribution is assigned to these $N+1$ boxes on the basis of experience. Using these *a priori* distributions, we can calculate the probability of getting a box of a particular composition, as in Example 5. It can be shown that for a large number of trials the results are independent of the *a priori* distribution, provided it satisfies certain mild restrictions.[5]

Opponents of the above approach point out that there is really only one box. Therefore, it is not logical to assign an *a priori* distribution. Methods of non-Bayesian inference[6] have been developed based on R.A. Fisher's theory of fiducial probability and on J. Neyman's theory of confidence intervals.

EXERCISES

1. A factory that produces ball bearings has two machines M_1 and M_2, which manufacture 80 percent and 20 percent of the ball bearings respectively. The probabilities of a standard bearing being produced by M_1 and M_2 are 0.9 and 0.8, respectively. What is the probability that the factory produces a standard bearing?

2. Suppose that a farmer has a mixture of corn seeds that includes 60 percent of type T_1, 30 percent of type T_2, and 10 percent of type T_3. The probabilities that an ear of corn will have more than 100 kernels are 0.75, 0.50, and 0.25 for types T_1, T_2, and T_3, respectively. What is the probability of the farmer getting an ear of corn with no more than 100 kernels?

3. Suppose that one coin in a million is two-headed and the remaining coins are fair. Suppose that a coin, selected at random, is tossed 30 times in succession. What is the probability that the coin is true if the result is heads each time?

4. The probability of a child belonging to blood group A, genotype AO, when the mother belongs to group O, is $\frac{1}{2}$ if the father is of genotype OA, 1 if the father is of genotype AA, and $\frac{1}{2}$ if the father is of genotype AB. The probabilities that a male belongs respectively to these three genotypes are 0.352, 0.069, and 0.015. If a child of an O mother is found to belong to genotype OA, what is the probability that its father has blood of type AB?

5. Suppose that it is equally likely for a family to have a boy or a girl, and that a family has two children. What is the probability that both children are girls if (a) the younger child is a girl, (b) at least one is a girl?

[5]The proof of this statement as well as further details on Bayesian inference can be found in *Mathematical Theory of Probability and Statistics*, by R. von Mises (New York: Academic Press Inc., 1964).

[6]H. Cramer, *Mathematical Methods of Statistics* (Princeton, N. J.: Princeton University Press, 1961).

6. Let a target be located on segment AB of a straight line. The line is divided into three subsegments: a central piece c and two equal end-pieces e and f. The exact position of the target is unknown, but we know the probabilities that it lies in one of the three segments: $P(c) = 0.8$ and $P(e) = P(f) = 0.1$, where c, e, and f each stand for the event that the target lies in the corresponding segment. The target may be damaged even if it lies in a segment at which we are not aiming, because of errors in firing. Let the probabilities of hitting the target be 0.6, 0.4, and 0.3 if it lies in segments e, c, and f, respectively. If the target is hit, what is the probability that it lies in each of the three segments?

7. A patient has one of three diseases D_1, D_2, and D_3 that have probabilities $\frac{5}{8}$, $\frac{2}{8}$, and $\frac{1}{8}$, respectively. To help diagnosis, a test is carried out; it yields a positive result with a probability of 0.2 for disease D_1, 0.3 for disease D_2, and 0.9 for disease D_3. What is the probability of each disease after testing?

8. Do Example 5 by applying Bayes' theorem in the following manner. Let the *a priori* probability of H_j be $1/(N+1)$ and calculate the *a posteriori* probability of H_j given that a card marked S was drawn. Use this probability as the new *a priori* probability for H_j. Repeat the procedure m times, use the *a posteriori* probability for H_j obtained in this manner in the total probability theorem, and find $P(B|A)$. (*Hint*: After k steps, the estimate for PH_j is j^k/t_k, where $t_k = 1^k + 2^k + \cdots + N^k$.)

9. Let $(\Omega, \mathscr{E}, P)$ be a probability space (see Section 1.4), and let $A \in \mathscr{E}$ such that $P(A) > 0$. We define a new field $\mathscr{E}_1$ of sets (see Section 1.5) whose elements are of the form BA, where $B \in \mathscr{E}$. On this field we define a new probability function P_1 by $P_1(E) = [P(AB)/P(A)]$, where $E = AB \in \mathscr{E}_1$. Show that $(A, \mathscr{E}_1, P_1)$ is a probability space; that is,

(a) $\mathscr{E}_1$ is a field.
(b) For any $E \in \mathscr{E}_1$, $P_1(E) \geq 0$ and $P_1(A) = 1$.
(c) If $E_1, E_2 \in \mathscr{E}_1$ and are disjoint, $P_1(E_1 + E_2) = P_1(E_1) + P_1(E_2)$. [Let $E_1 = AB_1$ and $E_2 = AB_2$; then the problem is equivalent to showing $P(B_1 + B_2|A) = P(B_1|A) + P(B_2|A)$.]

10. Prove that the following statements hold if $P(A) > 0$.

(a) $P(\Omega|A) = 1$. (if Ω is the sure event)
(b) $P(B|A) = 0$. (if $P(B) = 0$)
(c) $P(B_1 \cup B_2|A) = P(B_1|A) + P(B_2|A) - P(B_1 B_2|A)$.
(d) $P(B|A) + P(B'|A) = 1$.

3.3 INDEPENDENT EVENTS

We saw in the preceding section that $P(B|A)$ need not be the same as $P(B)$. However, in some situations the occurrence of the event A has no influence on the occurrence of the event B. Consider the experiment that consists of tossing a coin and then casting a die. Let A be the event that we obtain a head and let B be the event that the face bearing the number 6 turns up. Clearly, event B does not depend on whether event A did or did not take place. Accordingly, physical intuition would lead us to suspect that $P(B|A) = P(A)$. In other words, Equation 2.3 yields $P(AB) = P(A)P(B)$. Assuming that all possible outcomes are equally likely, the reader can easily verify that $P(AB) = \frac{1}{12}$, while $P(A) = \frac{1}{2}$ and $P(B) = \frac{1}{6}$. This leads us to frame the following formal definition which agrees with our intuition.

DEFINITION 1

Two events A and B are said to be *stochastically independent* or, in short, *independent* if and only if

$$P(AB) = P(A)P(B). \tag{1}$$

We can usually tell by experience when two events are independent and can calculate $P(AB)$ by assuming that Equation (1) holds.

We remark that the notion of independence is a symmetrical relation. In addition, if $P(B) > 0$, then $P(A|B)$ is well defined and Equation 2.2 implies $P(A|B) = P(A)$. Similarly, if $P(A) > 0$, then $P(B|A) = P(B)$.

The reader might feel that two mutually exclusive events are independent. However, for this to be true either $P(A)$ or $P(B)$ must be zero, since Equation (1) must hold and $P(AB) = 0$. For example, let us toss a coin twice. Let A be the event that we get exactly one head and B the event that we get two heads. Then $P(AB) = 0$, while $P(A) = \frac{1}{2}$ and $P(B) = \frac{1}{4}$.

If B does not depend on the occurrence of A, then it should not depend on the nonoccurrence of A (or A'). This leads to the following theorem.

LEMMA 1

If the events A and B are independent, then each pair of events $(A', B), (A, B')$, and (A', B') are independent.

Proof

From the proof of the total probability theorem 2.2,

$$P(A'B) + P(AB) = P(B).$$

By independence, $P(A'B) = P(B)[1 - P(A)] = P(B)P(A')$; that is, A' and B are independent. The independence of the remaining pairs can be proved in a similar manner.

The notion of independence can be extended to more than two events.

DEFINITION 2

The events A_i, where $i = 1, 2, \ldots, n$, are *mutually* (or *collectively*) *independent* if

$$P(A_{i_1}A_{i_2}A_{i_3}\cdots A_{i_m}) = PA_{i_1}PA_{i_2}PA_{i_3}\cdots PA_{i_m} \tag{2}$$

for every $m \leq n$ and for arbitrary distinct positive integers $i_1, i_2, i_3, \ldots, i_m \leq n$. [Equation (2) stands for $\binom{n}{2} + \binom{n}{3} + \cdots + \binom{n}{n}$ equations.]

In practice, if the events are *pairwise independent*—that is, $P(A_i A_j) = P(A_i)P(A_j)$—then usually they are independent. However, in general, pairwise independence is not a sufficient condition for events to be independent, as the following example shows. This fact was first discovered by S. Bernstein, a Russian mathematician.

EXAMPLE 1

The following combinations of symbols for apples, pears, and lemons appear on the face of a slot machine: (a, a, a), (p, p, p), (l, l, l), and (a, p, l). Assume that these four outcomes are equally likely. Let A, B, and C denote the events that the symbol for an apple, pear, and lemon is one of the symbols of the triplet appearing on the machine. Then $P(A) = P(B) = P(C) = \frac{1}{2}$ and $P(AB) = P(AC) = P(BC) = \frac{1}{4}$. Hence, the events A, B, and C are pairwise independent. However, they are not independent, since $P(ABC) = \frac{1}{4}$.

THEOREM 1

If the events A_i $(i = 1, 2, \ldots, n)$ are independent, then

$$\begin{aligned} P(A_{i_1}A_{i_2}\cdots A_{i_m}A'_{j_1}\cdots A'_{j_k}) \\ = P(A_{i_1})P(A_{i_2})\cdots P(A_{i_m})P(A'_{j_1})P(A'_{j_2})\cdots P(A'_{j_k}) \end{aligned} \tag{3}$$

for every $(m + k) \leq n$ and for arbitrary distinct positive integers $i_1, \ldots, i_m, j_1, \ldots, j_k < n$.

Proof

The proof is by a double induction on the number of complementary events appearing and on the value of $m + k$. Let (m, k) denote the fact that we are considering m of the A_i and k of the A'_i. For any value of $m + k$ the theorem is true for the situation $(m + k, 0)$ because of our definition of independence [replace m by $m + k$ in Equation (2)]. From Lemma 1, the theorem holds for (m, k) if $m + k = 2$.

Now assume that the theorem is true for $(m + 1, k - 1)$ and if the number of events under consideration is less than $m + k$. From these two facts we must show that it holds for (m, k). This will complete the proof, since from the cases $m+k=2$ and $(m, 0)$ it follows that the theorem is true for all $m + k$. We illustrate the induction procedure by showing that the theorem is true for all (m, k) such that $m + k = 3$. Since $(3, 0)$ is true and

$(m+1, k-1)$ implies that (m, k) is true, it follows that $(2, 1)$ is valid. By the same argument, $(2, 1) \to (1, 2)$ and $(1, 2) \to (0, 3)$.

Consider the following identity:

$$(4) \qquad P(A_{i_1} \cdots A_{i_m} A'_{j_1} \cdots A'_{j_k}) + P(A_{i_1} \cdots A_{i_m} A'_{j_1} \cdots A'_{j_{k-1}} A_{j_k}) = P(A_{i_1} \cdots A_{i_m} A'_{j_1} \cdots A'_{j_{k-1}}).$$

Since the number of events appearing in the right-hand side of Equation (4) is $m + k - 1$,

$$P(A_{i_1} \cdots A_{i_m} A'_{j_1} \cdots A'_{j_{k-1}}) = P(A_{i_1}) \cdots P(A_{i_m}) P(A'_{j_1}) \cdots P(A'_{j_{k-1}})$$

by the induction hypothesis. Also from the induction hypothesis,

$$P(A_{i_1} \cdots A_{i_m} A'_{j_1} \cdots A'_{j_{k-1}} A_{j_k}) = P(A_{i_1}) \cdots P(A_{i_m}) P(A'_{j_1}) \cdots P(A'_{j_{k-1}}) P(A_{j_k}),$$

since it is of the form $(m+1, k-1)$. Substituting these results in Equation (4) and simplifying as in the proof of Lemma 1 yields Equation (3).

Theorem 1 can be generalized; indeed, any two *Boolean functions* of a set of independent events are independent. By a Boolean function of the events $A_1, A_2, \ldots, A_n$, we mean any combination of the A_i and their complements using unions and intersections—e.g., $(A_1 A'_2 A_3 \cup A_4(A_5 A_6))A_7$. A Boolean function of events is an event, since the class of all events forms a field. A precise statement of this result is contained in the following theorem.

THEOREM 2

Let $A_1, A_2, \ldots, A_n$ be independent events. Let $F(A_1, A_2, \ldots, A_m)$ and $G(A_{m+1}, A_{m+2}, \ldots, A_n)$ be arbitrary Boolean functions. Then

$$P(F(A_1, A_2, \ldots, A_m) \cap G(A_{m+1}, A_{m+2}, \ldots, A_n)) = P(F(A_1, \ldots, A_m)) P(G(A_{m+1}, \ldots, A_n)).$$

Proof

We use the result that the functions can be written in the following normal forms:

$$(5) \qquad F(A_1, \ldots, A_m) \equiv H_k = B_1 \cup B_2 \cup \cdots \cup B_k$$

$$(6) \qquad G(A_{m+1}, \ldots, A_n) \equiv G_r = C_1 \cup C_2 \cup \cdots \cup C_r,$$

where each B_i consists only of the intersections of A_i or A'_i $(i = 1, \ldots, m)$, and each C_i consists of the intersections of A_i or A'_i $(i = m + 1, \ldots, n)$.

We will prove the theorem by a double induction. For the present, let us assume that the theorem is true for $k = 1$ and arbitrary r; this hypothesis will be proved later by induction. Under this assumption we will prove, by induction on k, that the theorem holds for arbitrary k. Hence we must demonstrate that if the theorem is true for $k - 1$, it is true for k.

Since $H_k G_r = H_{k-1} G_r \cup B_k G_r$,

$$P(H_k G_r) = P(H_{k-1} G_r) + P(B_k G_r) - P(H_{k-1} B_k G_r). \tag{7}$$

From the induction hypothesis, $P(H_{k-1} G_r) = P(H_{k-1})P(G_r)$ and $P(H_{k-1} B_k G_r) = P(H_{k-1} B_k)P(G_r)$, since $H_{k-1} B_k$ is of the form of Equation (5) with $k - 1$ factors. Also, since we have assumed that the theorem is true if we have only one factor B_i and for arbitrary r, $P(B_k G_r) = P(B_k)P(G_r)$. If we make all of these substitutions in Equation (7) and simplify, we obtain

$$P(H_k G_r) = P(H_{k-1} \cup B_k)P(G_r) = PH_k PG_r.$$

The only remaining task is to prove our first assumption that $P(B_1 G_r) = P(B_1)P(G_r)$; we will prove this by induction on r. For $r = 1$, the result follows from Theorem 1. We now assume that the theorem is true for $r - 1$ and prove that it is true for r. This follows from the observation that $B_1 G_r = B_1 G_{r-1} \cup B_1 C_r$, using reasoning similar to that in the preceding paragraph. We leave the details to the reader.

COROLLARY

Suppose that n independent events are divided into r disjoint classes $\mathscr{C}_i$. If F_i is a Boolean function of the events in $\mathscr{C}_i$, then

$$PF_1 F_2 \cdots F_r = PF_1 PF_2 \cdots PF_r.$$

Theorem 2 is not true for the arbitrary functions F and G, which map events into events, since they may map independent events into dependent ones.

EXAMPLE 2

What is the probability of getting a 6 or a 2 with one die and a 5 with the other when a pair of dice are thrown?

Let A and B be the events that 6 and 2 appear on the first die. Let C be the event that 5 appears on the second die. However, if we obtain a 6 or a 2 on the second die and a 5 on the first, the conditions of the problem are still satisfied. Hence the required answer is $2P((A \cup B)C)$. Since the events A, B, and C are independent, we have $P((A \cup B)C) = P(A \cup B)P(C)$, where $P(A \cup B) = \frac{2}{6}$ and $P(C) = \frac{1}{6}$. By enumerating all possible outcomes and counting the favorable cases, the reader can verify that the answer is $\frac{1}{9}$.

The preceding theorems show how, from a given set of events, we can form classes of independent events. For example, if A and B are independent events, then any event chosen from the class $\mathscr{C}_1 = \{\Omega, A, A', \phi\}$ is independent of any event chosen from the class $\mathscr{C}_2 = \{\Omega, B, B', \phi\}$. This follows from Lemma 1 and Exercise 1. Hence we are prompted to extend the notion of independence to classes of events.

DEFINITION 3

Let $\mathscr{C}_k$ be n classes of events defined on a space Ω. The classes $\mathscr{C}_k$ are *independent* if

$$P(A_1 A_2 A_3 \cdots A_n) = P(A_1)P(A_2)P(A_3) \cdots P(A_n)$$

for arbitrary events A_i selected from $\mathscr{C}_i$, where $i = 1, 2, \ldots, n$.

We can restate Theorem 2 in the above terminology as follows: The events A_i are independent if and only if the families $\mathscr{C}_i = \{\Omega, A_i, A'_i, \phi\}$ are independent, where $i = 1, 2, \ldots, n$.

The notion of independence for classes of events enables us to define independent random variables (see Exercise 14).

DEFINITION 4

Simple random variables v_i $(i = 1, 2, \ldots, n)$ are said to be *independent* if the partitions on which they are defined are independent.

An immediate consequence of this definition is the following multiplication property.

THEOREM 3

Let v_i $(i = 1, \ldots, n)$ be n independent simple random variables. Then

$$E(v_1 v_2 v_3 \cdots v_n) = (Ev_1)(Ev_2)(Ev_3) \cdots (Ev_n).$$

Proof

We need only prove the theorem for $n = 2$; the general case follows by induction, since the product of random variables is a random variable and $v_1 v_2 \cdots v_{n-1}$ is independent of v_n (see Exercise 9).

Let $v_1 = \sum_{r=1}^{m} x_r I_r$ and $v_2 = \sum_{s=1}^{n} y_s J_s$, where I_r and J_s are the indicators of A_r and B_s respectively. Then

$$E(v_1 v_2) = E\left(\sum_{r=1}^{m} \sum_{s=1}^{n} x_r y_s I_r J_s\right) = \sum_{r=1}^{m} \sum_{s=1}^{n} x_r y_s P(A_r B_s).$$

By independence, $P(A_r B_s) = P(A_r)P(B_s)$ and so

$$E(v_1 v_2) = \left[\sum_{r=1}^{m} x_r P(A_r)\right]\left[\sum_{s=1}^{n} y_s P(B_s)\right] = (Ev_1)(Ev_2).$$

The Bienaymé equality follows from this theorem.

THEOREM 4

If v_i are n independent random variables, then

$$\sigma^2 \sum_{i=1}^{n} v_i = \sum_{i=1}^{n} \sigma^2 v_i.$$

Proof

From Definition 3.2

$$\sigma^2 \sum_{i=1}^{n} v_i = E\left[\sum_{i=1}^{n} (v_i - \mu_i)\right]^2$$

$$= \sum_{i=1}^{n} E(v_i - \mu_i)^2 + E\left\{\sum_{\substack{i,j=1 \\ i \neq j}}^{n} (v_i - \mu_i)(v_j - \mu_j)\right\}.$$

However, if v_i and v_j are independent, then so are $v_i - \mu_i$ and $v_j - \mu_j$ (see Exercise 1).

$$E\{(v_i - \mu_i)(v_j - \mu_j)\} = E(v_i - \mu_i)E(v_j - \mu_j) = 0,$$

since $Ev_i = \mu_i$. (Note that only independence or pairs of random variables is required.) The Bienaymé equality follows from the definition of $\sigma^2 v_i$.

EXAMPLE 3

Let $s = (1/n)(r_1 + r_2 + \cdots + r_n)$ be the arithmetic mean of n independent random variables r_i, each having mean μ and variance σ^2. Then $Es = \mu$ and $\operatorname{Var} s = \sigma^2/n$.

$$Es = \frac{1}{n}(Er_1 + Er_2 + \cdots + Er_n) = \mu.$$

From Exercise 1(c) of Section 2.3,

$$\operatorname{Var}\left(\frac{1}{n} r_1\right) = \frac{\sigma^2}{n^2},$$

and from the Bienaymé equality,

$$\operatorname{Var} s = \sum_{i=1}^{n} \operatorname{Var}\left(\frac{1}{n} r_i\right) = \frac{\sigma^2}{n}.$$

This example is the formal basis of the intuitive belief that the arithmetic mean of a series of measurements is more accurate than any single measurement (the variance is reduced by a factor of $1/n$). We can regard each measurement as a random variable since it is influenced by a number of random factors. However, since the measurements are usually repeated under similar conditions, it is reasonable to assume that the mean and variance are the same for each measurement and that each measurement is independent of previous measurements.

If u and v are independent random variables, then $\rho(u, v) = 0$. The converse is not true. The correlation coefficient $\rho(u, v)$ can vanish even if v is a function of u.

EXAMPLE 4

Let $u = \pm 2, \pm 3$, with a probability of $\frac{1}{4}$ and $v = u^2$. The joint density function of u and v is given by $\frac{1}{4} = f(-2, 4) = f(2, 4) = f(-3, 9) = f(3, 9)$. Hence $\rho(u, v) = 0$ even though v depends on u.

Although $\rho(u, v)$ is not connected with general functional dependence between u and v, it is connected with the linear dependence of u and v.

THEOREM 5

$|\rho(u, v)| \leq 1$; the equality sign holds only if there exist constants $a = \pm\sigma v/\sigma u$ and b such that $v = au + b$ except for sets of points that have a probability of zero.

Proof

A direct calculation yields

$$\mathrm{Var}\left(\frac{u - Eu}{\sigma u} \pm \frac{v - Ev}{\sigma v}\right) = (1 \pm \rho(u, v))2.$$

The left-hand side cannot be negative and so $|\rho(u, v)| \leq 1$. For $\rho(u, v) = 1$ the variance is equal to zero. Chebyshev's inequality shows that the probability that $[(u - Eu)/\sigma u] - [(v - Ev)/\sigma v]$ is a constant is 1. Hence $v = au + b$. A similar argument applies when $\rho(u, v) = -1$.

EXERCISES

1. (a) Let Ω be the sure event, ϕ the impossible event, and A an arbitrary event. Show that Ω and A, ϕ and A are independent.

(b) Letting r_1 and r_2 be independent random variables, show that $r_i - c_i$ $(i = 1, 2)$ are independent random variables, where each c_i $(i = 1, 2)$ is a constant.

2. A rocket is tested by means of three independent tests that have the probabilities 0.85, 0.95, and 0.90 of detecting something wrong with a particular circuit. Find the probability that all of these tests will fail to detect a flaw.

3. Suppose that 60 percent of males and 65 percent of females of age 30 live to be 70 years old. What is the probability that a couple married when they are 30 years old will live to celebrate their fortieth anniversary?

4. The distribution of the four basic blood groups is O, 45 percent; A, 40 percent; B, 8 percent; and AB, 7 percent. What is the probability that three people selected at random are all of type A? none of type AB? of different types? at least one of type B?

5. Suppose that a neutron passing through plutonium releases one, two, or three other neutrons with probabilities of 0.4, 0.4, and 0.2 respectively. Each of these second-generation neutrons, in turn, releases one, two, or three third-generation neutrons with the same probabilities as those given above. Find the probabilities for the various possible numbers of third-generation neutrons.

6. Suppose that a machine has a straight shaft consisting of n pieces. Slight variations in the production process cause the length of each part to deviate from a standard length. Hence we can regard these lengths as random variables and assume that these random variables are independent. Let the mean length of part i be a_i and let its standard deviation be g_i, where $i = 1, 2, \ldots, n$. Find the mean value of the length of the shaft and the standard deviation. Let $n = 4$, $a_i = 10$ inches, and $g_i = 0.2$ inch; show that although the length of each part deviates by $g_i/a_i \times 100 = 2$ percent from its mean, the shaft only deviates by about 1 percent from its mean.

7. (a) Let $A_1, A_2, \ldots, A_n$ be independent events. Show that the probability that at least one of these events will occur (cf., Theorem 2.4.2) is

$$1 - (1 - P(A_1))(1 - P(A_2)) \cdots (1 - P(A_n)).$$

[*Hint:* The required probability is $P(A_1 \cup A_2 \cup \cdots \cup A_n) = 1 - P(A_1' A_2' \cdots A_n')$.]

(*b*) *The Problem of the Chevalier de Méré.* Which is more probable: throwing at least one ace with four dice or at least one double ace in 24 throws of a pair of dice?

8. Find the probability of curing a disease if 100 drugs that have independent effects are administered, each one having a probability of 0.004 of being successful.

9. Let $\mathscr{P}_1, \mathscr{P}_2, \ldots, \mathscr{P}_n$ be independent partitions of a space Ω. Show that the product of $\mathscr{P}_1, \mathscr{P}_2, \ldots, \mathscr{P}_{n-1}$ is independent of $\mathscr{P}_n$.

10. In the following figure, assume that the probability of each relay being closed is p and that each relay is open or closed independently of any other relay. Find the probability that current flows from A to B.

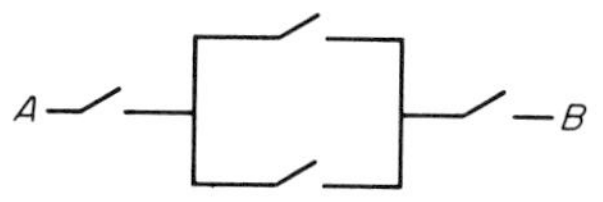

11. A single pair of genes determines a physical characteristic of a plant or animal by simple Mendelian inheritance. The color of flowers is an example. Let y and r represent yellow and red; flowers will be red if the plant has the color gene pair (r, r) and yellow if the color gene pair is (y, y). Yellow is dominant to red; if a plant has the color gene pair (y, r), it will have yellow flowers. The offspring get one gene from each parent and are equally likely to get either gene from each parent's pair. If (y, y) flowers are crossed with (r, r) flowers, all the resulting flowers will be (y, r), and therefore will be yellow.

(a) Show that if (y, r) flowers are crossed with (r, r) flowers, the probability is $\frac{1}{2}$ that the resulting flowers will be red and $\frac{1}{2}$ that they will be yellow.
(b) In crosses between (y, r) and (y, r), what proportion would be expected to be yellow? to be red?

12. Flowers may be smooth or wrinkled; this is a simple Mendelian characteristic. Smooth is dominant to wrinkled, so (s, s) and (s, w) flowers are smooth while (w, w) flowers are wrinkled. Find the possible outcomes and their associated probabilities if (y, r), (s, w) flowers are crossed with (r, r), (w, w) flowers.

13. Two machines M and N, operating independently, may have a number of breakdowns each day, with the probabilities indicated in the table below.

Number of breakdowns	0	1	2	3	4	5
M	0.1	0.2	0.3	0.2	0.11	0.09
N	0.45	0.1	0.1	0.1	0.1	0.15

Find the following probabilities:

(a) The total number of breakdowns is less than 3; less than 4.
(b) M has more breakdowns than N.
(c) M and N have the same number of breakdowns.
(d) M has twice as many breakdowns as N.
(e) The minimum (maximum) number of breakdowns of the two machines is 4, is less than 4.

14. Show that the random variables r_i are independent if and only if

$$f(x_1, x_2, \ldots, x_n) = f_1(x_1)f_2(x_2) \cdots f_n(x_n)$$

for all n-tuples of real numbers $(x_1, x_2, \ldots, x_n)$, where f is the *joint density function* of the random variable $(r_1, r_2, \ldots, r_n)$ and f_i is the *marginal density function* of the random variable r_i for $i = 1, 2, \ldots, n$.

3.4 CONDITIONAL PROBABILITY AND EXPECTATION

In view of Definition 1.1, a natural way to define the conditional probability $P(A|r = x)$ of an event A, given the event $(r = x)$, is

$$P(A|r = x) = \frac{P(A, r = x)}{P(r = x)}$$

if $P(r = x) > 0$, while $P(A|r = x)$ is undefined if $P(r = x) = 0$. This definition is suitable since we really only want to define $P(A|r = x)$ for real numbers x that could actually arise; that is, $P(r = x) > 0$.

In particular, we can define the *conditional density function* $f(y|x)$ of a random variable r_2, given a random variable r_1, by

$$f(y|x) = P(r_2 = y|r_1 = x) = \frac{f(x, y)}{f_1(x)},$$

where $f(x, y) = P(r_1 = x, r_2 = y)$ is the joint density function of (r_1, r_2) and $f_1(x) = P(r_1 = x)$ is the marginal density function of r_1.

The *conditional expectation* of r_2 given that $r_1 = x$, denoted by $E(r_2|r_1 = x)$, is defined as the mean of the conditional density function $f(y|x)$; that is,

(1) $$E(r_2|r_1 = x) = \sum_y yf(y|x),$$

where the summation is over all y such that $f(y|x) \neq 0$.

THEOREM 1

(2) $$E(r_2) = \sum_x E(r_2|r_1 = x)f_1(x). \qquad (f_1(x) > 0)$$

Proof

Replacing $E(r_2|r_1 = x)$ by its value given in Equation (1) yields

$$\sum_x E(r_2|r_1 = x)f_1(x) = \sum_x \sum_y yf(y|x)f_1(x)$$

$$= \sum_x \sum_y yf(x, y) = \sum_y y \sum_x f(x, y) = \sum_y yf_2(y) = E(r_2),$$

where $f_2(y) = P(r_2 = y)$. The second equality in this equation follows by replacing $f(y|x)$ by its defining value.

THEOREM 2

If r_1 and r_2 are independent random variables, then for all x such that $f_1(x) > 0$, we have $E(r_2|r_1 = x) = E(r_2)$.

Proof

If r_1 and r_2 are independent, then

$$f(y|x) = P(r_2 = y|r_1 = x) = P(r_2 = y) = f_2(y),$$

and so $E(r_2|r_1 = x) = \sum_y yf_2(y) = E(r_2)$.

THEOREM 3

For all x such that $f_1(x) > 0$,

$$E(\phi(r_1, r_2)|r_1 = x) = E(\phi(x, r_2)|r_1 = x)$$

for any function $\phi(\ ,\ )$ whose expectation $E\phi$ exists.

Proof

Define random variables u and v by $u = \phi(r_1, r_2)$ and $v = \phi(x, r_2)$; then

$$P(u = u_1, r_1 = x) = \sum_{\phi(x,y)=u_1} f(x, y)$$

$$P(v = v_1, r_1 = x) = \sum_{\phi(x,y)=v_1} f(x, y),$$

where the summations are over all y such that $\phi(x, y) = u_1$ and $\phi(x, y) = v_1$ respectively. Therefore, for all real numbers t and for x such that $f_1(x) > 0$,

$$P(u = t|r_1 = x) = P(v = t|r_1 = x).$$

Consequently,

$$E(u|r_1 = x) = \sum_t tP(u = t|r_1 = x) = \sum_t tP(v = t|r_1 = x)$$
$$= E(v|r_1 = x).$$

COROLLARY

Suppose that $\phi(x, y) = \phi_1(x)\phi_2(y)$ in this theorem; then

$$E(\phi(r_1, r_2)|r_1 = x) = \phi_1(x)E(\phi_2(r_2)|r_1 = x).$$

The simple proof is left to the reader.

The following properties of the conditional expectation can easily be verified by its definition (or by the obvious generalization to more than two random variables):

$$E(c|r_1 = x) = c \qquad (c \text{ is a constant})$$

$$E(\sum_{i=2}^{n} c_i r_i|r_1 = x) = \sum_{i=2}^{n} c_i E(r_i|r_1 = x). \quad (c_i \text{ are constants; } r_i \text{ are random variables})$$

The following example illustrates the use of these results as well as the use of difference equations in solving probability problems.

EXAMPLE 1

Consider n independent tosses of a coin that has a probability p of falling heads. Let s_n be the number of tosses in which the coin falls heads. Find Es_n.

Let r_i be a random variable that assumes the value 1 with a probability p if heads occurs on the ith trial, and 0 otherwise. Then the r_i $(i = 1, 2, \ldots, n)$ are independent random variables $Er_i = p$, and

$$s_n = r_1 + r_2 + \cdots + r_n.$$

Let $m_n = Es_n$; we shall derive a difference equation for m_n. By theorem 1,

$$Es_n = E(s_n|r_1 = 0)P(r_1 = 0) + E(s_n|r_1 = 1)P(r_1 = 1). \tag{3}$$

Using the facts

$$E(s_n|r_1 = 0) = E(r_1+r_2 + \cdots + r_n|r_1 = 0)$$
$$= Er_2 + Er_3 + \cdots + Er_n = Es_{n-1}$$

and

$$E(s_n|r_1 = 1) = E(1 + r_2 + \cdots + r_n|r_1 = 1)$$
$$= 1 + Er_2 + \cdots + Er_n = 1 + Es_{n-1},$$

Equation (3) reduces to

$$m_n = p(1 + m_{n-1}) + qm_{n-1} = m_{n-1} + p. \qquad (p + q = 1)$$

Since $m_1 = p$, it follows that $m_2 = 2p$, $m_3 = m_2 + p = 3p, \ldots, m_n = np$.

EXAMPLE 2

Chuck-a-luck is a gambling game played in the following manner. Three dice are rolled. A player may bet on any one of the numbers 1, 2, 3, 4, 5, and 6. If the player's number appears on one, two, or three of the dice, he receives respectively one, two, or three times his original stake plus his own money back; otherwise he loses his stake. What is the player's expected loss per unit stake?

Let l be a random variable representing the player's loss per unit stake. Let $r = 0$, 2, and 3 when the numbers on the three dice are different, when exactly two are alike, and when all three are alike, respectively. Then

$$E(l) = E(l|r = 0)P(r = 0) + E(l|r = 2)P(r = 2) + E(l|r = 3)P(r = 3).$$

Suppose that the player bets one unit on all six numbers and that the roll of the dice results in two of one number and one of a second, say 2, 2, and 6. The house takes in four units (the stakes on the numbers 1, 3, 4, and 5), and pays out two units for the bet on 2 and one unit for the bet on 6. Hence the house wins one unit or the player loses one unit. The loss per unit stake is $\frac{1}{6}$. A similar result is obtained for any other two repeated numbers. Therefore, $E(l|r = 2) = \frac{1}{6}$. The reader can easily verify that $P(r=2) = \frac{90}{216}$.

Using this reasoning, we can show that $E(l|r = 0) = 0$ and $E(l|r = 3) = \frac{2}{6}$. Since $P(r = 0) = \frac{120}{216}$ and $P(r = 3) = \frac{6}{216}$, we have

$$E(l) = \frac{120}{216} \times 0 + \frac{90}{216} \times \frac{1}{6} + \frac{6}{216} \times \frac{2}{6} = \frac{17}{216}.$$

It can easily be verified that

$$E(I_A|r = x) = P(A|r = x).$$

Consequently, we immediately obtain the following results from Theorems 1–3.

THEOREM 4

For any event A, random variables r_1 and r_2, function ϕ, and set of real numbers B, and for any x such that $f_1(x) \neq 0$,

1. $PA = \sum_x P(A|r_1 = x)f_1(x)$;
2. $P(r_2 \in B|r_1 = x) = P(r_2 \in B)$ if r_1 and r_2 are independent;
3. $P(\phi(r_1, r_2) \in B|r_1 = x) = P(\phi(x, r_2) \in B|r_1 = x)$.

EXAMPLE 3

Using the notation of Example 1, find the probability P_n that s_n is even:

$$P_n = P(s_n \text{ even}) = P(s_n \text{ even}|r_1 = 0)P(r_1 = 0) + P(s_n \text{ even}|r_1 = 1)P(r_1 = 1).$$

However,

$$P(s_n \text{ even}|r_1 = 0) = P(\sum_{i=2}^{n} r_i \text{ even}|r_1 = 0) = P(\sum_{i=2}^{n} r_i \text{ even}) = P_{n-1}.$$

Similarly,

$$P(s_n \text{ even}|r_1 = 1) = P(s_{n-1} \text{ odd}) = 1 - P_{n-1}.$$

Therefore,

$$P_n = qP_{n-1} + p(1 - P_{n-1}) = (q - p)P_{n-1} + p.$$

The solution of this difference equation is given (see Exercise 1) by

$$P_n = \frac{1}{2}(1 + (q - p)^n).$$

Further results on conditional probability and variance appear in the Exercises. These results can be generalized to spaces with a countably infinite number of points (see Chapter 5).

EXERCISES

1. (a) Show that the solution of the difference equation $u_n = au_{n-1}$ $(n = 1, 2, 3, \ldots)$, in which a is a constant, is $u_n = a^n u_0$ $(n = 0, 1, 2, 3, \ldots)$.

(b) Deduce that the difference equation $u_n = au_{n-1} + b$ $(n = 1, 2, 3, \ldots)$, in which a and b are given constants, has the solution

$$u_n = \left(u_0 - \frac{b}{1-a}\right) a^n + \frac{b}{1-a} \qquad (a \neq 1)$$

$$= nb + u_0. \qquad (a = 1)$$

[*Hint:* Set $u_n = v_n + (b/(1-a))$].

2. Consider n independent trials in each of which an event B occurs or does not occur with probabilities p and q, respectively. For example, B can denote the event that heads occur on each trial when a coin is tossed n times (cf., Example 1). Let w_r be the number of trials required until B occurs r times (the waiting time for r occurrences of B),[7] and let $m_r = Ew_r$.

(a) Show that $m_1 = 1/p$ by showing that m_1 satisfies the difference equation $m_1 = p + q(m_1 + 1)$.

(b) Find m_r by using the fact that

$$m_r = p(1 + m_{r-1}) + q(1 + m_r).$$

3. Cards in cereal boxes are numbered from 1 to n, and a set of one each is required for a prize. With one card per box, how many boxes on the average are required to make a complete set? Find an approximation for large n. [*Hint:* Use Exercise 2(a) and the fact that the mean of a sum of random variables is the sum of their means and

$$1 + \frac{1}{2} + \frac{1}{3} + \cdots + \frac{1}{n} \sim \log n + \frac{1}{2n} + \gamma,$$

where $\gamma = 0.57721 \cdots$ is Euler's constant.[8]]

4. A convict has been placed in a dungeon with three doors. One of the doors leads into a tunnel that returns him to the dungeon after one day's travel. Another door leads to a similar tunnel whose traversal requires three days. The third door leads to freedom. The convict is equally likely to choose each door (that is, each time he chooses a door he does not know what lies beyond). Find the mean number of days the convict will be imprisoned from the moment he first chooses a door to the moment he chooses the door leading to freedom.

5. Three players (denoted a, b, and c) take turns at playing a fair game (that is the probability of winning or losing is $\frac{1}{2}$) according to the following rules. First a

[7]Exercises 2–5 are examples of nonfinite probability spaces. A possibly infinite number of trials may be required.

[8]After Leonhard Euler, a Swiss mathematician (1707–1783).

plays b, while c is out. The winner of the match between a and b plays c. The winner of the second match then plays the loser of the first match. The game continues in this way until a player wins twice in succession, thus becoming the winner of the game.

(a) Find the probabilities that a, b, or c will win the game.
(b) Find the mean duration of the game.

6. Show that

$$\text{Var}\, r_2 = E(\text{Var}\,(r_2|r_1)) + \text{Var}\,(E(r_2|r_1)),$$

where the conditional variance of r_2, given r_1, is defined by

$$\text{Var}\,(r_2|r_1) = E[(r_2 - E(r_2|r_1))^2|r_1].$$

7. Let n be the number of female insects in a region, r_1 the number of eggs laid by an insect, and r_2 the number of eggs in the region. Find the mean and variance of r_2, assuming that n and r_1 are random variables.

8. The number of animals trapped in a day is a random variable n with a mean μ and a variance σ^2. Let r_2 be the number of mink trapped. Find the mean and variance of r_2, under both of the following assumptions:

(a) Each animal trapped has a probability p of being a mink.
(b) The event that a trapped animal is a mink is independent of whether any other trapped animal is a mink, also independent of the number n of animals that were trapped.

9. Two players A and B agree to play until one of them wins a certain number of games, the probabilities of A and B winning a single game being p and $q = 1 - p$. However, they are forced to quit when A has a games still to win and B has b games. How should they divide their total stake to be fair?

10. A and B have, respectively, $n + 1$ and n coins. If they toss their coins simultaneously, what is the probability that

(a) A will have more heads than B?
(b) A and B will have an equal number of heads?
(c) B will have more heads than A?

3.5 REPEATED INDEPENDENT TRIALS

Suppose that we perform a series of independent experiments, say n, whose outcomes depend upon chance. The results of these experiments can be represented as an n-tuples $(w_1, w_2, \ldots, w_n)$, where w_i is the outcome of the ith experiment. This set of all n-tuples forms a sample space Ω. Conceptually, we can forget all about the idea of repeating the experiment and regard the n-tuple as a point of a given space. The following examples should make this clear.

Consider the experiment that consists of first throwing a coin and then throwing a die. The outcomes can be represented as (h, i) and (t, i) for $i = 1, 2, \ldots, 6$. Alternatively, we could have thrown the die and coin simultaneously and the outcomes could be represented in the same manner by 2-tuples.

As another illustration, suppose that we toss a coin n times. The outcomes of this experiment are represented by n-tuples that could also be used to represent the result of tossing n identical coins simultaneously.

Two questions naturally arise: First, how do we generate the probability space corresponding to n repeated independent experiments? Conversely, given a probability space whose elements are n-tuples, how do we decide if it is generated by n repeated independent experiments?

Let us begin to answer these questions by considering the second of the above examples in more detail for the case of $n = 3$. Each of the three tosses is a separate experiment; consequently we have three probability spaces $(\Omega_i, \mathscr{E}_i, P_i)$ to consider. In this example, the three probability spaces are the same; $\Omega_i = \{h, t\}$, $P_i(\{h\}) = P_i(\{t\}) = \frac{1}{2}$, and the events $\mathscr{E}_i$ are all possible subsets of Ω_i.

The outcomes of the three throws can be represented by the elements of the Cartesian product $\Omega = \Omega_1 \times \Omega_2 \times \Omega_3$ of the three spaces Ω_1, Ω_2, and Ω_3. For example, the event A_1 that we get heads on the first trial is represented by

$$A_1 = \{(h, t, h), (h, t, t), (h, h, h), (h, h, t)\}.$$

The event A_2 that we get heads on the second trial is represented by all 3-tuples that have an h in the second position; the event A_3 that we get heads on the third trial is represented by all n-tuples that have an h in the third position.

Now the outcome of a toss is not influenced in any way by the preceding or following tosses. Hence if we had a probability function P defined on the events of the space Ω, we would expect that $P(A_1) = P_1(\{h\}) = \frac{1}{2}$, since A_1 should be equivalent to getting heads in one toss of a coin; similarly, we would expect $P(A_2) = P(A_3) = \frac{1}{2}$. Furthermore, we feel that all outcomes in Ω should be equally likely. If we assume this and calculate the probability of A_1, we obtain $\frac{1}{2}$, which strengthens our conviction that $P(A_1) = \frac{1}{2}$. In other words, for $A_1 = \{h\} \times \Omega_2 \times \Omega_3$,

$$P(A_1) = P_1\{h\} = P_1\{h\}P_2(\Omega_2)P_3(\Omega_3) = \tfrac{1}{2}.$$

Similarly, for any event $E_1 \in \mathscr{E}_1$ and $A = E_1 \times \Omega_2 \times \Omega_3$, we would expect $P(E_1) = P_1(E_1)$. We shall see that sets of the form A play an important role.[9]

Let $B = \Omega_1 \times E_2 \times \Omega_3$ and $C = \Omega_1 \times \Omega_2 \times E_3$. The events A, B, and C should be independent for independent trials, since an event occurring on any trial is not influenced by the previous trials or succeeding trials. For example, the events A_1, A_2, and A_3 are independent in the product space if we assume all outcomes to be equally likely. Accordingly,

[9] A set of the form $\Omega_1 \times \Omega_2 \times \cdots \times A_k \times \cdots \times \Omega_n$ is called a *cylinder set*.

$$(1) \qquad P(ABC) = P(A)P(B)P(C).$$

However, $ABC = E_1 \times E_2 \times E_3$, $P(A) = P_1(E_1)$, $P(B) = P_2(E_2)$, and $P(C) = P_3(E_3)$; hence Equation (1) becomes

$$(2) \qquad P(E_1 \times E_2 \times E_3) = P(A)P(B)P(C) = P_1(E_1)P_2(E_2)P_3(E_3).$$

Let us use these results as a guide in the general case. Suppose that we perform in succession n random experiments whose probability spaces are $(\Omega_i, \mathscr{E}_i, P_i)$. The space of these n experiments is $\Omega = \Omega_1 \times \Omega_2 \times \Omega_3 \times \cdots \times \Omega_n$. Let us assume that we have a probability space $(\Omega, \mathscr{E}, P)$; we can deduce some of its properties from the fact that it represents n independent trials. An event E is said to depend on the ith trial if the decision as to whether E occurs depends on the ith component of Ω—that is, on the results of the ith experiment. By convention it is assumed that the sure event Ω and the impossible event ϕ depend on every trial. This will be clarified by the following definition.

DEFINITION 1

Let $(\Omega, \mathscr{E}, P)$ be the probability space of n random experiments whose probability spaces are $(\Omega_i, \mathscr{E}_i, P_i)$. Then the event E *depends on the* i*th trial* if and only if

$$E = \Omega_1 \times \cdots \times \Omega_{i-1} \times E_i \times \Omega_{i+1} \times \cdots \times \Omega_n,$$

where $E_i \in \mathscr{E}_i$.

In addition, all sets of the form $E_1 \times E_2 \times \cdots \times E_n$ where E_i is an event[10]—that is, $E_i \in \mathscr{E}_i$—must be considered as events according to our previous discussion. Furthermore, from Equation (2),

$$(3) \qquad P(E_1 \times E_2 \times \cdots \times E_n) = P_1(E_1)P_2(E_2) \cdots P_n(E_n).$$

However, these two facts alone are not sufficient to determine the probability function P, since not all the events in $\mathscr{E}$ are Cartesian products of events in $\mathscr{E}_i$.

Returning to our previous example, let us denote by E the event that we get exactly one head in three tosses of a coin. Then $E = A_1 A_2' A_3' + A_1' A_2 A_3' + A_1' A_2' A_3$ and it cannot be written as a Cartesian product of events. However, it can be written as a union of Cartesian products—in fact, $E = (h \times t \times t) \cup (t \times h \times t) \cup (t \times t \times h)$, where we have denoted the set $\{h\}$ by h.

If $\{\Omega, \mathscr{E}, P\}$ is a probability space, then $\mathscr{E}$ must be a field (cf., Section 1.5); therefore, all finite sums of Cartesian product events, such as E, must be events. This suggests the following theorem and definitions.

[10] $E_1 \times E_2 \times \cdots \times E_n$ is called a *rectangular event* or *Cartesian product event.*

THEOREM 1

If $\mathscr{E}_1, \mathscr{E}_2, \ldots, \mathscr{E}_n$ are fields of events in $\Omega_1, \Omega_2, \ldots, \Omega_n$ respectively, then the class $\mathscr{E}$ of all finite disjoint unions $E_1 \times E_2 \times \cdots \times E_n$, where $E_i \in \mathscr{E}_i$ for $i = 1, 2, \ldots, n$, is a field of sets in $\Omega_1 \times \Omega_2 \times \cdots \times \Omega_n$.

Proof

First we will prove that any finite intersection of events is an event. It suffices to prove this result for two events, since the general result follows by induction. Let A_1 and A_2 be events—that is,

$$A_1 = B_1 + B_2 + \cdots + B_r$$
$$A_2 = C_1 + C_2 + \cdots + C_s,$$

where the B_i and C_i are Cartesian product events. Then $A_1 A_2 = \sum_{i=1}^{r} \sum_{j=1}^{s} B_i C_j$ is a disjoint union of the sets $B_i C_j$. Hence the proof reduces to showing that the intersection of two Cartesian product events is also a Cartesian product event. This follows from the identity

$$(E_1 \times E_2 \times \cdots \times E_n) \cap (F_1 \times F_2 \times \cdots \times F_n) = (E_1 \cap F_1) \times (E_2 \cap F_2) \times \cdots \times (E_n \cap F_n),$$

since $E_j \cap F_j \in \mathscr{E}_j$ for $j = 1, 2, \ldots, n$.

We now show that the complement of an event is an event. Since $A_1' = B_1' B_2' \cdots B_r'$, it follows from the above result that A_1' is an event if all the B_i' are events. The B_i' are events because

$$(E_1 \times E_2 \times \cdots \times E_n)' = E_1' \times E_2 \times \cdots \times E_n + E_1 \times E_2' \times \cdots \times E_n + \cdots + E_1 \times \cdots \times E_n' + E_1' \times E_2' \times \cdots \times E_n + \cdots + E_1' \times E_2' \times \cdots \times E_n',$$

where the E_i are complemented one at a time, two at a time, ..., n at a time, resulting in $\binom{n}{1} + \binom{n}{2} + \cdots + \binom{n}{n}$ summands. This is a disjoint union of rectangular events, since $E_1' \in \mathscr{E}_i$ for $i = 1, 2, \ldots, n$.

Using our previous results, we can easily show that the union of a finite number of events $A_1, A_2, \ldots, A_n$ is an event. The complements of these events are events; therefore, $A_1' A_2' \cdots A_n'$ is an event and $\bigcup_{i=1}^{n} A_i = (\bigcap_{i=1}^{n} A_i')'$.

The field $\mathscr{E}$ is called the *product field* of $\mathscr{E}_1, \mathscr{E}_2, \ldots, \mathscr{E}_n$ and will be denoted by $\mathscr{E}_1 \times \mathscr{E}_2 \times \cdots \times \mathscr{E}_n$.

The reader can easily verify that in the elementary case in which each field $\mathscr{E}_i$ is the class of all subsets of Ω_i, $\mathscr{E}$ is also the class of all subsets of $\Omega_1 \times \Omega_2 \times \cdots \times \Omega_n$ (see Exercise 6).

We must now extend our probability function, which we have defined only for rectangular events, to every event in $\mathscr{E}$. In general, an event is a disjoint sum of rectangular events. According to Axiom 4 of Section 1.4, this function must be additive; since its value on a rectangle is given by Equation (3), the following definition seems appropriate.

DEFINITION 2

Let $E \in \mathscr{E}$, where $\mathscr{E} = \mathscr{E}_1 \times \mathscr{E}_2 \times \cdots \times \mathscr{E}_n$— that is,

$$E = \sum_{i=1}^{r} E_{i1} \times E_{i2} \times \cdots \times E_{in},$$

where $E_{ij} \in \mathscr{E}_j$ for $j = 1, 2, \ldots, n$. The *product probability function* P, denoted by $P_1 \times P_2 \times \cdots \times P_n$, is defined by

$$P(E) = \sum_{i=1}^{r} P_1(E_{i1})P_2(E_{i2}) \cdots P_n(E_{in}). \tag{4}$$

Actually we have not shown that the function P is a probability function; this is the object of our next two theorems. We first must prove that the function P defined above is unambiguously defined. The difficulty arises from the fact that an event A does not have a unique representation —for example,

$$A = (E_1 + E_2) \times (E_3 + E_4) = E_1 \times E_3 + E_1 \times E_4 + E_2 \times E_3 + E_2 \times E_4.$$

THEOREM 2

The probability function P given in Definition 2 is unambiguously defined.

Proof

Suppose that

$$A = A_1 + A_2 + \cdots + A_r = B_1 + B_2 + \cdots + B_s,$$

where the A's and B's are rectangular events, are two representations of A. Therefore,

$$A_i = A_iB_1 + A_iB_2 + \cdots + A_iB_s, \qquad (i = 1, 2, \ldots, r)$$

and since A_iB_j $(j = 1, 2, \ldots, s)$ is a rectangular event, it follows from the definition of P that

$$\sum_{i=1}^{r} P(A_i) = \sum_{i=1}^{r}\sum_{j=1}^{s} P(A_iB_j) = \sum_{j=1}^{s} P(B_j).$$

Hence the function P is unambiguously defined for every $A \in \mathscr{E}$.

THEOREM 3

The probability function $P = P_1 \times P_2 \times \cdots \times P_n$ is a probability function on $\Omega = \Omega_1 \times \Omega_2 \times \cdots \times \Omega_n$ with $\mathscr{E} = \mathscr{E}_1 \times \mathscr{E}_2 \times \cdots \times \mathscr{E}_n$ as the class of events.

Proof

To show that P is a probability function, we must show that Axioms 3 and 4 for a finite probability space hold (see page 8). It is obvious from Equation (4) that $P(E) \geq 0$ for any $E \in \mathscr{E}$. From the same equation it easily follows that $P(\Omega) = 1$, since $\Omega = \Omega_1 \times \Omega_2 \times \cdots \times \Omega_n$ and $P_i(\Omega_i) = 1$.

It only remains to verify that if E and F are any two disjoint events, then $P(E + F) = P(E) + P(F)$; i.e., P is additive. However, this follows directly from Definition 2 for P, once it is noted that $E + F$ is the disjoint sum of the rectangular events that comprise E and F.

The above theorems are really special cases of more general theorems of measure theory (cf., Chapter 6), since a probability function is a special case of a measure and the product probability space corresponds to the product of measure spaces.

The importance of independent trials arises from the fact that they enable us to construct new probability spaces, as shown in the following theorem. The proof of this theorem follows immediately from Theorems 1–3.

THEOREM 4

Let $(\Omega_i, \mathscr{E}_i, P_i)$ be n probability spaces. Then $\mathscr{E} = \mathscr{E}_1 \times \mathscr{E}_2 \times \cdots \times \mathscr{E}_n$ is a field of events in $\Omega = \Omega_1 \times \Omega_2 \times \cdots \times \Omega_n$ and $P = P_1 \times P_2 \times \cdots \times P_n$ is a probability function on $\mathscr{E}$. The triplet $(\Omega, \mathscr{E}, P)$ is called the *probability space of* n *independent trials.*

Let us now turn to the converse problem of determining when a given probability space is a space of n independent trials. In practice, the answer to this question is often quite useful. It enables us to consider the outcomes as the result of n repeated experiments, which frequently simplifies calculations.

THEOREM 5

Let $(\Omega, \mathscr{E}, P)$ be a probability space for which the elements of Ω are n-tuples of the Cartesian product $\Omega_1 \times \Omega_2 \times \cdots \times \Omega_n$ and $\mathscr{E} = \mathscr{E}_1 \times \mathscr{E}_2 \times \cdots \times \mathscr{E}_n$. Let $\mathscr{C}_k$ $(k = 1, 2, \ldots, n)$ be the classes of events that depend only on the kth trial (see Definition 1). Then $(\Omega, \mathscr{E}, P)$ is a probability space of n independent trials if and only if the classes $\mathscr{C}_k$ are independent.

Proof

If the classes $\mathscr{C}_k$ are independent, let us define $P_i(E_i)$ for $i = 1, 2, \ldots, n$ and any $E_i \in \mathscr{E}_i$ by

$$P_i(E_i) = P(\Omega_1 \times \Omega_2 \times \cdots \times \Omega_{i-1} \times E_i \times \Omega_{i+1} \times \cdots \times \Omega_n).$$

It is not difficult to verify that P_i is a probability function on $\mathscr{E}_i$. Since P is additive, we need only show that Equation (3) holds. Equation (3) follows automatically because of the independence of the $\mathscr{C}_k$ and the identity

$$E_1 \times E_2 \times \cdots \times E_n = (E_1 \times \Omega_2 \times \cdots \times \Omega_n)(\Omega_1 \times E_2 \times \cdots \times \Omega_n) \cdots (\Omega_1 \times \Omega_2 \times \cdots \times E_n).$$

On the other hand, if $P = P_1 \times \cdots \times P_n$, the classes $\mathscr{C}_k$ are independent.

Events in the space Ω can be independent without being rectangular. For example, let E be the event that we get exactly one head or exactly one tail in three tosses of a coin. The reader can easily verify that E is not a rectangular event and that $P(E) = \frac{3}{4}$. Let F be the event that we get heads on the first throw; then $P(F) = \frac{1}{2}$. The events E and F are independent, since $P(EF) = \frac{3}{8}$.

Repeated trials are not always independent; they can be dependent. For dependent repeated trials, Equation (3) is replaced by

$$(5) \qquad P(E_1 \times E_2 \times \cdots \times E_n) = P(E_1)P(E_2|E_1)P(E_3|E_1, E_2) \cdots P(E_n|E_1, E_2, \ldots, E_{n-1});$$

because of Theorems 2 and 3 we can extend P to the whole space.

EXAMPLE 1

Let us consider the process of sampling without replacement from the elements of $\Omega = (1, 2, 3, \ldots, n)$. A sample size $r \leq n$ can be regarded as a succession of r experiments. We have n possible results in the first experiment, $n - 1$ in the second, $n - 2$ in the third, etc., since we remove an element each time. If at the ith stage we regard all outcomes as equally likely, then the probability of obtaining any of the remaining numbers is $1/(n - i)$.

We can regard the outcomes of the r successive experiments as elements of $\Omega_1 \times \Omega_2 \times \cdots \times \Omega_r$, where all $\Omega_i = \Omega$, and the events as all possible subsets. Then, using Equation (5), we have

$$P(i_1, i_2, \ldots, i_r) = \frac{1}{n(n-1)\cdots(n-r+1)}$$

if all the i_r are different, and 0 if any of the numbers i_r are the same. For instance, if $n = r = 3$, we have $P(1, 2, 3) = \frac{1}{6}$ and $P(1, 1, 3) = 0$. By enumerating all possible r-tuples and regarding all outcomes as equally likely, we obtain an equivalent result. Of course, r-tuples such as $(1, 1, \ldots, 3)$ do not really appear; they are assigned a probability of zero.

EXERCISES

1. Let p_i be the probability that a coin i $(i = 1, 2, 3)$ will turn up heads when it is tossed. Find the probability space corresponding to the experiment of tossing the coins in the order 1, 2, 3. What is the probability that the same side will turn up on all three coins?

2. Suppose that we have four urns which contain red and white balls denoted respectively by R and W. The composition of urns 1, 2, 3, and 4 is $4R$; $6W$; $3W$, $2R$; and $1R$, $5W$, respectively. One ball is drawn from each urn. What is the probability of drawing

(a) One white and three red balls?
(b) At least two red balls?
(c) More white than red balls?

3. A sample of size 2 is taken from the set $(1, 2, 3, \ldots, 6)$. If the procedure followed is sampling with replacement, find which of the following statements are true and prove your result. Let A_i be the event that the outcome of the ith draw is even, where $i = 1, 2$.

(a) A_i depends on the ith draw.
(b) A_1 and A_2 are mutually exclusive and independent.
Let B_1 be the event that the sum of the two numbers drawn is 7.
(c) B_1 depends on the first draw.
(d) A_1 and B_1 are independent.

4. An engineer is observing the output of a machine that produces the numbers 0 and 1 at random. He discovers that the probabilities of two successive observations are given by the table which follows. Does this probability space consist of two independent trials?

Event E	(1, 1)	(1, 0)	(0, 1)	(0, 0)
$P(E)$	$\frac{1}{3}$	$\frac{1}{6}$	$\frac{1}{6}$	$\frac{1}{3}$

5. Consider n spaces Ω_i, where the ith space contains n_i points. Assume that all outcomes in each of these spaces are equally likely. It is natural to assume that all n-tuples are equally likely in the space generated by n independent trials.

Under this assumption, prove that the probability of any cylinder event (say $E_1 \times \Omega_2 \times \cdots \times \Omega_n$) is the same as the probability of the event (E_1) in its own space (Ω_1).

6. Using the assumptions given in Exercise 5, verify that $\mathscr{E}_1 \times \mathscr{E}_2 \times \cdots \times \mathscr{E}_n$ is the class of all subsets of $\Omega_1 \times \Omega_2 \times \cdots \times \Omega_n$.

7. Let $(\Omega^n, P^n, \mathscr{E}^n)$ be the probability space generated by n independent trials. Let $(\Omega^r, P^r, \mathscr{E}^r)$ be the probability space generated by the first r trials, where $r \leq n$. Then the second space can be regarded as a subspace of the first space. [*Hint:* Show that P^n is an extension of P^r (cf., Section 3.3) and that they are both probability functions. Also show that $\mathscr{E}^r$ and $\mathscr{E}^n$ are fields and that $\mathscr{E}^r \subset \mathscr{E}^n$.]

8. Show that for the space generated by n independent trials, any $r \leq n$ rectangular events that depend on different trials are independent.

3.6 THE BINOMIAL DISTRIBUTION

In many practical situations we are concerned with repeated independent trials with only two possible outcomes for each trial. Moreover, the probabilities of these outcomes remain the same for each trial. A familiar example is provided by successive tosses of a coin; here the outcomes are heads or tails.

EXAMPLE 1

Suppose that a coin is tossed six times. Let us find the probability of the event B that we get heads, h, on the first four tosses and tails, t, on the remainder, under the assumption that $P(\{h\}) = p$.

In the previous section we saw that we can regard B as the set $\{(h, h, h, h, t, t)\}$. From Equation 5.4 we have $P(B) = p^4 q^2$, where $q = 1 - p$, since $B = \{h\} \times \{h\} \times \{h\} \times \{h\} \times \{t\} \times \{t\}$.

Further examples are: the occurrence or nonoccurrence of the number 5 when a die is thrown, the event that the next child born in a family will be a boy, and the event that a manufactured item is defective or nondefective.

DEFINITION 1

A probability space $(\Omega, \mathscr{E}, P)$ consists of repeated *Bernoulli trials* if it is the probability space of n independent trials where each of the spaces $(\Omega_i, \mathscr{E}_i, P_i)$ are the same, $\mathscr{E}_i = \{\Omega, B, B', \phi\}$, $P_i(B) = p$, and $P_i(B') = q$ for $i = 1, 2, \ldots, n$.

We recall that in order to specify a probability function P on $\mathscr{E}$ we need only define it on the Cartesian product events $E_1 \times E_2 \times \cdots \times E_n$, where E_i is either B or B'. The following lemma follows immediately from Equation 5.4.

LEMMA 1

If a probability space consists of n independent repeated Bernoulli trials, then the probability of any combinatorial event $E_1 \times E_2 \times \cdots \times E_n$ is

$$P(E_1 \times E_2 \times \cdots \times E_n) = p^k q^{n-k}, \tag{1}$$

where k is the number of occurrences of the event B.

An occurrence of the event B is commonly called a *success*, while the nonoccurrence of the event B is called a *failure*. Using this terminology, we can obtain an alternate representation of Bernoulli trials which is commonly employed in elementary courses. Let the spaces $\Omega = \{s, f\}$, where s and f represent success and failure respectively; also let $P(\{s\}) = p$ and $P(\{f\}) = q$. Then the space of n repeated trials consists of 2^n n-tuples $(Z_1, Z_2, \ldots, Z_n)$, where each Z_i is s or f. If k of the components are s and $n-k$ are f, we have $P(\{Z_1, Z_2, \ldots, Z_n\}) = p^k q^{n-k}$. The events are all possible subsets of the product space. It is obvious that these two formulations are the same.

Frequently, we are only interested in the probability of the number of successes that will occur in n repeated Bernoulli trials—for example, the number of heads in six tosses of a coin, fives in n throws of a die, boys in a given family, or defective items in a given lot. The number of successes s_n in n Bernoulli trials is a random variable. In fact, it can be written as the sum of n random variables (cf., Exercise 8)—namely,

$$s_n = I_1 + I_2 + \cdots + I_n. \tag{2}$$

Here I_k $(k = 1, 2, \ldots, n)$ is the indicator of the event A_k that B occurs on the kth trial, or

$$A_k = \Omega_1 \times \cdots \times \Omega_{k-1} \times B \times \Omega_{k+1} \times \cdots \times \Omega_n,$$

where $\Omega_i = B \cup B'$.

THEOREM 1

Let s_n be a random variable that gives the number of successes in n repeated Bernoulli trials. Then $P\{s_n = k\}$ has the probability distribution

$$b(k; n, p) = \binom{n}{k} p^k q^{n-k}. \tag{3}$$

The distribution $b(k; n, p)$ is called the *binomial distribution*[11] or *Bernoulli's formula*.

Proof

Let $x \in \Omega$. Then $s_n(x) = k$ if and only if $x \in E_1 \times E_2 \times \cdots \times E_n$, where k of the E_i are B and $n - k$ of the E_i are $B' = f$. In other words, the set $\{s_n = k\}$ is the sum of all the rectangular events $E_1 \times E_2 \times \cdots \times E_n$ in which k of the E_i are B. However, the number of terms in the sum $\binom{n}{k}$ is just the number of different ways of arranging k letters s and $n - k$ letters f in a row. According to Lemma 1, each of these terms has a probability $p^k q^{n-k}$; hence Equation (3) follows.

EXAMPLE 2

A marksman is shooting at a target. The probability that he will hit the target is $\frac{3}{4}$. Find the probability that he will hit the target 0, 1, 2, . . . , 6 times if he fires six shots.

We can describe the results of the six shots by a 6-tuple $(x_1, x_2, \ldots, x_6)$ whose ith component $x_i = s$ or f, depending on whether that target was hit or missed. Assuming that the six shots constitute six repeated Bernoulli trials, the probability of k hits is $b(k; 6, \frac{3}{4})$ from Theorem 1. Hence we obtain

$$0.0, 0.0, 0.03, 0.13, 0.30, 0.36, 0.18$$

as the probabilities for 0, 1, 2, . . . , 6 hits respectively.

The binomial distribution occurs frequently in applications of probability theory. Consequently, the values of $b(k; n, p)$ have tabulated. In constructing such tables only values of $b(k; n, p)$ for $p \leq 0.5$ must be calculated; the remaining values can be obtained by means of the following formula (cf., Exercise 10):

$$b(k; n, p) = b(n - k; n, 1 - p).$$

The name "binomial distribution" is quite appropriate, since the probability of k successes in n Bernoulli trials is the coefficient of $s^k f^{n-k}$ in the binomial expansion of $(ps + qf)^n$. In fact, if we do not simplify the resulting expansion, the terms represent symbolically the possible n-tuples and their probabilities. For instance, the outcomes and their probabilities when a coin is tossed twice are given by

$$(ph + qt)^2 = p^2(hh) + pq(ht) + pq(th) + q^2(tt).$$

[11] $\binom{n}{k} = \frac{n(n-1)\cdots(n-k+1)}{k!} = \frac{n!}{(n-k)!\,k!}.$

These are called *binomial coefficients* and are often denoted by C_k^n or ${}_nC_k$.

An examination of the results in Example 2 reveals that the probability of k hits increases with k, reaches a maximum at $k = 5$, and then decreases. In the general case, for fixed p and n, the same phenomenon is observed.

THEOREM 2

The terms $b(k; n, p)$ first increase, then decrease monotonically with increasing k. If $(n + 1)p$ is not an integer, the greatest value is attained at k_0,[12] where

$$(n + 1)p - 1 < k_0 \leq (n + 1)p. \tag{4}$$

However, if $k_0 = (n + 1)p$ is an integer, then $b(k_0 - 1; n, p) = b(k_0; n, p)$ are the greatest values.

Proof

Using Equation (3), the inequality

$$b(k + 1; n, p) > b(k; n, p) \tag{5}$$

can be simplified by canceling out the common factors appearing in both sides of inequality (5). This yields

$$\frac{p}{k + 1} > \frac{q}{n - k},$$

which reduces to

$$(n + 1)p - 1 > k \tag{6}$$

by setting $q = 1 - p$ and cross-multiplying. In a similar manner, we can see that $b(k+1; n, p) = b(k; n, p)$ if $k = (n + 1)p - 1$, and $b(k + 1; n, p) < b(k; n, p)$ if $(n + 1)p - 1 < k$.

If $(n + 1)p$ is not an integer, then $(n + 1)p - 1$ is not an integer; hence k cannot equal $(n + 1)p - 1$. The terms increase as long as inequality (6) holds. Let k_0 be the first integer that violates inequality (6)—that is, k_0 is given by Equation (4). Then we have $b(k_0 + 1; n, p) < b(k_0; n, p)$. However, $k_0 - 1$ satisfies inequality (6) and so from inequality (5) we have $b(k_0; n, p) > b(k_0 - 1; n, p)$. Hence the greatest value of the probability is attained at k_0. If $k > k_0$, the terms decrease.

If $(n + 1)p = k_0$ is an integer, we have $b(k + 1; n, p) = b(k; n, p)$ for $k = (n + 1)p - 1 = k_0 - 1$, which proves the last statement in the theorem.

The term $b(k_0; n, p)$ is called the *central term*, where k_0 is the most probable number of successes.

Suppose that the value of p remains fixed but we increase the number n

[12] k_0 is called the *mode*. In general, the mode is the value(s) of x such that $f(x)$ is maximum, where f is the density function.

of trials. Let us see what happens to the fraction k_0/n, the most probable frequency of successes. From Equation (4) we see that

$$\left(1+\frac{1}{n}\right)p-\frac{1}{n}<\frac{k_0}{n}\leq\left(1+\frac{1}{n}\right)p.$$

Therefore, as $n\to\infty$ we have $k_0/n\to p$. We conclude that for a large number of trials, the most probable frequency for the occurrence of an event is equal to the probability of the occurrence of the event in a single trial. Thus if we assume that the probability that a baby will be a boy is $\frac{1}{2}$, it is most probable that in 100 births, 50 will be boys. This does not mean that the probability of 50 boys being born is high. In fact, this probability is less than 0.08. The probability is only a relative maximum.

We feel intuitively that the most probable number of successes in a large number of trials is the mean. From the preceding results we see that k_0 is asymptotic to np. We can verify that this result is correct by formally calculating the mean value of the number of successes.

LEMMA 2

The mean of the binomial distribution is np and the variance is npq.

Proof

Using the representation of s_n given by Equation (2),

$$Es_n=EI_1+EI_2+\cdots+EI_n=np,$$

since

$$EI_k=PA_k=p. \qquad (k=1,2,\ldots,n)$$

The variance of the random variable I_k is

$$EI_k^2-p^2=p-p^2=pq.$$

Since the random variables I_k $(k=1,2,\ldots,n)$ are independent, Theorem 3.4 (the Bienaymé equality) yields $\sigma^2 s_n=npq$.

We have seen that the function $b(k;n,p)$ first increases, then attains its maximum value, and finally decreases as k goes from 0 to n. Consequently, the largest values of $b(k;n,p)$ will be attained for values of k near the most probable value k_0. For large values of n, the graph of the function $b(k;n,p)$ will have the form shown in Figure 2.

Although $b(k_0;n,p)$ may be small, the sum of the probabilities between the points A and B is quite close to unity—that is $\sum_{A<k<B} b(k;n,p)>1-\epsilon$ for some small $\epsilon>0$. Moreover, it can be seen from Figure 2 that the number of points that lie between A and B and whose coordinates are integers is small compared to the total number of trials. This fact was first discovered

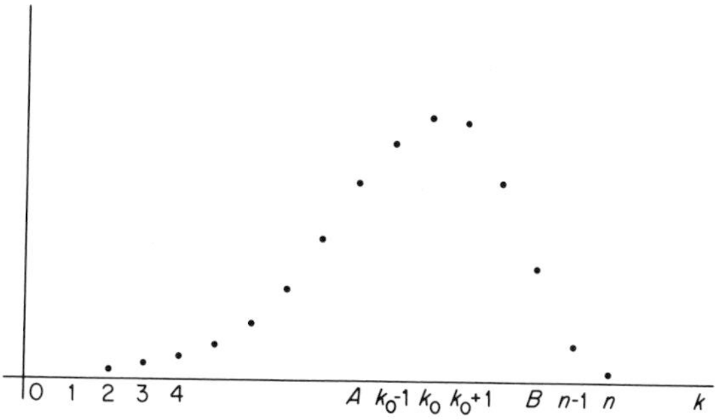

Figure 2

by Bernoulli in 1713 and is known as *Bernoulli's theorem*. We will now prove this result rigorously.

THEOREM 3

Given $\epsilon > 0$ and n Bernoulli trials,

$$P(|s_n - np| > \epsilon n) \to 0$$

as $n \to \infty$.

Proof

The proof is a simple consequence of Chebyshev's inequality, Theorem 3 of Chapter 2. Let $v = s_n - np$; then Ev^2 is the variance of s_n, which is npq. Accordingly,

(7) $$P(|s_n - np| \geq \epsilon n) < \frac{pq}{\epsilon^2 n},$$

which approaches zero as $n \to \infty$.

COROLLARY

Given $\epsilon > 0$ and n Bernoulli trials,

$$P\left(\left|\frac{s_n}{n} - p\right| \geq \epsilon\right) \to 0$$

as $n \to \infty$.

This corollary is known as the *Bernoulli* (or weak) *law of large numbers*. It states that the probability distribution of the values of the frequency s_n/n of the number of successes in n Bernoulli trials clusters about p as the number of trials becomes large. Earlier this theorem was assumed to be self-evident and was used in an intuitive manner as a basis for the early notions of probability. We refer the reader to the last paragraph in Section 1.2.

A large number of experiments have been carried out to verify the agreement between Bernoulli's theorem and actual experience. We cite a few coin-tossing experiments. The eighteenth-century French mathematician Buffon tossed a coin 4040 times, obtaining heads 2048 times for a relative frequency of 0.507. Karl Pearson, an English statistician (1875–1936), tossed a coin on one occasion 12,000 times and on another 24,000 times, obtaining 6019 and 12,012 heads, respectively. This leads to relative frequencies for heads of 0.5016 and 0.5005, respectively.

Are such small deviations from 0.5 a matter of luck? The reader will find after reading the next chapter that he can answer this question by applying Laplace's theorem (cf., Example 4.3.3).

EXAMPLE 3

Suppose that 400,000 candidates undergo a rigorous medical examination. Previous statistics show that the probability of passing is $\frac{3}{4}$. Obtain an estimate of the probability that between 296,000 and 304,000 will pass the test.

If we assume that we have a sequence of Bernoulli trials, then the answer is

$$\sum_{k=296,000}^{304,000} \binom{400,000}{k} \left(\frac{3}{4}\right)^{400,000} \left(\frac{1}{4}\right)^{400,000-k}. \tag{8}$$

However, because of the large numbers involved, a direct calculation of term (8) is tedious even with a table of the binomial distribution. Instead we can apply inequality (7), with $\epsilon = \frac{1}{100}$; hence

$$P(|s_n - 300,000| \geq 4000) < \left(\frac{3}{4}\right)\left(\frac{1}{4}\right)\frac{(100)^2}{4 \times 10^5} = 0.005.$$

There is a probability in excess of 0.995 that between 296,000 and 304,000 candidates will pass. In the next chapter we will see how to obtain more accurate estimates of sums of the form (8).

As an application of Chebyshev's inequality and the binomial distribution, we will prove the Weierstrass approximation theorem[13] stated below. The probabilistic approach is due to S. Bernstein.[14]

[13] After K. T. Weierstrass, a German mathematician (1815–1897).

[14] S. Bernstein, "Démonstration du théorème de Weierstrass, fondée sur le calcul des probabilités," *Comm. Soc. Math. Kharkow*, **2**, No. 13 (1912–1913), 1–2.

THEOREM 4

Let f be a continuous function on the interval $a \leq x \leq b$. Then for any $\epsilon > 0$ there is a polynomial p such that $|f(x) - p(x)| \leq \epsilon$ for all x in the given interval.

Since the function $f(a + (b - a)x)$ is continuous on $0 \leq x \leq 1$, there is no loss in generality if we prove the Weierstrass theorem for this interval only.

Before proceeding with a formal proof, it is instructive to look at the motivation behind this theorem. Imagine a loaded coin that turns up heads with a probability of x. The probability that exactly i heads occur in a game consisting of n consecutive tosses is given by the binomial distribution

$$\frac{n!}{i!(n-i)!} x^i(1-x)^{n-i}. \qquad (i = 0, 1, 2, \ldots, n)$$

Now suppose that the game pays off $f(i/n)$ dollars if exactly i of the n tosses are heads, where f is a continuous function. The expected value B_n of one game consisting of n tosses is

$$B_n(x) = \sum_{i=0}^{n} f\left(\frac{i}{n}\right)\binom{n}{i} x^i(1-x)^{n-i}. \tag{9}$$

If n is very large it is reasonable to expect about nx heads, since nx is the mean for the binomial distribution. This means that we can expect to receive $f(nx/n) = f(x)$. Hence $B_n(x)$ will be nearly the same as $f(x)$ for large values of n; that is, we expect the difference $|f(x) - B_n(x)|$ to approach zero as $n \to \infty$. However, $B_n(x)$ is a polynomial (a Bernstein polynomial) and so we have established the Weierstrass approximation theorem by an heuristic argument.

To proceed formally, we estimate the difference between $B_n(x)$ and $f(x) = \sum_{i=0}^{n} f(x)\binom{n}{i} x^i(1-x)^{n-i}$:

$$|B_n(x) - f(x)| \leq \sum_{i=0}^{n} \left|f(x) - f\left(\frac{i}{n}\right)\right|\binom{n}{i} x^i(1-x)^{n-i}$$

$$= \sum_{|x-i/n|\geq\delta} \left[\left|f(x) - f\left(\frac{i}{n}\right)\right|\binom{n}{i} x^i(1-x)^{n-i}\right] + \sum_{|x-i/n|<\delta} [\quad].$$

The second sum is less than ϵ by virtue of the uniform continuity of f, since we can find a $\delta = \delta(\epsilon)$ such that $|f(x) - f(i/n)| < \epsilon$ for $|x - (i/n)| < \delta$. The first sum is less than $2M[x(1 - x)/n\delta^2]$, since $|f| < M$ and

$$P(|i - nx| \geq \delta n) = P\left[\left|\frac{i}{n} - x\right| \geq \delta\right] \leq \frac{x(1-x)}{n\delta^2}$$

by virtue of Chebyshev's inequality. Since δ is independent of x, we see that $|B_n(x) - f(x)|$ can be made arbitrarily small by choosing n sufficiently large.

EXERCISES

1. (a) If 0.01 percent of patients afflicted with a certain disease die from it, what is the probability that just three will die in a group of 10,000?
(b) What is the mean number of deaths and the most probable number of deaths?

2. A manufacturing process yields radio tubes of which 10 percent are defective in the long run. Find the probability that in a sample of five tubes selected at random there will be

(a) No defectives.
(b) At least two defectives.
(c) Exactly two defectives.
(d) No more than two defectives.

3. If $\frac{2}{5}$ of people belong to blood group A, what is the probability that in a randomly selected sample of seven people, just four belong to group A?

4. Obtain an estimate of the number of times a coin must be tossed so that at least 95 percent of the time the frequency (sn/n) of heads will be within $1/1000$ of the probability p of getting heads.

5. Suppose that there are two players, say A and B, who need m and n points respectively to win a game. The probability that A will win a point is p. Assuming that the successive plays are Bernoulli trials, find the probability that A will win. [*Hint:* At most, $m + n - 1$ plays are necessary (cf., Section 2.2).]

6. A physicist sees a drunkard walking in a strange manner. After careful observation he discovers that this man walks five paces in a straight line and then halts. After each stop the drunkard continues his walk. However, he is just as likely to take the five steps backward as he is to take them forward. Show how the physicist can compute the probability that after 100 paces the drunkard will be

(a) Back at his starting point.
(b) Within five paces of his starting point.
(c) Exactly ten paces from his starting point.

7. (a) Show that $\sum_{k=0}^{n} b(k; n, p) = 1$.
(b) Find the mean and variance of the binomial distribution by evaluating $\sum_{k=0}^{n} k \binom{n}{k} p^k q^{n-k}$ and $\sum_{k=0}^{n} k^2 \binom{n}{k} p^k q^{n-k}$ directly.

8. Show that the I_k appearing in Equation (2) are random variables.

9. Plot $b(k; n, p)$ for $n = 20$ and $p = \frac{1}{2}$ and .3.

10. Show that $b(k; n, p) = b(n - k; n, 1 - p)$.

11. (a) Prove that the random variable s_n defined by Equation (2) can be written in the standard form

$$s_n = \sum_{k=0}^{n} kJ_k,$$

where J_k is defined by

$$J_k = \sum I_{j_1} I_{j_2} \cdots I_{j_k}(1 - I_{j_{k+1}}) \cdots (1 - I_{j_n}).$$

The summation is over all permutations of the subscripts $j_r = 1, 2, \ldots, n$ classified into two groups, one having k terms and the other $n - k$. [*Hint:* write

$$I_k = I_k \prod_{\substack{j=1 \\ j \neq k}}^{n} [I_j + (1 - I_j)];$$

expand it and substitute it in Equation (2).]
(b) Derive the binomial distribution by using the above representation of the s_n.

12. The following principle is the basis of the *sign test* of nonparametric statistics. Consider n pairs of numerical observations (x_i, y_i) obtained under similar conditions. If the observations are essentially the same except for random fluctuations, there should only be chance differences between any pair of observations. Accordingly, it is equally likely for the nonzero differences $d_i = x_i - y_i$ to be positive or negative. Let r be the number of occurrences of the less frequent sign.
(a) Show that

$$P(r = x) = \frac{m!}{x!(m - x)!}\left(\frac{1}{2}\right)^m,$$

where m is the number of $d_i \neq 0$.
(b) Twenty cows were paired according to equal weight and the cows of each pair were placed on different diets. The average daily gain (in ounces) over a period of time is given in the table which follows.

Pair	1	2	3	4	5	6	7	8	9	10
Diet *A* x_i	21	19	25	26	19	18	29	22	19	21
Diet *B* y_i	19	30	19	29	18	18	19	24	22	25

Apply the sign test to decide whether the diets produce significant mean gains in weight.

13. The mathematics department gives a multiple-choice comprehensive examination consisting of 25 questions. Each question has five answers listed. If a student knows none of the correct answers and resorts to guessing,

(a) What is his expected number correct and the standard deviation?
(b) What is a fair penalty to assign for an incorrect answer?
(c) If he passes the examination when he answers at least 13 questions correctly, what is the probability that he will pass by guessing?

3.7 THE MULTINOMIAL DISTRIBUTION

In the preceding section we considered n independent trials with only two outcomes for each trial. A natural generalization is to consider n repeated independent trials with $r > 2$ possible outcomes for each trial.

For instance, suppose that we throw a die five times. We might be interested in the probability of getting two 4's, two 5's and one 6. Here the following events can occur on each cast: $B_1 = \{1, 2, 3\}$, $B_2 = \{4\}$, $B_3 = \{5\}$, and $B_4 = \{6\}$. These form a partition of the spaces $\Omega_i = \{1, 2, 3, 4, 5, 6\}$, which for $i = 1, 2, 3, 4, 5$ are the spaces corresponding to the five throws. Let us assume that $P(B_i) = p_i$, where $i = 1, 2, 3, 4$; for a fair die, $P(B_1) = \frac{1}{2}$, while $P(B_2) = P(B_3) = P(B_4) = \frac{1}{6}$.

Now for each throw we must have a probability space $(\Omega_i, \mathscr{E}_i, P_i)$. At this point we do not know how to describe the fields $\mathscr{E}_1 = \mathscr{E}_2 = \mathscr{E}_3 = \mathscr{E}_4 = \mathscr{E}_5$. All that we know is that the field $\mathscr{E}_i$ contains the events B_1, B_2, B_3, B_4. Hence all finite intersections of these events or their complements and all finite unions of these intersections must belong to $\mathscr{E}_i$. This procedure can be applied to more than four events and to more general types of fields (cf., Chapter 6).

DEFINITION 1

The field *generated* by a class $\mathscr{C} = \{B_1, B_2, \ldots, B_r\}$ is the intersection of all fields containing $\mathscr{C}$.

This definition is not vacuous since the class of all subsets of Ω is a field containing $\mathscr{C}$. In the situation where the events B_i $(i = 1, 2, \ldots, r)$ are a partition of Ω, the field generated by $\mathscr{C}$ takes on a particularly simple form, that of all finite unions of sets of $\mathscr{C}$. The proof of this statement is simple and is left to the reader. The description of the field generated by $\mathscr{C}$ if the B_i are not disjoint is deferred until Chapter 6.

There still remains the task of defining a probability function P_i on the $\mathscr{E}_i$. However, it is just as easy to treat the more general case of $r > 4$ events. Suppose that a field $\mathscr{E}$ is generated by a class $\mathscr{C} = \{B_1, B_2, \ldots, B_r\}$ that forms a partition of Ω. In addition, assume that we have a nonnegative function P defined on the sets B_i by $P(B_i) = p_i$ for $i = 1, 2, \ldots, r$ where $\Sigma p_i = 1$. Then we can easily extend P to a probability function on $\mathscr{E}$. Since every element of $\mathscr{E}$ is of the form $B_{i_1} + B_{i_2} + \cdots + B_{i_k}$, where $i_1, i_2, \ldots, i_k$ are some of the integers $1, 2, \ldots, r$ and $k \leq r$, we merely set

$$(1) \qquad P(B_{i_1} + B_{i_2} + \cdots + B_{i_k}) = p_{i_1} + p_{i_2} + \cdots + p_{i_k}.$$

It is easy to see that P is a probability function on $\mathscr{E}$.

The probability space corresponding to the five repeated independent trials is just the product of the five spaces $(\Omega_i, \mathscr{E}_i, P_i)$. Similarly, the space corresponding to n repeated independent trials is the product of n similar spaces $(\Omega_i, \mathscr{E}_i, P_i)$. Here $\Omega_i = B_1 + B_2 + \cdots + B_r$, since there are r outcomes of each trial. Each $\mathscr{E}_i$ is the field generated by $\{B_1, B_2, \ldots, B_r\}$ and each P_i is defined by an equation similar to Equation (1), using the method described in the preceding paragraph.

Actually, this product space can be described in a simpler manner, which resembles the usual elementary treatment of this topic: It is just the space generated by the rectangular events of the form

$$E_1 \times E_2 \times \cdots \times E_n,$$

where each E_i is one of the events of $\mathscr{C}$ (cf., Theorem 5.1). This follows from the definition of $\mathscr{E}_1 \times \mathscr{E}_2 \times \cdots \times \mathscr{E}_n$ and from identities similar to

$$(B_{i_1} + B_{i_2} + \cdots + B_{i_k}) \times E_2 \times \cdots \times E_n$$
$$= B_{i_1} \times E_2 \times \cdots \times E_n + \cdots + B_{i_k} \times E_2 \times \cdots \times E_n,$$

which enable us to write products of a disjoint sum of events as a disjoint sum of products of these events. Our probability function is defined by

$$(2) \qquad P(E_1 \times E_2 \times \cdots \times E_n) = p_1^{k_1} p_2^{k_2} p_3^{k_3} \cdots p_r^{k_r},$$

in which $k_1, k_2, \ldots, k_r$ denote, respectively, the number of events E_i that are $B_1, B_2, \ldots, B_r$. For example, the probability of a particular occurrence of the event described in the second paragraph of this section (see page 80) is

$$P(B_2 \times B_2 \times B_3 \times B_3 \times B_4) = p_2^2 p_3^2 p_4.$$

In many situations we are only concerned with the probability that in n trials the outcome B_i will occur k_i times for any nonnegative integers k_i $(i = 1, 2, \ldots, r)$ that satisfy $k_1 + k_2 + \cdots + k_r = n$. For example, when six dice are rolled, what is the probability of getting each face once? In a class of 100 students, what is the probability of having 50 A's, 40 B's, and 10 C's if the probabilities of getting the grades A, B, and C are 0.1, 0.7, and 0.2 respectively? The number of occurrences of the events B_i in n independent trials is a random vector

$$v_r = (s_1, s_2, s_3, \ldots, s_r),$$

where the component s_i is a random variable that gives the number of occurrences of the event B_i in n trials. The distribution of v_r or $P(s_1 = k_1, s_2 = k_2, \ldots, s_r = k_r)$, where $k_i \geq 0$ and $k_1 + k_2 + \cdots + k_r = n$, is called the *multinomial distribution.* We will denote $P(s_1 = k_1, s_2 = k_2, \cdots, s_r = k_r)$ by $P_n(k_1, k_2, \ldots, k_r)$.

THEOREM 1

$$P_n(k_1, k_2, \ldots, k_r) = \frac{n!}{k_1!k_2! \cdots k_r!} p_1^{k_1} p_2^{k_2} \cdots p_r^{k_r}. \tag{3}$$

Proof

$x \in \{s_1 = k_1, s_2 = k_2, \ldots, s_r = k_r\}$ if and only if $x \in E_1 \times E_2 \times \cdots \times E_r$, where k_i of the E_i are B_i for $i = 1, 2, \ldots, r$. According to Equation (2), the probability of this event is $p_1^{k_1} p_2^{k_2} \cdots p_r^{k_r}$. The total number of such events is the same as the number of distinct permutations of the k_1 letters B_1, k_2 letters $B_2, \ldots, k_r$ letters B_r, where $k_1 + k_2 + \cdots + k_r = n$. This number is $n!/k_1!\, k_2! \cdots k_r!$; hence Equation (2) follows.

EXAMPLE 1

Let us find the answer to the problem posed on page 80. Under the assumption that the die is fair, we obtain

$$P_5(0, 2, 2, 1) = \frac{5!}{2!2!1!}\left(\frac{1}{6}\right)^5$$

upon applying Equation (3).

EXAMPLE 2

The probability that in a class of 100 students there will be 50 A's, 40 B's, and 10 C's, under the assumption that $PA = 0.1$, $PB = 0.7$, and $PC = 0.2$, is

$$P_{100}(50, 40, 10) = \frac{100!}{50!40!10!}(0.1)^{50}(0.7)^{40}(0.2)^{10}.$$

Equation (3) is called the multinomial distribution, since the right-hand side is the coefficient of $B_1^{k_1} B_2^{k_2} \cdots B_r^{k_r}$ in the expansion of

$$(p_1 B_1 + p_2 B_2 + \cdots + p_r B_r)^n.$$

The behavior of $P_n(k_1, k_2, \ldots, k_r)$ as a function of $k_1, k_2, \ldots, k_r$ for fixed n, r, and p_r is not as simple to determine as that of $b(k; n, p)$. However, we can determine bounds for the $k_1, k_2, \ldots, k_r$ that correspond to the maximum term.[15] To establish such bounds we need the following lemma.

[15]This result was discovered by P. A. P. Moran.

LEMMA 1

The term $P_n(k_1, k_2, \ldots, k_r)$ is maximal if and only if

$$p_i k_j \leq p_j(k_i + 1) \tag{4}$$

for every i and j.

Proof

Assume first that this term is maximal. Then it is larger than $P_n(k'_1, k'_2, k'_3, \ldots, k'_r)$. In particular, choosing $k'_s = k_s$ for $s \neq i, j$, $k'_i = k_i + 1$, and $k'_j = k_j - 1$ yields

$$\frac{1}{k_i!}\frac{1}{k_j!}p_i^{k_i}p_j^{k_j} \geq \frac{1}{(k_i+1)!(k_j-1)!}p_i^{k_i+1}p_j^{k_j-1},$$

which reduces to equation (4).

Suppose that Equation (4) holds. Using the same method as in the paragraph above we can show that if we decrease one of the k_r, say k_j, by 1 and increase another, say k_i, by 1, the resulting value of $P_n(k'_1, k'_2, \ldots, k'_r)$ is decreased. If we decrease k'_j by 1 and increase k'_i by 1, the resulting expression is decreased. In order to prove this we must show that

$$\frac{1}{(k_i+1)!}\frac{1}{(k_j-1)!}p_i^{k_i+1}p_j^{k_j-1} \geq \frac{1}{(k_i+2)!(k_j-2)!}p_i^{k_i+2}p_j^{k_j-2},$$

or equivalently,

$$(k_i + 1)p_j + (p_i + p_j) \geq p_i k_i,$$

which follows from Equation (4). By induction we can show that if k_i is increased by r and k_j is decreased by r the resulting term will be smaller than the original. However, any set of values $(k'_1, k'_2, \ldots, k'_r)$ can be obtained from $(k_1, k_2, \ldots, k_r)$ by successively increasing and decreasing pairs of the k_i, while preserving their sum $k_1 + k_2 + \cdots + k_r$. Since each change in a pair reduces the value of $P_n(k_1, k_2, \ldots, k_r)$, the result follows.

THEOREM 2

If $P_n(k_1, k_2, \ldots, k_r)$ reaches its maximum at $k_1, k_2, \ldots, k_r$, then

$$np_i - 1 \leq k_i \leq (n + r - 1)p_i \tag{5}$$

for $i = 1, 2, \ldots, r$.

Proof

If we sum over all j in Equation (4) and use the fact that $\sum_{j=1}^{r} p_j = 1$ and $\sum_{j=1}^{r} k_j = n$, we obtain

$$np_i \leq k_i + 1.$$

Summing over all i except $i = j$, we have

$$(1 - p_j)k_j \leq p_j(n - k_j + r - 1).$$

Simplifiying and combining these two inequalities, we obtain Equation (5).

If we increase the value of n while keeping the p_i fixed, we can see from Equation (5) that the most probable frequencies of occurrence for the events B_i approach p_i for $i = 1, 2, \ldots, r$. This leads us to conjecture that $Es_i = p_i$. We leave the verification of this fact to the reader.

EXERCISES

1. Suppose that we have a well-balanced cube with two red faces, three white faces, and one blue face. What is the probability that in a sequence of ten throws, four red faces, one white face, and five blue faces will show up?

2. As a result of a comprehensive examination, students are rated as superior, average, and inferior. The percentages falling into these categories are 5, 65, and 30, respectively. If the test is administered to nine persons, what is the probability that two will be rated superior, four average, and three inferior?

3. Let N elements be divided into r classes $B_1, B_2, \ldots, B_r$, which contain $N_1, N_2, \ldots, N_r$ elements respectively. If we draw a sample of size n with replacement, what is the probability that it will contain k_i elements from B_i, where $i = 1, 2, \ldots, r$?

4. A tin of mixed nuts is randomly composed from a stockpile of nuts in the following proportions: peanuts, 0.35; cashews, 0.10; walnuts, 0.30; filberts, 0.05; and pistachios, 0.20. What is the probability that a tin of 1000 nuts is mixed exactly in these proportions? What is the most probable composition of a tin of nuts?

5. (a) Show that the class of all finite sums of finite intersections of events $B_1, B_2, \ldots, B_r$ or their complements is a field.
(b) Show that this field is the intersection of all fields containing the B_i.
(c) If the B_i are a partition of Ω, then the class of all finite sums of the B_i is a field. (The empty set is included when no B_i are chosen.)
(d) Show that P defined by Equation (1) is a probability function on the field in Exercise 5(c).

6. (a) Write down an explicit expression for each s_i in terms of indicators in the multinomial theorem.
(b) Show that $Es_i = np_i$ and $\sigma^2 s_i = np_i(1 - p_i)$ for $i = 1, 2, \ldots, r$.

7. (a) Show that $\sum P_n(k_1, k_2, \ldots, k_r) = 1$, where the summation is over all r-tuples $k = (k_1, k_2, \ldots, k_r)$ satisfying $k_1 + k_2 + \cdots + k_r = n$.
(b) Verify Exercise 6(b) by finding $\sum k_i P_n(k_1, k_2, \ldots, k_r)$ and $\sum k_i^2 P_n(k_1, k_2, \ldots, k_r)$, where the summation is over all k satisfying $k_1 + k_2 + \cdots + k_r = n$.

8. Fill in the details necessary for deriving the multinomial distribution using the following alternate approach. The sample description space of a particular trial is $\Omega_i = \{S_1, S_2, \ldots, S_r\}$, where $P(\{S_j\}) = p_j$ and $\sum_{j=1}^{r} p_j = 1$. The sample space Ω of n independent trials consists of all n-tuples $(Z_1, Z_2, \ldots, Z_n)$, where each Z_i is one of the points S_j. The field of events consists of all possible subsets of Ω.

CHAPTER 4

SOME ELEMENTARY LIMIT THEOREMS

4.1 INTRODUCTION

A direct evaluation of $b(n; k, p)$, as given by equation 3.5.3, for large values of n and k is quite tedious. In practice it is often necessary to find the sum of a large number of such terms (c.f., Example 3.5.3), which is even more time-consuming. The approximations of these probabilities by means of the methods used in deriving the law of large numbers are usually too crude. Hence there is a need for asymptotic formulas which enable us to determine these probabilities to a high degree of accuracy. We will obtain such formulas for binomial and multinomial probabilities in the following sections.

4.2 THE LOCAL LIMIT THEOREM

As a prelude to finding an asymptotic formula for sums of binomial probabilities we will approximate the individual terms

$$b(k; n, p) = \frac{n!}{k!(n-k)!} p^k q^{n-k} \tag{1}$$

for large values of n. We recall (c.f. Section 3.5) that (1) represents $P\{s_n = k\}$, where s_n is a random variable which denotes the number of successes in n Bernoulli trials with probability p for success on each trial.

If we let $n \to \infty$ and keep p fixed, then $P\{|s_n - np| > n\epsilon\}$ approaches zero for every $\epsilon > 0$ by the law of large numbers. Accordingly only values of k such that $|k - np|n^{-1}$ tends to zero are important. This suggests the following transformation of variables: let

$$x_k = \frac{k - np}{\sqrt{npq}} \tag{2}$$

The factor $1/\sqrt{npq}$ is introduced for the sake of neatness in stating the final result.

Since n and k are large we can express the factorials in formula (1) by means of Stirling's[1] approximation

$$m! = (2\pi)^{1/2} m^{m+1/2} e^{-m} e^{\theta(m)} \tag{3}$$

where $|\theta(m)| \leq 1/12m$. This leads to

$$\begin{aligned} b(k; n, p) &= \frac{(2\pi)^{1/2} n^{n+1/2} e^{-n} e^{\theta(n)} p^k q^{n-k}}{(2\pi)^{1/2} k^{k+1/2} e^{-k} e^{\theta(k)} (2\pi)^{1/2} (n-k)^{n-k+1/2} e^{-(n-k)} e^{\theta(n-k)}} \\ &= \frac{1}{(2\pi)^{1/2}} \frac{n^{n+1/2}}{k^{k+1/2}(n-k)^{n-k+1/2}} p^k q^{n-k} e^{\theta} \\ &= \frac{1}{(2\pi)^{1/2}} \sqrt{\frac{n}{k(n-k)}} \left(\frac{np}{k}\right)^k \left(\frac{nq}{n-k}\right)^{n-k} e^{\theta} \end{aligned} \tag{4}$$

where $\theta = \theta(n) - \theta(k) - \theta(n - k)$. Using our previous bound for $\theta(m)$ yields

$$|\theta| \leq \frac{1}{12}\left(\frac{1}{n} + \frac{1}{k} + \frac{1}{n-k}\right).$$

Substituting for k from equation (2) the above inequality can be rewritten in the form

$$|\theta| \leq \frac{1}{12n}\left(1 + \frac{1}{p}\frac{1}{1 + x_k\sqrt{q/np}} + \frac{1}{q}\frac{1}{1 - x_k\sqrt{p/nq}}\right).$$

If we assume that $x_k/\sqrt{n} \to 0$, as $n \to \infty$, then θ tends to zero and the factor e^{θ} approaches 1. Alternatively we could assume that $a \leq x_k \leq b$, which implies $x_k/\sqrt{n} \to 0$ uniformly.

Similarly the factor $\sqrt{n/k(n-k)}$ can be replaced by

[1]James Stirling (1696–1770), an English mathematician. The original formula appears in *Methodus Differentialis,* 1730. H. Robbins, "A remark on Stirling's formula," *American Mathematical Monthly,* **62,** 26–29 (1955).

$$\frac{1}{\sqrt{npq}}\frac{1}{(1 + x_k\sqrt{q/np})(1 - x_k\sqrt{p/nq})}.$$

This expression can be approximated by $1/\sqrt{npq}$ for large values of n.

The only remaining task is to estimate the quantity

$$\left(\frac{np}{k}\right)^k \left(\frac{nq}{n-k}\right)^{n-k}. \tag{5}$$

In order to do this we consider the logarithm of the above expression—namely, $k \log(np/k) + (n-k)\log(nq/n-k)$, which can be written in the following form

$$\begin{aligned} -np(1 + \sqrt{q/np}\, x_k) \log (1 + \sqrt{q/np}\, x_k) - \\ nq(1 - \sqrt{p/nq}\, x_k) \log (1 - \sqrt{p/nq}\, x_k) \end{aligned} \tag{6}$$

upon substitution for k from equation (2). Since $x_k n^{-1/2}$ is small we can expand the logarithmic functions in power series. Using the Taylor expansion with a remainder

$$\log(1 + x) = x - \frac{x^2}{2} + \frac{\xi^3}{3!} \qquad (0 \le |\xi| \le x)$$

equation (6) becomes

$$-\frac{1}{2}x_k^2 + c\xi^3 n^{-1/2}, \tag{7}$$

where c is some constant. If we assume that $x_k^3/n^{1/2} \to 0$ then (7) can be approximated by $-\frac{1}{2}x_k^2$. Hence (5) is asymptotic to $\exp(-\frac{1}{2}x_k^2)$. Note that if we use our alternate assumption that $a \le x_k \le b$ the condition $x_k^3 n^{-1/2} \to 0$ is automatically satisfied.

Gathering together our various estimates yields[2]

$$b(k; n, p) \sim \frac{1}{\sqrt{2\pi\, npq}} e^{-x_k^2/2}. \tag{8}$$

We summarize the foregoing in the following theorem.

THEOREM 1

If $x_k^3 n^{-1/2} \to 0$ as $n \to \infty$ and $k \to \infty$ then (8) holds.

The reader can easily verify that if $x_k^3 n^{-1/2} \to 0$, then $x_k n^{-1/2} \to 0$ and so all of the assumptions made above are satisfied. Equation (8) is called the normal approximation to the binomial distribution.

[2]The sign $\sim$ indicates that the ratio of the two sides approaches 1 as $n \to \infty$.

EXAMPLE 1

Find the value of $b(55; 100, \frac{1}{2})$.

A direct evaluation of $\binom{100}{50}(\frac{1}{2})^{100}$ yields 0.04847. We can estimate this probability using (8) and tables of $e^{-1/2}x^2$. Here we have $\sqrt{npq} = 5$ and $x_k = 1$; hence (8) yields 0.04839.

Some care must be exercised in using equation (8) in practice. For example, for $p = \frac{1}{10}$ and $n = 10{,}000$ the relative error is about 0.30 for k around 1120. For k near 1150 the relative error exceeds $\frac{2}{3}$, while around 1180 it is nearly 1.4. Precise estimates for the error term are given in [1] and [2].

Using similar techniques the preceding theorem can be extended to the multinomial distribution. As in Section 3.6 we let the random variable s_i denote the number of occurrences of the event $B_i (i = 1, 2, \ldots, r)$ in n independent trials. The probability of B_i occurring on any trial is p_i. Then $P\{s_1 = k_1, s_2 = k_2, \ldots, s_r = k_r\}$, where $\sum_{i=1}^{r} k_i = n$, is given by the multinomial distribution

$$P_n(k_1, k_2, \ldots, k_r) = \frac{n!}{k_1!k_2!k_3! \cdots k_r!} p_1^{k_1} p_2^{k_2} \cdots p_r^{k_r}. \tag{9}$$

Before formulating the theorem it is convenient to introduce the following notation:

$$q_i = 1 - p_i \tag{10}$$

$$x_i = \frac{k_i - np_i}{\sqrt{np_i q_i}}. \tag{11}$$

THEOREM 2

$P\{s_1 = k_1, s_2 = k_2, \ldots, s_r = k_r\}$ satisfies the following asymptotic relation

$$P_n(k_1, k_2, \ldots, k_r) \sim \frac{\exp\left(-\frac{1}{2}\sum_{i=1}^{r} q_i x_i^2\right) n^{-(r-1)/2}}{(2\pi)^{(r-1)/2}\sqrt{p_1 p_2 \cdots p_r}}, \tag{12}$$

if $n \to \infty$ and $k_i \to \infty$ in such a way that $x_i^3 n^{-1/2} \to 0$.

Proof

Since the proof is similar to that of Theorem 1 we will omit many of the computations leaving the reader the task of filling in the details.

Applying Stirling's formula (3), Equation (9) reduces to

(13) $$P_n(k_1, k_2, \ldots, k_r) = \frac{1}{(2\pi)^{(r-1)/2}} \sqrt{\frac{n}{k_1, k_2, \ldots, k_r}} \prod_{i=1}^{r} \left(\frac{np_i}{k_i}\right)^{k_i} e^{\theta},$$

where $\theta = \theta(n) - \theta(k_1) - \theta(k_2) \cdots - \theta(k_r)$. As in the proof of Theorem (1), it is easy to see that

$$|\theta| \leq \frac{1}{12}\left(\frac{1}{n} + \frac{1}{k_1} + \frac{1}{k_2} + \cdots + \frac{1}{k_r}\right).$$

Substituting for the k_i from (11), the above inequality becomes

(14) $$|\theta| \leq \frac{1}{12n}\left(1 + \sum_{i=1}^{r} \frac{1}{p_i} \frac{1}{1 + x_i\sqrt{q/np_i}}\right).$$

By our assumption that $x_i/n^{-1/2} \to 0$, $|\theta|$ tends to zero as $n \to \infty$ and therefore $e^{\theta} \to 1$. Furthermore

$$\sqrt{\frac{n}{k_1, k_2, \ldots, k_r}} = \sqrt{\frac{n}{n^r p_1, p_2, \ldots, p_r}} \left(\prod_{i=1}^{r} (1 + x_i\sqrt{q_i/np_i})\right)^{-1/2}$$

which tends to

(15) $$\frac{1}{\sqrt{n^{r-1}}} \frac{1}{\sqrt{p_1, p_2, \ldots, p_r}}$$

since $x_i n^{-1/2} \to 0$.

Finally we consider the logarithm of the product term in (13):

(16) $$\log Q_n = \log \prod_{i=1}^{r} (np_i/k_i)^{k_i}$$

$$= -\sum_{i=1}^{r} np_i(1 + x_i\sqrt{q_i/np_i}) \log (1 + x_i\sqrt{q_i/np_i}).$$

Since each $x_i n^{-1/2}$ is small, all the logarithmic functions appearing in (16) can be expanded in a power series—namely,

$$\log Q_n = -\sum_{i=1}^{r} \frac{1}{2} q_i x_i^2 + c\,\frac{\xi_1^3 + \xi_2^3 + \cdots + \xi_r^3}{\sqrt{n}}$$

where $0 \leq |\xi_i| \leq |x_i|$ and c is some constant. It follows that $\log Q_n \to -\frac{1}{2}\sum_{i=1}^{r} q_i x_i^2$ and therefore

(17) $$Q_n \sim \exp\left(-\frac{1}{2}\sum_{i=1}^{r} q_i x_i^2\right).$$

Gathering together the estimates given by (14), (15), and (17) the theorem follows.

EXERCISES

1. Show that if $a_i \leq x_i \leq b_i (i = 1, 2, \ldots, r)$ then (12) holds uniformly in the x_i.

2. Using tables of the binomial distribution, plot the values of $\sqrt{npq}\, b(k; n, p)$ for $p = 0.1$ and $n = 100$ and compare with corresponding values of the function $\phi(x) = 1/\sqrt{2\pi}\, e^{-1/2x^2}$ (i.e., at points $x_k = k - np/\sqrt{npq}$). (For tables giving these values see [3] to [7].)

3. Show that when $r = 2$, Equation (12) reduces to (8).

4. Estimate the probability of each face turning up equally often when 6000 dice are tossed simultaneously.

5. Estimate the probability in the first part of Exercise 3.7.4.

4.3 THE DEMOIVRE-LAPLACE LIMIT THEOREM[3]

We will use the estimates for $b(k; n, p)$, derived in Section 4.2 to find an asymptotic formula for the probability that the number of successes in a sequence of Bernoulli trials, with probability p for success in each trial, lies between α and γ. However instead of $P\{\alpha \leq s_n \leq \gamma\}$ we shall find estimates for an equivalent sum $P\{a \leq s_n - np/\sqrt{npq} \leq c\}$ where $a = \alpha - np/\sqrt{npq}$ and $c = (\gamma - np)/\sqrt{npq}$. To be explicit

$$(1) \qquad P\left\{a \leq \frac{s_n - np}{\sqrt{npq}} \leq c\right\} = b(l; n, p) + b(l + 1; n, p) + \ldots + b(m; n, p)$$

where x_l and x_m defined by Equation 2.2 satisfy the following inequalities:

$$(2) \qquad a \leq x_l < a + \frac{1}{\sqrt{npq}} \qquad c - \frac{1}{\sqrt{npq}} < x_m \leq c.$$

An asymptotic formula for (1) was first discovered by DeMoivre for Bernoulli trials in which $p = q = \frac{1}{2}$ and was subsequently generalized by Laplace.

THEOREM 1

If s_n is the number of occurrences of an event in n independent Bernoulli trials, with probability p for success in each trial, then

[3]Abraham DeMoivre (1667–1754), a French mathematician, wrote *The Doctrine of Chance* in 1718. Pierre S. Laplace (1749–1827) published his *Théorie Analytique des Probabilités* in 1812.

(3) $$P\left\{a \le \frac{S_n - np}{\sqrt{npq}} \le c\right\} \sim \frac{1}{\sqrt{2\pi}} \int_a^c e^{-x^2/2}\,dx$$

if a and c vary so that $a^3 n^{-1/2} \to 0$ and $c^3 n^{-1/2} \to 0$ as $n \to \infty$.

Proof

The local limit theorem (2.1) enables us to write down the following approximation for the sum in (1)

(4) $$P\left\{a \le \frac{S_n - np}{\sqrt{npq}} \le c\right\} \sim \frac{1}{\sqrt{2\pi\, npq}} \{e^{-x_l^2/2} + e^{-x_{l+1}^2/2} + \cdots + e^{-x_m^2/2}\} = R_1$$

which holds uniformly for all the x_i since $a^3 n^{-1/2} \to 0$ and $c^3 n^{-1/2} \to 0$. Since the difference between successive x_i is $(npq)^{-1/2}$, the right hand side of (4) is an approximating sum for a Riemann integral.

In order to show that $R_1 \to 1/\sqrt{2\pi} \int_a^c e^{-x^2/2}\,dx$ we will approximate this integral by $1/\sqrt{2\pi} \int_{x_l - h/2}^{x_m + h/2} e^{-x^2/2}\,dx$, where $h = (npq)^{-1/2}$, and replace the latter by a sum. In fact, surrounding each point x_i^2 by an interval of length h and integrating $e^{-x^2/2}$ yields

$$R_2 = \frac{1}{\sqrt{2\pi}} \int_{x_l - h/2}^{x_m + h/2} e^{-x^2/2}\,dx = \frac{1}{\sqrt{2\pi}} \sum_{i=l}^{m} \int_{x_i - h/2}^{x_i + h/2} e^{-x^2/2}\,dx$$

$$= \frac{h}{\sqrt{2\pi}} \sum_{i=l}^{m} e^{-\xi_i^2/2},$$

by the mean value theorem for integrals, where

(5) $$x_i - h/2 < \xi_i < x_i + h/2.$$

We will first show that the sum $R_2 \to R_1$. Since $e^{-\xi_i^2/2} \le 1$,

(6) $$|R_2 - R_1| = \frac{h}{\sqrt{2\pi}} \left| \sum_{i=l}^{m} e^{-\xi_i^2/2}(1 - e^{-(1/2)(x_i^2 - \xi_i^2)})\right|$$

$$\le \frac{h}{\sqrt{2\pi}} \sum_{i=l}^{m} |1 - e^{-(1/2)(x_i^2 - \xi_i^2)}|.$$

From (5) it follows that

(7) $$x_i^2 - \xi_i^2 = (x_i - \xi_i)(x_i + \xi_i) \le (h/2)(2|x_i| + h/2).$$

Since the number of terms appearing in the sum in (6) is less than

$$1 + \frac{|c - a|}{h} \le \frac{|c| + |a|}{h} + 1,$$

[4]Note that $x_{i-1} + h/2 = x_i - h/2$.

(8) $$|R_2 - R_1| \le \frac{|c| + |a| + h}{\sqrt{2\pi}} \left|1 - e^{-(h/4)(2b+h/2)}\right|,$$

by using Equation (7) where $b = \max(|a|, |c|)$. Finally using the inequality $1 - e^{-x} \le x$ we see from (8) that $R_1 \to R_2$ since $a^3 h \to 0$ and $c^3 h \to 0$.

To complete the proof we show that the integral $R_2 \to \frac{1}{\sqrt{2\pi}} \int_a^c e^{-x^2/2}\, dx$. This follows from the inequality

(9) $$\left|R_2 - \frac{1}{\sqrt{2\pi}} \int_a^c e^{-x^2/2}\, dx\right| \le \frac{1}{\sqrt{2\pi}}\, h$$

which holds since the range of integration differs by at most h and $e^{-x^2/2} \le 1$.

COROLLARY

For every fixed $a < c$,

(10) $$P\left\{a \le \frac{s_n - np}{\sqrt{npq}} \le c\right\} \sim \frac{1}{\sqrt{2\pi}} \int_a^c e^{-x^2/2}\, dx.$$

If we want to approximate the sum $P\{a \le (s_n - np)/\sqrt{npq} \le c\}$ by an integral for large values of n, we must have some estimate of the difference between these two quantities. Analytic expressions for this remainder term can be found in [2]. The integral $\int_a^c e^{-x^2/2}\, dx$ must be evaluated numerically or by using the tables in [7] for the normal distribution function

(11) $$\Phi(x) = \frac{1}{\sqrt{2\pi}} \int_{-a}^{x} e^{-t^2/2}\, dt,$$

since the latter cannot be expressed in a closed form in terms of elementary functions. For most practical purposes, binomial probabilities can be computed to at least two-decimal accuracy by using the rules and tables appearing in [8]. Alternatively (see [3]–[6]), sums of binomial probabilities can be computed directly from tables of the binomial distribution function

$$F(x; n, p) = \sum_{k=0}^{x} \binom{n}{k} p^k q^{n-k}.$$

By a *distribution function*, we mean an increasing function of x which tends to 0 as $x \to -\infty$, and to 1 as $x \to \infty$, and which is continuous from the right. A *density function* is a nonnegative continuous function $f(x)$ such that $\int_{-\infty}^{\infty} f(x)\, dx = 1$.

Let us try to interpret the limit laws in terms of the axioms for a probability space and the definition of a simple random variable. As a first guess we can regard $e^{-x^2/2}/\sqrt{2\pi}$ as a generalization of the notion of probability density, since

$$\frac{1}{\sqrt{2\pi}}\int_{-\infty}^{\infty} e^{-x^2/2}\,dx = 1$$

is a property of it.[5] Now in order to speak of a probability density, we need a random variable. However, in this case the probability is distributed over a noncountable set of values, so the notion of a simple random variable cannot apply. Thus, if we want to interpret these limit theorems, we must modify our axioms so that they hold for noncountable probability spaces. This will be done in Chapter 5.

Let us now examine some typical problems that lead to the DeMoivre–Laplace limit theorem.

EXAMPLE 1

The probability of getting heads on a single toss of a coin is p. What is the probability that the relative frequency of the number of heads in n tosses will differ from p by no more than ϵ? Calculate this probability for $p = \frac{1}{2}$, $n = 10{,}000$, and $\epsilon = 1/100$.

Let s_n be the number of heads in n tosses. Then the required probability is obtained by using the DeMoivre–Laplace theorem—i.e.,

$$P\left\{\left|\frac{s_n}{n} - p\right| \le \epsilon\right\} = P\left\{-\epsilon\sqrt{\frac{n}{pq}} \le \frac{s_n - np}{\sqrt{npq}} \le \epsilon\sqrt{\frac{n}{pq}}\right\}$$
$$\doteq \frac{1}{\sqrt{2\pi}}\int_{-\epsilon\sqrt{n/pq}}^{\epsilon\sqrt{n/pq}} e^{-x^2/2}\,dx = \frac{2}{\sqrt{2\pi}}\int_{0}^{\epsilon\sqrt{n/pq}} e^{-x^2/2}\,dx.$$

For the given values of the constants, the probability is 0.96.

EXAMPLE 2

A well-balanced die is thrown 100 times. Let s_n represent the number of times a 1 appears. Find the bound ϵ such that

$$P\left\{\left|\frac{s_{100}}{100} - \frac{1}{6}\right| \le \epsilon\right\} = 0.8.$$

[5] Recall that $f(x) \ge 0$ is a density function if $\sum_x f(x) = 1$. Here the sum is replaced by an integral.

As in Example 1,

$$P\left\{\left|\frac{S_n}{n} - p\right| \leq \epsilon\right\} \doteq \frac{2}{\sqrt{2\pi}} \int_0^{60\epsilon/\sqrt{5}} e^{-x^2/2}\,dx = 0.8,$$

which gives us an equation for the determination of ϵ.

Using tables of the normal distribution function, we see that $60\epsilon\sqrt{5} = 1.3$ or $\epsilon = 0.049$.

EXAMPLE 3

What is the smallest number of tosses of a fair coin that must be made in order that the relative frequency of the number of heads will differ from $\frac{1}{2}$ by less than 0.01 with a probability of not less than 0.9?

The required answer is the smallest value of n that will satisfy the inequality

$$P\left\{\left|\frac{S_n}{n} - \frac{1}{2}\right| \leq 0.01\right\} \geq 0.9,$$

which can be written, using the DeMoivre–Laplace theorem, as

$$\frac{2}{\sqrt{2\pi}} \int_0^{\sqrt{n}/50} e^{-x^2/2}\,dx \geq 0.9.$$

Hence from a table of the normal distribution function, $\sqrt{n}/50 = 1.6$ or $n = .0098$.

We will now generalize Theorem 1 by finding an integral which is asymptotic to a sum of multinomial probabilities. The notation introduced in Section 4.2 will be employed.

An arbitrary event $\{s_1 = k_1, s_2 = k_2, \ldots, s_r = k_r\}$ can be represented by the point $(k_1, k_2, \ldots, k_r)$ in r-dimensional Euclidean space, where

$$k_1 + k_2 + \cdots + k_r = n. \tag{12}$$

Hence the outcomes of the n trials are represented by all the points on the hyperplane (12) whose coordinates are integers j_i with $0 \leq j_i \leq n$.

If we introduce the transformation of coordinates defined by Equations (2.10) and (2.11), the hyperplane (12) becomes

$$\sum_{i=1}^{r} \sqrt{np_i q_i}\, x_i = 0. \tag{13}$$

The points $(x_1, x_2, \ldots, x_r)$ where $x_i = (k_i - np_i)/\sqrt{np_i q_i}$, which correspond to the event $\{s_1 = k_1, s_2 = k_2, \ldots, s_r = k_r\}$, are called *lattice points*.

Let $P_n(R)$ denote the probability that the lattice points corresponding to the outcome of n trials lies in the region R. In other words,

$$P_n(R) = \sum_{(x_1, x_2, \ldots, x_r) \epsilon R} P\{s_1^* = x_1, s_2^* = x_2, \ldots, s_r^* = x_r\}, \tag{14}$$

where $s_i^* = (s_i - np_i)/\sqrt{np_i q_i}$ for $i = 1, 2, \ldots, r$.

The following theorem then holds.

THEOREM 2

Suppose that we have n independent trials. Each trial has r possible outcomes B_i, and $P(B_i) = p_i$, where $0 < p_i < 1$ and $\sum_{i=1}^{r} p_i = 1$, remains the same throughout the trials. Then the relation

$$P_n(R) \sim \left(\frac{q_1 q_2 \cdots q_r}{(2\pi)^{r-1} \sum_{i=1}^{r} p_i q_i} \right)^{1/2} \int_R e^{-(1/2)\Sigma q_i x_i^2}\, dv \tag{15}$$

holds uniformly in R as $n \to \infty$ for any region R whose boundary is integrable in any bounded portion of the hyperplane (13). Here dv denotes an infinitesimal element of volume of the region R and the integration is taken over the region R.

Proof

Let us first assume that the region R is bounded. Hence we can enclose R in a "parallelepiped" of volume C—that is, the intersection of $I_1 \times I_2 \times \cdots \times I_r$ and the hyperplane (13), where $I_j = (x_j : a_j \leq x_j \leq b_j)$ for $j = 1, 2, \ldots, r$. Since each x_j is contained in a bounded interval, the following asymptotic relation for the sum in Equation (14) holds uniformly in the x_j, by Theorem 2.2:

$$P_n(R) \sim \sum_{(x_1, x_2, \ldots, x_r) \epsilon R} \frac{\exp\left(-\frac{1}{2} \sum_{i=1}^{r} q_i x_i^2\right) n^{-(r-1)/2}}{(2\pi)^{(r-1)/2} \sqrt{p_1 p_2 \cdots p_r}} \tag{16}$$

As in the proof of Theorem 1, we surround each lattice point by a $(k-1)$-dimensional parallelepiped. To be explicit, suppose that $(x_1^0, x_2^0, x_3^0, \ldots, x_r^0)$ is a fixed lattice point and

$$J_i = \left(x_i : x_i^0 - \frac{1}{2} \frac{1}{\sqrt{np_i q_i}} \leq x_i \leq x_i^0 + \frac{1}{2} \frac{1}{\sqrt{np_i q_i}}\right); \qquad (i = 1, 2, \ldots, r),$$

then the parallelepiped is the intersection of $J_1 \times J_2 \times \cdots \times J_r$ and the hyperplane (13). The reader can verify that the volume Δv of this parallelepiped is

(17)
$$\Delta v = \frac{\sum_{i=1}^{r} p_i q_i}{n^{r-1} \prod_{i=1}^{r} p_i q_i}.$$

Hence if, by the use of Equation (17), we rewrite relation (16) as

(18)
$$P_n(R) \sim \left(\frac{q_1 q_2 \cdots q_r}{(2\pi)^{r-1} \sum_{i=1}^{r} p_i q_i}\right)^{1/2} \sum_{(x_1\, x_2, \ldots, x_r) \epsilon R} \exp\left(-\tfrac{1}{2} \sum_{i=1}^{r} q_i x_i^2\right) \Delta v = S(R),$$

we recognize that the sum $S(R)$ represents an approximation to a multiple Riemann integral.

We will show that $S(R)$ approaches the integral appearing in relation (15) by using the generalized mean value theorem (c.f., page 269 of Reference [13]). Let R_1 be the region obtained by surrounding each lattice point in R by the parallelepiped described in the preceding paragraph. Since R_1 is connected, replacing the integral in Equation (19) by a sum of integrals over the parallelepipeds and applying the mean value theorem yields

(19)
$$I(R_1) = \left(\frac{q_1 q_2 \cdots q_r}{(2\pi)^{r-1} \sum_{i=1}^{r} p_i q_i}\right)^{1/2} \int_{R_1} \exp\left(-\tfrac{1}{2} \sum q_i x_i^2\right) dv$$
$$= \left(\frac{q_1 q_2 \cdots q_r}{(2\pi)^{r-1} \sum_{i=1}^{r} p_i q_i}\right)^{1/2} \sum_{(\xi_1, \xi_2, \ldots, \xi_r) \epsilon R} \exp\left(-\tfrac{1}{2} \sum_{i=1}^{r} q_1 x_i^2\right) \Delta v,$$

where $(\xi_1, \xi_2, \ldots, \xi_r)$ is a point in the parallelepiped surrounding $(x_1, x_2, \ldots, x_r)$.

We will first show that $S(R)$ approaches the sum $I(R_1)$. Observing that $S(R_1) = S(R)$, we will have the desired result if we can show that $S(R_1) - I(R_1)$ is arbitrarily small as $n \to \infty$. However, this follows from the inequality

(20)
$$|S(R_1) - I(R_1)| \leq \left(\frac{q_1 q_2 \cdots q_r}{(2\pi)^{r-1} \sum_{i=1}^{r} p_i q_i}\right)^{1/2} \sum_{i=1}^{r} \frac{1}{4} \sqrt{\frac{q_i}{np_i}} \left(c_i + \frac{1}{\sqrt{np_i q_i}}\right) C,$$

where C is the volume of the parallelepiped enclosing R, and $c_i = \max(|a_i|, |b_i|)$. This inequality is derived by using Equation (17).

Next, we will show that $I(R_1) \to I(R)$. If a point belongs to $R_1 - R$, it lies in a parallelepiped whose center is in R, while if it belongs to $R - R_1$, it lies in some parallelepiped which has points in common with R but whose center is outside R. Hence in both cases its distance from the $(r-2)$-dimensional boundary of R is less than the diameter of a parallelepiped. Therefore, $I(R_1) - I(R)$ cannot exceed the product of the $(r-2)$-dimensional volume b of the boundary of R, the maximum value of the integrand, and the diameter of a parallelepiped surrounding a lattice point. Accordingly,

$$|I(R_1) - I(R)| \leq b\left(\frac{q_1 q_2 \cdots q_r}{(2\pi)^{r-1} \sum_{i=1}^{r} p_i q_i}\right)^{1/2} \left(\sum_{i=1}^{r} \frac{1}{np_i q_i}\right)^{1/2}, \tag{21}$$

since the diameter[6] is $\left(\sum_{i=1}^{r} 1/np_i q_i\right)^{1/2}$. The theorem follows from Equations (20) and (21).

Now let the region R be unbounded. Let S be an $(r-1)$-dimensional sphere of a radius ρ that will be determined below, with its center the origin. Then $R = RS + RS'$. Since RS is bounded, it follows from the result above that for $n \geq N_1$,

$$|P_n(RS) - I(RS)| < \frac{\epsilon}{2} \tag{22}$$

for an arbitrary $\epsilon > 0$. We will show that inequality (22) also holds for RS' if ρ is sufficiently large. Choose ρ so large that $I(S) > 1 - \epsilon/4$. This is possible since $I(X) = 1$, where X is the whole space. Since S is bounded, by choosing $n > N_2$, the difference between $I(S)$ and $P_n(S)$ can be made so small that $P_n(S) > 1 - \epsilon/4$. Therefore, both $P_n(S')$ and $I(S')$ are less than $\epsilon/4$. Since $P_n(RS') < P_n(S')$ and $I(RS') < I(S')$,

$$|P_n(RS') - I(RS')| < \epsilon/2. \tag{23}$$

The result follows from relations (22) and (23) with $n \geq max\,(N_1, N_2)$ and with $P_n(R) = P_n(RS) + P_n(RS')$ and $I(R) = I(RS) + I(RS')$.

EXERCISES

1. Do Example 3.6.3 by using the DeMoivre-Laplace theorem.

2. Deduce Bernoulli's theorem (3.6.3) from Theorem 1.

3. Use Theorem 1 to obtain an answer to Exercise 3.6.4.

4. Suppose that 12,000 tosses of a fair die are made. Find the probability of observing

[6] The diameter of a set is the supremum of all the distances between any two arbitrary points of the set. In the above, the distance between two points will be maximum if the difference between each coordinate is at its maximum, $(np_i q_i)^{-1/2}$.

(a) One thousand or more threes.
(b) Between 2100 and 2200 threes.

5. Let S_n be the number of successes in n independent repeated Bernoulli trials with a probability 0.2 for success. Calculate $P\{i \leq S_{10} \leq i+2\}$ for $i = 5, 6, \ldots, 49$ and find the percentage of error involved, by the use of

(a) The binomial probability law.
(b) The normal approximation.

6. Four theaters are competing for 800 patrons. Suppose that all the people arrive at the same time and choose a theater at random, independently of each other. How many seats should each theater have in order to be at least 90 percent certain of seating all of its customers?

7. In Theorem 2, x_r is really a function of $x_1, \ldots, x_{r-1}$ defined by Equation (13). Theorem 2 can then be formulated taking this dependence into account. Show that under the same hypothesis

$$P_n(R) \sim \frac{q_1 q_2 \cdots q_{r-1}}{(2\pi)^{r-1} p_r} \int_{R'} e^{-1/2 f(x_1 x_2 \cdots x_{r-1})}\, dx_1\, dx_2 \cdots dx_{r-1},$$

where

$$f(x_1 \cdots x_{r-1}) = \sum_{i=1}^{r-1} q_1\left(1 + \frac{p_i}{p_r}\right) x_i^2 + 2 \sum_{1 \leq i \leq j \leq r-1} x_i x_j \frac{(p_i q_i p_j q_j)^{1/2}}{p_r}$$

and R' is the projection of R on the coordinate hyperplane $x_r = 0$.

8. The probability of winning a game of craps is 0.495. If a player wins 25 games out of 50, should you be suspicious?

9. Suppose that two containers have the same volume and each contains 2.7×10^{22} molecules of a gas. The containers are connected so that a free exchange of molecules can occur between them. Find the probability that after 36 hours there will be at least $1 \times 10^{-10} N$ more molecules in one container than in the other where $N =$ the total number of molecules in both containers.

10. If a black mouse and a white mouse are mated, the probability that an offspring of the second generation will be pure white is $\frac{1}{4}$. Find the probability that among 400 such descendants more than 200 will be pure white.

4.4 POISSON'S THEOREM

It is frequently expedient to have an approximation for $b(k; n, p)$ for large n and small p, such that np is moderate. An examination of the remainder term reveals that the normal approximation (2.8) becomes poorer as p decreases from $\frac{1}{2}$ and finally breaks down for $p = 0$. An asymptotic formula covering this situation was discovered by Poisson (1781–1840), a French mathematician; it will be derived as a corollary to the following theorem.

THEOREM 1

Suppose that we have a sequence of n Bernoulli trials with the same probability p_n of success on each trial. If $p_n \to 0$ as $n \to \infty$, then

$$b(k; n, p_n) - \frac{\lambda_n^k}{k!} e^{-\lambda_n} \to 0, \tag{1}$$

where $\lambda_n = np_n$.

Proof

Substituting λ_n/n for p_n in

$$b(k; n, p_n) = \frac{n!}{k!(n-k)!} p_n^k (1 - p_n)^{n-k}$$

yields

$$b(k; n, p_n) = \frac{\lambda_n^k}{k!} \left(1 - \frac{\lambda_n}{n}\right)^n \frac{\left(1 - \frac{1}{n}\right)\left(1 - \frac{2}{n}\right) \cdots \left(1 - \frac{k-1}{n}\right)}{\left(1 - \frac{\lambda_n}{n}\right)^k}. \tag{2}$$

If $\{\lambda_n\}$ is unbounded, we can make both $b(k; n, p_n)$ and $(\lambda_n^k/k!)e^{-\lambda_n}$ *arbitrarily* small for sufficiently large n. The second fact is obvious, so we merely state that

$$\frac{\lambda_n^k}{k!} e^{-\lambda_n} < \frac{\epsilon}{2} \tag{3}$$

for an arbitrary $\epsilon > 0$ and fixed k if $\lambda_n > a$, where a is some constant. The first fact follows from the inequality $1 - x \le e^{-x}, 0 \le x \le 1$, which shows, using Equation (2), that

$$b(k; n, p_n) \le \frac{\lambda_n^k}{k!} e^{-(\lambda_n/n)(n-k)}$$

for $n > k$. Hence it follows that

$$b(k; n, p_n) < \frac{\epsilon}{2} \tag{4}$$

for $\lambda_n > b$. Now if $c = \max(a, b)$, then for all $n > k$ such that $\lambda_n > c$,

$$\left|b(k; n, p_n) - \frac{\lambda_n^k}{k!} e^{-\lambda_n}\right| < \frac{\epsilon}{4} \tag{5}$$

from relations (3) and (4).

It remains to consider those values of n for which $\lambda_n \le c$. Since $\lambda_n \le c$,

$$\left(1 - \frac{\lambda_n}{n}\right)^n \sim e^{-\lambda_n}$$

and

$$\lim_{n\to\infty} \frac{[1 - 1/n] \cdots [1 - (k-1)/n]}{[1 - (\lambda_n/n)]^k} = 1$$

for fixed k; it follows that $n \geq N(\epsilon)$,

$$|b(k; n, p_n) - \frac{\lambda_n^k}{k!} e^{-\lambda_n}| < \epsilon.$$

The result follows since either $\lambda_n \leq c$ or $\lambda_n > c$.

COROLLARY

Under the hypothesis of this theorem, if $n\, p_n \to \lambda$, then

(6) $$b(k; n, p_n) \to \frac{\lambda^k}{k!} e^{-\lambda}.$$

Under some circumstances, p may be so small that the Poisson approximation can be used. However, n may be so large that the normal approximation is also valid. This makes it likely that the normal density will also be an approximation to the Poisson distribution (c.f., page 177 of [9]).

Since $\sum_{k=0}^{\infty} (\lambda^k/k!)e^{-\lambda} = 1$, we can regard the function $(\lambda^k/k!)e^{-\lambda}$ as the probability distribution of some random variable. This random variable is not simple, because the probability is distributed over a countably infinite set of values. This again points out (c.f., the remarks following Theorem 3.1) that we must extend our definitions of random variable.

To use the asymptotic formula for calculating the value of $b(k; n, p)$, we must have some method of estimating the error involved. We will derive bounds for the ratio of the two quantities, which are quite good in most situations. From Equation (2),

(7) $$\frac{\lambda_n^k}{k!}\left(1 - \frac{\lambda_n}{n}\right)^{n-k} \geq b(k; n, p_n) \geq \frac{\lambda_n^k}{k!}\left(1 - \frac{k}{n}\right)^k \left(1 - \frac{\lambda_n}{n}\right)^n.$$

The left-hand side of the inequality follows from the fact that

$$\left(1 - \frac{1}{n}\right)\left(1 - \frac{2}{n}\right) \cdots \left(1 - \frac{(k-n)}{n}\right) \leq 1.$$

The right-hand side follows from the inequalities

$$\left(1 - \frac{\lambda_n}{n}\right)^{n-k} \geq \left(1 - \frac{\lambda_n}{n}\right)^n$$

[since $(1 - \lambda_n/n) \leq 1$], and

$$\left(1 - \frac{1}{n}\right)\left(1 - \frac{2}{n}\right) \cdots \left(1 - \frac{(k-1)}{n}\right) \geq \left(1 - \frac{k}{n}\right)^k.$$

Finally, Equation (7) and the inequality (c.f., Exercise 2)

$$e^{-t/(1-t)} < 1 - t < e^{-t} \qquad (0 < t < 1)$$

yield

$$(8) \qquad e^{k\lambda_n/n} > \frac{b(k; n, p_n)}{e^{-\lambda_n}(\lambda_n^k/k!)} > e^{-[k^2/(n-k)+\lambda^2{}_n/(n-\lambda_n)]}.$$

EXAMPLE 1

If the probability of a person having an accident is 0.01, what is the probability that there will be no accidents in a group of 100 people?

Assume that the event that one person has an accident is independent of the event that another person has an accident. Hence we have a sequence of 100 Bernoulli trials with the probability for an accident equal to 1/100. The required answer is $b(0; 100, 1/100) = 0.366$. The Poisson approximation (6), where $\lambda = 1$, is 0.368.

Since the Poisson distribution appears quite often in practice, tables of $e^{-\lambda}(\lambda^k/k!)$ as well as sums of such terms have been prepared (see Reference [10] and [11]). The reason for the frequent appearance of the Poisson distribution is given in Exercise 5.2.13.

EXERCISES

1. Show that $\sum_{k=0}^{\infty} (\lambda^k/k!)e^{-\lambda} = 1$.

2. Prove that $e^{-t/(1-t)} < 1 - t < e^{-t}$, for $0 < t < 1$. [*Hint:* Use the fact that $\log 1/(1-t) = t + t^2/2 + t^3/3 + \cdots$.]

3. If the probability of shooting down an airplane with a single rifle shot is 0.001, what is the probability of destroying the plane if 300 shots are fired simultaneously?

4. Let $p(k; \lambda) = (\lambda^k/k!)e^{-\lambda}$. For fixed λ, show that

(a) $p(k-1; \lambda) < p(k; \lambda)$ if $k < \lambda$.
(b) $p(k-1; \lambda) > p(k; \lambda)$ if $k > \lambda$.
(c) $p(k-1; \lambda) = p(k; \lambda)$ if $k = \lambda$.

Deduce that $p(k; \lambda)$ increases from $k = 0$ to $k_0 = [\lambda]$ and then decreases. If λ is an integer, then $p(k; \lambda)$ has two maximum values at $k_0 = \lambda$ and $k_0' = \lambda - 1$. (Compare this result with Theorem 3.6.2 for the binomial distribution.)

5. Suppose that there is a probability of 0.001 of manufacturing a defective ball bearing. If they are packed in boxes of 1000, what is the probability that there will be (*a*) no defectives, (*b*) more than 990 good ball bearings? How many ball bearings should be packed in a box so that the probability is at least 0.9 that the box will contain at least 1000 perfect ones? [*Hint:* Apply the Poisson approximation to each term in $n = 1000 + x$ trials. Since x is small, approximate $\lambda = 0.001n$ by $\lambda = 1$ and use the tables appearing in Reference [10].]

6. The number of oil tankers arriving at a refinery each day has a Poisson distribution with a parameter $\lambda = 3$. Present port facilities can service four tankers a day, the remainder being sent to another port.

(a) What is the expected number of tankers arriving per day?
(b) What is the most probable number of tankers arriving per day?
(c) What is the probability of having to send tankers away on a given day?
(d) How much must present facilities be increased to permit handling all tankers on about 90 percent of the days?
(e) What is the expected number of tankers serviced daily?
(f) What is the expected number of tankers turned away daily?

7. Glass vases are manufactured from molten glass containing small particles. If a single particle appears in a vase, the vase must be discarded. Assume that the particles are randomly scattered in the molten glass and that 100 pounds of molten glass always contains x particles.

(a) Under reasonable assumptions, find an approximation to the probability that there will be k particles in a vase if each vase is produced from α pounds of molten glass.
(b) For $x = 30$ and $\alpha = 1$, show that about 26 percent of the vases will be discarded.
(c) For $x = 30$ and $\alpha = 0.25$, show that only 7 percent of the vases will be discarded.

REFERENCES

1. Uspensky, J. V., *Introduction to Mathematical Probability*. New York: McGraw-Hill, 1937.
2. Feller, W., "On the normal approximation to the binomial distribution," *Annals of Math. Stat.*, **16** (1945), 319–329.
3. *Tables of the Binomial Probability Distribution*, Applied Mathematics Series, Vol. 6. Washington, D. C.: National Bureau of Standards, 1950. Here $n \leq 50$.
4. Romig, H. C., *50–100 Binomial Tables*. New York: Wiley, 1953.
5. *Tables of the Cumulative Binomial Probability Distribution*. Cambridge, Mass.: Harvard Computation Laboratory, 1955.
6. *Tables of the Cumulative Binomial Probabilities*, U. S. Army Ordnance Corps, ORDP 20-11. Washington, D. C.: Government Printing Office, 1952.
7. *Tables of Probability Functions*, Vol. 2. Washington, D. C.: National Bureau of Standards, 1942.
8. Hald, A., *Statistical Tables and Formulas*. New York: Wiley, 1952.
9. Feller, W., *An Introduction to Probability Theory and Its Applications*. New York: Wiley, 1960.
10. Molina, E. C., *Poisson's Exponential Binomial Limit*. Princeton, N. J.: Van Nostrand, 1942.
11. Kitagawa, J., *Tables of Poisson Distribution*. Tokyo: Baifukan, 1952.
12. Tukey, J. W., and F. Mosteller, "The use and usefulness of binomial probability paper," *J. Am. Stat. Ass.*, **44** (1949), 174.

CHAPTER 5

INFINITE PROBABILITY SPACES

5.1 INTRODUCTION

Until now we have dealt with probability spaces $(\Omega, \mathscr{E}, P)$ in which Ω has a finite number of points. However, there are many phenomena that require an infinite probability space for an adequate description. For instance, suppose that we want to find the probability that we will get at least one head when we toss a coin repeatedly. Since it is possible to get an arbitrary number of tails in a row, a typical point of a probability space Ω would consist of an infinite sequence of heads and tails. In fact, $\Omega = \times_{i=1}^{\infty} \Omega_i$, where each $\Omega_i = \{h, t\}$. Further examples of countable and noncountable spaces will be given below.

The introduction of an infinite number of events leads to certain problems in the definition of a probability function. These will be pointed out in the following section; however, their solution requires some knowledge of measure theory. Since these difficulties are of a theoretical rather than a practical nature, we will not resolve them until the next chapter. However, by assuming the existence of a probability function and proceeding intuitively, it is possible to solve many practical problems. This approach will be adopted below and some techniques for finding probabilities in infinite spaces, such as generating functions, will be described.

5.2 MODIFICATION OF THE PROBABILITY AXIOMS

Let us now examine in more detail the problem of finding the probability that we will get at least one head when we toss a coin repeatedly. Here $\Omega = \underset{i=1}{\overset{\infty}{\times}} \Omega_i$, where $\Omega_i = \{h, t\}$. Since most of us would agree that the statement "we will get at least one head" is an event, we should be able to express this statement in terms of sets of points. In fact, let $H_n = \underset{i=1}{\overset{\infty}{\times}} Y_i$, where $Y_i = \{t\}$ for $i = 1, 2, \ldots, n-1$, $Y_n = \{h\}$, and $Y_i = \{h, t\}$ for $i > n$. The English equivalent of H_n is "we will get a head on the nth toss for the first time and either heads or tails after the nth toss." The statement "we will get at least one head" is equivalent to the statement "we will get the first head on the first toss or second toss or third toss or $\cdots$," which is $\sum_{i=1}^{\infty} H_i$. We see that if H_i is an event, then $\sum_{i=1}^{\infty} H_i$ should be an event. Hence, in order to discuss infinite probability spaces we must revise Axiom 1 (see Section 1.4).

The class $\mathscr{E}$ of events in the above example is the class of all subsets of Ω.

We have seen that $P\left(\sum_{i=1}^{n} E_i\right) = \sum_{i=1}^{n} PE_i$ for arbitrary n. Hence it is natural, provided that we have an infinite number of events, to assume that $P\left(\sum_{i=1}^{\infty} E_i\right) = \sum_{i=1}^{\infty} PE_i$. It is easy to see that this assumption is reasonable. Most of us will agree that the probability of getting at least one head should be 1[1]; that is, $P\left(\sum_{i=1}^{\infty} H_i\right) = 1$. Further, since PH_n is just the probability of getting $n-1$ tails and then a head (we do not care what happens after the nth toss), we should set $PH_n = 1/2^n$.[2] In other words, if we only consider the first n trials, we will see that the above probability is $1/2^n$. Since $\sum_{n=1}^{\infty} PH_n = \sum_{n=1}^{\infty} 1/2^n = 1$, we see that $P\left(\sum_{n=1}^{\infty} H_n\right) = \sum_{n=1}^{\infty} PH_n$.

[1]The probability of getting at least one head in n tosses of a coin is 1 minus the probability of getting no heads; this is $1 - (\frac{1}{2})^n$.

[2]This seems clear intuitively. For further justfication, consider $n+1$ tosses. The probability of getting $n-1$ tails and a head on the first n tosses is $2/2^{n+1} = 1/2^n$, since on the last toss we can get heads or tails. For $n+2$ tosses there are four favorable cases, since we can get (h, h), (h, t), (t, h), or (t, t) on the last two tosses; once again the probability of getting $n-1$ tails and a head is $4/2^{n+2} = 1/2^n$. For $n+k$ tosses, $p = 2^k/2^{n-k} = 1/2^n$. Hence the probability of the above event remains the same for any $m \geq n$ tosses.

Therefore, in order to consider infinite probability spaces we must modify Axioms 1–4 of Section 1.4.

AXIOMS FOR A PROBABILITY SPACE

1. If $E_i \in \mathscr{E}$ $(i = 1, 2, 3, \ldots)$, then $\bigcup_{i=1}^{\infty} E_i \in \mathscr{E}$.
2. If $E \in \mathscr{E}$, then $E' \in \mathscr{E}$.
3. If $E \in \mathscr{E}$, then $PE \geq 0$; also $P\Omega = 1$.
4. If $\{E_i\}$ is a class of disjoint events, then

$$P\left(\sum_{i=1}^{\infty} E_i\right) = \sum_{i=1}^{\infty} PE_i.$$

A class of sets $\mathscr{E}$ satisfying Axioms 1 and 2 is called a *σ-field* (or *σ-algebra*). A set function P satisfying Axiom 4 is said to be completely additive or *σ-additive*.

A set of axioms for probability theory was first proposed by G. Bohlmann in 1908 [11]. These axioms involved two primitive concepts (event and probability) and the following postulates:

1. If A is an event, $PA \geq 0$.
2. $P\Omega = 1$.
3. If A and B are disjoint events, then $P(A + B) = PA + PB$.
4. If A is an event and $PA > 0$, then $P(B|A) = P(A + B)/PA$.

This paper immediately attracted the attention of many mathematicians such as U. Broggi ([2]), A. Padoa ([3]), and G. Peano ([4]). Later R. A. Fisher[3] ([5]) and R. von Mises ([6]) introduced an empirical approach, giving a statistical content to the concept of probability. In 1921, J. M. Keynes ([7]) gave a new axiomatization of probability theory based on conditional probability. The Keynes postulates are:

1. $P(A|H) \geq 0$.
2. $P(H|H) = 1$.
3. $P(A|H) + P(A'|H) = 1$.
4. $P(AB|H) = P(A|H)P(B|AH)$.

These efforts led to a new foundation of probability theory based on measure theory. The new foundation rose gradually during the 1920's under the influence of various authors. It reached the axiomatic basis that is used in modern probability theory in 1932, in A. N. Kolmogorov's famous paper ([8]). An analogous axiomatization was presented by H.

[3]Fisher's first statistical paper appeared in 1912.

Reichenbach ([9]) at almost the same time, but it did not stir up much interest.

We will assume that the spaces described in the examples and exercises appearing in this section satisfy the axioms for a probability space. At present, we cannot show that they are probability spaces. Later we will be able to prove these assertions by using measure theory. To see the source of the difficulty, let us consider the coin-tossing example. We have not proved that a probability function exists. Actually, we only know the value of P for rectangular events. If $E = \times_{i=1}^{\infty} E_i$ and n of the E_i are proper subsets of $\{h, t\}$ while the remainder are equal to $\{h, t\}$, we can set $PE = 1/2^n$ (see footnote 2). For example, letting $Y = \{h, t\}$,

$$P(\{h\} \times Y \times \{t\} \times Y \cdots) = \frac{1}{4}.$$

Now disjoint finite sums of rectangular events form a field $\mathscr{F}$ (see Exercise 1). If $F \in \mathscr{F}$, then $F = F_1 + F_2 + \cdots + F_n$ where the F_i ($i = 1, 2, \ldots, n$) are rectangular events. Since PF_i is known, we define $PF = PF_1 + PF_2 + \cdots + PF_n$. Hence we only know the value of P on a field $\mathscr{F} \subset \mathscr{E}$, where $\mathscr{E}$ is the class of all subsets of Ω. The problem is to extend P, which is defined on $\mathscr{F}$, to all of $\mathscr{E}$ so that the resulting extension is unique, nonnegative, and σ-additive. We shall prove that this is possible in Chapter 6. By assuming that P exists, we can calculate probabilities of events as indicated above and in the following examples.

EXAMPLE 1

What is the probability of getting an 8 before a 7 when a pair of dice is cast repeatedly?

Let E be the event that we get an 8 before a 7 and let E_i be the event that we get an 8 on the ith throw and neither an 8 nor a 7 on any of the previous throws; then $E = \sum_{i=1}^{\infty} E_i$. Since $PE_i = \frac{5}{36}(\frac{25}{36})^{i-1}$, we have

$$PE = P\left(\sum_{i=1}^{\infty} E_i\right) = \sum_{i=1}^{\infty} PE_i = \frac{5}{36}\sum_{i=1}^{\infty}\left(\frac{25}{36}\right)^{i-1} = \frac{5}{11}.$$

EXAMPLE 2

Suppose that two players A and B play a game. The probability that A will win a single game is a and the probability that B will win is b. The outcome of each game is independent of all the preceding games. Find the probability that A will win i games in a row before B will win j games in a row. (For brevity we will denote this event by "A wins.")

The space Ω of outcomes is $\mathop{\times}\limits_{i=1}^{\infty} Y_i$; here each $Y_i = \{l, \omega\}$, where ω indicates that A wins and l that A loses. We assume that we can define a probability function P on the set of events. Let A_n be the set in the n-dimensional space $\mathop{\times}\limits_{i=1}^{n} Y_i$, which denotes the event that A wins within the first n trials. In the infinite-dimensional space $\mathop{\times}\limits_{i=1}^{\infty} Y_i$, this event is the set of points $E_n = A_n \times Y^{(n)}$, where $Y^{(n)} = \mathop{\times}\limits_{i=n+1}^{\infty} Y_i$. From the definition of the events E_n, it follows that $E_i \supset E_{i+1} \supset E_{i+2} \supset E_{i+3} \cdots$. Since A_n is defined in a finite probability space, we can find the probability of this event, which we denote by p_n. We then set $P(E_n) = p_n$. Let E denote the event that A wins; then $E = \lim\limits_{n\to\infty} E_n = \bigcap\limits_{n=1}^{\infty} E_n$.

To find $p = P(E)$, we use the fact that $P(E) = P(\lim E_n) = \lim P(E_n) = \lim p_n$. The equality $P(\lim E_n) = \lim P(E_n)$ holds for monotone sequences. This result is a consequence of the σ-additivity of the set function P; we will defer the proof of this theorem until Chapter 6. Since $P(E)$ is unique, we could have used any other monotone sequence of events with known probabilities that approach E. In a similar manner, we can find the probability that B will win and the probability r that neither will win.

The quantities p_n are not easy to calculate; a method is given in Chapter 13, Example 8(b), of Reference [10]. We shall obtain the limits p, q and r by a simpler method.[4] Let x and y be the conditional probabilities of E given that the first trial results in A winning or losing respectively. Using the total probability theorem,

(1) $$p = ax + by.$$

Suppose that the first trial results in A's defeat. Then A can win only if he loses at most the next $j - 2$ games in a row. The probability that A will win given that he loses exactly k games in a row ($k = 0, \ldots, j - 2$) is ax, since he must win the next game and then win i games in a row under the hypothesis that he has already won a game. Applying the total probability theorem once more,

(2) $$y = xa(1 + b + b^2 + \cdots + b^{j-2}) = x(1 - b^{j-1}).$$

If the first trial results in A winning, an analogous argument yields

(3) $$x = a^{i-1} + by(1 + a + \cdots + a^{i-2}) = a^{i-1} + y(1 - a^{i-1}).$$

[4]We emphasize that the probability of an event is uniquely defined and is independent of the method used. This will be shown in the next chapter.

Solving Equations (2) and (3) for x and y and substituting the resulting expressions in Equation (1) leads to

$$p = a^{i-1}\frac{1 - b^j}{a^{i-1} + b^{j-1} - a^{i-1}b^{j-1}}.$$

By symmetry,

$$q = b^{j-1}\frac{1 - a^i}{a^{i-1} + b^{j-1} - a^{i-1}b^{j-1}}.$$

since $p + q = 1$, we see that $r = 0$; the probability that neither A nor B wins is zero.

The last two examples have the following feature in common. We considered only events E which are limits of monotone sequences of events E_n depending on a finite number of trials. Accordingly, the probability of each event E_n is known. Taking the limit of these probabilities, we find the probability of the event E. However, only by using some results of measure theory can we show that $P(E) = \lim P(E_n)$ and that the limits are independent of the particular sequence used.

Further examples of experiments with a countably infinite number of outcomes are provided by number theory problems. Let an infinite sequence of numbers $\{a_i\}$ be given, and let A be an event (a set of numbers having a given property). Denoting the number of points in $A_N = A \cap \{a_1, a_2, \ldots, a_N\}$ by $q(A_N)$, we define PA_N by $PA_N = q(A_N)/N$. If PA_N approaches a limit as $N \to \infty$, we define $PA = \lim_{N\to\infty} q(A_N)/N$. For example, if $a_i = i$ and A is the event that a number chosen at random is even, then $q(A_N) = [N/2]$ and $PA = \frac{1}{2}$. In general, this probability will depend on the way in which the numbers are arranged in the given sequence (see Exercise 4). In problems dealing with the positive integers, we assume that they are arranged in increasing order.

EXAMPLE 3

What is the probability that a positive integer selected at random is relatively prime to 6? That at least one of two integers selected at random is relatively prime to 6?

Let A be the event that an integer selected at random is relatively prime to 6. If B and C are the events that an integer selected at random is divisible by 2 and 3 respectively, then $A = (B \cup C)'$. Since $PB = \frac{1}{2}$ and $PC = \frac{1}{3}$,

$$P(B \cup C) = PB + PC - P(BC) = \frac{1}{2} + \frac{1}{3} - \frac{1}{6} = \frac{4}{6},$$

assuming that B and C are independent. Hence, $PA = \frac{1}{3}$.

Similarly, the probability that at least one of two integers selected at random is relatively prime to 6 is $\frac{1}{3} + \frac{1}{3} - \frac{1}{9} = \frac{5}{9}$.

Alternatively, if $N = 6Q + R$, then $q(A_N) = 2Q + r$, where $r = 0, 1,$ or 2, and $P(A_N) = (2Q + r)/(6Q + R) \to \frac{1}{3}$.

The concepts introduced for finite probability spaces can easily be generalized to probability spaces with a countably infinite number of outcomes by replacing n by ∞. For example, an (elementary) random variable r is defined by $r = \sum_{j=1}^{\infty} x_j I_j$, where I_j is the indicator of the event A_j and $\Omega = \sum_{j=1}^{\infty} A_j$. The expectation of r, Er, is defined by $Er = \sum_{j=1}^{\infty} x_j P(r = x_j) = \sum_{j=1}^{\infty} x_j f(x_j)$, where f is the density function of r. If A is an event and $\Omega = \sum_{i=1}^{\infty} E_i$, we have the total probability theorem

$$PA = P\left(\sum_{i=1}^{\infty} AE_i\right) = \sum_{i=1}^{\infty} P(A|E_i)PE_i.$$

We assume that the infinite analogs of Bayes' theorem, the conditional expectation theorems, etc., are valid. Their formulation and proofs are left to the reader. At this point the reader should rework the exercises in Section 3.4 and Exercise 4. 4. 6.

EXERCISES

1. (a) Show that $\mathscr{F}$ is a field.
(b) Prove that P is uniquely defined on $\mathscr{F}$ (cf., theorem 3.5.2).

2. Express the event T (getting no heads) when a coin is tossed repeatedly as a decreasing sequence of events; show $P(T) = 0$.

3. Suppose that in a trial there are three outcomes, A, B, and C, with probabilities p, q, and r, respectively. Find the probability that in a series of repeated independent trials, a run of a consecutive A's will occur before either a run of B's of length b or a run of C's of length c. (This problem can also be solved by using generating functions.)

4. Let the positive integers be arranged in the order 1, 3, 2, 5, 7, 4, 9, 11, 6, . . . ; that is, the first two odd numbers, then the first even number, then the next two odd numbers, then the next even number, and so on. Show that the probability that any number selected at random will be even is $\frac{1}{3}$.

5. A, B, and C, three players of equal skill, play a series of games. First A plays B, then the winner plays C, and so on, the loser always being eliminated. Find the probability that A will win two games in a row before B or C.

6. Consider a series of repeated independent trials on each of which an event E occurs or does not occur with probabilities p and $q = 1 - p$ respectively. Let r be the trial on which E occurs for the first time.

(a) Show that r has a geometric distribution; that is,

$$P(r = x) = pq^{x-1}. \qquad (x = 1, 2, 3, \ldots)$$

(b) Verify that $Er = 1/p$.

7. The game of craps is played with two dice in the following manner. A player throws the dice; he wins at once if the total for the first throw is 7 or 11 and loses at once if the total is 2, 3, or 12. Any other total is called his "point." If the first throw is a point, the player rolls the dice repeatedly until he either wins by throwing his point again or loses by throwing a 7. What is the probability of winning at craps?

8. A, B, and C fight a three-cornered pistol duel. All know that A's chance of hitting his target is 0.3, C's is 0.5, and B's is 1.0. They fire at their choice of target in succession in the order A, B, C, cyclically (but a hit man loses further turns and is no longer shot at). What should A's strategy be?

9. What is the probability that the square of an integer selected at random will end with the digit 1? That the cube of an integer selected at random will end with the digits 11?

10. (a) Prove that 2^n can begin with any sequence of digits.
(b) Let N be any r-digit number. What is the probability that the first r digits of the number 2^n represent the number N?

11. (a) Let n be a positive integer, and let p_n be the probability that two integers a, b chosen at random from the range $1 \leq a;\ b \leq n$, are relatively prime. Show that $\lim_{n\to\infty} p_n = p$ exists.

(b) Show that $1/p = 1 + 1/2^2 + 1/3^2 + 1/4^2 + \cdots$.

12. Show that the probability that an integer is prime is zero.

13. (a) In Chapter 4, the Poisson distribution was used as an approximation for the binomial distribution. However, the Poisson distribution plays an extremely important role in its own right, since it represents an appropriate probabilistic model for a large number of physical processes[5] (see [10]); the reason is that these processes satisfy the following axioms. Let $N(t)$ denote the number of events that have occurred in the interval $[0, t]$.

A_0. $N(0) = 0$.
A_1. The random variables $N(t_2) - N(t_1),\ N(t_3) - N(t_2), \ldots,\ N(t_n) - N(t_{n-1})$ are independent for all $t_1 < t_2 < \cdots < t_n$ for any n.

[5]These include such things as the distribution of weed seeds in grass seeds, raisins in bread, telephone calls coming into an exchange, flying bomb hits on London, deaths by horse kicks in the Prussian army, radioactive disintegration, flaws in material, animal litters in fields, bacterial colonies on Petri plates, and red (or white) corpuscles in a blood count.

A_2. $N(t_2 + h) - N(t_1 + h)$ has the same distribution as $N(t_2) - N(t_1)$ for all choices of indexes t_1 and t_2 and every $h > 0$.
A_3. The probability of at least one event happening in a period of time h is

$$p(h) = \lambda h + o(h) \qquad (\lambda > 0)$$

as $h \to 0$.
A_4. The probability of two or more events happening in a time h is $o(h)$.

Show that $P_k(t) = P(N(t) = k) = (\lambda^k t^k / k!)e^{-\lambda t}$ for $k = 0, 1, 2, 3, \ldots$. [*Hint:* Show that $P_0(t + h) = P_0(t)P_0(h) = P_0(t)\ [1 - p(h)]$ and, taking limits, show that $P_0'(t) = -\lambda P_0(t)$ and so $P_0(t) = e^{-\lambda t}$. Using the total probability theorem, prove that

$$P_k'(t) = -\lambda P_k(t) + \lambda P_{k-1}(t), \qquad (k = 1, 2, 3, \ldots)$$

subject to the initial condition

$$P_k(0) = 0. \qquad (k = 1, 2, 3, \ldots)$$

Solve by setting $Q_k(t) = P_k(t)e^{\lambda t}$].

(b) Often the Poisson process arises in a form where the time parameter is replaced by a suitable spatial parameter. Consider an array of points distributed in a Euclidean n space R^n. Let $N(R)$ denote the number of points (finite or infinite) contained in a region R of R^n. Let $V(R)$ denote the volume of the region R. Define a Poisson process $N(R)$ by rewriting the axiom with $N(t)$ replaced by $N(R)$ and t replaced by $V(R)$; show that

$$P(N(R) = k) = (\lambda^k V^k / k!)e^{-\lambda V}.$$

5.3 ON GAMBLING[6]

In this section we will show that there exists no successful system for gambling with games of chance. By a game of chance we mean the repeated performance of some experiment. This experiment has a finite number of outcomes, each of which has a definite probability of occurring. Each trial results in some outcome, and a player bets a constant sum on the outcome of the trials. Some examples of games of chance are roulette, tossing coins, and rolling dice. By definition, a system is a fixed set of rules that determines whether or not the player is to bet on a trial. The rules are such that the decision of whether or not to bet on a particular trial depends only on the outcomes of the preceding trials. For simplicity of exposition, we add the further requirement that the rules must insure an indefinite continuation of betting. For example, the player might decide to bet only on even-numbered trials or on trials determined by an infinite sequence of random numbers. In tossing coins the bettor may bet only after 1000 tails have occurred. (He

[6]This section need not be studied now, but it should be read before 7.9. A probabilistic treatment appears in Reference [10].

would then probably bet heads.) He might decide to make his first bet after the first tail, his second bet after a run of two tails, and his jth bet just after a run of j tails. He would then bet less and less frequently.

The mathematical counterpart of an experiment with a finite number of outcomes occurring randomly is a probability space $(Y, \mathscr{S}, F)$. Here $Y = \{y_1, y_2, \ldots, y_n\}$, $\mathscr{S}$ is the class of all subsets of Y, and F is a probability measure[7] defined on $\mathscr{S}$. The concept of repeated independent trials is represented by the probability space $(X, \mathscr{E}, P)$, where $X = \times_{i=1}^{\infty} Y_i$ and $Y_i = Y$ for all i. Here $\mathscr{E}$ is the class of all subsets of X and P is a probability measure on $\mathscr{E}$ determined by its value on rectangular events (as in Section 2); that is,

$$P(E_1 \times E_2 \times \cdots \times E_n \times Y_{n+1} \times Y_{n+2} \cdots) = \prod_{i=1}^{n} F(E_i).$$

The mathematical formulation of a rule requires a sequence of functions on X, —that is, f_1, $f_2(x_1)$, $f_3(x_1, x_2), \ldots$, where f_1 is either 0 or 1 and f_n (depending only on $x_1, \cdots x_{n-1}$) takes the same values. If the results of the first $n - 1$ trials are $x_1, x_2, \ldots, x_{n-1}$, the player bets on the nth trial if and only if $f_n(x_1 \cdots x_{n-1}) = 1$. Since the betting does not terminate, we will require the probability that a player will make an infinite number of bets to be 1. This is the probability of the set of points $x = (x_1, x_2, \ldots) \in X$ for which an infinite number of the $f_i = 1$; that is, $\limsup_{i\to\infty} f_i = 1$.

Since the amount wagered is the same at each trial, it will follow that a successful system is impossible if the use of the system does not change the probability of an event. To express this mathematically, suppose that the player bets on the j_kth trials ($k = 1, 2, 3, \ldots$) for a particular set of outcomes x. Then $f_i(x_1 \cdots x_{i-1}) = 1$ when i belongs to the sequence $\{j_k\}$. Set $b(n, x) = j_n$; that is, the nth bet is made on trial j_n. Let $\bar{x}_n = x_{b(n,x)}$; then the point $(\bar{x}_1, \bar{x}_2, \ldots)$ represents the results of the trials on which a bet is made. For example, consider the last system described in the first paragraph of this section. If

$$x = (t, h, h, h, t, t, t, t, \ldots),$$

then $b(1, x) = 2, b(2, x) = 7, b(3, x) = 8, b(4, x) = 9, \ldots.$

Still referring to the above example, let us investigate the meaning of the phrase "the system does not change the probability of an event." Suppose that a person using this system bets on heads each time he bets; let $\bar{E}_1 = (\bar{x}_1 = h)$. We will show that $P(\bar{x}_1 = h) = \frac{1}{2}$; in other words, the system is not helpful since $P(\bar{x}_1 = h) = P(x_1 = h)$. The event $(\bar{x}_1 = h) = \sum_{i=2}^{\infty} E_i$, where

$$E_i = (x\colon x_1 = h, \ldots, x_{i-2} = h, x_{i-1} = t, x_i = h);$$

[7]That is, a probability function.

hence $P(\bar{x}_1 = h) = \sum_{i=2}^{\infty} 1/2^i = \frac{1}{2}$. More generally, let $\bar{E}$ be any $\bar{x}$-event and E the corresponding x-event. The impossibility of a successful system means $P(\bar{E}) = PE$.

THEOREM 1

Let (f_i) be a sequence of functions on X such that:

1. $f_n(x) = 0$ or $f_n(x) = 1$.
2. $\limsup f_n(x) = 1$ for a set of points with a probability (measure) 1.
3. $f_1 \equiv 0$ or $f_1 \equiv 1$; if $n > 1$, $f_n(x)$ depends only on $x_1, \ldots, x_{n-1}$.
4. Let T be the transformation defined above, taking $x = (x_1, x_2, \ldots)$ into $\bar{x} = (\bar{x}_1, \bar{x}_2, \ldots)$ at every x at which condition 2 is true.

Let $\bar{E}$ be an event in X and $E = T^{-1}(\bar{E})$; then $P(\bar{x} \in \bar{E}) = P(x \in \bar{E})$.

Proof

The measure (or probability function) on X is determined by the probability of rectangular events (see Sections 3.5 and 6.7); hence it suffices to prove the theorem for $\bar{E}$ determined by events $\bar{E}_k$ for $k = 1, 2, \ldots, n$. We must show that

(1) $$P(\bar{E}) = P(\bar{x}_k \in \bar{E}_k, k = 1, \ldots, n) = P(x_k \in \bar{E}_k, k = 1, \ldots, n).$$

The last equality states that the probability of the event $\bar{E}$ is not changed because of the system.

We will prove Equation 1 by induction. For $n = 1$,

(2) $$P(\bar{E}) = P(\bar{x}_1 \in \bar{E}_1) = \sum_{j=1}^{\infty} P(b(1, x) = j, x_j \in \bar{E}_1).$$

The last equality follows from the fact that the first bet occurs on some trial and from the σ-additivity of P. Now the event $(x\colon b(1, x) = j) = (x\colon f_j(x_1, \cdots, x_{j-1}) = 1)$ and so it depends on the first $j - 1$ coordinates; consequently, it is independent of the event $(x_j \in \bar{E}_1)$. Since $P(x_j \in \bar{E}_1)$ is independent of j (it is just a set of some outcomes $y_1, y_2, \ldots, y_n$), then

(3) $$P(b(1, x) = j, x_j \in \bar{E}_1) = P(b(1, x) = j)P(x_1 \in \bar{E}_1).$$

If we substitute Equation (3) in Equation (2), noting that $\sum_{j=1}^{\infty} P(b(1, x) = j) = 1$, then Equation 1 will hold for $n = 1$.

Now suppose that Equation (1) is true for less than n sets $\bar{E}_k$. The event E can be decomposed into disjoint events determined by conditions of the form

(4) $$b(i, x) = k_i, x_{k_i} \in \bar{E}_1,\ i = 1, 2, \ldots, n - 1;\ b(n, x) = j, x_j \in \bar{E}_n,$$

where $k_1 < \cdots < k_{n-1} < j$. The first set of $n - 1$ conditions determines a set A which only depends upon the first $j - 1$ coordinates, as can be seen by expressing A in terms of the f_{k_i}. As in the preceding paragraph, we see that the event $(x_j \in \bar{E}_n)$ is independent of j and the events A and $(b(n, x) = j)$. Hence

$$P(E) = P(x_n \in \bar{E}_n) \sum_{j=n}^{\infty} P(x \in A, b(n, x) = j). \tag{5}$$

Since $\sum_{j=n}^{\infty} P(b(n, x) = j) = 1$, the last sum reduces to $P(A)$. Then Equation (5) becomes

$$P(\bar{E}) = P(x_n \in \bar{E}_n) P(\bar{x}_j \in \bar{E}_j, j = 1, \ldots, n - 1). \tag{6}$$

Using our induction hypothesis, we see that Equation (6) reduces to Equation (1), since

$$P(x_j \in \bar{E}_j, j = 1, \ldots, n - 1) = \prod_{j=1}^{n-1} P(x_j \in \bar{E}_j)$$

because of the fact that the trials are independent.

The above proof and formulation follow Doob [11]. In this paper a more general theorem is proved for arbitrary factor spaces. However, in this case the f_i are assumed to be measurable functions (see Chapter 7), since not all subsets are events.

Note that successful systems exist if the amount wagered on each trial can be arbitrarily large. For example, suppose that we have an infinite amount of capital and we choose heads every time in a coin-tossing game. We keep betting one dollar until we lose; on the next trial we double our bet and keep doubling until we win. If we win we stop playing (we bet zero dollars). In the last section we showed that the probability of getting at least one head is 1. Therefore, the probability that we will win is 1. However, this is not realistic, since the amount that can be wagered is fixed by house rules.

In some games of "chance" such as blackjack, the outcomes of the first $n - 1$ trials give us some information about the outcomes of the succeeding trials. Using this information, advantageous strategies can be found (see Reference [12]).

EXERCISES

1. Show that $\sum_{j=n}^{\infty} P(b(n, x) = j) = 1$.

2. Prove that if $\{E_i\}$ is a disjoint sequence of events and $\sum_{i=1}^{\infty} P(E_i) = 1$, then for any event E, $P(E) = \sum_{i=1}^{\infty} P(EE_i)$.

3. Show that for all the above rules, the set of points for which an infinite number of bets are made has a probability measure of 1. (*Hint:* Use the Borel–Cantelli theorem 12.1.3)

5.4 GENERATING FUNCTIONS

The method of generating functions is particularly suited to the study of random variables of the kind we have studied so far—that is, those that assume a countable number of values.

DEFINITION 1

Let r be a random variable and let $P(r = a_k) = p_k$ for $k = 0, 1, 2, 3, \ldots$. We define another random variable s such that $(s = k)$ if and only if $(r = a_k)$. Then the function

$$G(x) = E(x^s) = \sum_{k=0}^{\infty} p_k x^k$$

is called the (probability) *generating function*[8] of the random variable r.

Since $G(1) = 1$, a comparison with the geometric series shows that the series for $G(x)$ converges absolutely for $|x| \leq 1$. If $\{p_k\}$ is an arbitrary sequence of real numbers, and the series for $G(x)$ converges for $|x| < x_0$, then $G(x)$ is also called the *generating function* of the sequence $\{p_k\}$. Note that in the interval of convergence, $G(x)$ is uniquely determined by $\{p_k\}$ and vice versa.

For our purposes, we can assume that x is real, but in more advanced applications it is convenient to work with complex variables (cf., Exercise 3).

EXAMPLE 1

1. If $r = 1$ when we get a head tossing a fair coin and $r = -1$ when we get a tail, then $G(x) = \frac{1}{2} + \frac{1}{2}x$.
2. The generating function of the Poisson distribution $p(k; \lambda) = e^{-\lambda}(\lambda^k/k!)$ is

$$G(x) = \sum_{k=0}^{\infty} e^{-\lambda} \frac{(\lambda x)^k}{k!} = e^{-\lambda + \lambda x}.$$

[8]Generating functions are usually defined only for integral-valued nonnegative random variables with $a_k = k$; otherwise the generating function is not uniquely defined. For example, suppose that $P(r = -1) = p$ and $P(r = 1) = q$. If $a_0 = -1$ and $a_1 = 1$, then $G_1(x) = p + qx$, while for $a_0 = 1$ and $a_1 = -1$, $G_2(x) = q + px$.

In many problems the desired probability distribution is obtained by first finding the corresponding probability generating function $G(x)$ (see Section 5.5). The coefficients p_k are given by $p_k = G^{(k)}(0)/k!$. The explicit calculation of the kth derivative usually requires the decomposition of $G(x)$ into partial fractions. This technique is also useful in finding approximations (cf., Exercise 3).

THEOREM 1

Let r be a random variable such that $P(r = k) = p_k$ and let $G(x) = \sum_{k=0}^{\infty} p_k x^k$ be its generating function. Then

1. $Er = G'(1)$ if $Er < \infty$ and $G'(x) \to \infty$ as $x \to 1$ if Er diverges.
2. $\text{Var}\, r = G''(1) + G'(1) - (G'(1))^2$ if $\text{Var}\, r < \infty$ and $G''(x) \to \infty$ as $x \to 1$ in the case of an infinite variance.

Proof

1. By definition, $Er = \sum_{k=1}^{\infty} kp_k$. When $x = 1$, the derivative $G'(x) = \sum_{k=1}^{\infty} kp_k x^{k-1}$ reduces to Er. If $Er < \infty$, $G'(x)$ exists for $|x| < 1$ and the desired result follows from Abel's theorem. On the other hand, if Er diverges, then $G'(x) \to \infty$ as $x \to 1$, since all the terms are positive.
2. Let us assume that $\text{Var}\, r < \infty$. It is easy to see that $E(r(r-1)) = \sum_{k=2}^{\infty} k(k-1)p_k = G''(1)$. Since $\text{Var}\, r = Er^2 - (Er)^2$, we must add $Er - (Er)^2$; by using the result in the first part of this proof, we obtain the expression in the second part. In the case of infinite variance, Er^2 is divergent; hence $G''(x) \to \infty$ as $x \to 1$, since all the terms are positive.

THEOREM 2

Let r_1 and r_2 be nonnegative, integral-valued, mutually independent random variables. If their generating functions are $F(x) = \sum_{n=0}^{\infty} p_n x^n$ and $G(x) = \sum_{n=0}^{\infty} q_n x^n$ respectively, then their sum $r_1 + r_2$ has the generating function $F(x)G(x)$.

Proof

The event $\{r_1 + r_2 = k\}$ can be decomposed into the disjoint events $\{r_1 = 0,\ r_2 = k\}$, $\{r_1 = 1,\ r_2 = k-1\}, \ldots, \{r_1 = n, r_2 = k-n\}, \ldots, \{r_1 = k, r_2 = 0\}$. However, $P(r_1 = n, r_2 = k - n) = p_n q_{k-n}$ because of independence; therefore,

(1)
$$P(r_1 + r_2 = k) = \sum_{n=0}^{k} p_n q_{k-n},$$

which is the coefficient of x^k in the generating function of $r_1 + r_2$. The proof of the theorem is completed by noting that Equation (1) is also the coefficient of x^k in the power series for $F(x)G(x)$.

COROLLARY

Let s_1 and s_2 be nonnegative, integral-valued random variables. If $r_1 = x_0 + as_1$ and $r_2 = w_0 + as_2$ are independent random variables with generating functions $F(x)$ and $G(x)$ respectively, then their sum $r_1 + r_2$ has the generating function for $F(x)G(x)$.

Proof

The sum

$$r_1 + r_2 = x_0 + w_0 + a(s_1 + s_2).$$

Hence the random variable s associated with $r_1 + r_2$ is $s = s_1 + s_2$ and the result follows from the theorem.

DEFINITION 2

Suppose that $\{p_k\}$ and $\{q_k\}$ are any two sequences of numbers. The *convolution* of $\{p_k\}$ and $\{q_k\}$ is a sequence $\{r_k\}$ defined by $r_k = \sum_{n=0}^{k} p_n q_{k-n}$.

This operation will be denoted by $\{r_k\} = \{p_k\} * \{q_k\}$. Of course we can iterate this operation and form $\{r_k\} * \{s_k\} = (\{p_k\} * \{q_k\}) * \{s_k\}$.

Since multiplication of generating functions is associative and commutative, it follows directly that the convolution is also an associative and commutative operation. Of course these results can also be proved directly from the definition.

EXAMPLE 2

1. The number of successes in n Bernoulli trials is a random variable; it can be written as the sum of n independent random variables—namely,

$$s_n = I_1 + I_2 + \cdots + I_n,$$

where $I_j = 1$ if the desired event occurs on the jth trial and 0 otherwise (cf., Equation 3.6.2). The generating function of I_j is $(q + px)$, where $q = P(I_j = 0)$ and $p = P(I_j = 1)$. Hence the generating function of s_n is $(q + px)^n$.

2. The sum of s_n and s_m, two independent random variables that both have binomial distributions, is a random variable that also has a binomial distribution, since its generating function is $(q + px)^{m+n}$.
3. The above result also holds for Poisson random variables since, using the notation of part 2 of Example 1,

$$\{p(k; \lambda)\} * \{p(k; u)\} = \{p(k; \lambda + u)\}.$$

In Section 4.4 we showed that the Poisson distribution is a limiting form of the binomial distribution. We recall that the main hypothesis was that if $np_n = \lambda_n$, then $\lambda_n \to \lambda$ as $n \to \infty$. The generating function of the corresponding binomial distribution is $(q_n + p_n x)^n = (1 - \lambda_n(1 - x)/n)^n$. From the definition of $e = \lim_{n\to\infty} (1 + 1/n)^n$, the generating function of the Poisson distribution is $e^{-\lambda(1-x)}$. The following theorem shows that this result holds in general.

THEOREM 3

Suppose that for every fixed n the sequence $\{p_k(n)\}$ has the properties $p_k(n) \geq 0$ and $\sum_{k=0}^{\infty} p_k(n) = 1$. Then $\lim_{n\to\infty} p_k(n) = p_k$ for every k if and only if $G_n(x) \to G(x)$ for every x with $0 \leq x < 1$, where $G_n(x) = \sum_{k=0}^{\infty} p_k(n)x^k$ and $G(x) = \sum_{k=0}^{\infty} p_k x^k$.

Proof

Assume that $\lim_{n\to\infty} p_k(n) = p_k$. Then

$$(2) \qquad |G_n(x) - G(x)| \leq \sum_{k=0}^{S} |p_k(n) - p_k|x^k + \sum_{k=S+1}^{\infty} |p_k(n) - p_k|x^k.$$

For fixed x, where $0 \leq x < 1$ and $\epsilon > 0$, we choose s large enough so that $x^{S+1}/(1 - x) < \epsilon/4$. Since $|p_k(n) - p_k| \leq 2$, the second sum in Equation (2) is less than $\epsilon/2$. Chose n so large that $|p_k(n) - p_k| \leq \epsilon/2s$ for $k = 0, 1\ldots, s$; then the first sum appearing in Equation (2) is less than $\epsilon/2$, and it follows that $G_n(x) \to G(x)$.

Now assume that $G_n(x) \to G(x)$ as $n \to \infty$ for every x with $0 \leq x < 1$. In order to prove that $p_k(n) \to p_k$, we need the following lemma.

LEMMA 1

Let $\{p_k(n)\}$ be a double sequence of real numbers that are bounded with respect to n—that is, $|p_k(n)| < M_k$ for $k = 0, 1, 2, \ldots$. Then there exists an increasing sequence $\{n_n\}$ such that $p_k(n_n)$ converges for every k as $n \to \infty$.

Since $p_k(n) < 1$, we can always find a subsequence that will converge for every k. If $\lim_{n\to\infty} p_k(n) \neq p_k$ for every k, it is possible to extract two subsequences $\{\bar{p}_k(n_i)\}$ and $\{\bar{\bar{p}}_k(m_i)\}$ converging to $\{\bar{p}_k\}$ and $\{\bar{\bar{p}}_k\}$. Utilizing the first half of the proof, it follows that $G_{n_i}(x) \to G(x) = \sum_{k=0}^{\infty} p_k x^k$ and $\bar{\bar{G}}_{m_i}(x) \to \bar{\bar{G}}(x)$. Since the limiting function $G(x)$ is unique, we have arrived at a contradiction. The proof of the lemma remains.

Proof

The boundedness of the sequence $\{p_1(n)\}$ implies, by the Bolzano–Weierstrass theorem, the existence of a subsequence $\{p_1(n_1)\}$ which converges, where $n_1 = \{1_1, 2_1, 3_1, \ldots\}$. The sequence $\{p_2(n_1)\}$ is also bounded; hence there exists a subsequence $\{p_2(n_2)\}$ which converges. In general, having defined the sequence $\{p_r(n_r)\}$, we can extract a convergent subsequence from $\{p_{r+1}(n_r)\}$, which we denote by $\{p_{r+1}(n_{r+1})\}$. Note that $\{n_{r+1}\}$ is a subsequence of $\{n_k\}$ for $k = 0, 1, 2, \ldots, r$, where $\{n_0\}$ is identical to $\{n\}$. The last statement shows that the diagonal sequence $\{n_n\}$ is a subsequence of every $\{n_j\}$ except for a finite number of elements ($j = 0, 1, 2, 3, \ldots$). Accordingly, the sequence $\{p_k(n_n)\}$ $n \geq k$, being a subsequence of the convergent sequence $\{p_k(n_k)\}$, is convergent for all k.

Note that $\{p_k\}$ need not be a probability distribution even if all the hypotheses of Theorem 3 are satisfied. For example, if $p_k(n) = \delta_{nk}$, then $\lim_{n\to\infty} p_k(n) = 0$ for all k.

As an application of the continuity theorem 3 and the Bernstein polynomials, we will characterize generating functions. We require the following definition.

DEFINITION 3

The function f is *absolutely monotone* in (a, b) if $f^{(n)}(x) \geq 0$ for $n = 0, 1, 2, \ldots$ and $a < x < b$.

THEOREM 4

G is a probability generating function if and only if G is absolutely monotone for $0 < x < 1$ and $G(x) \to 1$ as $x \to 1$.

Proof

If G is a generating function, obviously G is absolutely monotone and $\lim_{x\to 1} G(x) = 1$.

Before proving the converse, we will introduce the following notation. The *difference operator* Δ is defined by

(3) $$\Delta u_n = u_{n+1} - u_n. \qquad (n = 1, 2, 3\ldots)$$

The k*th-order difference operator* Δ^k is defined recursively by the relation $\Delta^k = \Delta(\Delta^{k-1})$, where $\Delta^1 = \Delta$ and $\Delta^0 u_n = u_n$. For example,

$$\Delta^2 u_n = \Delta(\Delta u_n) = \Delta(u_{n+1} - u_n) = u_{n+2} - 2u_{n+1} + u_n.$$

The operator E is defined by

(4) $$Eu_n = u_{n+1}. \qquad (n = 1, 2, 3, \ldots)$$

The operator E^k is defined recursively by $E^k = E(E^{k-1})$, where $E^1 = E$ and $E^0 u_n = u_n$. Note that $E^k u_n = u_{n+k}$.

It is easy to verify that $\Delta = E - 1$; hence

(5) $$\Delta^k u_n = (E-1)^k u_n = \sum_{j=0}^{k} \binom{k}{j} (-\Delta)^{k-j} E^j u_n = \sum_{j=0}^{k} \binom{k}{j} (-1)^{k+j} u_{n+j}.$$

Similarly, the identity $u_n = (E - \Delta)u_n$ yields

(6) $$u_n = (E-\Delta)^k u_n = \sum_{j=0}^{k} \binom{k}{j} (-1)^{k-j} E^j u_n$$

$$= \sum_{j=0}^{k} \binom{k}{j} (-\Delta)^{k-j} \Delta^{k-j} u_{n+j}.$$

The operator Δ is frequently applied to sequences $u_n = f(x + nh)$ obtained from a function f by fixing a point x and $h > 0$. In such situations it is convenient to replace the operater Δ by the difference ratio $\underset{h}{\Delta} = h^{-1}\Delta$, where

$$\underset{h}{\Delta} f(x) = (f(x+h) - f(x))/h.$$

The kth-order operators $\underset{h}{\Delta}^k$ are again defined recursively by $\underset{h}{\Delta}^{k+1} = \underset{h}{\Delta}(\underset{h}{\Delta}^k)$, where $\underset{h}{\Delta}^1 = \underset{h}{\Delta}$ and $\underset{h}{\Delta}^0 f(x) = f(x)$. Equation 5 becomes

(7) $$\underset{h}{\Delta}^k f(x) = h^{-k} \sum_{j=0}^{k} \binom{k}{j} (-1)^{k+j} f(x + jh).$$

To illustrate the use of Equation (7), we rewrite the Bernstein polynomials equation (3.6.9) in this new notation. Expanding $(1 - x)^{n-i}$ by the binomial theorem, we see that the coefficient of x^j on the righthand side of Equation 3.6.9 is

$$\sum_k f\left(\frac{k}{n}\right) \binom{n}{k} \binom{n-k}{j-k} (-1)^{j-k} = \binom{n}{j} h^j \underset{h}{\Delta}^j f(0). \qquad (h = 1/n)$$

Hence 3.6.9 reduces to

(8) $$B_n(x) = \sum_{j=0}^{n} \binom{n}{j} (hx)^j \underset{h}{\Delta^j} f(0). \qquad (h = 1/n)$$

We are now prepared to complete the proof of Theorem 4. If G is absolutely monotone for $0 < x < 1$ and $G(x) \to 1$ as $x \to 1$, then obviously G is continuous for $0 < x \leq 1$ [if we set $G(1) = 1$]. Since $G' \geq 0, G$ is an increasing function and so $\underset{h}{\Delta} G \geq 0$. Similarly, $G^{(k)} \geq 0$ for $k = 1, 2, \ldots,$ implies $\underset{h}{\Delta} G^{(k)} \geq 0$. Hence, by induction, for all n $(h = 1/n)$, $\underset{h}{\Delta^k} G \geq 0$ $(k \leq n)$. It follows from Equation (8) that the coefficients x^j in B_n (for the function G) are all nonnegative. From Equation 3.6.9, these coefficients add to $B_n(1) = G(1) = 1$. Therefore, $B_n(x)$ is a probability generating function. By Theorem 3, $G = \lim_{n\to\infty} B_n$ is a generating function.

Generating functions are also useful in solving difference equations as illustrated in the next section (see Exercises 7 and 8 of Section 5.5).

EXERCISES

1. If $|c_k| < M$, show that $\sum_{k=0}^{\infty} c_k x^k$ converges for $|x| < 1$.

2. Let v be an integral-valued random variable such that $p(v = k) = p_k$ and $P(v > k) = g_k$. Let $\{p_k\}$ and $\{g_k\}$ have generating functions $P(x)$ and $G(x)$ respectively. Prove the following statements:

(a) $G(x) = [1 - P(x)]/(1 - x)$.
(b) $Ev = G(1)$.
(c) $\text{Var } v = 2G'(1) + G(1) - G^2(1)$.

3. Suppose that $G(x) = \sum_{n=0}^{\infty} p_n x^n$ is a rational function of the form $G(x) = U(x)/V(x)$, where $U(x)$ and $V(x)$ are relatively prime polynomials and $U(x)$ is of a lower degree than $V(x)$. Show that for large n we can approximate p_n by $\rho_1 x_1^{-n-1}$, where $\rho_1 = -[U(x_1)/V'(x_1)]$ and x_1 is the smallest root in absolute value of the denominator.

4. The generating function for a pair of integral-valued random variables (r, s) such that $P(r = i, s = j) = p_{ij}$ is defined as $G(x_1, x_2) = \sum_{i=0}^{\infty} \sum_{j=0}^{\infty} p_{ij} x_1^i x_2^j$. Show that

(a) The marginal distributions $P(r = i)$ and $P(s = j)$ have generating functions $G(x_1, 1)$ and $G(1, x_2)$ respectively.
(b) The random variable $r + s$ has a generating function $G(x, x)$.

(c) r and s are independent if and only if $G(x_1, x_2) = G(1, x_2)G(x_1, 1)$ for every x_1, x_2.

5. Consider part 1 of Example 2. Instead of assuming that the probabilities p and q are constant, assume that $p_j = P(I_j = 1)$ and $q_j = P(I_j = 0)$ depend upon j. Show that the limiting distribution of s_n is Poisson, provided that $\lim_{n\to\infty} (p_1 + p_2 + \cdots + p_n) = \lambda$.

6. Suppose that r_n is a sequence of independent random variables having the same distribution $P(r_n = i) = p_i$. Let $s_k = r_1 + \cdots + r_k$, where k is a random variable independent of the r_n and $P(k = i) = g_i$. Show that if $P(x)$ and $G(x)$ are the generating functions of s_n and k, then the generating function of s_k is $G(P(x))$.

7. An urn contains n balls altogether, among them a white balls. A series of drawings is held, and each time one ball is drawn it is replaced by a white ball. Find the probability $P_{x,r}$ that after r drawings there will be x white balls in the urn. (*Hint:* Show that

$$P_{x,r+1} = \frac{n - x + 1}{n} P_{x-1,r} + \frac{x}{n} P_{x,r}.)$$

8. Players $A_1, A_2, \ldots, A_n$ play a series of games in the following order: First A_1 plays with A_2. The loser is out and the winner plays the following player, A_3. The loser is out again and the next game is played with A_4, and so on. The probability of winning a single game is $\frac{1}{2}$ for each player. The series is won by the player who succeeds in winning over all his adversaries in succession. What is the probability that the series will stop exactly at the xth game? what is the probability that the series will stop before or at the xth game?

9. Show that the function f is absolutely monotone if and only if it possesses a power series with nonnegative coefficients converging for $0 < x < 1$.

5.5 RANDOM WALKS

This section illustrates the use of generating functions. For a detailed treatment of random walks the reader is referred to [13].

Random walk models serve as approximations to Brownian motion and diffusion processes. This connection with physics suggests the following description of a one-dimensional random walk. A particle is located initially at the point i of the x-axis. At times $n = 1, 2, 3, \ldots$, it moves a unit step to the right with a probability p or a unit step to the left with a probability q, where $p + q = 1$. If the particle can always move without any restriction, the random walk is called *unrestricted*. Frequently, there is an obstacle at some point, say the origin. When it strikes this barrier, the particle can either be absorbed (that is, the process terminates) or be reflected. Hence a particle located at point 1 can move to point 2 with a probability p, stay at 1 with a probability kq, or move to 0 and be absorbed with a probability $(1 - k)q$.

The barrier is called *elastic* if $k \neq 0$, *absorbing* if $k = 0$, and *reflecting* if $k = 1$. The process is more likely to continue if k is near 1. With two reflecting barriers the process never terminates.

We will consider the problem of finding the probability of absorption for a random walk with two absorbing barriers located at 0 and b. Actually, this probability is defined in the space of all points $x = (x_1, x_2, \ldots)$, where $x_n = x_{n-1} + 1$ if $x_{n-1} + 1 \leq b$ and b otherwise, or $x_n = x_{n-1} - 1$ if $x_{n-1} - 1 \geq 0$ and 0 otherwise. Here x_n is the position of the particle at time n. Let E_n denote the event that the particle is absorbed at one of the barriers in less than n trials. Then the event E, that the particle is absorbed, is the limit of the increasing sequence of events E_n and $P(E) = \lim P(E_n)$. As before, we assume that this probability is well defined. However, we use a different method to find p_i and q_i, the probability of absorption at b and 0 given that the particle is initially at i.

To find q_i we note that after the first step the particle is at $i - 1$ or $i + 1$; therefore, by the total probability theorem,

(1) $$q_i = pq_{i+1} + qq_{i-1}. \qquad (1 < i < b - 1)$$

If $i = 1$ or $b - 1$, the particle can be absorbed at the first step and so Equation (1) must be modified as follows:

(2) $$q_1 = pq_2 + q$$

(3) $$q_{b-1} = qq_{b-2}.$$

For completeness, we define $q_0 = 1$ and $q_b = 0$.

We will solve this set of equations by the method of generating functions[9]; let $G(x) = \sum_{i=0}^{b} q_i x^i$. Multiplying Equations (1), (2) and (3) by x^i, x^1, and x^{b-1} respectively and summing yields

(4) $$G(x) = \frac{p - (1 - pq_1)x + qq_{b-1}x^{b+1}}{qx^2 + p - x}$$

upon solving for $G(x)$. Since $G(x)$ is a polynomial, the numerator of Equation (4) must vanish at the zeros 1 and p/q of the denominator. This leads to the equations

$$pq_1 + qq_{b-1} = q$$

$$q_1 + q_{b-1}\left(\frac{p}{q}\right)^{b-1} = 1$$

for determining the two unknowns q_1 and q_{b-1}. Solving this system of equations yields

[9]Alternatively, set $q_i = x^i$ in Equation (1) and solve $x = px^2 + q$ for x. if the roots of this quadratic equation are 1 and r, then $q_i = A + Br^i$. Use Equations (2) and (3) to determine the unknown constants A and B.

$$q_1 = \frac{(q/p)^b - (q/p)}{(q/p)^b - 1}$$

$$q_{b-1} = \frac{(q/p)^b - (q/p)^{b-1}}{(q/p)^b - 1}.$$

Setting $q/p = \theta$ and substituting for q_1 and q_{b-1} in Equation (4) leads, after some simplification, to

(5) $$G(x) = \frac{\theta^b - 1 + (1 - \theta^{b+1})x + (\theta^{b+1} - \theta^b)x^{b+1}}{(1 - x)(1 - \theta x)(\theta^b - 1)}$$

$$= \frac{1}{\theta^b - 1}\left(\frac{\theta^b(1 - x^{b+1})}{1 - x} - \frac{1 - (\theta x)^{b+1}}{1 - \theta x}\right).$$

From Equation (5), if $p \neq q$, then

(6) $$q_i = \frac{(q/p)^b - (q/p)^i}{(q/p)^b - 1}.$$

If $p = q$, Equation (6) is indeterminate and q_i can be found by applying L'Hospital's rule at $\theta = 1$, which yields

(7) $$q_i = 1 - i/b.$$

By regarding the point b as the origin, we can obtain p_i from Equation (6) by interchanging q and p and replacing i by $b - i$. We leave it to the reader to verify that $p_i + q_i = 1$, so that the probability of nonabsorption is zero in the long run.

A critical examination of the fact that $p_i + q_i = 1$ shows that a new concept of convergence is required for random variables. We first express the position of the particle at the nth step by a sum of independent random variables r_k—namely,

$$s_n = \begin{cases} i + r_1 + r_2 + \cdots + r_n & (0 < s_{n-1} < b) \\ 0 & (s_{n-1} \leq 0) \\ b & (s_{n-1} \geq b) \end{cases}$$

where $r_k = \pm 1$ and represents the outcome of the kth trial. As $n \to \infty$, the random variables s_n approach a random variable s which represents the limiting position of the particle. However, at the point $x = (i + 1, i, i + 1, i, i + 1, \ldots)$, the random variables $s_1(x)$, $s_2(x)$, $s_3(x), \ldots$ assume the values i, $i + 1, i, \ldots$. Therefore, they do not converge in the ordinary sense (pointwise) to s. However, the set of all such points which do not have an infinite number of 0's and b's form an event that has a probability of 0. This concept of convergence will be investigated in more detail in Chapter 7.

A random walk can also be interpreted as a game of chance. At each trial a player A has a probability p of winning one dollar and a probability q of losing one dollar to B. A's capital is i and he decides to play until he has lost all of his money or has won $b - i$ dollars. Under this interpretation, Equation (6) and (7) represent the probability of his ruin. For example, from Equation (7) we see that a player who starts with i dollars and wants to win w dollars has the probability $i/(i + w)$ of doing so before he is ruined. If he is greedy—that is, if w is much greater than i—it is almost certain that he will be ruined.

Suppose that a more skillful player ($p > q$) is playing against a very rich adversary (b is much larger than i). The probability of his ruin is $(q/p)^i$ from Equation (6), and so the probability of his adversary's ruin is $1 - (q/p)^i$. Thus a skillful gambler, even with a small capital, stands less chance of being ruined than a less skillful gambler with a large amount of capital.

Let us now suppose that k dollars ($k > 1$) are staked at each trial. The gambler can be ruined by losing i/k times. Similarly, he can win $b - i$ dollars by playing successfully $(b - i)/k$ times. Therefore, the probability of ruin is the same as for a game that starts with i/k dollars and terminates with b/k dollars, with one dollar being staked each time. The probability of ruin $\bar{q}_i$, obtained from Equation (6), is

(8) $$\bar{q}_i = \frac{\theta^{b/k} - \theta^{i/k}}{\theta^{b/k} - 1}.$$

For $p > q$, $\theta^{1/k} > \theta$, since $\theta < 1$; for $p < q$, $\theta^{1/k} < \theta$. For $b > i$, Equation (8) can be rewritten as

(9) $$\bar{q}_i = 1 - \frac{z - 1}{z^l - 1},$$

with $l > 1$, by setting $z = \theta^{i/k}$. The function $(z - 1)/(z^l - 1)$ decreases as z increases for $z > 0$; thus $\bar{q}_i$ increases with z for $z > 0$. For $q > p$ we have $\bar{q}_i < q_i$, and for $p > q$ we have $\bar{q}_i > q_i$. In other words, if the stakes are increased while the initial capital is unaltered, the probability of ruin decreases for a player with $q > \frac{1}{2}$.

To find the mean number of steps E_i taken by a particle until it is absorbed, we assume that E_i is finite. The proof of this statement is left as an exercise. If the particle moves to the right (left) on the first step, the random walk continues as if the initial position had been $i + 1 \quad (i - 1)$. Since the particle has already taken one step, the expected number of steps is $E_{i+1} + 1 \quad (E_{i-1} + 1)$ under the condition that the particle moved to the right (left) on the first step. Hence the mean number of steps E_i (cf., Section 3.4) satisfies the difference equation

(10) $$E_i = pE_{i+1} + qE_{i-1} + 1 \qquad (0 < i < b)$$

with the boundary conditions

(11) $$E_0 = 0 \qquad E_b = 0.$$

This yields

$$E_i = \frac{i}{q-p} - \frac{b}{q-p}\left(\frac{1-\theta^i}{1-\theta^b}\right) \qquad (p \neq q) \tag{12}$$

$$= i(b-i). \qquad (p = q)$$

With the random walk model interpreted as a game of chance, E_i represents the average duration of the game.[10]

EXERCISES

1. Solve Equations (1)–(3) by the methods used for difference equations.

2. Show that a gambler's ultimate gain (or loss) is $b(1 - q_i) - i$ if $p \neq q$ and is 0 if $p = q = \frac{1}{2}$.

3. Find the probability of absorption for a random walk with an absorbing barrier at the origin only, using the method of generating functions. Check your answer by letting $b \to \infty$. (*Hint:* The generating function is analytic for $|x| < 1$.)

4. Verify that Equation (12) holds.

5. Find the probability of absorption q_{in} in n trials for a random walk with two absorbing barriers beginning at point i. [*Hint:* Write down a set of difference equations for the q_{in}. Reduce these to a set of difference equations in $G_i(x)$ by using the generating function $G_i(x) = \sum_{n=0}^{\infty} p_{in}x^n$.]

6. Suppose that A and B stake α and β respectively on a single game. The probabilities of winning a single game are p and q respectively. If A's fortune is a while B's fortune is b, find the probabilities that A or B will be ruined, in the sense that at a certain stage A's capital will become less than α or B's will become less than β.

7. If B's fortune is infinite (i.e., $b = \infty$), what is the probability that A will be ruined in the course of n games under the conditions described on page 126.

8. Rework Exercise 7 under the assumption that B's fortune is finite.

5.6 GEOMETRIC PROBABILITY

In geometric probability problems, the possible outcomes can be represented by points of a line segment, or of some plane figure or solid body. Since the number of outcomes is usually uncountable, we cannot define probability as

[10]One's intuition in such a game is not very good. Guess the average duration of a coin-tossing game if one player has one dollar and the other player has 1000 dollars and they bet one dollar at each play. From Equation (12) the average duration of the game will be 1000 plays!

the ratio of the favorable outcomes to the total number of outcomes. Nevertheless, we can still define the probability of an event in a natural way and calculate it by geometrical considerations.

As an illustration, let us consider what is meant by saying that a point chosen at random from the points of a square s lies inside a region r of the square.[11] If we stipulate that all points of s have the same probability p of being chosen, then $p = 0$, since there is an infinite number of such points. This interpretation does not enable us to solve the problem. Instead, let us assume that there exists a probability function P that satisfies the axioms, and let us try to determine this function. Since the point is certain to be in the square, we must have $P(S) = 1$, where S is the event that the point lies in s. Now decompose the square s into n^2 squares of equal area by dividing each side of s into n equal parts. We interpret the phrase "at random" to mean that the point has the same probability of lying inside any one of these squares. Let S_1 be the event that the point lies inside a particular square s_1. By our assumptions and the additivity of P, we have $n^2 P(S_1) = P(S) = 1$. Hence

$$P(S_1) = \frac{1}{n^2} = \frac{\text{area of } S_1}{\text{area of } S}. \tag{1}$$

Next, let us find the probability that the point lies inside a rectangle (event R) whose sides are parallel to the sides of the square s. By approximating the area of the rectangle by small squares of the type described above and by using the additivity of P, we can show that

$$P(R) = \frac{\text{area of rectangle}}{\text{area of } S}. \tag{2}$$

Finally, by approximating the region r by rectangles (as in Riemann integration), we can prove that Equation (2) continues to hold for the region r. By a similar procedure, we can show that the probability p that a point chosen at random from a hypercube s lies inside a k-dimensional region r that is contained in s is given by

$$p = \frac{\text{volume of } r}{\text{volume of } s}.$$

The following examples illustrate the use of this formula.

EXAMPLE 1

Suppose that the surface of a table is ruled into n squares. A coin of diameter d is tossed on the table. The squares are of width $a > d$. What is the probability that the coin will lie entirely within one of the squares?

[11]We assume that the region r has an area that can be evaluated by Riemann integration.

Let us assume that it is equally probable that the center of the coin will land in any of the n squares. Let A be the event that the coin lies entirely in a given square S if the center lies in S, and let B denote the event that the center lies in S. Then the probability that the coin lies in S is given by

$$P(AB) = \frac{1}{n} PA.$$

To calculate PA, we assume that the center falls randomly on S. The coin will lie entirely in S if and only if its center lands within a square r of width $a - d$ whose center coincides with that of the square S and whose sides are parallel to those of S. It follows from this that

$$PA = \frac{\text{area of } r}{\text{area of } S} = \frac{(a-d)^2}{a^2} = \left(1 - \frac{d}{a}\right)^2.$$

Therefore, the probability of the coin lying entirely within a given square is $(1/n)(1 - d/a)^2$. By adding up these probabilities for all n squares, we obtain

$$n\left(\frac{1}{n}\right)\left(1 - \frac{d}{a}\right)^2 = \left(1 - \frac{d}{a}\right)^2$$

as the probability that the coin will lie entirely within one of the squares.

EXAMPLE 2

On a line AB of length l, two points X and Y are selected at random. What is the probability that AX, XY, and YB can form a triangle?

Let the line AB lie along the x-axis, with X, Y, and B located at distances x, y, and l respectively from the origin. The variables x and y are independent and so are uniformly distributed over a square with side l. The three segments form a triangle if and only if the sum of the lengths of any two of the segments is greater than the length of the third. If $x < y$, then

$$\begin{aligned} y - x &< x + l - y \quad &\text{or}\quad y - x &< l/2 \\ x &< y - x + l - y \quad &\text{or}\quad x &< l/2 \\ l - y &< y - x + x \quad &\text{or}\quad y &> l/2. \end{aligned}$$

These last three inequalities mean that the point (x, y) lies in the triangle with vertex $(0, l/2)$, $(l/2, l/2)$, and $(l/2, l)$. The area of this triangle is $\frac{1}{8}l^2$. Hence the probability that the three segments can form a triangle if $y > x$ is $(\frac{1}{8}l^2)/l^2 = \frac{1}{8}$. By symmetry, a similar result holds for $x > y$ and so the required probability is $\frac{1}{4}$.

The reader is referred to References [14]–[20] for further details on geometric probability.

EXERCISES

1. Two persons agree to meet at a specified place between five and six o'clock. The one who arrives first waits 10 minutes for the other. What is the probability of their meeting if their arrival times are independent and can occur at random during the indicated hour?

2. *Bertrand's Paradox.* A chord of a circle is chosen at random.[12] Show that the probability that its length will exceed the length of a side of an inscribed equilateral triangle is

(a) One-half if the direction of the chord is fixed in advance. (Here the center of the chord is uniformly distributed on a diameter.)
(b) One-third if one end point of the chord is fixed in advance. (Here the other end point is uniformly distributed on the circumference.)
(c) One-fourth if the position of the chord is determined by its midpoint. (The midpoint is chosen at random inside the circle.)

3. How thick should a coin be to have a one-third chance of landing on its edge?

4. Suppose that n points are dropped at random on a unit interval. Let r_i be their coordinates, where $i = 1, 2, \ldots, n$.

(a) Find the distribution function F[that is, $F(x) = P(r \leq x)$] of $\max r_i \equiv r_{(n)}$.
(b) Find the distribution function of $\min r_i \equiv r_{(1)}$.
(c) Find the distribution function of $r_{(i)}$, where $r_{(i)}$ represents the ith largest coordinate.
(d) Show that the lengths of the $n + 1$ segments have identical distribution functions.

5. A handful of identical glass straws with one end marked red and the other marked yellow are dropped on the floor and break into three pieces each. What is the average length of the fragments with a red end?

6. (a) If a bar is broken in two at random, what is the average length of the smaller piece?
(b) What is the average ratio of the smaller length to the larger?

7. A bar is broken at random in two places. Find the average size of the smallest, the middle-sized, and the largest pieces. Generalize to n breaks.

8. What is the probability that the quadratic equation $x^2 + 2ax + b = 0$ has real roots?

9. *Buffon's Needle.* A set of parallel lines spaced $2a$ units apart is ruled on a floor

[12]Joseph Bertrand (1822–1900), a French mathematician, doubted that a definition of probability could be found if there were an infinite number of outcomes. This example was supposed to show that probability depends on the method of calculation. However, the word "random" is vague and we are really considering three different probability spaces.

(of infinite expanse). A needle of length $2l$ ($l < a$) is twirled and tossed on the floor. Show that the probability that the needle will intersect one of the lines is $2l/a\pi$.

10. Solve Exercise 9 if a closed convex curve of a diameter less than $2a$ is used instead of the needle.

11. Suppose that a needle of length $2l$ ($2l < 1$) is tossed on a grid with both horizontal and vertical rulings spaced one unit apart. What is the mean number of lines the needle crosses?

12. In Exercise 11, let the needle be of an arbitrary length. What is the mean number of crosses?

13. A segment AB is divided by a point C into two parts, part AC of length a and part CB of length b. Points X and Y are chosen at random in AC and CB respectively. What is the probability that a triangle can be formed from AX, XY, and BY?

14. A thin needle is thrown onto a table that has been ruled into congruent rectangles. Suppose that the needle is shorter than the smaller sides of the rectangles. Find the probability that the needle will be entirely contained in one of the rectangles.

15. On each of circles O_i with respective radii r_i, points M_i are taken at random. Suppose that $\sum_{i=1}^{\infty} r_i = \infty$ while $\sum_{i=1}^{\infty} r_i^2 < \infty$; then prove that the probability that the length of the vector

$$OM = O_1M_1 + O_2M_2 + \cdots O_nM_n > R$$

will tend to 0 as $R \to \infty$, no matter how large n is.

16. A piece is broken off at random from each of three identical rods. What is the probability that an acute-angled triangle can be formed from the three pieces?

5.7 PROBABILITIES IN R^n

In working with discrete random variables it is often necessary to evaluate cumbersome sums of probabilities. Frequently these sums are approximated by integrals (see Example 3.2.5). Sometimes these approximations take the form

$$P(c < r_n \leq d) \approx \int_c^d f(x)\,dx,$$

where $f(x) \geq 0$ has the property that $\int_{-\infty}^{\infty} f(x)\,dx = 1$; f is called a density function. An example of this is the normal approximation to the binomial distribution (see Theorem 4.3.1) or, more generally, to the multinomial distribution (see Theorem 4.3.2) in R^n, n-dimensional Euclidean space. Further examples are given below.

EXAMPLE 1

1. *Uniform Distribution.* Suppose that we have a random variable r_n such that $P(r_n = k/n) = 1/(n+1)$ for $k = 0, 1, 2, 3, \ldots, n$. In other words, we have $n + 1$ points lying in the interval $[0, 1]$ such that the probability of choosing any one of these points is $1/(n + 1)$. Let $a, b \in [0, 1]$ and $a < b$; then if $a \neq 0$ and $b \neq 1$, $(i - 1)/n < a \leq i/n$ and $(j - 1)/n < b \leq j/n$ for some integers i and j. Since $P(a < r_n \leq b) = 1/(n + 1)$ (the number of points in $(a, b]$),

$$\frac{j-1}{n+1} \leq P(a < r_n \leq b) \leq \frac{j-i+2}{n+1}.$$

Noting that $(j - i - 1)/(n + 1) \leq b - a \leq (j - i + 1)/(n + 1)$,

$$P(a < r_n \leq b) \approx (b - a) = \int_a^b dx \tag{1}$$

for sufficiently large n. This approximation yields the *uniform density* $f(x) \equiv 1$ on $[0, 1]$ and 0 elsewhere, since $\int_{-\infty}^{\infty} f\,dx = 1$.

2. *Exponential Distribution.* Some random processes exhibit a lack of memory for waiting times, the time between two successive occurrences of an event. By this we mean that the probability that the event will occur in a time interval of length Δt depends only on Δt. It does not depend on the time that has elapsed since the previous occurrence of the event. Under certain conditions, however, such things as the waiting times between arrivals of customers, successive collisions experienced by a molecule, and breakdowns of machines do possess the above property.

A mathematical model of such behavior can be formulated as follows. Suppose that the event under consideration can happen only at times $0, 1/n, 2/n, 3/n, \ldots$, and that it has not happened at time i/n. Then the probability of its occurrence at $(i + 1)/n$ is p, while the probability of its nonoccurrence is q. Both p and q are independent of i. In other words, we have an infinite sequence of Bernoulli trials that determine the waiting time. Let r_n be a random variable that represents the waiting time. Then $P(r_n = k/n) = q^{k-1}p$, since the event does not occur $k - 1$ times but does occur the kth time. $P(r_n > x) = q^k = p \sum_{i=k+1}^{\infty} q^{i-1}$, where k is the smallest integer such that $(k + 1)/n > x$. In an actual physical situation we could estimate λ, the number of occurrences of the event per unit time. The average waiting time $1/\lambda = E(r_n) = \sum_{k=1}^{\infty} (k/n)q^{k-1}p = 1/pn$, and so $p = \lambda/n$. Hence $P(r_n > x) \approx [1 - (\lambda/n)]^{nx}$, which approaches $e^{-\lambda x}$ as $n \to \infty$. The function $F(x) = 1 - e^{-\lambda x}$ is a distribution function and $f(x) = \lambda e^{-\lambda x}$ is the exponential density function. The reader can verify that $\int_0^\infty \lambda e^{-\lambda x}\, dx = 1$.

The random variable r_n can also be interpreted as the lifetime of an individual, a machine, or an atom. Physically this means that either we have an absence of aging or the probability distribution of the remaining lifetime is independent of age.

In the light of the above examples and Section 5.6, it would be convenient to be able to treat these limits as a density function of some random variable on some sample space. To make the proper interpretation, let us pose the analogous problem for a density function defined on a discrete set of points $\Omega = (x_1, \ldots, x_n)$. Suppose that we are given a function $f(x)$ such that $f(x_i) \geq 0$ and $\sum_{i=1}^{n} f(x_i) = 1$. We take Ω as our sample space and define a random variable r on Ω by $r(x) = x$. This random variable has a density function $f(x)$ and we have $P(c < r \leq d) = \sum_{c<x_i\leq d} f(x_i)$.

The procedure to follow, when f is a continuous function (except for a finite number of jumps) defined on the real line R, is now apparent. If $f(x) \geq 0$ and $\int_{-\infty}^{\infty} f(x)\, dx = 1$, then f is called a density function. We take R to be our sample space; as before, we consider the function $r(x) = x$ as a random variable on R with a density function $f(x)$ and $P(c < r \leq d) = \int_c^d f(x)\, dx$. The distribution function F of r is defined by

$$F(x) = P(r \leq x) = \int_{-\infty}^{x} f(y)\, dy.$$

To extend our theory of nondiscrete random variables, we first show that they can be approximated by discrete random variables. We need only consider a random variable r_n, which is defined on the discrete set of points $0, \pm 1/n, \pm 2/n, \ldots$, with probabilities $P(r_n = k/n) = F((k+1)/n) - F(k/n) = \int_{k/n}^{(k+1)/n} f(x)\, dx$. As n approaches infinity, $P(c < r_n \leq d)$ approaches $\int_c^d f(x)\, dx$. We then proceed by analogy with the theory already developed for discrete random variables. For example,

$$E(r_n^i) = \sum_{k=-\infty}^{\infty} \left(\frac{k}{n}\right)^i \left(F\left(\frac{k+1}{n}\right) - F\left(\frac{k}{n}\right)\right)$$

$$= \sum \left(\frac{k}{n}\right)^i f(\xi_k)\frac{1}{n} + \sum \left(\frac{k}{n}\right)^i \left(F\left(\frac{k+1}{n}\right) - F\left(\frac{k}{n}\right)\right),$$

where the first sum has been simplified by the mean value theorem with $(k+1)/n < \xi_k < k/n$, and the second finite sum corresponds to the fact that f is not continuous at a finite number of points. Hence, $E(r_n^i)$ is a Riemann sum approximating $\int_{-\infty}^{\infty} x^i f(x)\,dx$, assuming that the integral converges. This leads us to define $E(r^i) = \int_{-\infty}^{\infty} x^i f(x)\,dx$.

EXAMPLE 2

1. The kth moment of the exponential density function is given by $\lambda \int_0^{\infty} x^k e^{-\lambda x}\,dx = -x^k e^{-\lambda x}]_0^{\infty} + k\int_0^{\infty} x^{k-1}e^{-\lambda x}\,dx = k\int_0^{\infty} x^{k-1}e^{-\lambda x}\,dx$. Repeating the above procedure leads to the kth moment, $k!/\lambda^k$.
2. The kth moment about the mean of the normal density with parameters μ and σ is

$$\frac{1}{\sigma\sqrt{2\pi}} \int_{-\infty}^{\infty} (x-\mu)^k \exp\left[\frac{-(x-\mu)^2}{2\sigma^2}\right] dx = \sigma^k(2\pi)^{-1/2} \int_{-\infty}^{\infty} t^k e^{-t^2/2}\,dt,$$

upon setting $(x-\mu)/\sigma = t$. Hence the problem reduces to evaluating the integral

$$I = \int_{-\infty}^{\infty} t^k e^{-t^2/2}\,dt.$$

The function $f(t) = t^k e^{-t^2/2}$ is an odd function whenever k is odd; therefore, $\int_{-\infty}^{\infty} f(t)\,dt = 0$. Thus the kth moments about the mean are 0 for odd k. To compute the 0th moment, consider

$$I^2 = \left(\int_{-\infty}^{\infty} e^{-x^2/2}\,dx\right)\left(\int_{-\infty}^{\infty} e^{-y^2/2}\,dy\right) = \int_{-\infty}^{\infty}\int_{-\infty}^{\infty} e^{-(x^2+y^2)/2}\,dx\,dy$$

$$= \int_0^{2\pi}\int_0^{\infty} e^{-r^2/2} r\,dr\,d\theta = 2\pi.$$

Thus $I = (2\pi)^{1/2}$ and the 0th moment is 1, which verifies that the normal density is really a density function. To compute the higher even moments, say $k = 2n$, integrate

$$I_{2n} = \int_{-\infty}^{\infty} t^{2n} e^{-t^2/2}\, dt$$

by parts, setting $e^{-t^2/2} t\, dt = dv$ and $t^{2n-1} = u$. Then

$$I_{2n} = -t^{2n-1} e^{-t^2/2}\Big]_{-\infty}^{\infty} + (2n-1) \int_{-\infty}^{\infty} t^{2n-2} e^{-t^2/2}\, dt = (2n-1) I_{2n-2}.$$

By recursion,

$$I_{2n} = (2n-1)(2n-3) \cdots 3 \cdot 1 \cdot (2\pi)^{1/2},$$

and so the $2n$th moment about the mean is

$$(2n-1)(2n-3) \cdots 3 \cdot 1 \cdot \sigma^{2n}.$$

In Section 2.2 we saw that a random variable must satisfy certain conditions (i. e., be a "measurable" function) for a finite sample space. The extension of this concept will be postponed until Chapter 8. For the moment, we will assume that any continuous function on R is a random variable. We will see in the next section that this class of functions is insufficient for our purposes.

The definition of probability densities in R^n should now be evident. A *density function* f is a nonnegative integrable function whose integral over R^n equals unity. Actually, f may only be defined or nonzero on some subset S of R^n. For convenience, we will phrase the definitions as if this subset S is R^n. The necessary modifications are obvious. The corresponding *distribution function* F is defined by

(1) $$P(r_1 \le x_1, r_2 \le x_2, \ldots, r_n \le x_n) = F(x_1, x_2, \ldots, x_n)$$
$$= \int_{-\infty}^{x_1} \int_{-\infty}^{x_2} \cdots \int_{-\infty}^{x_n} f(x_1 \cdots x_n)\, dx_1 \cdots dx_n,$$

where the r_i are random variables defined by $r_i(x_1 \cdots x_n) = x_i$. It follows directly from Equation (1) and the additivity of the probability function that

(2) $$P(c_1 < r_1 \le d_1, \ldots, c_n < r_n \le d_n)$$
$$= \int_{c_1}^{d_1} \cdots \int_{c_n}^{d_n} f(x_1 \cdots x_n)\, dx_1 \cdots dx_n.$$

However, this is obvious if we pass to the limit from a discrete model. It follows from Equation (2) that we must define $P(r \in R)$, where $r = (r_1 \cdots r_n)$, for a sufficiently smooth region R by

$$P(r \in R) = \int \underset{R}{\cdots} \int f(x_1 \cdots x_n)\, dx_1 \cdots dx_n.$$

The proof is as in ordinary calculus. We approximate the region R by n-dimensional intervals and use the additivity of the probability function and the integral. A random variable is a continuous function on R^n [or on $S = (x : f(x) \neq 0)$].

The concepts introduced in Chapters 2 and 3 can now be reformulated for nondiscrete random variables. The *expectations* of r_i are defined by

$$\mu_i = Er_i = \int_{-\infty}^{\infty} \cdots \int_{-\infty}^{\infty} x_i f(x_1 \cdots x_n)\, dx_1 \cdots dx_n,$$

where f is their density function. The *variances* σ_i^2 and the *covariances* are given by

$$\sigma_i^2 = \text{Var}\,(r_i) = \int_{-\infty}^{\infty} \cdots \int_{-\infty}^{\infty} (x_i - \mu_i)^2\, f(x_1 \cdots x_n)\, dx_1 \cdots dx_n$$

$$\text{Cov}\,(r_i, r_j) = \int_{-\infty}^{\infty} \cdots \int_{-\infty}^{\infty} (x_i - \mu_i)(x_j - \mu_j) f(x_1 \cdots x_n)\, dx_1 \cdots dx_n.$$

The *correlation coefficients* $\rho_{ij} = \text{Cov}\,(r_i, r_j)/\sigma_i\sigma_j$. The *marginal distribution* F_i of r_i is given by

$$F_i(x_i) = \lim_{\substack{x_j \to \infty \\ j \neq i}} F(x_1, x_2, \ldots, x_n).$$

where x_i is fixed and all other x_j approach infinity. The *marginal density* $f_i(x)$ is defined by

$$f_i(x_i) = \int_{-\infty}^{\infty} \cdots \int_{-\infty}^{\infty} f(x_1 \cdots x_n)\, dx_1 \cdots dx_{i-1}\, dx_{i+1} \cdots dx_n.$$

EXAMPLE 3

1. *Uniform Distribution.* Suppose that R is a region of R^n such that $\int \underset{R}{\cdots} \int dx_1 \cdots dx_n$ is well defined. The n-tuple $r = (r_1, \ldots, r_n)$ is *uniformly distributed* in R if for any subset S of R we have

$$P(r \in S) = \frac{\text{volume } S}{\text{volume } R}.$$

For example, if R is the closed n-dimensional interval $(x : a_i \leq x_i \leq b_i)$, the density function of r is given by

$$f(x_1, \ldots, x_n) = \frac{1}{(b_1 - a_1)(b_2 - a_2) \cdots (b_n - a_n)} \qquad (x \in R)$$
$$= 0. \qquad (x \in R')$$

It is easily seen that the marginal density functions of r_i are uniform distributions on the line. Furthermore, $\mu_i = (b_i + a_i)/2$ and $\sigma_i^2 = \frac{1}{12}(b_i - a_i)^2$.
2. *Normal Density in* R^2. The density function of (r_1, r_2) is

$$f(x_1, x_2) = \frac{1}{2\pi\sigma_1\sigma_2\sqrt{1-\rho^2}} \exp\left[-\frac{1}{2(1-\rho^2)}\frac{x_1^2}{\sigma_1^2} - 2\rho\frac{x_1x_2}{\sigma_1\sigma_2} + \frac{x_2^2}{\sigma_2^2}\right],$$

where $-1 < \rho < 1$ and $\sigma_i > 0$. The reader can verify that $Er_i = 0$, $\text{Var}\, r_i = \sigma_i^2$, and $\text{Cov}\,(r_1, r_2) = \rho$ by the technique of completing the square and making substitutions of the form $U = x_2/\sigma_1 - 2\rho/\sigma_2$, $V = x_2$.

The random variables $r_1, r_2, \ldots, r_n$ with a joint density function $f(x_1 \cdots x_n)$ are *independent* if and only if

$$f(x_1 \cdots x_n) = f_1(x_1)f_2(x_2) \cdots f_n(x_n).$$

Under these circumstances, their distribution function is the product of their corresponding marginal distribution functions. If $n > 2$, this definition can be generalized, because we can have independent sets of random variables—for example, (r_1, r_2, r_3), $(r_4, \ldots, r_i)$, $(r_{i+1}, \ldots, r_n)$ are independent if

$$f(x_1, x_2, \ldots, x_n) = f_1(x_1, x_2, x_3)f_2(x_4, \ldots, x_i)f_3(x_{i+1}, \ldots, x_n).$$

It is not difficult to see that any decomposition of $r_1, \ldots, r_n$ into sets leads to independent sets for the uniform distribution whose density function is defined in part 1 of Example 3.

The concept of *convolution*, defined for discrete random variables in Definition 5.4.2., can easily be extended to continuous random variables. Suppose that u and v are independent random variables with density functions f and g which both vanish for negative values of the argument. The density h of the sum $u + v$ is given by

(3) $$h(s) = \int_0^s f(s-t)g(t)\,dt = \int_0^s g(s-t)f(t)\,dt.$$

The proof follows by approximating u and v by their discrete analogs u_n and v_n. Alternatively, we note that $P(u + v \le s)$ is the probability of the set of points in R^2 bounded by $x + y = s$, $x = 0$, and $y = 0$. Since u and v are independent, their joint density function is $f(x)g(y)$. The distribution function $H(s)$ of $u + v$ is obtained by integrating over this region; therefore,

(4) $$H(s) = \int_0^s \int_0^{s-y} f(x)g(y)\,dx\,dy = \int_0^s F(s-y)g(y)\,dy.$$

Finally, the validity of Equation (3) is established by differentiating Equation

(4). If we allow u and v to assume negative values, Equation (3) must be replaced by

$$h(s) = \int_{-\infty}^{\infty} f(s - t)g(t)\,dt.$$

We leave the proof to the reader. As in Section 5.4, we define the convolution of two functions f and g by Equation (3), denoting this operation by $h = f * g$. By an abuse of language, we often call h the convolution of f and g. The addition of random variables is associative and commutative. Since the convolution represents the resulting density, it follows that the operation of convolution is also associative and commutative. A direct proof of this fact is more difficult.

EXAMPLE 4

Frequently we are interested in the number of breakdowns, arrivals, etc., in an interval of time t for processes of the type described in part 2 of Example 1.2. In order to find the probability distribution of this quantity (cf., Exercise 4), it is expedient to know the probability distribution for the time of n occurrences of the event under consideration. In abstract terms, we have n independent random variables $r_1, r_2, \ldots, r_n$ (representing the waiting time) and we want to find the density f_n of $s_n = r_1 + r_2 + \cdots + r_n$. Applying Equation (3), we see that the density of s_2 is given by

$$f_2(x) = \lambda^2 \int_0^x e^{-\lambda(x-t)}e^{-\lambda t}\,dt = \lambda^2 x e^{-\lambda x}.$$

Since $s_3 = s_2 + r_3$, we can repeat the above calculation to find the density of s_3—namely,

$$f_3(x) = \lambda \int_0^x f_2(x - t)e^{-\lambda t}\,dt = \lambda\frac{(\lambda x)^2}{2!}e^{-\lambda x}. \tag{5}$$

By repeating this procedure a few more times, the reader should have no difficulty in guessing that

$$f_n(x) = \lambda\frac{(\lambda x)^{n-1}}{(n-1)!}e^{-\lambda x}. \qquad (x > 0) \tag{6}$$

We can prove that Equation (6) is correct by induction. Assume that this equation is true for some $n \geq 1$; writing $s_{n+1} = s_n + r_n$, the density f_{n+1} can easily be found from the density of s_n and r_n as in Equation (5). The detailed calculation is left to the reader. The distribution function is

$$F_n(x) = 1 - \sum_{k=0}^{n-1}\frac{(\lambda x)^k}{k!}e^{-\lambda x}. \tag{7}$$

Finally, let us consider the concept of conditional probability. By definition, the conditional probability of the event $A = (r_n \leq x_n)$, given the event $B = (x_i < r_i \leq x_i + h, i = 1, 2, \ldots, n-1)$, is $P(AB)/PA$. Since the probability of a set of points is obtained by integrating the density $f(x_1, x_2, \ldots, x_n)$ of $(r_1, \ldots, r_n)$ over the set,

$$\text{(8)} \qquad P(r_n \leq x_n | x_i < r_i \leq x_i + h) = \frac{\int_{x_1}^{x_1+h} \cdots \int_{x_{n-1}}^{x_{n-1}+h} \int_{-\infty}^{x_n} f\, dt_1 \cdots dt_n}{\int_{x_1}^{x_1+h} \cdots \int_{x_{n-1}}^{x_{n-1}+h} g\, dt_1 \cdots dt_{n-1}}.$$

where $g(t_1, \ldots, t_{n-1}) = \int_{-\infty}^{\infty} f(t_1, \ldots, t_n)\, dt_n$. Suppose that $f_1(x_1, \ldots, x_{n-1}) > 0$. Dividing the numerator and denominator of the fraction appearing in Equation (8) by h^{n-1}, it is easy to see that as $h \to 0$, Equation (8) approaches

$$\text{(9)} \qquad \int_{-\infty}^{x_n} \frac{f(x_1, \ldots, x_{n-1}, t_n)}{g(x_1, \ldots, x_{n-1})}\, dt_n.$$

We call this limit the *conditional distribution* of r_n, given that $r_i = x_i$ for $i = 1, 2, \ldots, n-1$; we will denote it by $F(x_n | x_1, \ldots, x_{n-1})$. The conditional density $f(x_n | x_1, \ldots, x_{n-1})$ is obtained by differentiating Equation (9). The *conditional expectation of* r_n, given that $r_i = x_i$ for $i = 1, 2, \ldots, n-1$, is then defined by

$$\text{(10)} \qquad E(r_n | r_1 = x_1, \ldots, r_{n-1} = x_{n-1}) = \frac{1}{g(x_1, \ldots, x_{n-1})} \int_{-\infty}^{\infty} t_n f(x_1, \ldots, x_{n-1}, t_n)\, dt_n.$$

The *conditional variance* is defined as an ordinary variance using the conditional mean and the conditional density function. The conditional density, distribution, and expectation are functions of $x_1, \ldots, x_{n-1}$ and so they are random variables. However, they are not defined for any x such that $g(x_1, \ldots, x_{n-1}) = 0$. Since the density f is a continuous function, the set of all such x has a probability of zero. Hence the values assigned to these random variables on this set are immaterial for most purposes. If we arbitrarily set them equal to zero, the resulting random variables are well defined. We then have the following analog of the total probability theorem for continuous random variables:

$$\text{(11)} \qquad E(f(x_n | r_1, \ldots, r_{n-1})) = f_n(x_n),$$
$$E(F(x_n | r_1, \ldots, r_{n-1})) = F_n(x_n),$$

where the expectation is taken with respect to the random variables $r_1, \ldots, r_{n-1}$, which have a density function $g(x_1, \ldots, x_{n-1})$. The proof follows from Equation (9). Similarly, we have

$$E(E(r_n|r_1, \ldots, r_{n-1})) = Er_n.$$

Bayes' theorem (Theorem 3.2.3) can also be extended to continuous random variables.

EXAMPLE 5

1. For the normal density defined in part 2 of Example 2, the conditional density of r_2 for a given r_1 is normal, with $E(r_2|r_1) = \rho(\sigma_2/\sigma_1)\, r_1$ and $\mathrm{Var}(r_2|r_1) = \sigma_2^2(1 - \rho^2)$.
2. Let $n(t)$ be a random variable representing the number of cars arriving at a toll booth in a time interval of length t and let

(12) $$P(n(t) = n) = e^{-\lambda t}\frac{(\lambda t)^n}{n!}.$$

The time it takes to pay the toll is a random variable T with a density function $f(t)$. To find the probability p_n of n arrivals during the time it takes to pay the toll, we note that $P(n(t) = n|t < T \leq t + dt)$ is given by Equation (12) upon setting $n = 2$, $r_2 = n(t)$, and $r_1 = T$ (cf., part 1 of Theorem 3.4.4). Applying the results of Equation (11),

$$p_n = \int_0^\infty e^{-\lambda t}\frac{(\lambda t)^n}{n!} f(t)\, dt.$$

EXERCISES

1. (a) If the random variable r has an exponential distribution, show that

(1) $$P(r > t + h|r > h) = P(r > t).$$

(b) Let $F(t) = P(r > t)$. Verify that

(2) $$F(t + h) = F(t)F(h)$$

if Equation (1) holds.

(c) Prove that the only nonnegative function bounded in every finite interval that satisfies Equation (2) is $F(t) = e^{-\alpha t}$ for some α. [*Hint:* Let $G(t) = e^{\alpha t}F(t)$, where $\alpha = -\log F(1)$. Then $G(t)$ satisfies Equation (2). Show that $G(m/n) = 1$ for any integers m and n. Show that for any real number t, $G(t) = G(t')$, where $0 < t' \leq 1$. If $G(t) \equiv 1$, there is a t' such that $G(t') = c > 1$, since $G(1 - t') = c^{-1}$. Show that G is unbounded on (0, 1).]

2. Derive Equation (6) of Example 4 by assuming that each r_i has a geometric distribution and passing to the limit.

3. (a) *Bayes' Theorem.* Show that the density function of r_2 under the hypothesis $r_1 = x$ is given by

$$f(y|x) = \frac{f(x|y)f_2(y)}{E[f(x|r_2)]}$$

if the conditional density function $f(x|y)$ of r_1 given r_2 and the marginal density function $f_2(y)$ of r_2 are known.
(b) A skew coin with a probability p of heads determined by a density function $f(p)$ is thrown n times. Find the probability that $p \leq p_0$ if k heads were obtained.

4. Let $n(t)$ denote the number of occurrences of an event in a time interval of length t for the type of process described in part 2 of Example 1. Show that

$$P(n(t) = n) = \frac{(\lambda t)^n}{n!}e^{-\lambda t}.$$

[*Hint:* Use the fact that $(s_{n+1} \leq t) = (n(t) > n)$; cf., Example 4].

5. When bones are matched in anthropology, problems of the following type arise. Suppose that r_1, r_2, and r_3 are three independent characteristic properties; find $P(r_1 > a,\ r_2 < b, r_3 > c)$. (Express your answer in terms of density functions.)

6. Verify the following properties of the expectation operator E, where the r_i denote random variables and the c_i are constants:

(a) $Ec = c$. (c is *a* constant)
(b) $E(c_1r_1 + c_2r_2 + \cdots + c_nr_n) = c_1Er_1 + c_2Er_2 + \cdots + c_nEr_n$.
(c) If the r_i are independent, then $E(r_1r_2 \cdots r_n) = (Er_1)(Er_2) \cdots (Er_n)$.

7. (a) Show that if r_i are independent random variables,

$$\text{Var}\left(\sum_{i=1}^{n} r_i\right) = \sum_{i=1}^{n} \text{Var}\, r_i.$$

(b) $\text{Var}\,(ar + b) = a^2 \text{ Var}\, r$. ($a$ and b are constants; r is a random variable)

8. (a) Let r be a random variable with $Er = \mu$ and $\text{Var}\, r = \sigma^2$. Show that if $s = h(r)$, where h is at least twice differentiable at $x = \mu$, then $Es \approx h(\mu) + h''(\mu)\sigma^2/2$ and $\text{Var}\, s \approx (h'(\mu))^2\sigma^2$.
(b) Generalize the above result to the n-dimensional random variable $(r_1, r_2, \ldots, r_n)$ and show that

$$Eh(r_1 \cdots r_n) \approx h(\mu_1, \mu_2, \ldots, \mu_n) + \sum_{i=1}^{n} \frac{\partial^2 h}{\partial x_i^2}\sigma_i^2$$

$$\text{Var}\, h(r_1 \cdots r_n) \approx \sum_{i=1}^{n} \left(\frac{\partial h}{\partial x_i}\right)^2 \sigma_i^2.$$

9. Suppose that (r_1, r_2) is uniformly distributed over the semicircle $S: y = \sqrt{1 - x^2}$.

(a) Show that their density function $f(x, y) = 2/\pi$ if $(x, y) \in S$.

(b) Verify that

$$f_1(x) = (2/\pi)\sqrt{1 - x^2} \qquad (-1 \leq x \leq 1)$$

and

$$f_2(y) = (2/\pi)\sqrt{1 - y^2} \qquad (-1 \leq y \leq 1)$$

(c) Prove that

$$f(x|y) = 2(1 - y^2)^{-1/2} \qquad (-\sqrt{1 - y^2} \leq x \leq \sqrt{1 - y^2})$$
$$f(y|x) = (1 - x^2)^{-1/2} \qquad (0 \leq y \leq \sqrt{1 - x^2})$$

and show that $E(r_2|x) = \frac{1}{2}\sqrt{1 - x^2}$ and $E(r_1|y) = 0$.

10. Suppose that it costs C_1 dollars to produce one gallon of petroleum. If the petroleum distills at a temperature of less than 200°C., the product is naphtha, which sells for C_2 dollars per gallon. If it is distilled at a temperature greater than 200°C., it is refined oil, which sells for C_3 dollars per gallon. Find the expected net profit (per gallon) under the assumption that the distilling temperature is a random variable uniformly distributed over [150, 300].

11. Assume that the scores in a Graduate Record Examination are normally distributed with $\mu = 500$ and $\sigma = 100$. If 1000 people take the examination and we wish 800 of them to pass, what should be the lowest score permitted for passing?

12. The intelligence quotient (IQ) of a person is defined by

$$\text{IQ} = \frac{\text{MA}}{\text{CA}} \times 100,$$

where CA and MA denote the chronological and mental ages of the person. The mental age is determined by a test. For adults, CA is taken to be 16. For example, a person 12 years old with an MA of 16 would have an IQ of 133, while an adult with the same MA would have an IQ of 100. It has been found that IQ values are distributed according to a normal distribution with a mean of 100 and a standard deviation of 16.

(a) Show that 68.3 percent of the population have IQ's between 84 and 116.

(b) Show that only 2.3 percent have IQ's above 132, while only about three in a million have IQ's above 172.

(c) Suppose that a television program is designed for people with an MA of 12. What percentage of the potential adult viewers will be lost if the program is redesigned for an MA of 16. (Assume that only adults with MA $\geq$ MA of the television program will watch it.)

12. The annual rainfall at a certain locality is normally distributed with $\mu = 29.5$ inches and $\sigma = 2.5$ inches. How many inches of rain (annually) are exceeded about 5 percent of the time?

14. The orbital life of a satellite is an exponentially distributed random variable with an expected lifetime equal to two years. If three such satellites are launched simultaneously, what is the probability that at least two will still be in orbit after three years?

15. The reliability $R(t)$ of a component (or system) at time t is defined as $R(t) = P(l > t)$, where l is the life length of a component. The density function f of the time to failure of the system is $f(t) = -R'(t)$.

(a) Show that if El is finite, the mean time to failure is $El = \int_0^\infty R(t)\,dt$.

(b) Show that if n components, functioning independently, are connected in series, as in the figure below, and if the ith component has a reliability

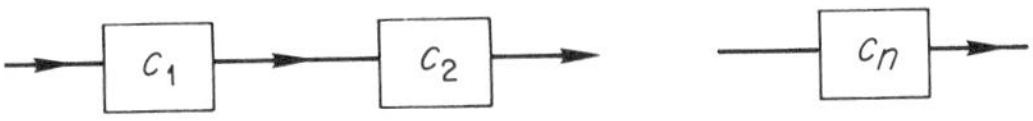

$R_i(t)$, then the reliability of the entire system $R(t)$ is given by $R(t) = R_1(t)R_2(t) \cdots R_n(t)$.

(c) Show that if the components are connected in parallel, as in the figure below,

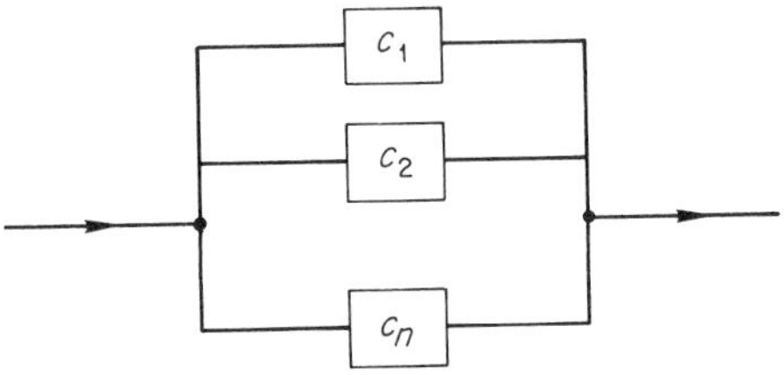

then $R(t) = 1 - \prod_{i=1}^{n} (1 - R_i(t))$. In particular, if $R_i(t) = e^{-\lambda_i t}$ for $i = 1, 2$, show that the expected time to failure is $1/\lambda_1 + 1/\lambda_2 - (\lambda_1 + \lambda_2)^{-1}$.

5.8 TRANSFORMATION OF VARIABLES

A metal cylinder of a height h is formed from a strip of a width l. Both h and l are random variables, because of minute variations in the cutting operation. Hence the volume $v = hl^2/4\pi$ will also be a random variable. Given the joint density function of (h, l), it should be possible to find the density function of v. In this section, we will be concerned with problems of this general nature.

Let us begin by considering discrete random variables.

EXAMPLE 1

Let r have the Poisson density function

$$f(x) = \frac{\lambda^x e^{-\lambda}}{x!}. \qquad (x = 0, 1, 2, \ldots)$$

Find the density function of $s = 2r + 1$.

To find the density function of s by the change-of-variable technique, let $y = 2x + 1$. Let $X = (x : x = 0, 1, 2, \ldots)$, so that X is the set where $f > 0$. The transformation defined by $y = 2x + 1$ maps the space X into the space $Y = (y : y = 1, 3, 5, \ldots)$. Because there is a one-to-one correspondence between the points of X and Y, the event $(s = y)$ or $(2r + 1 = y)$ can occur if and only if the event $r = (y - 1)/2$ occurs. Hence

$$g(y) = P(s = y) = P(r = (y - 1)/2) = \frac{\lambda^{(y-1)/2}e^{-\lambda}}{((y-1)/2)!} \qquad (y = 1, 3, 5, \ldots)$$

and 0 otherwise.

This example suggests the following generalization. Let r be a random variable of the discrete type whose density function is f. Let $X = (x : f(x) > 0)$ and let $y = \phi(x)$ define a one-to-one transformation that maps X onto Y. We can solve $y = \phi(x)$ for x in terms of y; let $x = \psi(y)$ denote the inverse transformation. Consider the random variable $s = \phi(r)$. If $y \in Y$, the event $(s = y)$ or $(\phi(r) = y)$ occurs if and only if the event $(r = \psi(y))$ occurs. Therefore, the density function of s is

$$g(y) = P(s = y) = P(r = \psi(y)) = f(\psi(y)) \qquad (y \in Y)$$

and 0 otherwise.

The generalization to n-dimensional random variables is straightforward. We illustrate the procedure for a two-dimensional discrete random variable with a density function f. Let $x = (x_1, x_2)$ and let $X = (x : f(x_1, x_2) > 0)$. Let $y_1 = \phi_1(x_1, x_2)$ and $y_2 = \phi_2(x_1, x_2)$ define a one-to-one transformation that maps X onto Y, and let $x_1 = \psi_1(y_1, y_2)$ and $x_2 = \psi_2(y_1, y_2)$ be the inverse transformation. The joint density function of the new random variables $s_1 = \phi_1(x_1, x_2)$ and $s_2 = \phi_2(x_1, x_2)$ is given by

$$g(y_1, y_2) = f(\psi_1(y_1, y_2), \psi_2(y_1, y_2)), \qquad (y_1, y_2) \in Y$$

and 0 elsewhere.

Often the transformation $y = \phi(x)$ is not one-to-one and several values of x lead to the same value of y, as the following example illustrates.

EXAMPLE 2

Suppose that the random variable r assume the values $-2, -1, 0, 1$, and 2 with probabilities $\frac{1}{6}, \frac{1}{6}, \frac{1}{2}, \frac{1}{12}$, and $\frac{1}{12}$ respectively. Let $y = x^2$ and $s = r^2$. The possible values of s are 0, 1, and 4 assumed with probabilities $\frac{1}{2}, \frac{1}{4}$, and $\frac{1}{4}$, since $(s = 1)$ or $(r^2 = 1)$ if and only if $(r = x)$ and $x \in A = (x : x^2 = 1)$. In other words, $(s = 1)$ if and only if $(r = 1)$ or $(r = -1)$ and so

$$g(1) = P(s = 1) = P(r = 1) + P(r = -1) = f(1) + f(-1),$$

where f and g denote the density functions of r and s. Similarly,

$$g(4) = \sum_{x \in A_4} f(x) = f(2) + f(-2),$$

where $A_4 = (x : x^2 = 4)$.

The generalization of the procedure described in the above example is

$$g(y) = P(s = y) = \sum_{x \in A_y} f(x) \qquad (y \in Y)$$

and 0 elsewhere, where $A_y = (x : \phi(x) = y)$.

The most frequently encountered case arises when r is a continuous random variable and ϕ is a continuous function. We first assume that ϕ is strictly monotone and is differentiable.

THEOREM 1

Let r be a continuous random variable with a density function f, where $f(x) > 0$ for $a < x < b$. Suppose that $y = \phi(x)$ is a strictly monotone function of x which is differentiable for all x. Then the random variable $s = \phi(r)$ has a density function g given by

$$g(y) = f(\phi^{-1}(y))\left|\frac{dx}{dy}\right|.$$

If ϕ is increasing, then g is nonzero for $\phi(a) < y < \phi(b)$. If ϕ is decreasing, then g is nonzero for $\phi(b) < y < \phi(a)$.

Proof

Assume that ϕ is a strictly increasing function. Then

$$G(y) = P(s \leq y) = P(\phi(r) \leq y) = P(r \leq \phi^{-1}(y)) = F(\phi^{-1}(y)).$$

Differentiating $G(y)$ with respect to y yields

$$G'(y) = \frac{dF}{dx}\frac{dx}{dy} = f(x)\frac{dx}{dy},$$

where $x = \phi^{-1}(y)$. For a decreasing function of y,

$$G'(y) = -f(x)\frac{dx}{dy}.$$

EXAMPLE 3

Let $f(x) = 3x^3$ for $0 < x < 1$, and let $y = e^{-x}$. Then $x = -\log y$ and $dx/dy = -1/y$; therefore, $g(y) = -3(\log y)^3/y$ for $e^{-1} < y < 1$.

If $y = \phi(x)$ is not a monotone function of x, we cannot apply Theorem 1. Instead we must use the method described above for discrete random variables. The following example illustrates the procedure.

EXAMPLE 4

Suppose that

$$f(x) = \frac{1}{2}(x + 1) \qquad (-1 < x < 1)$$

and 0 elsewhere. Let $\phi(x) = x^2$. We obtain the density function of $s = r^2$ as follows:

$$\begin{aligned} G(y) &= P(s \leq y) = P(r^2 \leq y) = P(-\sqrt{y} \leq r \leq \sqrt{y}) \\ &= F(\sqrt{y}) - F(-\sqrt{y}), \end{aligned}$$

where F is the distribution function of the random variable r. Hence

$$g(y) = G'(y) = (2\sqrt{y})^{-1}(f(\sqrt{y}) + f(-\sqrt{y})).$$

Theorem 1 can be generalized to n-dimensional random variables. We denote the point $(x_1, x_2, \ldots, x_n)$ by x and $(r_1, \ldots, r_n)$ by r.

THEOREM 2

Let $(r_1, \ldots, r_n)$ be an n-dimensional random variable with a density function f, and let $X = (x : f(x) > 0)$. Suppose that $y_i = \phi_i(x_1, \ldots, x_n)$ $(i = 1, 2, \ldots, n)$ defines a one-to-one transformation that maps X onto Y. Let the first partial derivatives of the inverse functions be continuous and let the Jacobian

$$J = \begin{vmatrix} \dfrac{\partial x_1}{\partial y_1} & \dfrac{\partial x_1}{\partial y_2} & \cdots & \dfrac{\partial x_1}{\partial y_n} \\ \dfrac{\partial x_2}{\partial y_1} & \dfrac{\partial x_2}{\partial y_2} & \cdots & \dfrac{\partial x_2}{\partial y_n} \\ \cdots & \cdots & \cdots & \cdots \\ \dfrac{\partial x_n}{\partial y_1} & \dfrac{\partial x_n}{\partial y_2} & \cdots & \dfrac{\partial x_n}{\partial y_n} \end{vmatrix}$$

not vanish identically in Y. Then the random variables $s_i = \phi_i(r_1, \ldots, r_n)$ have a density function g given by

$$(1) \qquad g(y) = f(\phi_1^{-1}(y), \phi_2^{-1}(y), \ldots, \phi_n^{-1}(y))|J| \qquad (y \in Y)$$

and 0 elsewhere.

Proof

Let $A \subset X$ and let B be the image of A under the given one-to-one transformation. Then $((r_1, r_2, \ldots, r_n) \in A)$ if and only if $((s_1, s_2, \ldots, s_n) \in B)$. Hence

$$P(s \in B) = P(r \in A) = \iint_A \cdots \int f(x_1, x_2, \ldots, x_n)\, dx_1 dx_2 \cdots dx_n.$$

Changing variables in this last integral yields

$$P(s \in B) = \int_A \cdots \int f(x_1 \cdots x_n)\, dx_1 \cdots dx_n$$

$$= \iint_B \cdots \int f(\phi_1^{-1}(y), \ldots, \phi_n^{-1}(y))|J|\, dy_1 \cdots dy_n.$$

Since

$$P(s \in B) = \iint_B \cdots \int g(y)\, dy_1 \cdots dy_n$$

for every $B \subset Y$ (for which the integral is defined), we obtain Equation (1) upon equating the last two integrals.[13]

EXAMPLE 5

1. Let r_1 and r_2 be independent random variables with $P(r_1 \leq 0) = 0$. Find the density function of the random variable $r_2/r_1 = s_2$.

Let f_1 and f_2 be the density functions of r_1 and r_2. By Theorem 7, the transformation of variables $x_1 = y_1, x_2 = y_1 y_2$ transforms the density function $f_1 f_2$ of (r_1, r_2) into the density function g of (s_1, s_2), where

$$g(y_1 y_2) = f_1(y_1) f_2(y_1 y_2) y_1. \qquad (y_1 > 0,\ -\infty < y_2 < \infty)$$

The marginal density function of s_2 is the required density function, since $s_2 = r_2/r_1$. Hence the density function h of r_2/r_1 is

$$h(y_2) = \int_0^\infty f_1(y_1) f_2(y_1 y_2) y_1\, dy_1. \qquad (-\infty < y_2 < \infty)$$

2. Let $r_1, \ldots, r_n$ be normally distributed and independent, and let each have a mean 0 and a variance 1. Find the density function of the random variable

$$s = r_1^2 + r_2^2 + \cdots + r_n^2.$$

This is the so-called χ^2 (chi square) random variable.

It is easier to proceed from first principles than to apply Theorem 2. The joint probability density function of $(r_1, r_2, \ldots, r_n)$ is

$$\left(\frac{1}{\sqrt{2\pi}}\right)^n \exp\left(-\frac{x_1^2 + \cdots + x_n^2}{2}\right).$$

[13]This can be proved for a region Y and continuous f and g by using the mean value theorem.

Hence

$$P(s \le y) = \int\limits_{|x|^2 \le y} \left(\frac{1}{\sqrt{2\pi}}\right)^n \exp\left(-\frac{x_1^2 + \cdots + x_n^2}{2}\right) dx_1 \cdots dx_n$$

$$= \int_0^{\sqrt{y}} \left(\frac{1}{\sqrt{2\pi}}\right)^n A_t \exp\left(\frac{-t^2}{2}\right) dt,$$

where A_t is the area of the surface bounding the n-ball of radius t. The area function is the derivative of the volume function and so it has the form kt^{n-1}. Therefore,

$$P(s \le y) = c \int_0^{\sqrt{y}} t^{n-1} \exp\left(-\frac{t^2}{2}\right) dt,$$

where c denotes all constant multipliers. Let the density function of s be g; then

$$P(s \le y) = \int_0^y g(u)\, du,$$

where $u = t^2$, since $t = \sqrt{x_1^2 + \cdots + x_n^2}$. Setting $t = u^{1/2}$ in the integral involving t leads to

$$g(u) = cu^{(n/2)-1} \exp(-u/2) \qquad (u > 0)$$

and 0 otherwise. Since $\int_0^\infty g(u)\, du = 1$, $c^{-1} = 2^{n/2}\Gamma(n/2)$, where

$$\Gamma(x) = \int_0^\infty y^{x-1} e^{-y}\, dy$$

is the gamma function. The integer n is the number of degrees of freedom.

If the transformation is not one-to-one, we cannot apply Theorem 2. However, we can generalize the reasoning applied in Example 4 as follows.

EXAMPLE 6

Let $r_1, r_2, \ldots, r_n$ be n independent continuous random variables having a common density function[14] f that is positive, provided that $a < x < b$. Let s_1 be the smallest of these r_i, s_2 the next smallest, . . . , s_n the largest.

[14]This is known in statistics as a random sample.

Then s_i $(i = 1, 2, \ldots, n)$ is called the *ith-order statistic* of the random sample $r_1, \ldots, r_n$. Show that the joint density function g of $(s_1, s_2, \ldots, s_n)$ is given by

$$g(y_1, y_2, \ldots, y_n) = (n!)f(y_1)f(y_2) \cdots f(y_n) \qquad (a < y_1 < y_2 < \cdots < y_n < b)$$

and 0 elsewhere.

For convenience, we shall prove this only for the case of $n = 3$, but the argument can easily be generalized. With $n = 3$, the joint density function of (r_1, r_2, r_3) is $f(x_1)f(x_2)f(x_3)$. The set X, where $f(x_1)f(x_2)f(x_3) > 0$, is the union of six mutually exclusive sets:

$$X_1 = (x : a < x_1 < x_2 < x_3 < b)$$
$$X_2 = (x : a < x_1 < x_3 < x_2 < b)$$
$$X_3 = (x : a < x_2 < x_1 < x_3 < b)$$
$$X_4 = (x : a < x_2 < x_3 < x_1 < b)$$
$$X_5 = (x : a < x_3 < x_1 < x_2 < b)$$
$$X_6 = (x : a < x_3 < x_2 < x_1 < b)$$

because a probability such as $P(a < r_1 = r_2 < b, a < r_3 < b) = 0$. There are six of these sets, since there are $3! = 6$ permutations of the symbols x_1, x_2, x_3. The functions $y_1 = \min(x_1, x_2, x_3)$, $y_2 =$ middle in size (x_1, x_2, x_3), and $y_3 = \max(x_1, x_2, x_3)$ define a many-to-one transformation on X which maps each X_i into the same set $Y = (y : a < y_1 < y_2 < y_3 < b)$. However, the transformation restricted to each X_i is one-to-one and so it has a unique inverse. For example, the inverse functions for points in X_1 are $x_1 = y_1$, $x_2 = y_2$, $x_3 = y_3$. Let J_i be the corresponding Jacobians; it is easy to verify that $|J_i| = 1$. Let A be any set in X and let B be its image under the given transformation; then

$$P(s \in B) = P(r \in A) = \int_A f(x_1)f(x_2)f(x_3)\, dx_1 dx_2 dx_3$$
$$= \sum_{i=1}^{6} \int_{AX_i} f(x_1)f(x_2)f(x_3)\, dx_1 dx_2 dx_3.$$

Changing variables on each set AX_i yields

$$\int_{AX_i} f(x_1)f(x_2)f(x_3)\, dx_1 dx_2 dx_3 = \int_B f(y_1)f(y_2)f(y_3)\, dy_1 dy_2 dy_3,$$

from which the result follows.

The above method can be applied to generalize Theorem 2 (see Exercise 13).

EXERCISES

1. Let r have a density function $f(x) = (\frac{1}{2})^x$ for $x = 1, 2, 3, \ldots$, and $f(x) = 0$ otherwise. Find the density function of $s = r^3$.

2. Let (r_1, r_2) have a density function $f(x_1, x_2) = x_1x_2/36$ for $x_1 = 1, 2, 3$ and $x_2 = 1, 2, 3$, and $f(x_1, x_2) = 0$ otherwise. Let $s_1 = r_1r_2$ and $s_2 = r_2$. Find the marginal density of s_1.

3. If the density function of r is $f(x) = 2xe^{-x^2}$ for $0 < x < \infty$, and $f(x) = 0$ otherwise, determine the density function of $s = r^2$.

4. Suppose that r is uniformly distributed over $(-1, 1)$. Find the density function of the following random variables:

(a) $s = \sin(\pi/2)r$.
(b) $s = 4 - r^2$.
(c) $s = |r|$.
(d) $s = \cos(\pi/2)r$.

5. Let r_1, r_2 be a random sample from the normal distribution with $\mu = 0$ and $\sigma = 1$. Show that the marginal density function of $s_1 = r_1/r_2$ is the Cauchy density

$$g_1(y_1) = [\pi(1 + y_1^2)]^{-1}. \qquad (-\infty < y_1 < \infty)$$

6. Suppose that the radius of a ball bearing is a continuous random variable with a density function $f(x) = 6x(1 - x)$ for $0 < x < 1$. Find the density function of the volume and the surface area of the sphere.

7. The speed s of a molecule in an ideal gas at equilibrium is distributed according to the Maxwell-Boltzmann law: $f(x) = 4\pi^{-1/2}a^{3/2}x^2e^{-ax^2}$ for $x > 0$, where $a = m/2kT$, and k, T, and m denote Boltzmann's constant, the absolute temperature, and the mass of the molecule respectively. Derive the density function of the kinetic energy $r = ms^2/2$ of the molecule.

8. (a) In part 2 of Example 5, show that if the variances of the r_i are σ^2 instead of 1, the density function becomes

$$f(x) = (2^{n/2}\sigma^n\Gamma(n/2))^{-1}x^{(n/2)-1}\exp(-x/2\sigma^2). \qquad (x > 0)$$

(b) Show that the random variable $[(r_1^2 + r_2^2 + \cdots + r_n^2)/n]^{1/2}$ has the density function

$$f(x) = \frac{2(n/2)^{n/2}}{\sigma^n\Gamma(n/2)}x^{n-1}\exp\left(-\frac{nx^2}{2\sigma^2}\right). \qquad (x > 0)$$

9. Derive the density function of the Student's t-distribution, where $t = r/[(1/n)\sum_{i=1}^{n} r_i^2]$ and $r, r_1, \ldots, r_n$ are normal and independent, with a mean 0 and a variance σ^2. [*Hint:* Derive the density function of the denominator first and then use part 1 of Example 5. The answer is

$$h(u) = \frac{\Gamma(n+1)}{\Gamma(n\sqrt{n\pi}/2)}\left(1 + \frac{u^2}{n}\right)^{-(n+1)/2}.\Big]$$

10. The F-random variable is the quotient $F = r/s$, where $r = (r_1^2 + r_2^2 + \cdots + r_m^2)/m$ and $s = (s_1^2 + \cdots + s_n^2)/n$. Let $r_1, \ldots, r_m, s_1, \ldots, s_n$ be independent and normal, with a mean 0 and a variance 1. Verify that the density function of F is

$$h(u) = \left(\Gamma\left(\frac{m+n}{2}\right)/\Gamma\left(\frac{m}{2}\right)\Gamma\left(\frac{n}{2}\right)\right)\left(\frac{n}{m}\right)^{n/2} u^{(n/2)-1}\left(1 + \frac{n}{m}u\right)^{-(m+n)/2}.$$

11. Suppose that r is a continuous random variable and $s = \phi(r)$ is a discrete random variable. Describe how to find the density function of s.

12. Let r_i be independent random variables and let $s_i = \phi_i(r_i)$, a function of r_i alone. Show that s_i are independent random variables.

13. Suppose that the transformation in Theorem 2 is not one-to-one. Assume, however, that $X = \sum_{j=1}^{k} X_j$ and that the transformation restricted to X_i is one-to-one, with each X_i mapped on Y. Let

$$x_i = \psi_{ij}(y) \qquad (i = 1, 2, \ldots, n;\ j = 1, 2, \ldots, k)$$

denote the k groups of n inverse functions and let J_i be the Jacobians of these transformations. Assume that each J_i satisfies the hypothesis stated for J in Theorem 2. Show that

$$g(y_1, y_2, \ldots, y_n) = \sum_{j=1}^{k} |J_j| f(\psi_{1j}(y), \ldots, \psi_{nj}(y)).$$

5.9 THE NEED FOR MEASURE THEORY

In the preceding section we defined probability by means of density functions. We did not point out the connection with the probability axioms in Section 5.2, nor did we define rigorously the concept of random variables. Hence we must reconcile the intuitive and rigorous approaches.

Suppose that we are given a probability density function f on the real line. Let us try to determine the class of events $\mathscr{E}$. We saw that it was sensible to speak about a random variable r lying between two real numbers a and b when $r(x) = x$. Hence the point set $[a, b]$ should be an event.[15] Let $\mathscr{B}_0$ be the class of all such intervals; then $\mathscr{B}_0 \subset \mathscr{E}$. However, $\mathscr{B}_0 \neq \mathscr{E}$, since countable unions of sets in $\mathscr{B}_0$ do not necessarily belong to $\mathscr{B}_0$. Let $\mathscr{B}_1$ be the class of all sets that are finite or countable unions of sets in $\mathscr{B}_0$. Then $\mathscr{B}_1 \subset \mathscr{E}$, but $\mathscr{B}_1 \neq \mathscr{E}$, because it is not closed under complementation—for example, $\phi \notin \mathscr{B}_1$. Since the sets in $\mathscr{B}_1$ are events, the complements of these sets must be events; therefore, we form the class $\mathscr{B}_2$, which consists of such sets and of all sets in $\mathscr{B}_1$. The above process must be repeated indefinitely, since we always get events if we take countable unions and complements of events.

[15]Here any type of interval (open, closed, half-open) is considered.

The class of sets $\mathscr{B}$ generated in this manner is called the *Borel class* and its elements are *Borel sets*. The class $\mathscr{B}$ is suitable for a class of events since it is a σ-field. A similar construction can be carried out for n-dimensional intervals. Hence we have determined which sets should be considered as events in the spaces of Section 5.8. It remains to define a probability function P on these sets.

We know that $P(a < r \leq b) = \int_a^b f(x)\,dx$, $P(r > b) = \int_b^\infty f(x)\,dx$, and $P(r \leq a) = \int_{-\infty}^a f(x)\,dx$. Let $\mathscr{F}$ be the class of sets consisting of finite disjoint unions of intervals of the following type: $(a, b]$, (b, ∞), or $(-\infty, a]$. Then $\mathscr{F}$ is a field and $\mathscr{F} \subset \mathscr{B}$. Since P is defined on these three types of intervals, we can define P on $F = F_1 + \cdots + F_n \in \mathscr{F}$ by $PF = PF_1 + PF_2 + \cdots + PF_n$. We then are faced with the problem that appeared in Section 5.2. We have P defined on a field $\mathscr{F} \subset \mathscr{B}$ and we must show that P can be uniquely extended to a nonnegative, σ-additive set function on $\mathscr{B}$. We will show that this is possible in Chapter 6, where we will prove a more general result. This generality is necessary for arbitrary probability space. For example, in Exercise 4 of Section 5.7 we considered the number of occurrences of an event. For this process a sample point is represented by a step function of t with unit jumps at the moment of occurrence of the event. Our sample space is not simply a Euclidean space but a space of functions.

Suppose that r is a random variable defined on the real line such that $r(x) = x$. In Section 5.7 we defined a random variable to be $s = \phi(r)$, where ϕ is a continuous function. Since s is a random variable, it should have a probability distribution. In other words, the set of points $(s \leq y)$ should be an event (cf., Theorem 8.1), since the probability function is defined on Borel sets. We will defer the full proof of this statement until Chapter 8. Let us assume for the present that the function ϕ has a unique inverse, say $x = \psi(y)$. This will be true if ϕ is strictly monotone. If ψ is increasing, then $(s \leq y)$ if and only if $(r \leq \psi(y))$; this set has a probability measure $F(\phi(y))$. The proof that $P(s \leq y)$ is defined in the situation where ϕ is decreasing is left to the reader. The density of the random variable is $f(\psi(y))\psi'(y)$. Note that the density would not exist if ψ were not differentiable; however, $P(s \leq y)$ is well-defined. Hence density functions are not satisfactory for a general theory of probability. We will return to this point later. In order to study a function s on a probability space, we will see that it is important that the set of points $(s \leq y)$ be an event for every real y. A function with this property is called a random variable (or a measurable function). We will show that the class of measurable functions properly contains the class of continuous functions.

In many instances it is necessary to integrate such functions—for example, in calculating Es. Since these functions can be discontinuous and since the density function may not exist, we must generalize the concept of integration based on the fact that $P(s \leq y)$ is well-defined. Example 1 illustrates some of these points.

EXAMPLE 1

Suppose that we try to find the probability of choosing a rational number from the real numbers in the interval [0, 1]. Since there are many more real numbers than rationals, we would expect the probability of this event E to be zero. Let us try to verify this fact using our probability axioms. Since the rationals are countable, we may order them in a sequence $r_1, r_2, r_3, \ldots$. Now cover each r_k with an interval I_k of length ϵ^k, where $\epsilon < 1$. Actually, we consider $I_k \cap [0, 1]$, since some I_k may protrude. The event $E \subset A = I_1 \cup I_2 \cup \cdots$, and so $PE \leq PA$. However, some of the intervals overlap; hence $P(A) \leq \sum_{k=1}^{\infty} PI_k = \epsilon + \epsilon^2 + \epsilon^3 + \cdots = \epsilon/(1 - \epsilon)$. The proof of the last inequality is left as an exercise for the reader (cf., Exercise 1.4.6). Finally, $PE \leq \epsilon/(1 - \epsilon)$, which can be made arbitrarily small. Hence our conjecture is verified; $PE = 0$.

Let I be the indicator function of the set E of rational numbers. Then I is a measurable function according to the definition given above, since E is a Borel set of measure zero. We observed earlier (cf., Lemma 2.3.1) that the expected value of I is $PE = 0$. Assuming a uniform distribution, the expected value is given by $\int_0^1 I\,dx$. However, this integral does not exist in the Riemann sense. Since every interval contains a rational number, $\sup I = 1$ on every interval. Accordingly the upper sums equal 1; the lower sums are 0. In Chapter 7, we will give the definition of integration by which this expectation exists. Moreover, this more general concept of integration will enable us to treat discrete and continuous density functions in the same manner. For example, in calculating expectations, we use two different expressions: $\sum x_i f(x_i)$ and $\int f(x)\,dx$. These are both particular cases of a more general integral.

As a final example, we will consider a problem which gives rise to a continuous distribution with no density. To do this, we will define the *Cantor ternary set*. Divide the interval [0, 1] into three equal parts and remove the set $(1/3, 2/3)$, the interior of the middle part. This removes the set of numbers whose ternary expansion as an infinite decimal begins with 1, except for $0.100 \cdots = 0.0222 \cdots$. Next, subdivide each of the two remaining parts into three equal parts, and remove the interiors of the middle parts of each of them. Any number lying in a "middle part" has a 1 in its ternary expansion. Hence the set of remaining numbers does not have a 1 in the second place of its ternary expansion except for $0.01 = 0.00222 \cdots$ and $0.21 = 0.20222 \cdots$. At the nth step we remove 2^{n-1} "middle" intervals; this process is repeated indefinitely. The remaining set of points C is called the Cantor ternary set. These points are represented by the infinite decimals

$$0.\ a_1a_2a_3 \cdots a_n \cdots$$

in the scale of 3, where each a_i is 0 or 2.

The points that do not belong to the set C lie in open intervals. Let $0.a_1a_2 \cdots a_n \cdots$ be a number that does not belong to C; then some of the a_i will be 1. Suppose that a_r is the first integer equal to 1. Then a_i for $i > r$ cannot be all 0 or all 2 since $0.a_1 \cdots a_{r-1}100 = 0.a_1 \cdots a_{r-1}022 \cdots$, while $0.a_1 \cdots a_{r-1}122 \cdots = 0.a_1 \cdots a_{r-1}2$, and both of these numbers belong to C. Hence the given number lies in an open interval bounded by the above numbers or

$$0.a_1a_2 \cdots a_{r-1}1 < 0.a_1a_2 \cdots a_{r-1}1a_{r+1} \cdots < 0.a_1a_2 \cdots a_{r-1}2.$$

No number in this open interval belongs to C, since it has the form $0.a_1a_2 \cdots a_{r-1}1a_{r+1}a_{r+2} \cdots$ and since a_i for $i > r$ are not all 0 or 2. We will call this interval the *associated interval* of $0.a_1a_2 \cdots a_{r-1}1$.

The Cantor set has measure zero.[16] We will prove this statement by showing that C' has measure (or total length) 1. Now there are 2^{r-1} intervals, each of length $1/3^r$, of the type described in the preceding paragraph, since we have two choices (0 or 2) for each of the members a_i where $i = 1, \ldots, r-1$. Hence the total length of the intervals comprising C' is $\sum_{r=1}^{\infty} 2^{r-1}/3^r = 1$.

The set C is not enumerable. This can be proved by expressing the real numbers in the interval [0, 1] in binary notation and mapping a number in C onto the binary number

$$0.\frac{a_1}{2}\frac{a_2}{2} \cdots \frac{a_n}{2} \cdots.$$

The details of the proof are left to the reader.

We are now in a position to give an example of a continuous distribution function that has no density.

EXAMPLE 2

Let $F(x)$ be a function which for $x \in C$ and $x = 0.a_1a_2 \cdots a_n \cdots$ in ternary notation takes the value $0.(a_1/2)(a_2/2) \cdots (a_n/2) \cdots$ in binary nota-

[16] C is a set that can be covered by a finite number or a denumerable sequence of intervals such that the sum of the individual lengths is arbitrarily small. It is clear intuitively that C is of measure zero if and only if C' is of measure 1. This can be proved rigorously using the methods of Chapter 6.

tion. If $x \in C'$, it belongs to some associated interval with a left-hand end point $y \in C$; then the value of $F(x)$ is defined to be $F(y)$. In other words, on the interval $[\frac{1}{3}, \frac{2}{3}]$, $F(x) = \frac{1}{2}$, on $[\frac{1}{9}, \frac{2}{9}]$, $F(x) = \frac{1}{4}$, on $[\frac{7}{9}, \frac{8}{9}]$, $F(x) = \frac{3}{4}$, etc. We define $F(x) = 0$ if $x < 0$ and $F(x) = 1$ if ≥ 1.

The function $F(x)$ is continuous and never decreasing; $\lim_{x \to -\infty} F(x) = 0$ and $\lim_{x \to \infty} F(x) = 1$. Therefore, it is a distribution function. Almost every x belongs to an interval where this function is constant and so its derivative is equal to zero almost everywhere (recall that the measure of C is zero).

We can interpret the above probabilistically. Suppose that at the nth trial of an infinite sequence of coin tossings the player receives $2/3^n$ if he gets heads and nothing if he gets tails. Let r be a random variable representing his total gain. Then the values of r lie in the Cantor set. If we represent heads by 1 and tails by 0, then if $r = 0.a_1 a_2 \cdots a_n \cdots$, the outcomes of this game are represented by $(a_1/2), (a_2/2), (a_3/2), \ldots, (a_n/2), \ldots$. $P(r \leq 0.a_1 a_2 \cdots a_n \cdots)$ is obtained in the following manner. Suppose that $a_{i_1} = a_{i_2} = \cdots = a_{i_n} = \cdots = 2$. This means that we obtained a head on trials number i_k for $k = 1, 2, 3, \ldots$. To obtain less money we must get fewer heads. In other words, some $a_{i_n} = 0$. Hence the set of all such points x (which are expressed in binary notation as a sequence) such that $r(x) \leq 0.a_1 a_2 \cdots$ can be divided into disjoint sets A_n, where

$$A_n = (x : x_i = a_i/2 \ (i < i_n),\ x_{i_n} = 0,\ x_i = 0 \text{ or } 1 \ (i > i_{n+1})).$$

Then $P(A_n) = (a_{i_n}/2)2^{-i_n}$, since the first i_n coordinates are fixed and the remainder are arbitrary. Hence $P(r \leq 0.a_1 a_2 \cdots a_n \cdots) = \sum_{n=1}^{\infty} (a_n/2)2^{-n} = 0.(a_1/2)(a_2/2) \cdots (a_n/2) \cdots$ in binary notation. If x is not in the C set, $P(r \leq x)$ is the value of the associate interval containing x as above. Under this interpretation, it is obvious that the distribution function of the random variable r is $F(x)$ defined in the last paragraph.

We cannot find Er, since there is no density function. However, intuitively, the expectation should exist. In fact, a moment's reflection will convince the reader that $r = 2 \sum_{1}^{\infty} 3^{-n} r_n$, where r_n is a sequence of mutually independent random variables assuming the values 0 and 1 with a probability of $\frac{1}{2}$. Proceeding formally, $Er = 2 \sum_{1}^{\infty} 3^{-n} Er_n = \sum_{1}^{\infty} 3^{-n} = \frac{1}{2}$. The generalization of the integral given in Chapter 7 will justify this calculation.

REFERENCES

1. Bohlmann, G., "Die Grundbegriffe der Wahrscheinlichkeitsrechnung und ihrer Anwendung auf Lebensversicherung," *Proc 4th Int. Cong. Math.*, Vol. 3 (1908).
2. Broggi, U., "Il teorema della probabilità composta e la definizione descritiva di probabilità," *Rend. Circ. Mat. Palermo,* **28** (1909).
3. Padoa, A., "Frequenza previsione probabilità," *Atti. R. Acc. delle Scienze,* Torino (1912).
4. Peano, G., "Sulla definizione di probabilità," *Rend. R. Acc. Lincei,* Rome (1912).
5. Fisher, R. A., *Statistical Methods for Research Workers*, Edinburgh: Boyd, 1925.
6. von Mises, R., "Grundlagen der Wahrscheinlichkeitsrechnung," *Math. Zeitschrift* (1919).
7. Keynes, J. M., *A Treatise on Probability*, London: Macmillan, 1921.
8. Kolmogorov, A. N., *Grundbegriffe der Wahrscheinlichkeitsrechnung.* Berlin, 1933.
9. Reichenbach, "Axiomatik der Wahrscheinlichkeitsrechnung," *Math. Zeitschrift* (1932).
10. Feller, W., *An Introduction to Probability Theory and Its Applications,* Vol. I. New York: Wiley, 1960.
11. Doob, J. L., "Note on Probability," *Ann. of Math.*, **37** (1936), 363–367.
12. Thorp, E., *Beat the Dealer*, New York: Random House, 1962.
13. Spitzer, F. L., *Principles of Random Walk*, Princeton, N.J.: Van Nostrand, 1964.
14. Edwards, J., *Treatise on Integral Calculus*, Vol. II. New York: Chelsea, 1954.
15. Kendall, M. G., and P. A. P. Moran, "Geometrical Probability," *Griffin's Stat. Mon.* **10** (1963).
16. Czuber, E., *Geometrische Wahrscheinlichkeiten und Mittelwerte*, Leipzig, 1884.
17. Poincaré, H., *Calcul des Probabilités*, Paris, 1912.
18. Czubér, E., *Wahrscheinlichkeitsrechnung*, Leipzig, 1908.
19. Crofton, W., "On the theory of local probability applied to straight lines drawn at random in a plane, the method used being also extended to the proof of *certain* new theorems in integral calculus," *Philos. Trans.*, **158** (1868).
20. Crofton, W., "Probability," *Encyclopaedia Britannica*, 9th ed.

CHAPTER 6

THEORY OF MEASURE

6.1 INTRODUCTION

In this chapter we will try to settle some of the problems raised in the preceding chapter, such as the extension of a probability function. These questions could be answered by using the properties of the probability function as described in the axioms. However, it is expedient to introduce the terminology of measure theory, especially as the difficulty of the proofs is not increased.

Although we have tried to motivate measure theory from probability theory, the subject originated in the study of functions of a real variable. Around 1882, Hankel introduced the notion of the *content* of a set in order to try to measure the portion of an interval occupied by the discontinuities of a function. The foundations of the theory of content were laid by Cantor and Stolz (1884) when they extended these ideas to R^n. This theory was not satisfactory for nonclosed sets; to correct this defect, Peano (1887) and Jordan (1892) introduced the notions of *inner* and *outer content*. This theory was generalized to a more satisfactory form by Borel (1898) and Lebesgue (1902) and is now called the *theory of Lebesgue measure*. Studies of integration based upon these concepts have been carried out by a number of others, among whom we may mention J. Radon [1], P. J. Daniell [2], O. Nikodyn [3], A. Kolmogorov [4], and B. Jessen [5].

6.2 FUNDAMENTAL CONCEPTS IN PROBABILITY AXIOMS

For the sake of completeness and ease of reference, we will summarize some of the notions that were introduced earlier. Unless otherwise specified, all sets are subsets of X.

DEFINITION 1

A *field*[1] is a nonempty class of sets $\mathscr{F}$ such that:

1. If $A, B \in \mathscr{F}$, then $A \cup B \in \mathscr{F}$.
2. If $A \in \mathscr{F}$, then $A' \in \mathscr{F}$.

The reader can easily verify by induction that if $E_i \in \mathscr{F}$, then $\bigcup_{i=1}^{n} E_i \in \mathscr{F}$. From part 2 of the definition, it follows that $\left(\bigcup_{i=1}^{n} E_i\right)' \in \mathscr{F}$; an application of DeMorgan's law shows that $\bigcap_{i=1}^{n} E_i' \in \mathscr{F}$. Since $\mathscr{F}$ is not empty, we have $X = E \cup E' \in \mathscr{F}$ and $X' = \phi \in \mathscr{F}$.

DEFINITION 2

A *σ-field*[2] is a nonempty class $\mathscr{S}$ of sets such that:

1. If $E_i \in \mathscr{S}$, then $\bigcup_{i=1}^{\infty} E_i \in \mathscr{S}$.
2. If $E \in \mathscr{S}$, then $E' \in \mathscr{S}$.

A *relative field* (*σ-field*) in $E \subset X$ is defined exactly as above except that A' denotes $E - A$.

EXAMPLE 1

1. All subsets of the spaces Ω_1 and Ω_2 defined in Section 1.3 are fields. More generally, the class of all subsets of any set X is a field.
2. The class of events $\mathscr{E}$ specified in Definition 2.2.1 is a field.
3. The class $\mathscr{F} = (X, A, A', \phi)$ is a field.
4. Any field containing a finite number of elements is a σ-field. However, if it has an infinite number of members, it is not necessarily a σ-field. Any σ-field is a field.
5. Let $\mathscr{F}$ be the class of finite sums of all types of rectangles (open, closed,

[1]Also called an *algebra*.

[2]Also known as an *additive class*, a *Borel field*, or a *σ-algebra*.

open on right, etc.) with edges parallel to the x- and y-axes. It is not difficult to verify that $\mathscr{F}$ is a field. However, $\mathscr{F}$ is not a σ-field; $C = [(x, y) : x^2 + y^2 \leq 1]$ does not belong to $\mathscr{F}$ because it is impossible to square a circle. The well-known technique of finding the area of a figure as a limiting sum of rectangles shows that C belongs to a σ-field containing $\mathscr{F}$. This can be proved rigorously using Theorem 6.3.1.

The following theorems completely describe the structure of fields with a finite number of elements.

THEOREM 1

Let $X = A_1 + A_2 + \cdots + A_n$; then the field $\mathscr{F}$ containing the class $\mathscr{P} = \{A_1, A_2, \ldots, A_n\}$ consists of all possible finite sums that can be formed from the sets of $\mathscr{P}$.

The proof follows directly from the definition of a field provided that we interpret the sum formed by choosing no sets of $\mathscr{P}$ as the empty set. Note that $\mathscr{F}$ consists of 2^n elements.

In order to prove that every finite field is constituted as described in Theorem 1, we will require the following definition.

DEFINITION 3

A set A of a field $\mathscr{F}$ is called an *atom* if $\phi \neq B \in \mathscr{F}$ and $B \subset A$ then $B = A$.

THEOREM 2

Any field containing a finite number of elements is of the form described in Theorem 1.

Proof

Suppose that $\mathscr{F} = (B_i : i \in I)$, where I is a finite index set. Then every subset A of the form $A = \bigcap_{i \in I} C_i$, where $C_i = B_i$ or B_i' for $i \in I$, is either empty or an atom of $\mathscr{F}$. By construction, two nonempty sets A are necessarily disjoint. Furthermore, every $F \in \mathscr{F}$ (particularly X) is a finite sum of the nonempty sets A which it contains. Hence the class $\mathscr{P}$ of all sets of the form A satisfies the conditions of Theorem 1; obviously, $\mathscr{F}$ contains the sets A.

The remaining notion in the axioms is that of a probability function. This function is a set function—that is, a function f whose domain of definition is a class $\mathscr{C}$ of sets (the events $\mathscr{E}$ for a probability function). We generalize the concept of a probability function by letting the range of the function be the extended real-number system. Examples of set functions are furnished by the number of signs on a stretch of road, the number of banks in each area of a city, and the number of births in a certain time interval as well as by Example 2.

EXAMPLE 2

1. Let X be the set of nonnegative integers. For any $A \subset X$ and $A \neq \phi$, we define $f(A)$ as the smallest integer in A and $f(\phi) = 0$.
2. Let X be any set. Define $f(A)$ to be the number of points in A if A is a finite set; otherwise, $f(A) = \infty$.
3. Suppose that g is an arbitrary bounded point function on a finite set X. Then $f\{x_1, \ldots, x_n\} = \sum_{i=1}^{n} g(x_i)$ is a set function.
4. The probability function on the class of events is a set function.
5. $f(A) = 0$ if A is a finite set and $f(A) = \infty$ if A is an infinite set, where A is any subset of a countable infinite set X.

The probability that at least one event will occur is the sum of the probabilities of the individual events if the events are mutually exclusive. This additive property is also possessed by some set functions.

DEFINITION 4

A real-valued set function f defined on a class of sets $\mathscr{C}$ is *finitely additive* if for every finite disjoint class $\{E_i\}$ of sets of $\mathscr{C}$ whose sum is in $\mathscr{C}$,

$$f\left(\sum_{i=1}^{n} E_i\right) = \sum_{i=1}^{n} f(E_i);$$

it is *σ-additive* (or countably additive) if for any $E_i \in \mathscr{C}$ such that $\sum_{i=1}^{\infty} E_i \in \mathscr{C}$,

$$f\left(\sum_{i=1}^{\infty} E_i\right) = \sum_{i=1}^{\infty} f(E_i).$$

The function defined in part 1 of Example 2 is not additive; those appearing in parts 2–4 are σ-additive. Part 5 is additive but not σ-additive since, if A is any infinite set, $f(A) = \infty$, while $\sum_{x \in A} f(\{x\}) = 0$. A σ-additive function is finitely additive if $\phi \in \mathscr{C}$. However, if $\phi \notin \mathscr{C}$ this might not be true.

DEFINITION 5

If $|f(A)| < \infty$, then f is said to be *finite on* A. If $A \in \mathscr{C}$ is the countable union of sets in $\mathscr{C}$ on each of which f is finite, then f is said to be *σ-finite*[3] *on* A. Finally, if f is finite (σ-finite) on every set in $\mathscr{C}$, then f is said to be *finite* (*σ-finite*).

The set function f is finite in parts 1, 3, and 4 of Example 2 and σ-finite in part 5. In part 2, if X is a set of arbitrary cardinality, then f is not necessarily σ-finite.

Since we are allowing f to be infinite, we must exclude expressions of the form $\infty - \infty$ in order for $\sum f(E_i)$ to be meaningful. Suppose further that f is additive and that it is defined on a field that has sets A and B such that $f(A) = \infty$ and $f(B) = -\infty$. Then $f(X) = f(A) + f(A') = \infty$ and $f(X) = f(B) + f(B') = -\infty$, since $f(A') \neq -\infty$ and $f(B') \neq \infty$, because of the restriction imposed above. Since f is single valued, one of these values must be excluded. *From this point on, we assume that if* f *is an additive set function on a field* $\mathscr{F}$, *then it never takes the value* $-\infty$.

EXERCISES

1. (*a*) Show that (X, ϕ) is a field.
(*b*) Does a field always contain the difference of any two of its sets?
(*c*) Show that a class of sets closed under finite intersections and complements is a field.

2. Which of the following classes of sets are fields? Which are σ-fields?

(*a*) $\mathscr{C} = (A, A', X, \phi)$.
(*b*) X is an uncountable set; $\mathscr{C}$ is the class of countable subsets of X.
(*c*) X is an uncountable set; $\mathscr{C}$ is the class of all finite subsets of X.
(*d*) X is an uncountable set; $\mathscr{C}$ is the class of all sets that either are countable or have countable complements.

3. Consider the class $\mathscr{R} = (EF : F \in \mathscr{F})$, where $\mathscr{F}$ is a field and E is a fixed subset of X. This class is not a field, but it has the following properties:

(*a*) If $R_1, R_2 \in \mathscr{R}$, then $R_1 \cup R_2 \in \mathscr{R}$.
(*b*) If $R_1, R_2 \in \mathscr{R}$, then $R_1 - R_2 \in \mathscr{R}$.

Such a class of sets is called a *ring*. Show that the class of all finite unions of semiclosed intervals $(a, b] = (x : a < x \leq b)$, where $b < \infty$ and $a > -\infty$, of the real line is a ring.

[3]Some authors only require that A be contained in a countable union of sets of $\mathscr{C}$ on each of which f is finite. These definitions are equivalent if $\mathscr{C}$ is a field.

4. (*a*) Show that if $E, F \in \mathscr{R}$, then $E \Delta F$ and $E \cap F$ belong to $\mathscr{R}$.
(*b*) Prove that if a ring contaings X it is a field.
(*c*) Prove that every field is a ring.

5. If we replace Exercise 3(*a*) by the following: $R_i \in \mathscr{R}$ implies $\bigcup_{i=1}^{\infty} R_i \in \mathscr{R}$, the resulting class is called a *σ-ring*. Show that if $\mathscr{R}$ is a σ-ring and $R_i \in \mathscr{R}$, then $\bigcap_{i=1}^{\infty} R_i \in \mathscr{R}$.

6.3 FIELDS AND MONOTONE CLASSES

To find the probability of an event, it is often necessary to consider a monotone sequence of events whose probability is known, and which approach the given event. This was illustrated in the last chapter by several examples. We assumed, because of the intuitive interpretation, that the limit of an increasing sequence of events is an event. We will demonstrate below that the foregoing is true for σ-fields. Classes that contain monotone limits of their sets are also important in measure theory and have acquired a special name.

DEFINITION 1

A nonempty class of sets $\mathscr{M}$ is *monotone* if, for every monotone sequence $\{E_i\}$ of sets in $\mathscr{M}$, $\lim E_i \in \mathscr{M}$.

THEOREM 1

A σ-field is a monotone class; a monotone field is a σ-field.

Proof

Suppose that $E_1 \subset E_2 \subset \cdots \subset E_n \cdots$, where $E_i \in \mathscr{S}$ and $\mathscr{S}$ is a σ-field. Then $\lim E_i = \bigcup_{i=1}^{\infty} E_i \in \mathscr{S}$ by definition. An analogous proof shows that a decreasing sequence of sets $E_i \in \mathscr{S}$ belongs to $\mathscr{S}$ and so $\mathscr{S}$ is a monotone class.

To prove the second assertion, we need only show that if $E_i \in \mathscr{M}$, then $\bigcup_{i=1}^{\infty} E_i \in \mathscr{M}$, since $\mathscr{M}$ is already a field. However, $\bigcup_{i=1}^{n} E_i \in \mathscr{M}$ forms an increasing sequence of sets as n increases. Since $\mathscr{M}$ is a monotone class, the limit $\bigcup_{i=1}^{\infty} E_i \in \mathscr{M}$.

This theorem enables us to prove the statement made in part 5 of Example 2.1 and supplies us with an example of a monotone class.

EXAMPLE 1

1. A σ-field is a monotone class. In particular, all subsets of X and the events $\mathscr{E}$ in a probability space are monotone classes.
2. All spheres (open and closed) having their center at the origin in n-space, plus the whole space X and the set consisting of the single point that is the origin, comprise a monotone class that is not a σ-field.

The class of all subsets of X is a field (σ-field, monotone class). Therefore, there exists at least one field (σ-field, monotone class) containing any arbitrary class $\mathscr{C}$ of sets of X. Accordingly, the following definition is meaningful.

DEFINITION 2

The field $\mathscr{F}(\mathscr{C})$ [σ-field $\mathscr{S}(\mathscr{C})$, monotone class $\mathscr{M}(\mathscr{C})$] generated by any class $\mathscr{C}$ of sets is the minimal field (σ-field, monotone class) containing $\mathscr{C}$.

EXAMPLE 2

1. Let $\mathscr{C} = \{A_1, A_2\}$. Then $\mathscr{F}(\mathscr{C})$ is comprised of the sets

$$X, \phi, A_1, A_2, A_1', A_2', A_1 \cup A_2, A_1 \cup A_2', A_1' \cup A_2, A_1' \cup A_2',$$
$$A_1 A_2, A_1 A_2', A_1' A_2, A_1' A_2'.$$

2. The Borel class described in Section 5.9 is generated by the class of all intervals (open, closed, half-open, degenerate, etc.).

The following theorem describes the construction of the field generated by a class of sets.

THEOREM 2

Let $\mathscr{C}$ be an arbitrary class of subsets of X. Form successively the classes $\mathscr{C}_i$ where

1. $\mathscr{C}_1$ consists of all $A \subset X$ such that A or $A' \in \mathscr{C}$.
2. $\mathscr{C}_2$ consists of finite intersections of sets in $\mathscr{C}_1$.
3. $\mathscr{C}_3$ consists of finite sums of sets belonging to $\mathscr{C}_2$.

Then $\mathscr{C}_3$ is the field generated by $\mathscr{C}$.

Proof

We first show that if $A, B \in \mathscr{C}_3$, then $AB \in \mathscr{C}_3$; that is, $\mathscr{C}_3$ is closed under the operation of intersection. Now A is of the form $A = \sum_{i=1}^{m} A_i$ where $A_i \in \mathscr{C}_2$, and B is of the form $B = \sum_{j=1}^{n} B_j$ where $B_j \in \mathscr{C}_2$; hence

$$AB = \sum_{i=1}^{m}\sum_{j=1}^{n} A_i B_j \in \mathscr{C}_3$$

for $A_i B_j \in \mathscr{C}_2$, since $\mathscr{C}_2$ is closed under the operation of intersection.

To show that $\mathscr{C}_3$ is closed under the operation of complementation, let $A = \sum_{i=1}^{n} A_i \in \mathscr{C}_3$ where $A_i \in \mathscr{C}_2$; then $A' = \bigcap_{i=1}^{n} A_i'$. If $A_i' = \mathscr{C}_3$, the result would follow by the closure of $\mathscr{C}_3$ under the operation of intersection. Thus it must be proved that $B \in \mathscr{C}_2$ implies $B' \in \mathscr{C}_3$. To this end, suppose that $B = B_1 B_2 \cdots B_n$, where $B_i \in \mathscr{C}_1$. Note that the sets of the form $C_1 C_2 \cdots C_n$, where $C_i = B_i$ or B_i', form a partition of X; that is, $X = \sum C_1 C_2 \cdots C_n$. Hence $X - B = \sum C_1 \cdots C_n$, where the summation is over all possible combinations of the C_i except for $C_i = B_i$, where $i = 1, 2, \ldots, n$. Since $\mathscr{C}_1$ is closed under the operation of complementation, it follows that all $C_1 C_2 \cdots C_n \in \mathscr{C}_2$ and finally that $B' \in \mathscr{C}_3$.

The reader unacquainted with the theory of partial ordering might find the above definition unpalatable. The concept of the smallest class of an infinite number of classes will perhaps be clarified by Lemma 1.

LEMMA 1

The field $\mathscr{F}(\mathscr{C})$ [σ-field $\mathscr{S}(\mathscr{C})$, monotone class $\mathscr{M}(\mathscr{C})$] is the intersection of all fields [σ-fields, monotone classes] that contain $\mathscr{C}$.

Proof

We only have to show that $\bigcap_{\alpha} \mathscr{F}_\alpha$ is a field if $\mathscr{F}_\alpha$ are fields containing $\mathscr{C}$, since such a field is the smallest. Suppose that $F_i \in \bigcap_{\alpha} \mathscr{F}_\alpha$; then for every α, F_i is an element of the field $\mathscr{F}_\alpha$ and so $\bigcup_{i=1}^{n} F_i \in \mathscr{F}_\alpha$. Hence $\bigcap_{i=1}^{n} F_i \in \bigcap_{\alpha} \mathscr{F}_\alpha$. A similar proof holds for the complement of a set; $F \in \bigcap_{\alpha} \mathscr{F}_\alpha$ implies $F' \in \bigcap_{\alpha} \mathscr{F}_\alpha$. The proofs of the bracketed statements are left as exercises for the reader (c.f., Exercise 1).

In most of the examples in Chapter 5, we knew the probability of any event in a certain field $\mathscr{F}$. Then we took monotone limits of these events. Theorem 1 shows that these limits belong to the σ-field of events $\mathscr{E}$. We would suspect that all events generated by $\mathscr{F}$ are expressible as such limits, since we should be able to calculate their probability. In other words, $\mathscr{M}(\mathscr{F}) = \mathscr{S}(\mathscr{F})$. Theorem 3 shows that this result is true for any field.

THEOREM 3

The monotone class $\mathscr{M}$ and the σ-field $\mathscr{S}$ generated by the same field $\mathscr{F}$ coincide.

Proof

From the first part of Theorem 1 we have $\mathscr{S} \supset \mathscr{M}$, since $\mathscr{M}$ is the minimal monotone class containing $\mathscr{F}$. To obtain the inverse inclusion, we need only show that $\mathscr{M}$ is a field, since the second part of Theorem 1 shows that a monotone field is a σ-field. Then $\mathscr{M} \supset \mathscr{S}$, since $\mathscr{S}$ is the minimal σ-field containing $\mathscr{F}$.

To show that $\mathscr{M}$ is a field, it suffices to prove that if $A \in \mathscr{M}$, then $A' \in \mathscr{M}$, and if $A, B \in \mathscr{M}$, then $A \cap B \in \mathscr{M}$. This requirement can be put in a more symmetrical form:

(1) If $A, B \in \mathscr{M}$, then AB, $A'B$, and AB' belong to $\mathscr{M}$.

Note that this requirement implies that $A' \in \mathscr{M}$, since $A'X = A'$ and $X \in \mathscr{M}$ (for $\mathscr{M} \supset \mathscr{F}$).

For every fixed $A \in \mathscr{M}$, let $\mathscr{M}_A$ be the class of all $B \in \mathscr{M}$ that satisfy (1). If $A \in \mathscr{F}$, then (1) is satisfied for every $B \in \mathscr{F}$, since $\mathscr{F}$ is a field and $\mathscr{M} \supset \mathscr{F}$; hence $\mathscr{M}_A \supset \mathscr{F}$. Moreover, $\mathscr{M}_A$ is a monotone class, since if the sequence $B_n \in \mathscr{M}_A$ is monotone, $B = \lim B_n$ belongs to $\mathscr{M}$ and so do the limits of the monotone sequences AB_n, AB_n', and $A'B_n$. Since $AB = \lim AB_n$, $AB' = \lim AB_n'$, and $A'B = \lim A'B_n$ all belong to $\mathscr{M}$, then $B \in \mathscr{M}_A$ by definition. Now $\mathscr{M} \supset \mathscr{M}_A \supset \mathscr{F}$; since $\mathscr{M}$ is the minimal monotone class containing $\mathscr{F}$, it follows that $\mathscr{M} = \mathscr{M}_A$ for every $A \in \mathscr{F}$.

Since the conditions imposed upon the pairs A and B are symmetrical, $A \in \mathscr{M}_B$ for every $A \in \mathscr{F}$ and *any* $B \in \mathscr{M}$; hence $\mathscr{M}_B \supset \mathscr{F}$. Reasoning as in the preceding paragraph, we see that $\mathscr{M}_B$ is monotone and, accordingly, $\mathscr{M}_B = \mathscr{M}$. This last equality, which is just a restatement of (1), completes the proof.

COROLLARY

If a monotone class contains a field $\mathscr{F}$, then it contains the σ-field generated by $\mathscr{F}$.

Let $E\mathscr{F} = (EF : F \in \mathscr{F})$. If $\mathscr{F}$ is a field, $E\mathscr{F}$ is a field in E (recall that $A' = E - A$). $\mathscr{S}_E(E\mathscr{C})$ denotes the intersection of all σ-fields in E that contain $E\mathscr{C}$.

LEMMA 2

Let $\mathscr{F}$ be a class of sets; then $\mathscr{S}_E(E\mathscr{F}) = \mathscr{S}(\mathscr{F}) \cap E$ for any set $E \subset X$.

Proof

We can prove that $\mathscr{S}(\mathscr{F}) \cap E \equiv \mathscr{S}$ is a σ-field in E as follows: Let $B_i = A_i \cap E \in \mathscr{S}$, where $A_i \in \mathscr{S}(\mathscr{F})$; then $\cup B_i = \cup A_i \cap E$ and $E - B_i = A_i' \cap E$ both belong to $\mathscr{S}$ since $\mathscr{S}(\mathscr{F})$ is closed under union and complementation. Consequently, $\mathscr{S}(\mathscr{F}) \cap E \supset \mathscr{S}_E(E\mathscr{F})$, the smallest σ-field containing $E\mathscr{F}$ in E.

To prove the reverse inclusion, we look for a σ-field in X that contains $\mathscr{S}(\mathscr{F})$ and whose intersection with E is $\mathscr{S}_E(E\mathscr{F})$. A natural candidate for such a field is $\mathscr{C} = \mathscr{S}_E(E\mathscr{F}) \cup \mathscr{S}(\mathscr{F})E'$. A typical element C_i of $\mathscr{C}$ has the form $C_i = A_i \cup B_i E'$, where $A_i \in \mathscr{S}_E(E\mathscr{F})$ and $B_i \in \mathscr{S}(\mathscr{F})$. It is obvious that $C_i \in \mathscr{C}$ implies $\cup C_i = (\cup A_i) \cup (\cup B_i)E' \in \mathscr{C}$.

The difficult part in verifying that $\mathscr{C}$ is a σ-field is to prove that $\mathscr{C}$ is closed under complementation. To accomplish this we need the fact that $\mathscr{C}$ is closed under intersection. The intersection of two typical elements $(A_1 \cup B_1 E')(A_2 \cup B_2 E') = A_1 A_2 \cup B_1 B_2 E'$ since $A_i \subset E$, and so $A_i E' = \phi$. The result follows since $A_1 A_2 \in \mathscr{S}_E(E\mathscr{F})$ and $B_1 B_2 \in \mathscr{S}(\mathscr{F})$.

The following inclusions will also be helpful: $\mathscr{C} \supset \mathscr{S}(\mathscr{F})E' \cup E = \mathscr{S}(\mathscr{F}) \cup E, \mathscr{C} \supset \mathscr{S}_E(E\mathscr{F})$, and $\mathscr{C} \supset \mathscr{S}(\mathscr{F})E'$. These follow directly from the definition of $\mathscr{C}$ when the classes are replaced by particular elements of each class—namely, $E \in \mathscr{S}_E(E\mathscr{F})$, $\phi \in \mathscr{S}(\mathscr{F})$, and $\phi \in \mathscr{S}_E(E\mathscr{F})$.

Now let $C = A \cup BE' \in \mathscr{C}$, where $A \in \mathscr{S}_E(E\mathscr{F})$ and $B \in \mathscr{S}(\mathscr{F})$. Then $C' = A'B' \cup A'E = A'EB' \cup A'E \cup A'B'E' = A'E(B' \cup E) \cup B'E'$, where $A'B'E' = B'E'$, since $E' \subset A'$. However, $A'E = E - A \in \mathscr{S}_E(E\mathscr{F})$ and $B' \cup E \in \mathscr{S}(\mathscr{F}) \cup E$; therefore, each belongs to $\mathscr{C}$. Since $\mathscr{C}$ is closed under union and intersection and $B'E' \in \mathscr{S}(\mathscr{F})E'$, the proof of the closure of $\mathscr{C}$ under complementation follows.

Knowing that $\mathscr{C}$ is a σ-field enables us to prove that $\mathscr{C}$ contains $\mathscr{S}(\mathscr{F})$, since $\mathscr{C} \supset E\mathscr{F} \cup E'\mathscr{F} = \mathscr{F}$. Hence $\mathscr{C} \cap E \supset \mathscr{S}(\mathscr{F}) \cap E$. The proof of the reverse inclusion is completed when it is noted that $\mathscr{C} \cap E = \mathscr{S}_E(E\mathscr{F})$.

EXERCISES

1. Let a c-class be a class of sets closed under some set operation or operations. For example, let the c-class consist of all semiclosed intervals $(ab]$ of the real line. This class is closed under intersection. Show that the intersection of an arbitrary number of c-classes is a c-class.

2. Show that a monotone ring is a σ-ring.

3. Prove that if $\mathscr{R}$ is a ring, $\mathscr{M}(\mathscr{R})$ is a σ-ring generated by $\mathscr{R}$.

4. Let $\mathscr{C}$ be an arbitrary class of sets and let $\mathscr{F} = \mathscr{F}(\mathscr{C})$ be the field generated by $\mathscr{C}$. Show that $\mathscr{S}(\mathscr{C}) = \mathscr{S}(\mathscr{F})$.

5. Describe the field generated by each of the following classes:

(*a*) The class $\mathscr{C}$ of all semiclosed intervals $(ab]$ of the real line.
(*b*) The class $\mathscr{C}$ of all finite subsets of X, where X is an uncountable set.
(*c*) $\mathscr{R}$, where $\mathscr{R}$ is a ring.

6. Describe the σ-field and monotone field generated by the classes where:

(*a*) X is an uncountable set; $\mathscr{C}$ is the class of all finite subsets of X.
(*b*) X is an uncountable set; $\mathscr{C}$ is the class of all countable subsets of X.
(*c*) X is any set; P is a one-to-one transformation of X onto itself. A subset of X is invariant if, whenever $x \in A$, $P(x) \in A$ and $P^{-1}(x) \in A$. $\mathscr{C}$ is the class of all invariant sets.
(*d*) X and Y are any two sets; F is a function with a domain X and a range contained in Y. $\mathscr{C}$ is the class of sets of the form $F^{-1}(A)$, where $A \subset Y$.
(*e*) X is R^2. Let a subset C of X be called a *cylinder* if, whenever $(x, y) \in$ C, $(x, \bar{y}) \in$ C for every real number $\bar{y}$. Let $\mathscr{C}$ be the class of cylinders.
(*f*) X is as in part (*e*); $\mathscr{C}$ is the class of sets that can be covered by countably many horizontal lines.

6.4 PROPERTIES OF SET FUNCTIONS

Before turning our attention to measures, we will study some general properties of set functions defined on a σ-field $\mathscr{S}$. Many of these properties have been encountered previously in our study of probability. For example, the following lemma is a slight generalization of Exercise 1.4.3; therefore its proof should offer no difficulty to the reader.

LEMMA 1

Let f be additive and $B \subset A$; then[4]

$$f(A) = f(B) + f(A - B).$$

Because set functions that always take the value of infinity are not very interesting, we will exclude such functions from our study and assume that there is at least one set B such that $f(B)$ is finite.

LEMMA 2

If f is additive, then $f(\phi) = 0$.

[4]Whenever we write $f(B)$, we have assumed that $B \in \mathscr{S}$.

Proof

By additivity, $f(B + \phi) = f(B) + f(\phi)$; the result follows by canceling $f(B)$, which is finite.

The sum of a convergent series of terms of varying sign may depend on the order of the terms. This cannot occur for sums of values of a set function, since we have excluded the value $-\infty$.

THEOREM 1

If f is σ-additive, (A_n) is a sequence of disjoint sets, and $\sum f(A_n)$ is conditionally convergent, then $\sum f(A_n)$ is absolutely convergent.

Proof

Set $A_n^+ = A_n$ if $f(A_n) \geq 0$ and $A_n^+ = \phi$ otherwise; similarly, set $A_n^- = A_n$ if $f(A_n) \leq 0$ and $A_n^- = \phi$ if $f(A_n) > 0$. Then $f(\sum A_n^+) = \sum f(A_n^+)$ and $f(\sum A_n^-) = \sum f(A_n^-)$; the terms of each series are of constant sign. Since $\sum f(A_n)$ is conditionally convergent, either both $\sum f(A_n^+)$ and $\sum f(A_n^-)$ converge or both diverge. The last series converges, since the value $-\infty$ is excluded. Hence both series converge, and so $\sum f(A_n)$ is absolutely convergent.

LEMMA 3

If f is additive, $f(A)$ is finite, and $A \supset B$, then $f(B)$ is finite.

COROLLARY

1. If f is additive and $f(X)$ is finite, then f is finite.
2. If $f \geq 0$, f is additive, and $A \supset B$, then $f(A) \geq f(B)$.

The proofs, which follow from Lemma 1, are left as an exercise (see Exercise 1).

DEFINITION 1

A set function is *subadditive* if for any $E_j \in \mathscr{C}$,

$$f\left(\bigcup_{j=1}^{n} E_j\right) \leq \sum_{j=1}^{n} f(E_j).$$

It is *σ-subadditive* if the last inequality is valid, with n replaced by ∞.

This property of the probability function has been used repeatedly (c.f., Exercise 1.4.6 and Example 5.9.2); Lemma 4 justifies these applications.

LEMMA 4

If f is additive (σ-additive) and $f \geq 0$, then f is subadditive (σ-subadditive).

The result follows from the fact that a union can be written as a disjoint sum, by using the additivity of f and part 2 of the corollary to Lemma 3.

By analogy with continuity on the real line, we can define continuity for set functions.

DEFINITION 2

A function f is said to be *continuous from below* if $f(\lim A_n) = \lim f(A_n)$ for every increasing sequence A_n. It is *continuous from above* if $f(\lim A_n) = \lim f(A_n)$ for every decreasing sequence A_n such that $f(A_n)$ is *finite* for some n. If f is continuous from above and below it is said to be *continuous*.

Of course, continuity at ϕ reduces to continuity from above at ϕ. The following example shows that we cannot drop the restriction that $f(A_n) < \infty$ for some n and still have f continuous from above.

EXAMPLE 1

Let $A_n = (n, \infty)$; then $\lim A_n = \phi$. If $f(A_n) =$ length of A_n, then $f(A_n) = \infty$ for all n and $\lim f(A_n) \neq f(\lim A_n)$.

In Chapter 5, we used the fact that $P(\lim E_n) = \lim P(E_n)$. The following theorem, which holds for any σ-additive set function, justifies this usage.

THEOREM 2

A σ-additive set function f is finitely additive and continuous.

Proof

Since f is σ-additive, it is finitely additive (ϕ belongs to the domain of f).

We will first show that f is continuous from below. Let $A_n \uparrow$; then we have

$$\lim A_n = A_1 + (A_2 - A_1) + (A_3 - A_2) + \cdots + (A_n - A_{n-1}) + \cdots.$$

Therefore,

$$\begin{aligned} f(\lim A_n) &= f(A_1) + f(A_2 - A_1) + \cdots + f(A_n - A_{n-1}) + \cdots \\ &= \lim\,(f(A_1) + f(A_2 - A_1) + \cdots + f(A_n - A_{n-1})) = \lim f(A_n). \end{aligned}$$

The last equality follows from the additivity of f.

Now if $A_n \downarrow$ and $f(A_N)$ is finite, then $A_N - A_n \uparrow$ for $n \geq N$ and, since $f(\lim A_n)$ is finite by Lemma 3, the foregoing result yields

$$f(A_N) - f(\lim A_n) = f[\lim (A_N - A_n)] = \lim f(A_N - A_n)$$
$$= f(A_N) - \lim f(A_n).$$

This shows that f is continuous from above, since we can cancel the finite term $f(A_N)$.

THEOREM 3

If a set function f is additive and continuous from below, then it is σ-additive.

Proof

Since f is additive, we have, for all n, $\sum_{k=1}^{n} f(E_k) = f\left(\sum_{k=1}^{n} E_k\right)$. Taking limits, and noting that $\sum_{k=1}^{n} E_k$ increases, yields $\lim f(\sum_{k=1}^{n} E_k) = f(\lim \sum_{k=1}^{n} E_k)$; it follows that

$$\lim \sum_{k=1}^{n} f(E_k) = f(\lim \sum_{k=1}^{n} E_k),$$

which states that f is σ-additive.

THEOREM 4

If a set function f is additive, finite, and continuous at ϕ, then it is σ-additive.

Proof

The σ-additivity of f follows from

$$f(\sum E_k) = f\left(\sum_{k=1}^{n} E_k\right) + f\left(\sum_{k=n+1}^{\infty} E_k\right) = \sum_{k=1}^{n} f(E_k) + f\left(\sum_{k=n+1}^{\infty} E_k\right)$$

and

$$f\left(\sum_{k=n+1}^{\infty} E_k\right) \to f(\phi) = 0.$$

A continuous real function attains its extrema on a closed bounded set. An analogous result holds for a continuous σ-additive set function f on a σ-field. The proof of this statement requires the concept of positive and negative sets.

DEFINITION 3

A set P is *positive* (*negative*) with respect to a σ-additive set function f defined on a σ-field $\mathscr{S}$ if for every $E \in \mathscr{S}, EP \in \mathscr{S}$ and $f(EP) \geq 0$ ($f(EP) \leq 0$).

Of course, positive and negative sets exist—for example, the empty set. The above definition implies that f is defined on such sets (Let $E = X$), and that it has the same sign on every subset of such sets on which it is defined. Thus it follows that the intersection and difference of two negative (positive) sets is negative (positive). Since any union can be written as a disjoint union, we see that the union of negative (positive) sets is a negative (positive) set.

LEMMA 5

If f is σ-additive on a σ-field $\mathscr{S}$ and $f(E) \leq 0$ ($f(E) \geq 0$), there exists a negative (positive) subset N (P) of E such that $f(N) \leq f(E)$ ($f(P) \geq f(E)$).

Proof

If E is negative, let $N = E$. Now suppose that E is not negative; thus there exists at least one $P \subset E$ such that $f(P) > 0$. The set of all such P can be divided into mutually disjoint classes $\mathscr{C}_n$ defined by

$$\mathscr{C}_1 = \left(P : f(P) \geq \frac{1}{2}\right) \qquad \mathscr{C}_n = \left(P : \frac{1}{n} > f(P) \geq \frac{1}{n+1}\right).$$

$$(n = 2, 3, \ldots)$$

Choose some P_1 from the first nonempty class $\mathscr{C}_j$ (at least one such class exists by hypothesis). Since $f(E - P_1) = f(E) - f(P_1) < f(E) \leq 0$, the argument just applied to E can be repeated for $E - P_1$. Either $E - P_1$ is negative and the theorem follows or $E - P_1$ contains a set $\bar{P}$ such that $f(\bar{P}) > 0$. In the latter situation, we choose a P_2 from the same class from which P_1 was chosen or, if that class is empty, from the first nonempty class $\mathscr{C}_k$ that follows. This process is repeated; either it comes to a halt, in which case the theorem follows, or we obtain an infinite sequence (P_i).

From any class $\mathscr{C}_k$ we can only choose a finite number of P_i. Suppose that $\mathscr{C}_k$ contained an infinite number of P_i; then $f(\sum P_i) = \infty$, since $f(P_i) \geq 1/(k+1)$ and the P_i are disjoint. This is impossible, since for every subset $A \subset E, f(A) < \infty$ by Lemma 3. Therefore, no $\mathscr{C}_k$ contains any P disjoint from every P_i of our sequence (by construction). Thus if $A \in \mathscr{S}$ and A is a subset of $N = E - \sum_{i=1}^{\infty} P_i$ then $f(A) \leq 0$. It is now clear that $N \in \mathscr{S}, N$ is negative, and $f(N) < f(E)$. The proof is similar for the existence of a positive subset P when $f(E) \geq 0$.

THEOREM 5

If f is σ-additive on a σ-field $\mathscr{S}$, then there exist sets C and D of $\mathscr{S}$ such that $f(C) = \sup f$ and $f(D) = \inf f$.

Proof

We will prove that the set D exists and leave the proof of the existence of C as an exercise. Using Lemma 5, it is only necessary to consider $\inf f(N)$ over negative sets, since $f(\phi) = 0$. Let $N_i \in \mathscr{S}$ be a sequence of negative sets such that $\lim f(N_i) = \inf f$. The set $D = \bigcup_{k=1}^{\infty} N_k$ is a negative set that belongs to $\mathscr{S}$. Since $f(D) \leq f(N_i)$ for all values of i, it follows that $f(D) = \inf f$.

COROLLARY

If f, whose domain is a σ-field, is σ-additive, then f is bounded below.

DEFINITION 4

The disjoint sets A and B form a *Hahn decomposition* of X with respect to f if $X = A + B$, where A is positive and B is negative.

LEMMA 6

If f is σ-additive on a σ-field, a Hahn decomposition of X exists.

Proof

Let $B = D$ be the set on which f is minimal and $A = D'$. The result follows if we can show that D' is positive. Suppose that for some $E \in \mathscr{S}$ we have $f(ED') < 0$; then

$$f(D + ED') = f(D) + f(ED') < f(D),$$

which is impossible.

The Hahn decomposition of X is not unique, as Example 1 illustrates.

EXAMPLE 1

E	X	ϕ	A	B	A'	B'	$A \cup B$	$A' \cup B$	$A \cup B'$	$A' \cup B'$	$A'B'$	AB'	$A'B$	AB
$f(E)$	4	0	−6	−6	10	10	−6	4	4	10	10	0	0	−6

$X = A + A'$ and $X = B + B'$ are Hahn decompositions; $X = (A' \cup B) + AB'$ is not a Hahn decomposition since $A' \cup B$ contains the negative set B.

An examination of this example reveals that $f(AE) = f(BE)$ and $f(B'E) = f(A'E)$ for every set E. This result holds in general.

LEMMA 7

If $X = A_1 + B_1 = A_2 + B_2$ are two Hahn decompositions of X, then for every $E \in \mathcal{S}, f(A_1E) = f(A_2E)$ and $f(B_1E) = f(B_2E)$, where the A_i are the positive sets.

Proof

Since $X = A_1 + B_1$, $EA_2 = EA_1E_2 + EA_2B_1$. If we note that $EA_2B_1 \subset A_2$, it follows that $f(EA_2B_1) \geq 0$ since A_2 is a positive set; similarly, $f(EA_2B_1) \leq 0$. The last two inequalities imply that $f(EA_2B_1) = 0$ and so $f(EA_2) = f(EA_1A_2)$. By symmetry, $f(EA_1) = f(EA_1A_2)$, which shows that $f(EA_2) = f(EA_1)$. The remainder of the theorem can be proved in a similar manner.

EXERCISES

1. Prove Lemma 3 and its corollary.
2. Find some examples of σ-subadditive set functions that are not additive.
3. Prove that if $f > 0$ and if it is σ-additive, then it also is σ-subadditive.
4. Finish the proof of Theorem 5.
5. Let $X = A_1 + B_1 = A_2 + B_2$ be two Hahn decompositions of X, where A_i are the positive sets. Show that $f(A_i) = f(A_1A_2)$ and $f(A_iB_j) = 0$.
6. Suppose that $(A_\alpha + B_\alpha)$ are all the Hahn decompositions of X, where the A_α are positive. Let $N = (\cup A_\alpha) \cap (\cup B_\alpha)$. Show that $f(N) = 0$.
7. Show that the Hahn decomposition theorem (Lemma 6) remains valid if f is defined on a σ-ring (with suitable modification of the definition of positive and negative sets).

6.5 MEASURE

We will begin by defining the term "measure."

DEFINITION 1

A nonnegative, σ-additive, set function μ defined on a field $\mathcal{F}$ is called a *measure*. If a nonnegative set function is only additive, it is sometimes called a *content*.

Since a measure is a set function, the following results are immediate consequences of the theorems in Section 6.4. The reader should try to derive these results without referring back to the corresponding theorems. All sets appearing in the theorems below belong to $\mathscr{F}$.

1. If $B \supset A$, then $\mu B = \mu(B - A) + \mu A$.
2. If $B \supset A$, then $\mu B \geq \mu A$.
3. If $\mu X < \infty$, then μ is finite.
4. $\mu(\phi) = 0$.
5. $\mu\left(\bigcup_{j=1}^{\infty} E_j\right) \leq \sum_{j=1}^{\infty} \mu(E_j)$.
6. If $A_n \uparrow A$, then $\mu A_n \to \mu A$.
7. If $A_n \downarrow A$ and μA_n is finite for some n, then $\mu A_n \to \mu A$.

The following theorem is a generalization of statements 6 and 7.

THEOREM 1

Let μ be a measure defined on a σ-field $\mathscr{S}$, and let $A_n \in \mathscr{S}$. Then $\liminf \mu A_n \geq \mu(\liminf A_n)$; if μ is finite, then $\limsup \mu A_n \leq \mu(\limsup A_n)$.

Proof

Let $B_n = \bigcap_{i=n}^{\infty} A_i$; then $B_n \uparrow$ and $\lim B_n = \liminf A_n$. Using statement 6, $\mu(\liminf A_n) = \mu(\lim B_n) = \lim \mu B_n$. However, $\mu B_n \leq \mu A_n$ and it follows that $\lim \mu B_n \leq \liminf \mu A_n$, which completes the proof.

The second statement is left as an exercise for the reader.

Actually, we can discuss measures without having to consider general set functions, because any σ-additive set function on a σ-field can be expressed as the difference of two measures, as the following theorem shows. This also explains why a *signed measure* is defined sometimes as an extended σ-additive set function and at other times as the difference of two measures.

THEOREM 2

Let f be σ-additive on a σ-field $\mathscr{S}$ and let $f^+(E) = f(ED')$ and $f^-(E) = -f(ED)$, where D is the set on which f is minimal. Then f^+ and f^- are measures and $f(E) = f^+(E) - f^-(E)$ for every $E \in \mathscr{S}$. Furthermore, if f is finite or σ-finite, then so are f^+ and f^-; in fact, one of these measures is always finite.

Proof

From Lemma 4.7 it follows that f^+ and f^- are unambiguously defined—that is, we could use any Hahn decomposition to define them. They are clearly nonnegative and, since

$$f^+(\sum E_n) = f(\sum E_n D') = \sum f(E_n D') = \sum f^+(E_n),$$

they are σ-additive. The equation $f = f^+ - f^-$ follows from the definition of f^+ and f^-. If every $E \in \mathscr{S}$ is the union of sets on which f is finite, the same is true of f^+ and f^-. The fact that f takes on only one infinite value implies that one of f^+ or f^- is finite. In fact, according to our convention, f^- is always finite.

The measures f^+ and f^- are called the *upper* and *lower variations* of f. The set function $|f|$ defined by $|f| = f^+ + f^-$ is called the *total variation* of f. The representation of f as the difference of its upper and lower variations is called the *Jordan decomposition* of f.

One of the main questions left unanswered in Chapter 5 involved the possibility of extending a probability function defined on a field $\mathscr{F}$ to the σ-field generated by $\mathscr{F}$. Here we will study the more general problem of extending a measure. We will show that a measure μ on a field $\mathscr{F}$ can be extended to a measure on the minimal σ-field over $\mathscr{F}$. This extension of μ is unique and is σ-finite if μ is σ-finite. We will prove this extension theorem by means of an intermediate extension that preserves the properties of measure given in the definition below.

DEFINITION 2

A nonnegative set function μ° on the class $\mathscr{S}(X)$ of all subsets of the space X is called an *outer measure* if it is σ-subadditive, if it is increasing,[5] (i.e., if $A \subset B$, then $\mu^\circ A \le \mu^\circ B$), and if $\mu^\circ(\phi) = 0$.

Before showing how outer measures arise, we will give some examples of these functions.

EXAMPLE 1

1. Let X be an arbitrary set and define μ° on the class of all subsets of X by $\mu^\circ E = I_E(x_0)$, where x_0 is a fixed point of X. In general, any measure μ, defined on $\mathscr{S}(X)$, is also an outer measure.
2. $X = \{x, y\}$, $\mu^\circ(\phi) = 0$, $\mu^\circ(x) = \mu^\circ(y) = 1$, and $\mu^\circ X = 1$. Note that if we set $\mu^\circ(x) = 2$, μ° will not be an outer measure, since it is not monotone.
3. For any $E \subset X$, $\mu^\circ(E) = 1$ if $E \ne \phi$ and $\mu^\circ \phi = 0$.
4. Let X be the space of 16 points arranged in a square array of four rows each having four points, and let $\mu^\circ E$ be the number of columns that contain

[5] A synonym for "increasing" is "monotone."

at least one point of E. The outer measure μ° is not additive; consider the set E composed of two disjoint sets E_1 and E_2, each of which has one point P_1 in the same column ($P_1 \neq P_2$). Then $\mu^\circ E = 1 = \mu^\circ E_1 = \mu^\circ E_2$.

The extension of μ to some set not in $\mathscr{F}$ must in some way involve the sets of $\mathscr{F}$, since we only know the value of μ on such sets. As a first attempt, we extend μ to countable unions of sets in $\mathscr{F}$. To preserve σ-additivity, we write such a union as a disjoint sum, say $\sum E_i$ and define $\mu(\sum E_i)$ as $\sum \mu E_i$. However, the class of all countable unions of sets in $\mathscr{F}$ is not a σ-field, since it may not be closed under complementation. Hence, in general, μ has not been extended to a σ-field containing $\mathscr{F}$. To define μ on a wider class of sets, we might try to approximate such sets by unions $\cup F_i$ of sets in $\mathscr{F}$. Since we want the closest possible approximation, we attach the number $\inf \sum \mu F_i$ for all possible coverings (see Theorem 3). Since $X \in \mathscr{F}$, every subset will have some number assigned to it; this leads to the same value of μ for sets of $\mathscr{F}$ and countable sums of such sets. In general, the resulting function will no longer be a measure. The situation is described as follows.

THEOREM 3

If μ is a measure on a field $\mathscr{F}$ and if for every $E \subset X$

$$\mu^\circ E = \inf\left(\sum_{n=1}^{\infty} \mu E_n : E \subset \bigcup_{n=1}^{\infty} E_n,\ E_n \in \mathscr{F}\right),$$

then μ° is an extension of μ to an outer measure on $\mathscr{S}(X)$. If μ is (finite) σ-finite, then μ° is (finite) σ-finite.

Proof

If $E \in \mathscr{F}$, then $E \subset E$, which implies that $\mu^\circ E \leq \mu E$. Now if $E_n \in \mathscr{F}$ and $E \subset \bigcup_{n=1}^{\infty} E_n$, then $E = \bigcup_{n=1}^{\infty} EE_n$. Accordingly, $\mu E \leq \sum \mu(EE_n) \leq \sum \mu E_n$; therefore, $\mu E \leq \mu^\circ E$. This proves that μ° is an extension of μ.

Let (A_j) be a countable class of sets. From the definition of μ°, it follows that for every A_j there is a covering (A_{jk}) in $\mathscr{F}$ such that $\sum_{k=1}^{\infty} \mu A_{jk} \leq \mu^\circ A_j + \epsilon/2^j$, for arbitrary $\epsilon > 0$. Since $\cup A_j \subset \bigcup_{j,k} A_{jk}$, it follows that

$$\mu^\circ(\cup A_j) \leq \sum_{j,k} \mu A_{jk} \leq \sum_j \mu^\circ A_j + \epsilon.$$

The arbitrariness of ϵ implies σ-subadditivity.

Finally, suppose that μ is σ-finite and let $E \in \mathscr{S}(X)$. Then by definition, $X = \bigcup_{j=1}^{\infty} A_j$, where $A_j \in \mathscr{F}$ and $\mu A_j < \infty$. It follows that $E = \bigcup_{i=1}^{\infty} EA_i$ and

$\mu^\circ EA_i \le \mu^\circ A_i = \mu A_i < \infty$, since it is obvious that μ° is monotone. Hence μ° is σ-finite.

Although we have succeeded in extending our set function, we have not obtained a measure; instead, we have only obtained an outer measure. In general, outer measures are not countably or even finitely additive (cf., part 4 of Example 1). To satisfy the requirement of additivity, we must single out those sets A which split every other set additively.

DEFINITION 3

A set A is called μ°*-measurable* if for every $D \subset X$,

$$\mu^\circ D = \mu^\circ(AD) + \mu^\circ(A'D).$$

Since $D = AD + A'D$ and the outer measure is σ-subadditive, we have $\mu^\circ(D) \le \mu^\circ(AD) + \mu^\circ(A'D)$; therefore, to prove that a set A is μ°-measurable, we need only show that

$$\mu^\circ(D) \ge \mu^\circ(AD) + \mu^\circ(A'D).$$

The latter statement is often taken as the definition of μ°-measurability. A few examples may help to illuminate this concept.

EXAMPLE 2

1. In part 1 of Example 1, μ° is a measure; therefore, every set is μ°-measurable.
2. Only X and ϕ are μ°-measurable in part 2 of Example 1.
3. In part 4 of Example 1, a set E is μ°-measurable if and only if $x \in E$; when this is true, the entire column that includes x is contained in E.
4. In general, X and ϕ are μ°-measurable; in addition, if E is μ°-measurable, so is E'.

Since μ°, the extension of μ, is additive on $\mathscr{F}$, we would expect the sets of $\mathscr{F}$ to be μ°-measurable.

LEMMA 1

If $A \in \mathscr{F}$, then A is μ°-measurable.

Proof

By definition of the outer measure of a set $D \subset X$, we can find a covering $E_j \in \mathscr{F}$ such that

$$\mu^\circ(D) + \epsilon \ge \sum_{j=1}^{\infty} \mu E_j = \sum_{j=1}^{\infty} \mu(E_j A) + \sum_{j=1}^{\infty} \mu(E_j A'). \tag{1}$$

Since μ° is an extension of μ, monotone, and σ-subadditive, $\sum_{j=1}^{\infty} \mu(E_j A) \geq \mu^\circ(\bigcup_{j=1}^{\infty} E_j A) \geq \mu^\circ(DA)$. A similar result holds for $\sum \mu(E_j A')$; the following inequality is obtained from Equation (1):

$$\mu^\circ D + \epsilon \geq \mu^\circ(DA) + \mu^\circ(DA').$$

The result follows from the arbitrariness of ϵ.

THEOREM 4

If μ° is an outer measure, then the class $\mathscr{F}^\circ$ of μ°-measurable sets is a σ-field and μ° is a measure on $\mathscr{F}^\circ$.

Proof

First we must prove that $\mathscr{F}^\circ$ is a field. If $A \in \mathscr{F}^\circ$, then $A' \in \mathscr{F}^\circ$, since the definition of μ°-measurability is symmetrical in A and A'. Thus we need only show that $\mathscr{F}^\circ$ is closed under the operation of intersection. Let $A, B \in \mathscr{F}^\circ$; from the definition of μ°-measurability,

$$\mu^\circ(D) = \mu^\circ(AD) + \mu^\circ(A'D) \tag{2}$$

for an arbitrary set D. Since B is also μ°-measurable,

$$\mu^\circ(AD) = \mu^\circ(ADB) + \mu^\circ(ADB') \tag{3}$$

$$\mu^\circ(A'D) = \mu^\circ(A'DB) + \mu^\circ(A'DB'). \tag{4}$$

Substituting (3) and (4) in (2) and using the fact that μ° is subadditive yields

$$\begin{aligned} \mu^\circ(D) &\geq \mu^\circ(ABD) + \mu^\circ(ADB' \cup A'DB \cup A'DB') \\ &= \mu^\circ(ABD) + \mu^\circ((AB)'D). \end{aligned}$$

Finite additivity is proved as follows. Let $A, B \in \mathscr{F}^\circ$ and let A and B be disjoint; then

$$\mu^\circ(A + B) = \mu^\circ((A + B)A) + \mu^\circ((A + B)A') = \mu^\circ A + \mu^\circ B.$$

Since $\mu^\circ(A) \geq \mu^\circ(\phi) = 0$, μ° is a content on $\mathscr{F}^\circ$.

Next we show that if $A_n \in \mathscr{F}^\circ$ are disjoint, then $A = \sum A_n \in \mathscr{F}^\circ$. Since $B_n = \sum_{k=1}^{n} A_k \in \mathscr{F}^\circ$, we have $\mu^\circ B_n D = \sum_{k=1}^{n} \mu^\circ(A_k D)$. Therefore, as $B'_n \supset A'$,

$$\mu^\circ(D) = \mu^\circ(E_n D) + \mu^\circ(B'_n D) \geq \sum_{k=1}^{n} \mu^\circ(A_k D) + \mu^\circ(A'D). \tag{5}$$

Letting $n \to \infty$ gives

$$\mu^\circ(D) \geq \sum_{n=1}^{\infty} \mu^\circ(A_n D) + \mu^\circ(A'D) \geq \mu^\circ(AD) + \mu^\circ(A'D),$$

by using the σ-subadditivity of μ°. This last inequality implies that $\sum_{n=i}^{\infty} A_n \in \mathscr{F}^\circ$, and since any union of sets can be written as a disjoint union of sets, it follows that $\mathscr{F}^\circ$ is a σ-field.

Finally, we will show that μ° is σ-additive on $\mathscr{F}^\circ$. First, replace D by A in equation (5) and let $n \to \infty$; then $\mu^\circ(A) \geq \sum_{n=1}^{\infty} \mu^\circ A_n$. The reverse inequality is always true; thus $\mu^\circ(\sum A_n) = \sum \mu^\circ A_n$.

We are now in a position to prove an extension theorem formulated by Carathéodory.[6] Note that a probability measure satisfies all of the hypotheses of the theorem.

THEOREM 5

A measure μ on a field $\mathscr{F}$ can be extended to a measure on the minimal σ-field over $\mathscr{F}$. If μ is finite (σ-finite), then the extension is unique and is finite (σ-finite).

Proof

By Lemma 1, we see that $\mathscr{F}^\circ \supset \mathscr{F}$; since $\mathscr{F}^\circ$ is a σ-field, it contains $\mathscr{S}(\mathscr{F})$. The first part of the theorem follows from Theorem 4 by restricting μ° to $\mathscr{S}(\mathscr{F})$. Note that the existence of the extension of μ follows without the assumption of σ-finiteness. The fact that μ is finite (σ-finite) is a consequence of Theorem 3.

We first prove uniqueness under the assumption that μ is finite. Suppose that μ_1 is another measure on $\mathscr{S}(\mathscr{F})$ such that $\mu_1(E) = \mu(E)$ whenever $E \in \mathscr{F}$. Let $\mathscr{M}$ be the class of all sets E in $\mathscr{S}(\mathscr{F})$ for which the measures coincide. If (E_n) is a monotone sequence of sets in $\mathscr{M}$, then by properties 6 and 7 (since $\mu_1 E_n = \mu E_n$ is finite),

$$\mu_1(\lim E_n) = \lim (\mu_1 E_n) = \lim \mu E_n = \mu(\lim E_n)$$

and so $\lim E_n \in \mathscr{M}$. Hence $\mathscr{M}$ is a monotone class; since $\mathscr{M} \supset \mathscr{F}$, $\mathscr{M} \supset \mathscr{S}(\mathscr{F})$ as a consequence of the corollary to Theorem 3.3.

In the σ-finite case we proceed as follows. Let $X = \sum_{i=1}^{\infty} E_i$, where $E_i \in \mathscr{F}$ and $\mu E_i < \infty$. Since μ is finite on $E_i\mathscr{F}$, we can prove that the measure is unique on each of the classes $E_i \cap \mathscr{S}(\mathscr{F})$ by the procedure used above. We merely take the space to be E_i and our field to be $E_i\mathscr{F}$. Then the extension μ° is unique on the σ-field $\mathscr{S}_{E_i}(E_i\mathscr{F})$ generated by $E_i\mathscr{F}$, which by Lemma 3.2 is $\mathscr{S}(\mathscr{F}) \cap E_i$.

[6]Constantin Carathéodory (1873–1950), a Greek mathematician.

The following example shows that, in general, the extension is not unique if the measure μ is not finite.

EXAMPLE 3

Let $\mathscr{F}$ be a field such that all sets in $\mathscr{F}$ except ϕ have a countable (not finite) number of elements and $\mathscr{S}(\mathscr{F})$ consists of all subsets of a countable space X. We define $\mu_2 E = 2\mu_1 E$. If E has an infinite number of points, $\mu_1 E = \infty$; otherwise, $\mu_1 E$ is the number of points. Then μ_1 and μ_2 agree on $\mathscr{F}$ but not on $\mathscr{S}(\mathscr{F})$.

The field $\mathscr{F}$ defined here is generated by the class of all intervals $(x : a < x \leq b;\ x \text{ is rational})$, where $-\infty < a < b < \infty$ and the space X is composed of the rational numbers.

DEFINITION 4

The measure μ is called *complete* if the conditions $E \in \mathscr{S}, F \subset E$, and $\mu E = 0$ imply that $F \in \mathscr{S}$.

If $F \in \mathscr{S}$ and $F \subset E$, we have $\mu F \leq \mu E$; therefore, $\mu F = 0$ if $\mu E = 0$. Conversely, if we define the measure of any subset of a set of measure zero as zero we can obtain a complete measure known as the *completion* of μ.

THEOREM 6

Let μ be a measure on a σ-field $\mathscr{S}$, and $\bar{\mathscr{S}}$ (*completion of* $\mathscr{S}$) be the class of all sets of the form $E \cup N$, where $E \in \mathscr{S}$ and N is a subset of a set of measure zero of $\mathscr{S}$. Then $\bar{\mathscr{S}}$ is a σ-field and the set function $\bar{\mu}$ defined by $\bar{\mu}(E \cup N) = \mu(E)$ is a complete measure.

Proof

$\bar{\mathscr{S}}$ is closed under the formation of countable unions, since $\mathscr{S}$ is a σ-field and a countable union of sets of measure zero is a set of measure zero. The proof that $\mathscr{S}$ is closed under complementation is more difficult. We first note that the intersection of two sets of $\bar{\mathscr{S}}$ belong to $\bar{\mathscr{S}}$ and that $\mathscr{S} \subset \bar{\mathscr{S}}$. Now let $M \in \mathscr{S}, \mu M = 0$, and $N \subset M$; then $N' = M' + (M - N)$. It follows by definition that $N' \in \bar{\mathscr{S}}$ for $M - N$ is a subset of a set of measure zero and $M' \in \mathscr{S}$. Finally, $(E \cup N)' = E'N' \in \bar{\mathscr{S}}$, since $\bar{\mathscr{S}}$ is closed under intersection.

We must prove that $\bar{\mu}$ is uniquely defined. Suppose that $F = E \cup N = E_1 \cup N_1$, where $E, E_1 \in \mathscr{S}$ and N and N_1 are each contained in sets of measure zero. Then $\mu E_1 = \mu EE_1 + \mu(E_1 - E)$, since $E_1 - E \in \mathscr{S}$. Since

$E_1 - E \subset F - E \subset N$, which is contained in a set of measure zero, then $\mu(E_1 - E) = 0$. Hence $\mu E_1 = \mu EE_1$; by symmetry, $\mu E = \mu EE_1$. Therefore, $\bar{\mu}F = \mu E_1 = \mu E$ is uniquely defined.

The σ-additivity of $\bar{\mu}$ follows from the definition by using the fact that a countable union of sets of measure zero is a set of measure zero. The remaining properties of measure are easy to verify.

Suppose that F is a subset of a set of $\bar{\mu}$ measure zero. Then $F \subset E \cup N$ and $\bar{\mu}(E \cup N) = \mu E = 0$; hence F is a subset of a set of μ measure zero. The completeness of $\bar{\mu}$ is a consequence of the fact that $\bar{\mathscr{S}}$ contains all the subsets of sets of measure zero in $\mathscr{S}$.

The extension theorem (Theorem 5) provides us automatically with extensions to complete measures, as Theorem 7 shows.

THEOREM 7

Suppose that μ° is an outer measure on $\mathscr{S}(X)$ and $\mathscr{F}^\circ$ is the class of all μ°-measurable sets. Then every set of outer measure zero belongs to $\mathscr{F}^\circ$ and the restriction of μ° to $\mathscr{F}^\circ$ is a complete measure.

Proof

If $E \subset X$ and $\mu^\circ E = 0$, then for every $A \subset X$, $\mu^\circ A \geq \mu^\circ AE + \mu^\circ AE'$, since $\mu^\circ AE = 0$ and $\mu^\circ A \geq \mu^\circ AE'$. This means that $E \in \mathscr{F}^\circ$.

Now let $E \in \mathscr{F}^\circ$, $\mu^\circ E = 0$, and $F \subset E$. Then $\mu^\circ F = 0$ by the monotonicity of μ° and $F \in \mathscr{F}^\circ$ by the first part of this theorem.

Let us denote the domain of $\bar{\mu}$, the completion of the extension of μ to $\mathscr{S}(\mathscr{F})$, by $\bar{\mathscr{S}}(\mathscr{F})$. Theorem 7 implies that $\bar{\mathscr{S}}(\mathscr{F}) \subset \mathscr{F}^\circ$ and that $\bar{\mu}$ and μ° coincide on $\bar{\mathscr{S}}(\mathscr{F})$. Actually, it can be shown that $\bar{\mathscr{S}}(\mathscr{F}) = \mathscr{F}^\circ$ (see Exercise 14).

EXERCISES

1. Let f be a σ-additive set function whose domain is a σ-field. Show that $f^+(A) = \sup_{B \subset A} f(B)$ and $f^-(A) = -\inf_{B \subset A} f(B)$. [*Hint:* $B \subset A$ implies that $f(B) \leq f(BD') \leq f(BD') + f((A - B)D') = f(AD')$.]

2. Let f be an additive set function on a field $\mathscr{F}$. We define f^+ and f^- as in Exercise 1, where $A, B \in \mathscr{F}$. If f is bounded, prove that it is a signed content. [*Hint:* Use the definition, showing first that $f^\pm$ are bounded and additive and that $\sup(A + B) \geq \sup A + \inf B$].

3. Show that a countable union of sets of measure zero is a set of measure zero.

4. Complete the proof of Theorem 1.

5. Is the extension theorem (Theorem 7) valid for σ-finite signed measures?

6. (a) Show that if $\bar{\mu}$ is the completion of μ and $\mu E = 0$, then $\bar{\mu} E = 0$.
(b) Prove that if F is a subset of a set of measure zero, then $\bar{\mu} F = 0$.
(c) Verify that $\bar{\mu} E_1 = \bar{\mu} E_2$ if and only if $\bar{\mu}(E_1 \Delta E_2) = 0$.

7. Formulate and prove a corresponding extension theorem for rings.

8. Two sets E and F are almost equal, denoted by $E \sim F$, if and only if $\mu(E - F) = \mu(F - E) = 0$.

(a) Show that if $E \sim F$, then $\mu^\circ E = \mu^\circ(EF) = \mu^\circ F$.
(b) Verify that if $E \sim F$ and $F \sim H$, then $E \sim H$.

9. A set $O = \bigcup_1^\infty E_n$, where $E_n \in \mathscr{F}$ is a field, is called a σ-set. A set $F = O' = \bigcap_1^\infty E'_n$, where $E'_n \in \mathscr{F}$ is a field, is called a δ-set.

(a) Verify that a σ-set O can be written in the form $\sum_{n=1}^\infty A_n$, with $A_n \in \mathscr{F}$.
(b) Show that the union of any number of σ-sets is a σ-set and that finite intersections of σ-sets are σ-sets.

10. (a) Show that if A is an arbitrary set, then

$$\mu^\circ A = \inf (\mu O : A \subset O;\ O \text{ is a } \sigma\text{-set}).$$

(b) Deduce that if A is an arbitrary set, then

$$\mu^\circ A = \inf \mu E$$

over all measurable sets $E \supset A$.

11. A set $O_\delta = \bigcap_{n=1}^\infty O_n$, where all O_n are σ-sets, is called a $\sigma\delta$-set. Similarly, a set $F_\sigma = \bigcup_1^\infty F_n$, where all F_n are δ-sets, is called a $\delta\sigma$-set.

(a) Prove that all $\delta\sigma$- and $\sigma\delta$-sets are measurable.
(b) Show that if $\mu^\circ A < \infty$, there exists a $\sigma\delta$-set $O_\delta \supset A$ such that $\mu(O_\delta) = \mu^\circ(A)$. Moreover, if A is measurable, then $A \sim O_\delta$.
(c) Show that if A is measurable and μ is σ-finite, then the restriction $\mu^\circ A < \infty$ in Exercise 11(b) can be removed.

12. For the remaining problems, assume that μ is σ-finite.

(a) Show that if A is measurable and $\epsilon > 0$, there exists a σ-set O and a δ-set F such that $F \subset A \subset O$, $\mu(O - A) < \epsilon$, and $\mu(A - F) < \epsilon$.
(b) Deduce that if A is measurable, there exists a $\sigma\delta$-set O_δ and a $\delta\sigma$-set F_σ such that $F_\sigma \subset A \subset O_\delta$ and $F_\sigma \sim A \sim O_\delta$.

13. Prove that a set A is measurable if and only if, given $\epsilon > 0$, there exist two measurable sets A_1 and A_2 such that $A_1 \subset A \subset A_2$ and $\mu(A_2 - A_1) < \epsilon$.

14. Verify the last assertion of this section. [*Hint:* Use Exercise 10(b).]

6.6 LEBESGUE–STIELTJES MEASURES

A set of axioms for infinite probability space was formulated in Chapter 5. In the same chapter we showed that it was possible to solve many practical problems using density or distribution functions. To reconcile the intuitive and axiomatic treatments, we must show that a distribution function defines a probability measure on a certain class of sets and, conversely, that a measure defines a distribution function. The results of Section 6.5 will enable us to carry out this task.

Let $\mathscr{P}$ be the class of all intervals $\{a, b\}$, where $-\infty < a \leq b < \infty$. An interval can be open, half-open, closed, etc. In other words, the brace "}" is either "]" or ")." Since $\mathscr{P}$ is closed under intersection, then by Theorem 3.2, the field $\mathscr{F} = \mathscr{F}(\mathscr{P})$ generated by $\mathscr{P}$ is the class of all finite sums of elements of $\mathscr{P}$ and of intervals of the form $(-\infty, a\}\{b, \infty)$.

DEFINITION 1

The elements of $\mathscr{S} = \mathscr{S}(\mathscr{F})$, the σ-field generated by $\mathscr{F}$, are called *Borel sets*. The elements of $\bar{\mathscr{S}}$, the completion of $\mathscr{S}$, are called *Lebesgue–Stieltjes sets.*

DEFINITION 2

Lebesgue–Stieltjes measures are complete measures, defined on Lebesgue–Stieltjes sets, which always assign finite values to finite intervals. If $\mu(ab] = b - a$, μ is called a *Lebesgue measure.*

Sometimes a measure is called a Lebesgue–Stieltjes measure if it is defined only on Borel sets and if it has the above properties. We will show that Lebesgue–Stieltjes measures are determined by distribution functions. In this chapter, the term "distribution function" is used in a wider sense than it is in probability theory. We do not require that $F(\infty) = 1$ and $F(-\infty) = 0$.

DEFINITION 3

A real-valued function defined on R that is increasing and continuous from the right is called a distribution function.

LEMMA 1

A nonnegative, σ-additive, set function μ defined on $\mathscr{P}$ such that $\mu(a, b] < \infty$ determines a distribution function up to a constant.

Proof

Let c be a fixed real number. Define a real-valued function F by

$$F(x) = \begin{cases} \mu(c, x] & (x \geq c) \\ -\mu(x, c] & (x < c) \end{cases}$$

Suppose that $b \geq a > c$; then $(c, a] + (a, b] = (c, b]$. Hence

(1) $$F(b) = F(a) + \mu(a, b]$$

by the additivity of μ and the definition of F. It is not difficult to verify that Equation (1) remains valid for $b > c > a$ and $c > b > a$. Therefore, F is an increasing function. Setting $b = a + h$ and using the continuity of μ, we find that F is continuous from the right. Hence F is a distribution function.

Let $a > c$ be another fixed real number. Define a function G as in the preceding paragraph; then $F(x) - G(x) = \mu(c, a]$. This completes the proof that μ determines a distribution function up to a constant.

Conversely, a distribution function F determines a set function μ on $\mathscr{P}$ defined by

(2) $$\begin{aligned} \mu(a, b] &= F(b) - F(a) \qquad & \mu(a, b) &= F(b^-) - F(a) \\ \mu[a, b] &= F(b) - F(a^-) \qquad & \mu[a, b) &= F(b^-) - F(a^-). \end{aligned}$$

Obviously, μ is a nonnegative, additive, and real-valued function on $\mathscr{P}$. Actually, μ is σ-additive.

LEMMA 2

The set function μ is σ-additive on $\mathscr{P}$.

Proof

By Theorem 4.4, we need only show that if $A_n \downarrow \phi$ in $\mathscr{P}$, then $\mu A_n \downarrow 0$. By the continuity of F and the additivity of μ, there exists a closed interval $C_n \subset A_n$ such that $\mu A_n \leq \mu C_n + \epsilon 2^{-n}$. Since $\bigcap C_n \subset \bigcap A_n = \phi$, there exists an integer m (by compactness) such that $\bigcap_{n=1}^{m} C_n = \phi$. This implies that

$$A_m = \bigcap_{n=1}^{m} A_n = \bigcap_{n=1}^{m} A_n - \bigcap_{n=1}^{m} C_n = \bigcup_{i=1}^{m} \left(\bigcap_{n=1}^{m} A_n - C_i \right) \subset \bigcup_{i=1}^{m} (A_i - C_i).$$

The finite additivity and subadditivity of μ show that

$$\mu A_m \leq \sum_{i=1}^{m} \mu(A_i - C_i) \leq \sum_{i=1}^{m} (\mu A_i - \mu C_i) \leq \epsilon.$$

Letting $\epsilon \to 0$, we obtain $\lim \mu A_n = 0$.

The set function μ can be uniquely extended from $\mathscr{P}$ to $\mathscr{F}$, the field generated by $\mathscr{P}$, as follows. First, every unbounded interval can be represented as the sum of a countable number of elements of $\mathscr{P}$; this representation is not unique. Consequently, every $F \in \mathscr{F}$, being a finite sum of elements of $\mathscr{P}$ and unbounded intervals, can be expressed as a countable sum of elements of $\mathscr{P}$. If $F = \sum_{i=1}^{\infty} A_i$, where $A_i \in \mathscr{P}$, we define

$$\mu F = \sum_{i=1}^{\infty} \mu A_i.$$

To show that μ is properly defined on $\mathscr{F}$, we must verify that if $F = \sum_{i=1}^{\infty} B_i$, then $\sum \mu B_i = \sum \mu A_i$. From the formula

$$A_i = A_i F = \sum_{j=1}^{\infty} A_i B_j$$

and the σ-additivity of μ on $\mathscr{P}$, it follows that $\mu A_i = \sum_{j=1}^{\infty} \mu(A_i B_j)$. Therefore, $\sum_{i=1}^{\infty} \mu A_i = \sum_{i=1}^{\infty} \sum_{j=1}^{\infty} \mu(A_i B_j)$. Similarly, it can be shown that $\sum \mu B_i = \sum\sum \mu(A_i B_j)$ and so $\sum \mu A_i = \sum \mu B_i$.

LEMMA 3

The set function μ on $\mathscr{P}$ defined by Equation (2) can be uniquely extended to a σ-finite measure on $\mathscr{F}$.

Proof

The σ-finiteness and all other properties, except for σ-additivity, obviously hold. To prove that the set function is σ-additive, let $F = \sum_{i=1}^{\infty} F_i$, where $F_i \in \mathscr{F}$. Each F_i can be expressed as a countable sum of elements of $\mathscr{P}$. Thus, for the results of the paragraph preceding this lemma, the measure of the resulting sum equals the sum of the measures. This immediately implies σ-additivity.

THEOREM 1

The formula

$$F(x) = \begin{cases} \mu(c, x] & (x \geq c;\ -\infty < c < \infty) \\ -\mu(x, c] & (x < c;\ -\infty < c < \infty) \end{cases}$$

establishes a one-to-one correspondence between Lebesgue–Stieltjes measures μ and distribution functions F defined up to a constant (i.e., F and $F + K$ correspond to μ).

Proof

The restriction of a Lebesgue–Stieljes measure μ to $\mathscr{P}$ determines a distribution function according to Lemma 1. Conversely, a distribution function determines a σ-finite measure on $\mathscr{F}$ by Lemma 3. The result follows by applying the extension theorem (5.5) and completing the resulting measure.

EXERCISES

1. A right semiclosed interval of dimension n is the set $(x: a_i < x_i \leq b_i)$, where $-\infty < a_i \leq b_i < \infty$ and $x = (x_1, x_2, \ldots, x_n)$. If $a = (a_1, \ldots, a_n)$ and $b = (b_1, \ldots, b_n)$, then the right semiclosed interval defined above will be denoted by $(a, b]$. The other types of n-dimension intervals are defined similarly. Let $\mathscr{P}$ be the class of all intervals and let $\mathscr{S}$ be the σ-field generated by $\mathscr{P}$. The elements of $\mathscr{S}$ are called (n-dimensional) Borel sets. Show that the Borel class can also be defined as the minimal σ-field containing the class of all open intervals (a, b). Repeat this problem for $(a, b]$, $[a, b)$, and $[a, b]$.

2. The partial difference operator of step $h_i \geq 0$, denoted by Δ_{h_i} or $\Delta_{x_i}^{x_i+h_i}$, is defined by

$$\Delta_{h_i} F(x_1, x_2, \ldots, x_n) = F(x_1, \ldots, x_{i-1}, x_i + h_i, x_{i+1}, \ldots, x_n) - F(x_1, \ldots, x_{i-1}, x_i, x_{i+1}, \ldots, x_n).$$

(a) Show that the partial difference operators are commutative and associative.
(b) Verify that $\Delta_{a_i}^{b_i} \Delta_c^{x_i} F = \Delta_{a_i}^{b_i} F$.
(c) Show that

$$\begin{aligned}\Delta_h F &\equiv \Delta_{h_1} \Delta_{h_2} \cdots \Delta_{h_n} F \\ &= F(x_1 + h_1, x_2 + h_2, \ldots, x_n + h_n) - F(x_1, x_2 + h_2, \ldots, x_n + h_n) \\ &\quad - F(x_1 + h_1, x_2 + h_2, \ldots, x_n) + \cdots \\ &\quad + F(x_1, x_2, x_3 + h_3, \ldots, x_n + h_n) + \cdots \\ &\quad + F(x_1 + h_1, x_2 + h_2, \ldots, x_{n-1}, x_n) + \cdots \\ &\quad + (-1)^n F(x_1, x_2, \ldots, x_n),\end{aligned}$$

where there are $\binom{n}{i}$ elements of the type

$$(-1)^i F(x_1, x_2, \ldots, x_i, x_{i+1} + h_{i+1}, \ldots, x_n + h_n).$$

3. A distribution function on R^n is a real-valued function whose nth-order differences at any point are nonnegative [i.e., $\Delta_{b-a} F(a) \geq 0$, using the notation of Exercise 2(c)], and which is continuous on the right in each variable. Show that all lower-order differences are nonnegative.

4. Extend Theorem 1 to R^n.

5. Show that nonmeasurable sets exist with respect to the Lebesgue measure. (In most elementary examples, the events are the class of all subsets of the space Ω. The above result shows that we can have a probability space in which the events are not the class of all subsets.) (*Hint:* See Reference [6].)

6.7 PRODUCT MEASURES

A probability space $(\Omega, \mathscr{E}, \mathscr{P})$ was defined in Section 1.4. This concept can be generalized.

DEFINITION 1

A *measurable space* $(X, \mathscr{S})$ is a set X and a σ-field $\mathscr{S}$ of subsets of X. A subset E of X is *measurable* if and only if it belongs to the σ-field $\mathscr{S}$.

Note that the fact that a set is measurable has nothing to do with a measure μ. By abuse of notation we often denote a measurable space, the doublet $(X, \mathscr{S})$, by the single letter X. This will also be done for the measure spaces defined below; in most instances, this will cause no confusion.

DEFINITION 2

A *measure space* $(X, \mathscr{S}, \mu)$ is a measurable space $(X, \mathscr{S})$ and a measure μ on $\mathscr{S}$.

The measure space X is called *(totally) finite*, *σ-finite*, or *complete* depending on whether the measure μ is (totally) finite, σ-finite, or complete.

In Section 3.5 we saw that the outcomes of repeated trials could be represented as the Cartesian product of sets. However, the Cartesian product of the corresponding fields was not a field, although if we took finite sums of such events we obtained a field (cf., Theorem 3.5.1). In the case of an infinite number of trials, we still consider finite sums of rectangles (cf., Section 5.2).

Our intuition told us how to assign values to such rectangles and the resulting calculations confirmed our belief. Nevertheless, in the infinite-dimensional situation we did not prove the existence of a probability function. It is natural to inquire whether the same process can be applied to products of measure spaces to prove the existence of a "product measure." Since the treatment of finite product measure spaces differs very little from the method used for infinite spaces, we will study the latter.

Let $(X_i, \mathscr{S}_i, \mu_i)$ be a collection of measure spaces for $i \in I$, an arbitrary index set. We assume that $\mu_i(X_i) = 1$ in order to define a product measure[7]

[7]If $\mu_i(X_i) \neq 1$, we would have an infinite product [c.f., Equation (1)], which might not converge. For finite products, X_i can be arbitrary.

on $X = \underset{i \in I}{\times} X_i$. As we do with infinite products of probability spaces (c.f., Section 5.2), we begin by defining measurable rectangles. We drop $i \in I$ when no confusion is possible.

DEFINITION 3

A *rectangle* is a set of the form $\times A_i$, where $A_i \subset X_i$ for all i and $A_i = X_i$ for all but a finite number of values of i.

Suppose that $I_N = (i_1, i_2, \ldots, i_N)$ is the finite set of indexes for which $A_i \neq X_i$; we denote this rectangle by $(\underset{i \in I_N}{\times} A_i)(\underset{i \notin I_N}{\times} X_i)$. If $I_N = (1, 2, \ldots, N)$, then $\underset{i \notin I_N}{\times} X_i$ will be denoted by $X^{(N)}$.

DEFINITION 4

A *measurable rectangle* is a rectangle $\times A_i$ for which each A_i is a measurable subset of X_i.

Let $\mathscr{R}$ be the class of all measurable rectangles and let $\mathscr{F}$ be the class of all finite sums of elements of $\mathscr{R}$.

THEOREM 1

$\mathscr{F}$ is a field.

Proof

Let $B = \times B_i$ and $A = \times A_i$, and let M be the maximum integer for which $A_i \neq X_i$ and $B_i \neq X_i$. If $I_M = (1, 2, \ldots, M)$, then $AB = (\underset{i \in I_M}{\times} A_i B_i)(\underset{i \notin I_M}{\times} X_i)$, which is a measurable rectangle. It follows by induction that the intersection of a finite number of measurable rectangles is a measurable rectangle. Since elements of $\mathscr{F}$ are finite sums of rectangles, $\mathscr{F}$ is closed under finite intersections.

The complement of a rectangle is a rectangle, since

$$[(\underset{i \in I_N}{\times} A_i)(\underset{i \notin I_N}{\times} X_i)]' = (\times A_i)'(\times X_i);$$

the result follows from Exercise 1. Therefore $\mathscr{F}$, being closed under finite intersections, is closed under complementation and is a field.

DEFINITION 5

The minimal σ-field over $\mathscr{F}$ is called the *product σ-field*; it is denoted by $\times \mathscr{S}_i$.

To define the product measure μ, we first assign its value on the class $\mathscr{R}$ of measurable rectangles by

$$(8) \qquad \mu(\underset{i \in I_N}{\times} A_i)(\underset{i \notin I_N}{\times} X_i) = (\prod_{i \in I_N} \mu_i A_i)(\prod_{i \notin I_N} \mu_i X_i) = \prod_{i \in I_N} \mu_i A_i.$$

Clearly $\mu X = 1$, where $X = \times X_i$; μ can be uniquely extended to $\mathscr{F}$ by defining the measure of a sum of rectangles as the sum of the measures of its rectangles. The reader should have no difficulty in verifying that μ is finitely additive and uniquely defined. Theorem 2 will show that μ is a measure on $\mathscr{F}$; that by the extension theorem (5.5), μ is uniquely determined on the product σ-field $\times \mathscr{S}_i$ and is denoted by $\times \mu_i$. Thus the triplet $(\times X_i, \times \mathscr{S}_i, \times \mu_i)$ is a measure space called the *product measure space.*

Before proceeding with the proof of the σ-additivity of μ, we will introduce some helpful notation.

DEFINITION 6

Let $A \subset X$ and let x_i be fixed points in X_i for $i \in I_N$, where $I_N = (1, 2, \ldots, N)$. Then the set of all $(x_i, i \notin I_N)$ such that $(x_i, i \in I) \in A$ is called the *section of A at* $(x_1, x_2, \ldots, x_n)$ and is denoted by $A(x_1, x_2, \ldots, x_N)$ (see Figure 3).

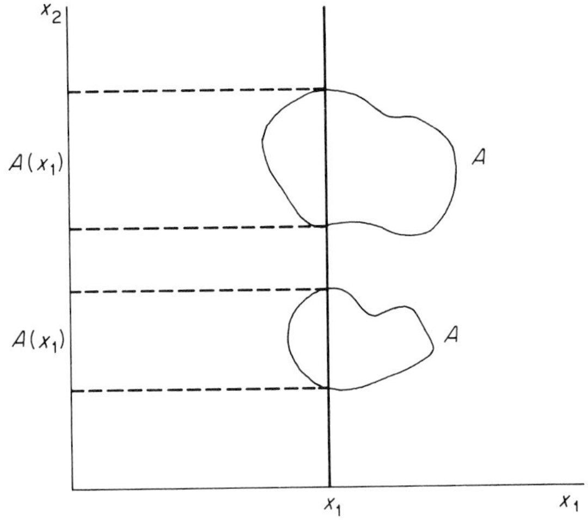

Figure 3

The reader should note that $A(x_1 \cdots x_N) \subset \underset{i \notin I_N}{X} X_i$. The analogs of $\mathscr{F}$ and μ in the spaces $X^{(N)}$ will be denoted by $\mathscr{F}^{(N)}$ and $\mu^{(N)}$ respectively.

It is not difficult to verify that for a measurable rectangle R, every section of the form $R(x_1 \cdots x_n)$ belongs to $\mathscr{F}^{(n)}$. Moreover,

$$(E_1 + E_2)(x_1, \ldots, x_n) = E_1(x_1 \cdots x_n) + E_2(x_1 \cdots x_n)$$

for any disjoint sets E_1 and E_2. From these two facts it follows that if $F \in \mathscr{F}$, then $F(x_1 \cdots x_n) \in \mathscr{F}^{(n)}$.

THEOREM 2

The product measure μ is σ-additive on $\mathscr{F}$.

Proof

To prove that μ is σ-additive it suffices to show, by Theorem 4.4, that μ is continuous at ϕ. However, it is easier to prove the following contrapositive statement: If (A_n) is a decreasing sequence of sets in $\mathscr{F}$ such that $\mu A_n > \epsilon > 0$ for $n = 1, 2, 3 \ldots$, then $\cap A_n \neq \phi$. Since every A_n is the sum of a finite number of rectangles, each of which depends on a finite number of indexes, the set of all indexes involved in defining the sequence is countable. By interchanging the indexes if necessary, we can restrict our attention to the product space $X = \overset{\infty}{\underset{n=1}{\times}} X_n$.

Let $B_n \subset X_1$ be defined by $B_n = (x_1 : \mu^{(1)} A_n(x_1) > \epsilon/2)$. Then as a consequence of the relation $A_n = (B_n \times X^{(1)}) \cap A_n + (B_n' \times X^{(1)}) \cap A_n$ (see Exercise 2),

$$\mu_1 B_n + \frac{\epsilon}{2}(1 - \mu_1 B_n) \geq \mu A_n > \epsilon; \tag{9}$$

hence $\mu_1 B_n > \epsilon/2$. Since (B_n) is a decreasing sequence of measurable sets in X_1 and since μ_1, being σ-additive, is continuous from above at ϕ, it follows that there exists at least one point $\bar{x}_1 \in X_1$ such that $\mu^{(1)}(A_n(\bar{x}_1)) > \epsilon/2$ for all n. If not, then $\cap B_n = \phi$, which implies that $\mu_1(\lim B_n) = 0$. Since $(A_n(\bar{x}_1))$ is a decreasing sequence of sums of rectangles in $X^{(1)}$, the argument just applied to $X, (A_j)$, and ϵ may be repeated for $X^{(1)}$, $(A_j(\bar{x}_1))$, and $\epsilon/2$. We obtain a point $\bar{x}_2 \in X_2$ such that $\mu^{(2)} A_n(\bar{x}_1, \bar{x}_2) > \epsilon/2^2$ for all n. Continuing in this manner, we obtain a sequence $(\bar{x}_m)$ such that $\bar{x}_m \in X_m$, where $m = 1, 2, 3, \ldots$, and $\mu^{(m)}(A_n(\bar{x}_1, \ldots, \bar{x}_m)) \geq \epsilon/2^m$.

To prove that the point $(\bar{x}_1, \bar{x}_2, \ldots, \bar{x}_m, \ldots)$ belongs to every A_n, we choose m so large that every rectangle comprising A_n can be written in the form $R \times X^{(m)}$, where $R \subset \overset{m}{\underset{i=1}{\times}} X_i$. This is always possible since there are only a finite number of factors differing from X_i. The fact that $\mu^{(m)} A_n(\bar{x}_1 \cdots \bar{x}_m) > 0$ implies that A_n contains at least one point $(x_1, x_2, \ldots, x_n, \ldots)$ such that $x_i = \bar{x}_i$ for $i = 1, 2, \ldots, m$ and x_i is arbitrary otherwise. Since every rectangle in A_n is of the form $R \times X^{(m)}$, we see that $(\bar{x}_1, \bar{x}_2, \ldots)$ itself belongs to A_n.

If the set of indexes involved is finite, say N, then the above argument still applies and we can find points $\bar{x}_1, \ldots, \bar{x}_{N-1}$ such that $\mu^{(N-1)} A_n(\bar{x}_1, \ldots, \bar{x}_{N-1}) > 0$. However, the last part of the proof requires a slight modification. Since the $A_n(\bar{x}_1, \ldots, \bar{x}_{N-1})$ form a decreasing sequence of sets in X_N and since μ_N is continuous at ϕ, there is a point $\bar{x}_N$ common to all of these sets; hence $(\bar{x}_1, \bar{x}_2, \ldots, \bar{x}_N) \in \bigcap_{n=1}^{\infty} A_n$.

EXERCISES

1. (a) Show that $(A_1 \times A_2 \times \cdots \times A_n)' = \sum B_1 \times \cdots \times B_n$, where each $B_i = A_i$ or A_i' but $B_i \neq A_i$ for all i.
(b) Verify that $(\sum A_i) \times B = \sum (A_i \times B)$.

2. (a) Prove that Equation (2) is valid.
(b) Why are the B_n measurable?

3. Prove the following statements:

(a) Let X_0 be a measurable subset in the measure space $(X, \mathscr{S}, \mu)$ and let $\mathscr{S}_0$ be the class of all measurable subsets of X_0. For E in $\mathscr{S}_0$ we define $\mu_0(E) = \mu E$; then $(X_0, \mathscr{S}_0, \mu_0)$ is a measure space.
(b) Let $X_0 \subset X$ and let $(X_0, \mathscr{S}_0, \mu)$ be a measure space. Let $\mathscr{S} = (E: EX_0 \in \mathscr{S}_0, E \subset X)$ and define $\mu E = \mu_0(EX_0)$. Then $(X, \mathscr{S}, \mu)$ is a measure space.
(c) Let X_0 be a measurable subset of X. Define a new measure μ_0 on all $E \in \mathscr{S}$ by the equation $\mu_0 E = \mu E X_0$. Then $(X, \mathscr{S}, \mu_0)$ is a measure space.

4. Generalize (where possible) the results and definitions of this section to rings and σ-rings.

5. Show that the Lebesgue measure can be considered as the completion of a product measure.

6. For a product distribution function F—that is, for

$$F(x_1, \ldots, x_n) = \prod_{i=1}^{N} F_k(x_k), \qquad (x_k \in E_k = R)$$

where each F_k on E_k is a distribution function—show that $\mu_F = \mu_1 \times \mu_2 \times \cdots \times \mu_N$. Here μ_k is the measure on the Borel field in E_k defined by means of the relation $\mu_k(a, b] = F_k(b) - F_k(a)$.

7. Prove that any set of distribution functions $(F_t, t \in I)$, where $\lim_{x\to\infty} F(x) = 1$ and $\lim_{x\to-\infty} F(x) = 0$, determines a product probability on the Borel field in the product space $\times E_t$.

8. A set $(\underset{i \in I_N}{\times} A_i)(\underset{i \in I - I_N}{\times} X_i)$, where $I_N = (1, 2, \ldots, N)$, is called a cylinder in $\times X_i$ with a base $\times A_i$ Show that the class of all cylinders whose bases are measurable generates the product σ-field.

9. Review Chapter 5 and cite the relevant theorems developed in this chapter that justify the assumptions made in Chapter 5.

REFERENCES

1. Radon, J., "Theorie und Anwendungen der absolut additiven Mengenfunktionen," *S.-B. Akad. Wiss. Wien*, **122** (1913), 1295–1438.
2. Daniell, P. J., "Integrals in an infinite number of dimensions," *Ann. of Math.*, **20**, No. 2 (1918), 281–288.
3. Nikodym, O., "Sur une généralisation des intégrals de M. Radon," *Fund. Math.*, **10** (1930), 131–179.
4. Kolmogorov, A., "Untersuchungen über den Integralbegriff," *Math. Ann.*, **103** (1930), 654–696.
5. Jessen, B., "Abstrakt Maal-og Integralteori," *Mat. Tidsskr. B.* (1934), pp. 73–84.
6. Burkill, J., *The Lebesgue Integral*. New York: Cambridge University Press, 1963.

CHAPTER 7

INTEGRATION

7.1 INTRODUCTION

It was pointed out in Section 5.9 that the definition of random variables as continuous functions and of expectations as their Riemann–Darboux integrals is not satisfactory for a general theory of probability. In this chapter we will define the class of measurable functions that includes continuous functions and that can serve as random variables. Moreover, we will generalize the concept of the integral for these more general functions. The resulting integral is more satisfying since it reduces to the Riemann integral for continuous random variables and to summation for discrete random variables. In addition, it will enable us to calculate the expectation of the random variables introduced in Section 5.9. Apart from this, the general definition of integration is simpler than that of Riemann integration in that it is defined for a certain class of unbounded functions. It is permissible to interchange the operations of integration and take limits in situations where it was not allowed earlier. Consequently, we can integrate certain sequences of functions term by term. The reason for this is to be found in the σ-additivity of measure.

As mentioned in Section 6.1, these concepts were developed in the study of functions of a real variable. In fact, the examples in Section 5.9 were formed by merely restating them in the language of probability theory. Accordingly, we will use the terminology of real-variable theory, because of its greater applicability, until the next chapter.

7.2 MEASURABLE FUNCTIONS

We can lead to the definition of a measurable function by examining our notions of random variables. The simple type of random variable introduced in Section 2.2 was a linear combination of indicator functions of events. In Section 5.7, we defined random variables as continuous functions; in particular, for the identity function $v(x) = x$, $(v \leq x)$ had to be an event. The simple random variables mentioned in Section 2.2 also have this property; i.e., $(v \leq x)$ belongs to a certain σ-field $\mathscr{E}$. However, we also saw that continuous random variables could be defined as limits of simple random variables. We will show that both of these properties can be used to characterize measurable functions.

Although random variables were defined on probability spaces, the probability function played no role in determining whether a function was a random variable (cf., example on page 16). Consequently, we need only consider a measurable space $(X, \mathscr{S})$ and an extended real-valued function f defined on X.

DEFINITION 1

An extended real function f on a measurable space $(X, \mathscr{S})$ is said to be *measurable* if the set $(x\colon f(x) \leq c)$ is measurable for every extended real number c.

Of course, if f is real valued, c must only take values in the real number system. The same class of measurable functions is obtained if any one of the sets $(f > c)$, $(f \geq c)$, or $(f < c)$ is substituted for $(f \leq c)$ in the above definition[1] (cf., Exercise 1). If X is the real line and $\mathscr{S}$ is the class of Borel sets, then the measurable functions are called *Borel measurable functions.* If $\mathscr{S}$ is the class of Lebesgue sets, then the measurable functions are called *Lebesgue measurable functions.*

A simple example of a measurable function is the indicator function of a measurable set. It is obvious that the random variables defined in the previous chapters are measurable functions.

DEFINITION 2

A function f defined on a measurable space X is called *elementary*[2] if $f = \sum_{i=1}^{\infty} y_i I_i$, where the I_i are the indicator functions of disjoint measurable

[1] $(f \leq c) \equiv (x\colon f(x) \leq c)$.
[2] The zero function can be written as I_ϕ.

sets E_i and the y_i are nonzero real numbers. If there are only a finite number of y_i, then f is called a *simple function.*

It is obvious that the sum and the product of two simple functions are also simple functions (cf., Theorem 2.3.1).

LEMMA 1

Simple and elementary functions are measurable functions.

This result follows directly from the definition of a measurable function. To prove that every measurable function is the limit of a sequence of simple functions, we observe that every real-valued function f can be written in the form

$$f = f^+ - f^-, \tag{1}$$

where

$$f^+ = \max(f, 0) \qquad f^- = -\min(f, 0). \tag{2}$$

The f^+ and f^- are called the *positive part* and *negative part* of f, respectively.

LEMMA 2

If f is an extended, real-valued, measurable function, then f^+ and f^- are measurable.

Proof

Let $N = (x: f(x) \leq 0)$; then $(f^- \leq c) = (f \geq -c) \cap N$ is a measurable set. A similar proof shows that f^+ is measurable.

THEOREM 1

Every extended, real-valued, measurable function f is the limit of a sequence $\{f_n\}$ of simple functions.

Proof

Since the difference of two simple functions is a simple function, we need only show that f^+ and f^- can be approximated by a sequence of simple functions. Hence it is sufficient, by virtue of Lemma 2, to prove the theorem for a nonnegative measurable function f. We define

$$\begin{aligned} f_n(x) &= \frac{i-1}{2^n} \qquad \text{if } \frac{i-1}{2^n} \leq f(x) < \frac{i}{2^n} \qquad (i = 1, 2, \ldots, 2^n n) \\ &= n \qquad \text{if } f(x) \geq n \end{aligned} \tag{3}$$

for $n = 1, 2, 3, \ldots$ and every $x \in X$. Since f is measurable, the sets $((i-1)/2^n \leq f < i/2^n)$ and $(f \geq n)$ are measurable. Clearly, f_n is a nonnegative simple function, and the sequence $\{f_n\}$ is increasing. If f is bounded, then $f < N$ for some N. From (3) we see that for all $n \geq N$, $0 \leq f(x) - f_n(x) \leq 1/2^n$, and so the convergence is uniform. Suppose that f is unbounded; if $f(x) < N$ for some N, the preceding argument shows that $f_n(x) \to f(x)$. If $f(x) = \infty$, then $f_n(x) = n$ for every n. The functions f_n still converge to f but not uniformly.

COROLLARY 1

A nonnegative measurable function is the limit of an increasing sequence of nonnegative simple functions.

COROLLARY 2

A bounded measurable function is the limit of a uniformly convergent sequence of simple functions.

These corollaries are the byproducts of the proof of the theorem. Corollary 2 can be modified for unbounded measurable functions by using the fact that a measurable function $f \geq 0$ is the limit of a uniformly convergent sequence of elementary functions f_n, where

$$f_n = \sum_{i=1}^{\infty} \frac{i-1}{2^n} I_{((i-1)/2^n \leq f < 1/2^n)} + \infty I_{(f=\infty)}.$$

In order to prove the converse of Theorem 1 we require Theorem 2.

THEOREM 2

If $\{f_n\}$ is a sequence of extended, real-valued, measurable functions, then $\sup f_n$ and $\inf f_n$ are measurable functions.

Proof

The equations $(\sup f_n \leq c) = \bigcap_{n=1}^{\infty} (f_n \leq c)$ and $(\inf f_n < c) = \bigcup_{n=1}^{\infty} (f_n < c)$ imply the theorem since the class of measurable sets is a σ-field.

COROLLARY

If $\{f_n\}$ is a sequence of extended, real-valued, measurable functions, then $\liminf f_n$ and $\limsup f_n$ are measurable.

Proof

The measurability of these functions is a consequence of the relations $\limsup f_n = \inf_{n \geq 1} \sup_{m \geq n} f_m$ and $\liminf f_n = \sup_{n \geq 1} \inf_{m \geq n} f_m$.

As a consequence of the corollary, the limit of a measurable sequence of functions is measurable. In particular, the limit of a sequence of simple functions is measurable. Hence, Theorem 1 implies the following theorem.

THEOREM 3

An extended real-valued function is measurable if and only if it is the limit of a sequence of simple (elementary) functions.

Another equivalent definition of measurability is contained in the following theorem.

THEOREM 4

An extended, real-valued function f is measurable if and only if the inverse images of all extended Borel sets[3] are measurable.

Proof

Assume that the function is measurable according to Definition 1. We recall that f^{-1} preserves all set operations and the measurable sets form a σ-field. Therefore, the class of all sets whose inverse images are measurable is a σ-field. This σ-field contains the intervals $[-\infty, c]$ and so it contains the minimal σ-field generated by these intervals, which is the Borel field.

Conversely, suppose that the inverse images of all extended Borel sets are measurable. In particular, the inverse images of sets of the form $[-\infty, c]$ are measurable and so f is measurable.

Note that we could have used any class $\mathscr{C}$ that generates the extended Borel field in the definition of measurability (cf., Exercise 1). The class $\mathscr{C}$ of semiclosed intervals $(a, b]$, where $-\infty < a < b < \infty$, cannot be used, since it does not generate the extended Borel field. However, if we add the two Borel sets $\{\infty\}$ and $\{-\infty\}$ to the σ-field generated by $\mathscr{C}$, we generate the extended Borel field. To show that an extended real-valued function f is measurable, we need only prove that $f^{-1}(M)$, where M is an ordinary Borel set, $f^{-1}(\infty)$, and $f^{-1}(-\infty)$ are measurable sets. Therefore, if we

[3]An *extended Borel set* is a member of the σ-field generated by the class of all intervals of the extended real line. The end points of an interval can be ∞ or $-\infty$.

know that $f^{-1}(\pm\infty)$ are measurable, we can show that f is measurable by showing that its restriction to finite values is an ordinary measurable function (cf., Theorem 5).

The next theorem shows that certain combinations of measurable functions yield measurable functions.

THEOREM 5

Let f and g be extended measurable functions and let c be a real number. Then the following are measurable functions:

1. cf.
2. $c + f$.
3. $f + g$.
4. fg.
5. $|f|$.
6. $\max(f, g)$.
7. $\min(f, g)$.

We assume that functions 3 and 4 are well defined; that is, expressions such as $\infty - \infty$ and $0 \times \infty$ do not appear.

Proof

Let $\{f_n\}$ and $\{g_n\}$ be sequences of simple functions whose limits are f and g respectively. Theorem 3 and the fact that cf_n, $c + f_n$, $f_n + g_n$, and $f_n g_n$ are simple functions yield the proof of the measurability of functions 1–4.

The proof of the measurability of function 5 follows from Lemma 2 upon noting that $|f| = f^+ + f^-$.

The relation

$$[\max(f, g) = \pm\infty] = [(f = \infty) \cup (g = \infty)] \cup [(f = -\infty) \cup (g = -\infty)]$$

reduces the proof of 6 to showing that it is true for the finite-valued measurable functions f and g [see Exercise 2(a)]. However, this follows from the identity

$$\max(f, g) = \frac{1}{2}(f + g + |f - g|).$$

The proof of 7 follows from a similar argument upon noting that

$$\min(f, g) = \frac{1}{2}(f + g - |f - g|).$$

Some examples of Borel measurable functions are given below. These functions map the real line into the real line; the measurable sets are Borel sets.

EXAMPLE 1

1. $f(x) = c$ is a measurable function.
2. $f(x) = x$ is a measurable function, since any interval is transformed into an interval of the same type.
3. From Theorem 5, we see that a polynomial is a measurable function.
4. Let $f(x)$ be a continuous function. Let $f_n(x) = f(x)$ where $|x| \leq n$ and zero otherwise. By the Weierstrass approximation theorem, $f_n(x)$ can be approximated by a polynomial on $[-n, n]$. Hence, from Theorem 2 and the previous result, $f_n(x)$ is a measurable function. Another application of Theorem 2 implies that $\lim_{n\to\infty} f_n = f$ is a measurable function.

Alternatively, this result can be proved from the fact that a continuous function can be approximated by a step function (i.e., a simple function).

5. A monotone function is measurable. This follows from the fact that such a function can be decomposed into a step function and a continuous function (cf., Theorem 9.2.6). Alternatively, we can approximate it by elementary functions.
6. For a nonmeasurable function, see Example 7.3.1.

These results can be generalized to functions defined on R^n.

Part (3) of Theorem 5 can be reinterpreted as follows. Let $h(x, y) = x + y$ and let f and g be measurable; then $h(f, g)$ is a measurable function. Actually, this result can be generalized to read that a continuous function of a measurable function is also a measurable function (see Exercise 7). An application of Theorem 2 yields the result that limits of continuous functions, when applied to measurable functions, lead to measurable functions. The smallest class closed under passages to the limit containing all continuous functions is called the *Baire class*. The members of the Baire class are the *Baire functions*. Since the class of measurable functions is closed under passages to the limit and contains the continuous functions, Baire functions are measurable functions.

THEOREM 6

A Baire function of a measurable function is measurable.

This theorem can be proved by transfinite induction by starting with the fact that if f is the limit of a sequence of continuous functions and g is a measurable function then $f(g)$ is a measurable function.

Encouraged by this result, we might conjecture that a measurable function of a measurable function is measurable. However, it is possible to construct an example of a Lebesgue measurable function of a measurable function that is *not* measurable [1]. This shows that not all measurable functions are Baire functions.

Using the results of Theorem 4, we can generalize the definition of a measurable function. Let f be a function on the measurable space $(X, \mathscr{S})$ to the measurable space $(Y, \mathscr{T})$. Then f is called measurable if and only if $f^{-1}(T) \in \mathscr{S}$ for every $T \in \mathscr{T}$. Using this definition, with $Y = R^n$ and with $\mathscr{T}$ the extended Borel class, the results of this section can easily be extended for functions on a measurable space to an n-dimensional space. However, it is not necessary to verify this directly, as the following theorem shows.

THEOREM 7

A function $f = (f_1, \ldots, f_n)$ into R^n is measurable if and only if its components $f_1, \ldots, f_n$ are measurable.

Proof

If $f = (f_1, \ldots, f_n)$ is measurable, then the sets $(f_j \leq c_j) = f_j^{-1}[-\infty, c] = f^{-1}(M)$, where $M = (x\colon -\infty \leq x_k \leq \infty, k \neq j; -\infty \leq x_j \leq c_j)$, are measurable; hence f_j is measurable for every $j \leq n$.

Conversely, suppose that all the f_j are measurable; then for $c = (c_1, c_2, \ldots, c_n)$, the sets

$$(f \leq c) = (f_1 \leq c_1, f_2 \leq c_2, \ldots, f_n \leq c_n) = \bigcap_{j=1}^{n} (f_j \leq c_j)$$

are measurable, so that f is measurable (cf., Exercise 11).

Finally, we say that a complex-valued function is measurable if its real and imaginary parts are extended, real-valued, measurable functions.

EXERCISES

1. (a) Show that the same class of functions is obtained if the sign $\leq$ is replaced by $<, \geq$, or $>$. [*Hint:* $(f < c) = \bigcup_{n=1}^{\infty} (f \leq c - 1/n)$.]
(b) Show that the same class of functions is obtained if c is restricted to belong to an everywhere-dense set of numbers.

2. (a) Verify that if f is an extended, real-valued, measurable function, the set of points $(f = c)$ is a measurable set.
(b) Show that if f and g are measurable functions, the set of points $(f = g)$ is a measurable set.
(c) Show that the set of points of convergence of a sequence of extended, real-valued, measurable functions is measurable.

3. Is a Borel measurable function of a measurable function a measurable function?

4. Show that $f + g$ is a measurable function when f and g are measurable functions by verifying that $(f + g > c) = \bigcup_r (f > r)(c - g < r)$, where r is a rational number.

5. Prove that a set is measurable if and only if its indicator function is measurable.

6. If $|f|$ is measurable, does f have to be measurable?

7. Show that a continuous function of a measurable function is measurable under the assumption that the resulting function is well defined.

8. The set of points of discontinuity of an arbitrary function (even a nonmeasurable one) is F_σ. [*Hint:* Consider the oscillation.] Deduce that there is no function that is continuous at all the rational points and discontinuous at the irrational points.

9. Show that there exists a function continuous at the irrationals and discontinuous at all the rationals. [*Hint:* If $x = p/q$ for rational x, set $f(x) = 1/q$; otherwise, set $f(x) = 0$].

10. Let $X = \sum_{n=1}^{\infty} F_n$, where F_n is a measurable set. Show that if a function is measurable on each F_n, then it is measurable in the whole space.

11. Prove that $f: X \to R^n$ is measurable if and only if $(x: f(x) \leq c)$ is measurable for all c in R^n.

7.3 CONVERGENCE

In our study of probability theory, we came across sequences of random variables that did not converge in the sense of classical analysis. Nevertheless, it still made sense to speak about the "limit" of these random variables. For example, for a sequence of repeated independent trials we defined the random variable s_n, the number of successes in n trials. In Theorem 3.6.3 we saw that $\lim_{n\to\infty} P(|s_n/n - p| < \epsilon) = 1$ for every $\epsilon > 0$. This type of convergence is known as convergence in probability; it will be studied later. In Example 1 of Section 5.7, another kind of convergence (convergence in distribution) was encountered. The study of this type of convergence will be deferred until Chapter 8. Instead of using a probability function and random variables, we can study these concepts for measurable functions by using an arbitrary measure. However, before carrying out this program, let us recall the ordinary concept of convergence.

We say that f_n *converges* to a finite-valued f on S if, for every $x \in S$ and every $\epsilon > 0$, there is an integer $N = N(x, \epsilon)$ such that $|f_n(x) - f(x)| < \epsilon$ for every $n \geq N$. If f is an extended real-valued function, we will also require that $f_n(x) > 1/\epsilon$ if $f(x) = \infty$ and $f_n(x) < -1/\epsilon$ if $f(x) = -\infty$ for all $n \geq N$. We denote pointwise convergence by $f_n \to f$. If the same $N = N(\epsilon)$ can be chosen for every $x \in S$, then the convergence is *uniform* and we write $f_n \to f(u)$.

A sequence $\{f_n\}$ of real-valued functions is a *Cauchy sequence* or *fundamental* on S if for every $\epsilon > 0$ there exists an integer $N = N(x, \epsilon)$ with the property that $|f_n(x) - f_m(x)| < \epsilon$ whenever $n \geq N$ and $m \geq N$. We recall Cauchy's theorem, which states that if the f_n are finite, then $\lim f_n$ exists and is finite if and only if $\{f_n\}$ is a fundamental sequence. A fundamental sequence is *uniformly fundamental* if $N = N(\epsilon)$. The Cauchy criterion also holds for uniform convergence.

For the remainder of this section, we will assume that the underlying space X is a measure space $(X, \mathscr{S}, \mu)$. Then we can slightly generalize the foregoing definitions of convergence by requiring that the functions converge on $X - A$, where $\mu A = 0$.

DEFINITION 1

A sequence $\{f_n\}$ is said to *converge a.e.* (almost everywhere) to f if $f_n \to f$, except on a set of measure zero. We denote this by $f_n \to f$ (a.e.). It is *fundamental* a.e. (we write $f_m - f_n \to 0$ (a.e.) or $f_{n+i} - f_n \to 0$ (a.e.)) if it is a Cauchy sequence outside a set of measure zero.

By the Cauchy criterion, it is clear that if a sequence converges to a finite limit function a.e., then it is fundamental a.e. and conversely. In fact, if the functions f_n are assumed to be only *finite a.e.* (finite except on a set of measure zero), the same result holds. This is a consequence of the fact that a countable union of null sets is a null set. Moreover, if $f_n \to f$ (a.e.) and $f_n \to g$ (a.e.), then $f = g$ (a.e.); that is, the limit function is determined uniquely except on a set of measure zero. The relation $f = g$ (a.e.) is an equivalence relation and so the class of all functions on our measure space is split into equivalence classes. As long as we are concerned with convergence a.e., these functions are defined only up to an equivalence and we can replace a function that is finite a.e. and measurable a.e. (the limit a.e. of a sequence of simple functions) by a finite and measurable function respectively.

EXAMPLE 1

Let $X = [0, 2]$ and $\mathscr{S} = \{X, \phi, [0, 1), [1, 2]\}$, with $\mu[0, 1) = 1$ and $\mu[1, 2] = 0$. Suppose that $f = 1$ on $[0, 1)$, $\frac{1}{2}$ on $[1, 1\frac{1}{2})$, and 0 on $[1\frac{1}{2}, 2]$. Then f is not a measurable function since $f^{-1}(\{0\}) = [1\frac{1}{2}, 2] \notin \mathscr{S}$. Nevertheless, the measurable functions $f_n(x) = 1 - 1/n$ on $[0, 2]$ converge to f a.e.

In general, since each f_n is the limit of a sequence of simple functions, there exists a sequence of simple functions that approaches f a.e. (see the proof of Theorem 7.5.1). Hence, we could say that f is measurable a.e. By modifying f on a set of measure zero, we can always obtain a measurable function. In the above example, we could define $f = 1$ on $[1, 2]$ or $f = \frac{1}{2}$ on $[1, 2]$, etc., and then the f_n would approach a measurable function a.e. However, if the measure is complete, then f is measurable (see Exercise 9).

DEFINITION 2

A sequence $\{f_n\}$ is said to *converge uniformly almost everywhere* to f if $f_n \to f$ uniformly except on a set of measure zero. We write this as $f_n \to f$ (u.a.e.). The sequence is *uniformly fundamental almost everywhere*, written as $f_m - f_n \to 0$ (u.a.e.), if it is uniformly fundamental except on a set of measure zero.

By applying the Cauchy criterion, we can easily verify that if the functions f_n are finite a.e., then $f_n \to f$ (u.a.e.) if and only if $f_n - f_m \to 0$ (u.a.e.).

In the sequel, we will refer to several different kinds of convergence, using terminology similar to that employed above. Thus, suppose that we define a new kind of convergence of a sequence $\{f_n\}$ to a limit f by specifying the sense in which f_n is to be near f for large n. Then the notion of a fundamental sequence in the new sense means that for large m and n the differences $f_m - f_n$ are to be near to zero in the specified sense of nearness.

EXAMPLE 2

1. The space $X = [0, 1]$, the measurable sets are the Borel sets of this interval, and the measure μ is the Lebesgue measure. Let $f_n = x^n$ and $f = 0$. Then $f_n \to f$ (a.e.), since we have pointwise convergence on the set $0 \leq x < 1$ and $\mu(\{1\}) = 0$. Note that the sequence $\{f_n\}$ does not converge pointwise to this function.
2. The above sequence does not converge u.a.e. to zero. At the points $x_n = 1 - 1/n$, we have $f_n(x_n)$ close to e^{-1} for sufficiently large n. Also, if $x > x_n$, then $f_n(x) > f_n(x_n)$. Therefore, it is not true that there exists an M such that $n \geq M$ implies $|f_n(x)| < \epsilon$ for arbitrary $\epsilon > 0$ and every x. Moreover, the set of points where this inequality does not hold is not of measure zero.
3. The space is the Borel line $[0, \frac{1}{2}]$ and the measure is the Lebesgue measure. Let $f_n(x) = x^n$ for $0 \leq x < \frac{1}{2}$ and 1 for $x = \frac{1}{2}$. Then $f_n(x) \to 0$ (u.a.e.) since $\mu(\{\frac{1}{2}\}) = 0$. This sequence does not converge pointwise to zero.

If a sequence converges pointwise a.e., it may not converge u.a.e. However, in the examples above, by removing a small neighborhood of the point which causes the trouble, we obtain a function that converges uniformly on the remainder of the set. Egoroff's theorem shows that this result holds in more general situations.

THEOREM 1

Let E be a measurable set such that $\mu(E) < \infty$. Suppose that a sequence of a.e. finite-valued measurable functions f_n converges a.e. on E to a finite-valued measurable function f. Then for every $\epsilon > 0$, there exists a measurable subset F of E such that the sequence $\{f_n\}$ converges uniformly to f on $E - F$ and $\mu F < \epsilon$.

Proof

Assume first that the sequence $\{f_n\}$ converges to f everywhere on E and that every f_n is finite on E. Since $|f_i - f|$ is a measurable function, the sets

$$E_k(m) = \bigcap_{i=k}^{\infty} \left(|f_i(x) - f(x)| < \frac{1}{m}\right)$$

are measurable and $E_k(m) \subset E_{k+1}(m)$ for all k. Since the squence $\{f_n\}$ converges to f on E, $\lim_{k\to\infty} E_k(m) \supset E$ for every $m = 1, 2, 3, \ldots$. Hence $\lim_{k\to\infty} \mu(E - E_k(m)) = 0$, so that by the continuity of μ there exists a positive integer $n = n(m, \epsilon)$ such that $\mu(E - E_n(m)) < \epsilon/2^m$. Let $F = \bigcup_{m=1}^{\infty} (E - E_n(m))$; then $F \subset E$, F is a measurable set, and $\mu F \leq \sum_{m=1}^{\infty} \mu(E - E_n(m)) < \epsilon$. For $x \in E - F = \bigcap_{m=1}^{\infty} EE_n(m)$ and for $k \geq n$, $|f_k(x) - f(x)| \leq 1/m$, since $x \in E_k(m)$. Since m is arbitrary, this proves the uniform convergence on $E - F$.

If the sequence $\{f_n\}$ converges to f everywhere except on a set E_0 of measure zero and if f_n is finite except on a set E_n, of measure zero, then $T = \bigcup_{i=0}^{\infty} E_i$ is a set of measure zero. Applying these results to $\bar{E} = E - T$, we obtain a set $\bar{F}$; it is not difficult to see that the set $F = \bar{F} + T$ fulfills the stated requirements.

The following example shows that Egoroff's theorem does not hold if $\mu E = \infty$.

EXAMPLE 3

Let X be the set of positive integers, let $\mathscr{S}$ be the class of all subsets of X, and let $\mu(E)$ be the number of points in E for $E \in \mathscr{S}$. Let f_n be the indicator of $\{1, 2, \ldots, n\}$. Then f_n converges pointwise to $f \equiv 1$, but the convergence is not uniform on X. There are no subsets of X (except ϕ) of arbitrarily small measure.

A new type of convergence appears in Egoroff's theorem.

DEFINITION 3

A sequence of a.e. finite, measurable functions f_n *converges almost uniformly* to the measurable function f, written as $f_n \to f$ (a.u.), if for every $\epsilon > 0$ there is a measurable set F with the property that $\mu F < \epsilon$ and such that the sequence $\{f_n\}$ converges to f uniformly on F'.

If $f_n \to f$ (u.a.e.), then $f_n \to f$ (a.u.); however, the converse is not true, as the following example illustrates.

EXAMPLE 4

Let the measure space be the Borel interval $(0, 1)$ with the Lebesgue measure and let $f_n = 1$ on $(0, 1/n)$ and $f_n = 0$ elsewhere. Then $f_n \to f \equiv 0$ everywhere; if we take $F = (0, \epsilon)$, the sequence $\{f_n\}$ will converge to f uniformly on F'. However, we do not have $f_n \to f$ (u.a.e).

The most we can prove in the converse direction is convergence a.e.

THEOREM 2

Let E be a measurable set such that $\mu E < \infty$ and let $\{f_n\}$ be a sequence of functions that converges to f almost uniformly on E. There exists a sequence $\{E_i\}$ of measurable subsets of E such that on each E_i the convergence is uniform and $\mu(E - E_i) \to 0$ $(i \to \infty)$. Moreover, $f_n \to f$ (a.e.) on E even if $\mu E = \infty$.

Proof

By definition of convergence a.u, there exists for every $i = 1, 2, \ldots$ a measurable set $F_i \subset E$ such that $\mu(F_i) < 1/i$ and such that $f_n \to f(u)$ on $E_i = E - F_i$. Since $\mu E_i = \mu E - \mu F_i$, $\lim \mu E_i = \mu E$ and so $\mu(E - E_i) \to 0$. If $x \in \bigcup_{i=1}^{\infty} E_i$, then $x \in E_i$ for some i so that $f_n \to f$. Moreover, $\mu(E - \cup E_i) \leq \mu F_i \leq 1/i$ and so $\mu(E - \cup E_i) = 0$.

A detailed examination of the set of points of convergence of a sequence $\{f_n\}$ leads to the following criterion for convergence a.e.

THEOREM 3

The sequence $\{f_n\}$ of finite measurable functions converges to a finite measurable function f a.e. on a measurable set E if and only if

(1) $$\mu\left(\bigcap_{n=1}^{\infty}\bigcup_{m=n}^{\infty} EA_m(\epsilon)\right) = 0$$

for every $\epsilon > 0$, where

$$A_m(\epsilon) = (|f_m - f| \geq \epsilon).$$

Proof

The sequence of real numbers $\{f_n(x)\}$ for a fixed x does not converge to the real number $f(x)$ if and only if there is an $\epsilon > 0$ such that x belongs to $A_m(\epsilon)$ for an infinite number of values of m. If we use the definition of the limit superior,[4] the set S of points of divergence is given by

(4) $$S = \bigcup_{\epsilon>0}\bigcap_{n=1}^{\infty}\bigcup_{m=n}^{\infty} A_m(\epsilon) = \bigcup_{k=1}^{\infty}\left(\bigcap_{n=1}^{\infty}\bigcup_{m=n}^{\infty} A_m\left(\frac{1}{k}\right)\right) \equiv \bigcup_{k=1}^{\infty} S\left(\frac{1}{k}\right).$$

Now if $\{f_n\}$ converges a.e. to f on E, the set of points of divergence ES has measure zero. However, if $\mu ES = 0$, we see from (2) that (1) holds, since $ES \supset \bigcap_{n=1}^{\infty}\bigcup_{m=n}^{\infty} EA_m(\epsilon)$. Conversely, assume that (1) holds. Then, *a fortiori*, $\mu(ES(1/k)) = 0$ for every k; the result follows from (2) since $\mu(ES) \leq \sum_{k=1}^{\infty} \mu ES(1/k)$.

COROLLARY 1

If μ is finite or E is of finite measure, the above criterion becomes

$$\lim_{n\to\infty} \mu\left(E \cap \bigcup_{m=n}^{\infty} A_m(\epsilon)\right) = 0.$$

The proof follows directly from Theorem 6.4.2, which requires that μ be finite in order for it to be continuous from above.

The following inclusion:

$$\bigcup_{m=n}^{\infty} (|f_m - f_n| \geq \epsilon/2) \supset (\sup_{m\geq n} |f_m - f_n| \geq \epsilon) \supset \bigcup_{m=n}^{\infty} (|f_m - f_n| \geq \epsilon)$$

and Corollary 1 imply Corollary 2 (cf., Exercise 3).

COROLLARY 2

If the measure μ is finite, then $\{f_n\}$ converges to a finite-limit function a.e. if and only if

[4] $\limsup A_n = \lim_n \downarrow (\sup_{m\geq n} A_m) = \bigcap_{n=1}^{\infty}\bigcup_{m=n}^{\infty} A_m.$

$$\lim_{n\to\infty} \mu(\sup_{m\geq n} |f_m - f_n| \geq \epsilon) = 0.$$

The desire to study convergence in probability for sequences of random variables, as mentioned at the beginning of this section, motivates the following definition.

DEFINITION 4

A sequence $\{f_n\}$ of a.e. finite measurable functions is said to *converge in measure* to the measurable function f, written as $f_n \to f(\mu)$ if $\lim_{n\to\infty} \mu(|f_n - f| \geq \epsilon) = 0$ for every $\epsilon > 0$. It is *fundamental in measure* if $\mu(|f_n - f_m| \geq \epsilon) < \delta$ for arbitrary positive ϵ and δ whenever m and n are greater than $N = N(\epsilon, \delta)$.

We may easily obtain the following result from Theorem 3.

THEOREM 4

If a sequence of a.e. finite measurable functions f_n converges a.e. to a finite function f on a set E of finite measure, then it converges in measure on E. Similarly, if it is fundamental a.e., then it is fundamental in measure.

On sets of infinite measure, convergence a.e. does not imply convergence in measure. In Example 3 the $f_n \to f$ (a.e.); however, the statement that $\mu(|f_n - f| \geq \epsilon) < \delta$ for $n \geq N(\epsilon, \delta)$ is not true for arbitrary $\delta > 0$. The assumption of finiteness is not needed in the following theorem.

THEOREM 5

Almost uniform convergence implies convergence in measure.

Proof

We must show that $\mu(|f_n - f| \geq \epsilon) < \delta$ for $n \geq N$ if $f_n \to f$ (a.u.). However, by the definition of a.u. convergence, there exists an F with the property that $\mu F < \delta$; for $n \geq N$ we have $|f_n - f| < \epsilon$ on F'. Hence if $x \in (|f_n - f| \geq \epsilon)$, then $x \in F$ and so $\mu(|f_n - f| \geq \epsilon) \leq \mu F < \delta$.

However, convergence in measure does not imply a.u. convergence, as the following example illustrates.

EXAMPLE 5

Let the measure space be [0, 1] with the Lebesgue measure, and let the sequence $\{g_n\}$ be $f_{11}, f_{21}, f_{22}, f_{31}, f_{32}, f_{33}, \ldots$, with $f_{nk} = 1$ on $[(k-1)/n, k/n]$

and $f_{nk} = 0$ elsewhere. Then $g_n \to 0(\mu)$, but the sequence does not even converge pointwise.

THEOREM 6

If $f_n \to f(\mu)$, then $\{f_n\}$ is fundamental in measure; moreover, if $f_n \to g(\mu)$, then $f = g$ (a.e.).

Proof

The relation

$$(|f_n - f_m| \geq \epsilon) \subset (|f_m - f| \geq \epsilon/2) \cup (|f_n - f| \geq \epsilon/2)$$

proves the first assertion. The fact that $f = g$ (a.e.) is proved in an analogous manner.

The converse, that $f_n - f_m \to 0(\mu)$ implies the existence of a measurable function f such that $f_n \to f(\mu)$, is more difficult to prove. We need the following preliminary result.

THEOREM 7

Every sequence $\{f_n\}$ of measurable functions which is fundamental in measure contains an almost uniformly convergent subsequence.

Proof

From the definition of the phrase "fundamental in measure," we can find an integer $m(k)$ such that $m \geq m(k)$, and $n \geq m(k)$ implies that $\mu(|f_n - f_m| \geq 1/2^k) < 1/2^k$ for every integer $k > 0$. Let us assume for the moment that $m(k) < m(k+1)$ for $k = 1, 2, 3, \ldots$, so that the sequence $\{f_{m(k)}\}$ is an infinite subsequence of $\{f_n\}$. We shall show that this subsequence is a.u. fundamental. Since the sequence $\{f_{m(k)}\}$ cannot be fundamental on the sets $A_k = (|f_{m(k)} - f_{m(k+1)}| \geq 1/2^k$, we will consider what happens for $x \notin \bigcup_{m=k}^{\infty} A_m$.

If $k \leq i \leq j$, we have $|f_{m(i)}(x) - f_{m(j)}(x)| \leq \sum_{s=1}^{\infty} | f_{m(s)}(x) - f_{m(s+1)}(x)| < 1/2^{i-1}$, since $|f_{m(s)} - f_{m(s+1)}| < 1/2^s$. Therefore, our subsequence is uniformly fundamental on $\left(\bigcup_{m=k}^{\infty} A_m\right)'$. Since $\mu\left(\bigcup_{m=k}^{\infty} A_m\right) \leq \sum_{m=k}^{\infty} \mu A_m < 1/2^{k-1}$, this subsequence is a.u. fundamental by definition. The theorem follows from the fact that an a.u. fundamental sequence is an a.u. convergent sequence (see Exercise 4).

The assumption that the $m(k)$ are increasing is no restriction, since we can always satisfy this condition. We define $n(1) = m(1)$ and $n(k) = \max(m(k), n(k-1)+1)$; then $n(1) < n(2) < n(3) \cdots$ and the required subsequence is $\{f_{n(k)}\}$.

COROLLARY

Every sequence $\{f_n\}$ of measurable functions which is fundamental in measure contains an a.e. convergent subsequence.

The proof follows by Theorem 2.

THEOREM 8

If a sequence of measurable functions f_n is fundamental in measure, then there exists a measurable function f such that $f_n \to f$ in measure.

Proof

By Theorem 7, we can extract an almost uniformly convergent subsequence $\{f_{n(k)}\}$. Suppose that $f_{n(k)} \to f$ (a.u.); then by Theorem 5, $f_{n(k)} \to f(\mu)$. We shall prove that $f_n \to f(\mu)$. The result follows from the relation

$$(|f_n - f| \geq \epsilon) \subset (|f_n - f_{n(k)}| \geq \epsilon/2) \cup (|f_{n(k)} - f| \geq \epsilon/2).$$

By hypothesis, the measure of the first term on the right-hand side of the relation is arbitrarily small if n and $n_{(k)}$ are sufficiently large. The second term also approaches zero, since $f_{n(k)} \to f(\mu)$.

Figure 4 summarizes the relations among the various modes of convergence for finite measurable functions on sets of finite measure. The dotted arrows indicate that the result is true only for a subsequence. The numbers refer to the appropriate theorems; $\mu E = \infty$ signifies that the measure of E does not have to be finite.

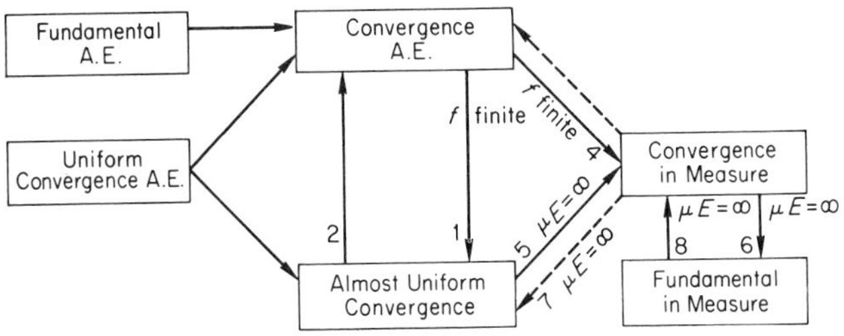

Figure 4

EXERCISES

1. If $f_n = f$ (a.e.) and the f_n are finite a.e., is it true that f is finite a.e.?

2. Given $f_n \to f(\mu)$, prove that $f = g$ (a.e.).

3. (a) Prove that $f_{n+v} - f_n \to 0$ (a.e.) if and only if, for every $\epsilon > 0$,

$$\mu\left[\bigcap_{n=1}^{\infty} \bigcup_{v=1}^{\infty} (|f_{n+v} - f_n| \geq \epsilon)\right] = 0.$$

In other words, $\{f_n\}$ is fundamental a.e.
(b) For finite μ, show that the above result can be stated as

$$\lim_{n\to\infty} \mu\left[\bigcup_{v}^{\infty} (|f_{n+v} - f_n| \geq \epsilon)\right] = 0.$$

(c) For finite μ, show that $f_n \to f$ (a.e.) and f is finite if and only if the equation in Exercise 3(b) holds.

4. Prove that $f_n \to f$ (a.u.) if and only if $f_n - f_m \to 0$ (a.u.). [*Hint:* If $\{f_n\}$ is a.u. fundamental, $f_n \to g_k$ on F_k with $\mu F_k' < 1/k$ and $F_k \uparrow$. Use the g_k to construct a measurable function g.]

5. Without using Theorem 7, show directly that a sequence of measurable functions that is fundamental in measure contains an a.e. convergent subsequence.

6. *Lusin's Theorem.* Let A be a measurable set of σ-finite measure and let $\{f_n\}$ be a sequence of a.e. finite measurable functions that converges a.e. on A to a finite measurable function f. Show that there exists a sequence $\{A_i\}$ of measurable sets such that the sequence $\{f_n\}$ converges uniformly on each A_i and such that $\mu(A - \bigcup_{i=1}^{\infty} A_i) = 0$. [*Hint:* Prove that one can assume $\mu A < \infty$ and then apply Theorem 2.]

7. If $f_n \to f$ (a.u.), does $f_n \to f$ (a.u.) on every measurable subset of X?

8. Suppose that μ is finite and $\{f_n\}$ and $\{g_n\}$ are sequences of finite measurable functions converging in measure to f and g respectively. Prove that the indicated sequences approach the given limits in measure:

(a) $\alpha f_n + \beta g_n \to \alpha f + \beta g$, where α and β are real constants.
(b) $|f_n| \to |f|$.
(c) $f_n^2 \to f^2$ if $f = 0$ (a.e.).
(d) $f_n g \to fg$. [*Hint:* Find a constant k such that if $A = (|g| \leq k)$, then $\mu A' < \delta$, and consider convergence on A and A'.]
(e) $f_n^2 \to f^2$. [*Hint:* Apply Exercise 8(c) to $\{f_n - f\}$.]
(f) $f_n g_n \to fg$. [*Hint:* $fg = \frac{1}{4}((f + g)^2 - (f - g)^2)$.]
(g) $f_n^{-1} \to f^{-1}$ if f_n and f are not equal to zero anywhere.

9. Let $\{f_n\}$ be a sequence of measurable functions and let $f_n \to f$ (a.e.); show that f is a measurable function if the measure μ is complete.

10. *Lusin's Theorem.* Let $E \subset R^n$. A finite-valued function f on E is measurable on E if and only if there exists for every $\epsilon > 0$ a closed set $F \subset E$ on which f is continuous and $\mu^\circ(E - F) < \epsilon$.

7.4 INTEGRALS

The definition of the expectation of a simple random variable was given in Definition 2.3.1. The integral of a nonnegative simple function is defined in exactly the same way. The only novelty is that, in this section, we are working with a measure space $(X, \mathscr{S}, \mu)$ that remains fixed instead of with a probability space.

NONNEGATIVE SIMPLE FUNCTIONS

DEFINITION 1

The *integral* on X of a nonnegative simple function $f = \sum_{j=1}^{n} x_j I_{A_j}$, denoted by $\int f\,d\mu$ or $\int_X f\,d\mu$, is defined by

$$\int f\,d\mu = \sum_{j=1}^{n} x_j \mu A_j.$$

If $\int f\,d\mu < \infty$, the function f is said be *integrable*.

The integral is unambiguously defined;[5] if $f = \sum_{i=1}^{m} y_i J_i$, it follows as in Section 2.3 that $\int f\,d\mu = \sum_{i=1}^{m} y_i \mu B_i$, where B_i are the sets whose indicators are J_i. The proof of the following theorem, which is similar to that of Theorem 2.3.1, should offer no difficulty to the reader.

THEOREM 1

If α and β are real numbers and f and g are integrable functions, then $\alpha f + \beta g$ is an integrable function and

$$\alpha\int f\,d\mu + \beta\int g\,d\mu = \int (\alpha f + \beta g)\,d\mu.$$

DEFINITION 2

The *indefinite integral* over a measurable set E, denoted by $\int_E f\,d\mu$, is defined by

[5]Our definition of simple functions precludes the appearance of the expression $0 \times \infty$; in addition, $\int 0\,d\mu = 0$.

$$\int_E f\,d\mu = \int I_E f\,d\mu,$$

where I_E is the indicator of the set E.

Using the explicit expression $\int_E f\,d\mu = \sum_{i=1}^{n} x_i\,\mu(EA_i)$, the reader should have no difficulty in verifying that

$$\int_E f\,d\mu \leq \int_X f\,d\mu.$$

Hence if f is integrable, the indefinite integral is finite. Moreover, the indefinite integral is countably additive on the class of measurable sets; that is, if $A, A_i \in \mathscr{S}$ and $A = \sum_{i=1}^{\infty} A_i$, then

$$\int_A f\,d\mu = \sum \int_{A_i} f\,d\mu.$$

From the fact that $\int f\,d\mu \geq 0$, we can easily prove the following result.

THEOREM 2

If f and g are each integrable and $f \geq g$, then $\int f\,d\mu \geq \int g\,d\mu$.

Proof

If $f \geq g$, then $f = g + h$, where $h \geq 0$ is simple. Hence $\int f\,d\mu = \int g\,d\mu + \int h\,d\mu \geq \int g\,d\mu$.

NONNEGATIVE MEASURABLE FUNCTIONS

Since every nonnegative measurable function f is the limit of an increasing sequence of nonnegative simple functions (see Corollary 1 of Theorem 2.1), we can define the integral of f as follows.

DEFINITION 3

The integral on X of a nonnegative measurable function f is defined by

$$\int f\,d\mu = \lim_{n\to\infty} \int f_n\,d\mu,$$

where f_n is an increasing sequence of nonnegative simple functions converging to f. The function f is *integrable* if $\int f\,d\mu < \infty$.

Since the sequence f_n is increasing, it follows from Theorem 2 that the sequence $\int f_n \, d\mu$ is also increasing; hence it has a limit (which may be infinite). Therefore, to justify the definition we need only show that the limit is independent of the particular choice of the sequence f_n.

LEMMA 1

Let f be a nonnegative measurable function and let f_n and g_n be two sequences of nonnegative simple functions such that $f_n \uparrow f$ and $g_n \uparrow f$ as $n \to \infty$; then

$$\lim \int f_n \, d\mu = \lim \int g_n \, d\mu.$$

Proof

Let p be a fixed positive integer and let $\epsilon > 0$. Set $A_n = (f_n \geq g_p - \epsilon)$; then $A_n \uparrow X$ as $n \to \infty$. By the additivity of the integral,

$$\begin{aligned} \int f_n \, d\mu &= \int_{A_n} f_n \, d\mu + \int_{A'_n} f_n \, d\mu \\ &\geq \int_{A_n} (g_p - \epsilon) \, d\mu + \int_{A'_n} f_n \, d\mu \geq \int (g_p - \epsilon) \, d\mu \\ &\quad + \int_{A'_n} f_n \, d\mu - \int_{A'_n} g_p \, d\mu, \end{aligned}$$

the inequality being a consequence of Theorem 2. Now if $M = \max g_p$, the last inequality becomes

$$\int f_n \, d\mu \geq \int (g_p - \epsilon) \, d\mu - M\mu A'_n.$$

If $\mu X < \infty$, taking limits in the above inequality yields $\lim \int f_n \, d\mu \geq \int (g_p - \epsilon) \, d\mu$, since $A'_n \downarrow \phi$. Therefore, letting $\epsilon \to 0$,

$$\lim_{n \to \infty} \int f_n \, d\mu \geq \int g_p \, d\mu. \tag{1}$$

Now suppose that $\mu X = \infty$ and $\min g_p = m > 0$. Then

$$\int f_n \, d\mu \geq \int_{A_n} (g_p - \epsilon) \, d\mu \geq (m - \epsilon) \, \mu A_n \to \infty$$

and so Equation (1) holds trivially. Finally, if $\min g_p = 0$ we apply the previous results on the set $B = (g_p > 0)$, obtaining

$$\lim_{n \to \infty} \int_B f_n \, d\mu \geq \int_B g_p \, d\mu = \int g_p \, d\mu,$$

from which (1) follows. Since (1) is valid for arbitrary p, it follows that

$\lim_{n\to\infty} \int f_n \, d\mu \geq \lim_{p\to\infty} \int g_p \, d\mu$. The desired result follows upon observing that the reverse inequality is also valid.

Theorem 1 remains valid for the nonnegative measurable functions f and g. To verify that this extension holds, let f_n and g_n be nonnegative simple functions that increase to f and g respectively. Since

$$\alpha \int f_n \, d\mu + \beta \int g_n \, d\mu = \int (\alpha f_n + \beta g_n) \, d\mu,$$

the result follows by passage to the limit.

The indefinite integral of a nonnegative measurable function f is defined as in Definition 2. Again, if f is integrable, the indefinite integral is finite. We will defer the proof of the σ-additivity of the indefinite integral until after we have proved Lebesgue's monotone convergence theorem; this will be done in the next section. However, finite additivity is a simple consequence of Theorem 1. Setting $\alpha = \beta = 1, f = I_A h$, and $g = I_B h$ yields

$$\int_A h \, d\mu + \int_B h \, d\mu = \int_{A+B} h \, d\mu$$

for any integrable function h and disjoint measurable sets A and B. The general result is proved by induction.

Theorem 2 also remains valid since it is a consequence of Theorem 1.

MEASURABLE FUNCTIONS

Since every measurable function f can be uniquely decomposed into its positive and negative parts, which are both nonnegative measurable functions by Lemma 2.2, the following definition is justified.

DEFINITION 4

The integral on X of a measurable function f is defined by

$$\int f \, d\mu = \int f^+ \, d\mu - \int f^- \, d\mu,$$

where $f^+ = \max(f, 0)$ and $f^- = \min(f, 0)$, provided that at least one of the integrals $\int f^+ \, d\mu$ or $\int f^- \, d\mu$ is finite. If $\int f \, d\mu$ is finite—that is, if both of the last two integrals are finite—then f is said to be *integrable* on X.

The indefinite integral is defined as in Definition 2. The finite additivity of the indefinite integral follows from Definition 4, which defined the integral and finite additivity for nonnegative measurable functions.

We will now show that Theorem 1 can be extended to measurable functions. First we assume that $\alpha = \beta = 1$ and we decompose X into the following six sets:

$$A = (f \geq 0, g \geq 0, f + g \geq 0), B = (f \geq 0, g < 0, f + g \geq 0),$$
$$C = (f \geq 0, g < 0, f + g < 0), \ldots, F = (f < 0, g < 0, f + g < 0).$$

On the set B we have $f^+ = f, g^- = -g$, and $(f + g)^+ = f + g$, while the remaining parts of the functions are zero. Hence, by applying Definition 4 and Theorem 2 to the nonnegative functions $(f + g)I_B$ and $(-g)I_B$, we have

$$\int_B f\, d\mu = \int_B (f + g)\, d\mu + \int_B (-g)\, d\mu = \int_B (f + g)\, d\mu - \int_B g\, d\mu.$$

Since $\int_B g\, d\mu$ is finite, we obtain

$$\int_B f\, d\mu + \int_B g\, d\mu = \int_B (f + g) d\mu.$$

A similar proof holds for the other sets; Theorem 1 follows from the finite additivity of the indefinite integral.

Suppose that $\alpha \neq 1$. For $\alpha \geq 0, (\alpha f)^+ = \alpha f^+$ and $(\alpha f)^- = \alpha f^-$; for $\alpha < 0, (\alpha f)^+ = \alpha f^-$ and $(\alpha f)^- = \alpha f^+$. From these facts and from Theorem 1 with $\beta = 0$ applied to nonnegative functions, it follows that $\int \alpha f d\mu = \alpha \int f\, d\mu$ from Definition 4. Combining these last two results, we obtain

$$\int (\alpha f + \beta g)\, d\mu = \int (\alpha f)\, d\mu + \int (\beta g)\, d\mu = \alpha \int f\, d\mu + \beta \int g\, d\mu.$$

Theorem 2, being a consequence of Theorem 1, remains valid for measurable functions.

We will now prove some elementary properties of the integral.

THEOREM 3

If f is a measurable function and if $\mu A = 0$, then $\int_A f\, d\mu = 0$.

Proof

This is obvious for simple functions. The general result follows from the definitions of the integral.

THEOREM 4

If $0 \leq f \leq g$ (a.e.), f is measurable, and g is integrable, then f is integrable.

Proof

By Theorem 3, there is no loss of generality in assuming that $f \leq g$. Let f_n be an increasing sequence of nonnegative simple functions that converges to f. Since $f_n \leq f \leq g$, it follows by Theorem 2 that $\int f_n \, d\mu \leq \int g \, d\mu$. Thus $\lim \int f_n \, d\mu \leq \int g \, d\mu < \infty$ and so f is integrable.

COROLLARY 1

A measurable function f is integrable if and only if $|f|$ is integrable.

Proof

Suppose that f is integrable; then by definition, f^+ and f^- are both integrable. Since $|f| = f^+ + f^-$, it follows from Theorem 1 that $|f|$ is integrable. Conversely, suppose that $|f|$ is integrable. Since f^+ is measurable and $0 \leq f^+ \leq |f|$, it follows from the theorem that f^+ is integrable. A similar argument applies to f^-; hence f is integrable.

COROLLARY 2

If f is measurable, g is integrable and $|f| \leq g$ (a.e.), then f is integrable.

THEOREM 5

If f is integrable, then $|\int f \, d\mu| \leq \int |f| \, d\mu$.

Proof

Since $\int f^+ \, d\mu$ and $\int f^- \, d\mu$ are both nonnegative, we have

$$\left|\int f \, d\mu\right| = \left|\int f^+ \, d\mu - \int f^- \, d\mu\right| \leq \max\left(\int f^+ \, d\mu, \int f^- \, d\mu\right)$$
$$\leq \int f^+ \, d\mu + \int f^- \, d\mu = \int |f| \, d\mu.$$

Finally, suppose that f is defined and measurable outside a set N of measure zero.[6] Let $\bar{f}$ be any measurable function such that $f = \bar{f}$ on N'. The integral of f is defined by setting $\int f \, d\mu = \int \bar{f} \, d\mu$, provided that $\int \bar{f} \, d\mu$ exists. By Theorem 3, the integrals of such functions $\bar{f}$ whose restrictions are f on N' are equal. Hence, $\int f \, d\mu$ is unambiguously defined. We leave the proofs of Theorems 1 through 4 for a.e. defined and a.e. measurable functions as exercises for the reader.

[6] $(f \leq c) \cap N' \in \mathscr{S}$.

EXERCISES

1. Prove the following statements under the assumption that f is integrable:

(a) f is measurable.
(b) f is finite a.e.
(c) $\mu(|f| \geq \epsilon) < \infty$.

2. Show that if $\lim \int |f_n - f| \, d\mu \to 0$ as $n \to \infty$, then $f_n \to f(\mu)$.

3. (a) Show that if $f \geq 0$ (a.e.), then $\int f \, d\mu = 0$ if and only if $f = 0$ (a.e.).
(b) Show that if $f = g$ (a.e.) and g is integrable, then $\int f \, d\mu = \int g \, d\mu$.

4. Prove that if f is integrable, then $\int_A f \, d\mu = 0$, where A is a set of measure zero.

[*Hint:* $I_A f = 0$ (a.e.).]

5. If $f > 0$ (a.e.) on a measurable set A and if $\int_A f \, d\mu = 0$, show that $\mu A = 0$.

[*Hint:* Let $E_0 = (f(x) > 0)$; then $E_0 = \bigcup_{n=1}^{\infty} E_n$, where $E_n = (f(x) \geq 1/n)$. Show that $\int_{AE_n} f \, d\mu = 0$ and hence $\mu AE_n = 0$.]

6. Prove that if $\int_E f \, d\mu = 0$ for every measurable set E, then $f = 0$ (a.e.).

[*Hint:* Let $E = (f(x) > 0)$ and apply Exercise 5].

7. Show that the product of two elementary functions is integrable if one of them is integrable and the other is bounded a.e.

8. Give an example to show that we cannot choose arbitrary sequences of simple functions $f_n \to f \geq 0$ to define the integral, even if the convergence is uniform.

9. Show that if the measure space consists of a finite number of points and μ is finite, then the theory of integration reduces to summation.

10. Find the probability of getting a rational number.

11. *Chebyshev's Inequality*. Show that if $f \geq 0$ on A and $m > 0$, then

$$\mu(x\colon x \in A, f(x) \geq m) \leq \frac{1}{m} \int_A f \, d\mu.$$

12. *First Mean Value Theorem*. Show that if f is measurable, if $m \leq f < M$ (a.e.) on A, and if $g \geq 0$ is integrable over A, then there exists a real number k such that $m \leq k \leq M$ and

$$\int_A f(x)g(x) \, d\mu = k \int g \, d\mu.$$

13. Let the measure space be the Borel line with the Lebesgue measure and suppose that f is integrable over $[a, b]$. Prove that

(a) $F(x) = \int_{[a,x]} f\,d\mu$ is defined for $a \leq x \leq b$.

(b) $(F(y) - F(x))/(y - x) = 1/(y - x) \int_{[x,y]} f\,d\mu$ for $a \leq x < y \leq b$.

(c) $F'(x_0) = f(x_0)$ for any point x_0 at which $f(x)$ is continuous and $x_0 \in [a, b]$.

14. Under the assumptions of Exercise 13, show that if $\int_{[a,c]} f\,d\mu = 0$ for any $c \in [a, b]$, then $f = 0$ (a.e.) on $[a, b]$. [*Hint:* Show that the class $C = (A\colon A \subset [a, b], \int_A f\,d\mu = 0)$ is a σ-field and apply Exercise 6.]

15. It is known that if f is continuous and satisfies the functional equation $f(x+y) = f(x) + f(y)$, then $f(x) = cx$ for some constant c. Prove that if f is integrable and satisfies such an equation, then $f(x) = cx$.

7.5 LIMIT THEOREMS

A sufficient condition for taking the limit under the integral sign in Riemann integration is the uniform convergence of the sequence involved. In this section we will generalize this result. The following theorem is known as *Lebesgue's monotone convergence theorem.*

THEOREM 1

If an increasing sequence $\{f_n\}$ of nonnegative measurable functions approaches f a.e., then $\lim_{n\to\infty} \int f_n\,d\mu = \int f\,d\mu$.[7]

Proof

We can assume that the f_n are integrable; otherwise the theorem holds trivially. Since each f_n is measurable, we can find an increasing sequence $\{f_{nk}\}$ of nonnegative simple functions such that $\lim_{k\to\infty} f_{nk} = f_n$. These sequences can always be chosen so that $f_{nk+1} \geq f_{nk}$. We now proceed to extract a sequence $\{g_n\}$ from the $\{f_{nk}\}$ that converges to f. Let $g_n = \max(f_{1n}, f_{2n}, \ldots, f_{nn})$; then for $i \leq n$,

$$f_{in} \leq g_n \leq f_n. \tag{1}$$

The g_n are simple integrable functions; they are increasing, since $f_{in+1} \geq f_{in}$. Letting $n \to \infty$ in (1) yields

$$f_i \leq \lim_{n\to\infty} g_n \leq \lim_{n\to\infty} f_n; \tag{2}$$

[7]The integral can be infinite.

taking limits as $i \to \infty$ in (2) yields $\lim_{n\to\infty} g_n = \lim_{n\to\infty} f_n = f$. From the definition of the integral, $\lim_{n\to\infty} \int g_n \, d\mu = \int f \, d\mu$. Using this last fact, integrating inequality (1), and taking limits as $n \to \infty$ gives

$$\int f_i \, d\mu \leq \int f \, d\mu \leq \lim_{n\to\infty} \int f_n \, d\mu, \tag{3}$$

since $\lim_{n\to\infty} \int f_{in} \, d\mu = \int f_i \, d\mu$ by definition. Letting $i \to \infty$ in (3) yields $\int f \, d\mu = \lim_{n\to\infty} \int f_n \, d\mu$, the desired result.

COROLLARY 1

The hypothesis that $f_n \geq 0$ can be dropped if $|\int f_i \, d\mu| < \infty$ for some i.

The proof is an immediate consequence of the theorem applied to the sequence $f_n - f_i$.

COROLLARY 2

If $f_n \geq 0$ and $\sum_{n=1}^{\infty} \int f_n \, d\mu < \infty$, then the series $\sum_{n=1}^{\infty} f_n$ converges a.e. and

$$\int \sum_{n=1}^{\infty} f_n \, d\mu = \sum_{n=1}^{\infty} \int f_n \, d\mu.$$

Corollary 2 follows from the theorem when it is observed that $0 \leq \sum_{k=1}^{n} f_k$ increases to $\sum_{k=1}^{\infty} f_k$. This result also holds for arbitrary f_n if we assume that $\sum_{k=1}^{\infty} \int |f_n| \, d\mu < \infty$.

COROLLARY 3

Let A and A_i be measurable sets and suppose that $A = \sum_{i=1}^{\infty} A_i$; then for any measurable h,

$$\sum_{i=1}^{\infty} \int_{A_i} h \, d\mu = \int_A h \, d\mu.$$

Proof

By definition of the integral, we need only prove the result for $h \geq 0$. Then Corollary 3 follows immediately from the theorem, with $f_n = \sum_{k=1}^{n} hI_{A_k}$ and $f = hI_A$.

COROLLARY 4

The theorem also holds for a decreasing sequence provided that some f_n is integrable.

The following theorem derived by Fatou and Lebesgue is an extension of the previous result.

THEOREM 2

Let f be an integrable function and let f_n be measurable functions.

1. If $f_n \geq f$ (a.e.), then $\int \liminf f_n \, d\mu \leq \liminf \int f_n \, d\mu$.
2. If $f_n \leq f$ (a.e.), then $\int \limsup f_n \, d\mu \geq \limsup \int f_n \, d\mu$.

Proof

We will first assume that the f_n are nonnegative and prove part 1 of the theorem. Since $f_n \geq \inf_{k \geq n} f_k$ and $\inf_{k \geq n} f_k$ increases to $\liminf f_n$, then $\liminf \int f_n \, d\mu \geq \lim \int \inf_{k \geq n} f_k \, d\mu = \int \liminf f_n \, d\mu$, where the least equality follows by Theorem 1.

The proof of parts 1 and 2 of the theorem follows in the general case by applying the first part of the proof to the sequences $\{f_n - f\}$ and $\{f - f_n\}$ respectively.

COROLLARY 1

If g is integrable and $g \leq f_n$ a.e., where the f_n are measurable functions increasing to f, then $\int f_n \, d\mu \to \int f \, d\mu$.

COROLLARY 2

Let g and h be integrable functions and let f_n be measurable functions such that $g \leq f_n \leq h$ and $f_n \to f$ (a.e.). Then $\int f_n \, d\mu \to \int f \, d\mu$.

The proofs of these corollaries follow directly from the theorem and are left to the reader.

The following result is known as *Lebesgue's dominated convergence theorem.*

THEOREM 3

If $|f_n| \leq g$ a.e. with g integrable and if f_n converges in measure to f [or $f_n \to f$ (a.e.)], then f is integrable and the sequence $\{f_n\}$ converges to f in the mean; that is $\lim \int |f_n - f| \, d\mu = 0$.

Proof

The theorem holds under the assumption that $f_n \to f$ (a.e.). Since under this hypothesis $|f_n - f| \to 0$ (a.e.) and $|f_n - f| \leq 2g$, then by Corollary 2 of Theorem 2, $\int |f_n - f|\, d\mu \to 0$.

To prove the result if $f_n \to f(\mu)$, we will show that $\limsup \int |f_n - f|\, d\mu = 0$. To do this we choose a subsequence $\{|f_{n'} - f|\}$ such that $\int |f_{n'} - f|\, d\mu \to \limsup \int |f_n - f|\, d\mu$. Since $|f_{n'} - f| \to 0(\mu)$, we can find a subsequence $\{|f_{n''} - f|\}$ of this sequence that approaches zero a.e. (by the corollary to Theorem 3.7). Then, since every subsequence of a convergent sequence approaches the same limit,

$$\limsup \int |f_n - f| d\mu = \lim \int |f_{n''} - f|\, d\mu = \int \lim |f_{n''} - f|\, d\mu = 0,$$

where the last relation follows from the first part of the proof.

COROLLARY

If f_n are measurable functions such that $|f_n| \leq M$, where M is a constant, and $f_n \to f$ (a.e.) [or $f_n \to f(\mu)$], then $\int_A f_n\, d\mu \to \int_A f\, d\mu$ for any set A of finite measure.

The proof of the corollary follows directly from the theorem.

We recall that convergence a.e. does not imply convergence in measure for sets of infinite measure; this is demonstrated by Example 3.3. However, Theorem 3 shows that if a sequence of measurable functions converges a.e. and is bounded by an integrable function, then it converges in measure.

EXERCISES

1. Let $\{f_n\}$ be an increasing sequence of measurable functions on a measurable set A, and suppose that $\int_A f_n\, d\mu \leq k$ for some constant k and all n. Prove that $f = \lim f_n$ exists a.e. on A and $\int_A f_n\, d\mu \to \int_A f\, d\mu$.

2. (a) Let f_n be elementary integrable functions such that $f_n - f_m \to 0(m)$ (that is, approaches zero in the mean) and $f_n \to f(\mu)$. Define $\rho(f, g) = \int |f - g|\, d\mu$. Show that $\rho(f, f_n) \to 0$ as $n \to \infty$. [*Hint:* Observe that $|f_n - f_m|$

$\to |f - f_m|(\mu)$ and $\{|f_n - f_m|\}$ is a mean fundamental sequence for fixed m—i.e., $\rho(f_n, f_m) < \epsilon$ for $n, m \geq N$. Then apply the technique used to prove the last part of Theorem 3.]

(b) Deduce from Exercise 2(a) that to every integrable function f there corresponds an elementary integrable function f such that $\rho(f, g) < \epsilon$ for an arbitrary $\epsilon < 0$.

3. Show that if $\{f_n\}$ is a mean fundamental sequence of integrable functions, then there exists an integrable function f such that $\rho(f_n, f) \to 0$ as $n \to \infty$. [*Hint:* Use Exercise 2(b) to find an elementary g_n such that $\rho(f_n, g_n) < 1/n$.]

4. Show that if f is integrable, then $\int_A |f|\, d\mu \to 0$ as $\mu A \to 0$. [*Hint:* Define $f_n = f$ if $|f_n| < n$ and $f_n = n$ if $|f| \geq n$. Then use the fact that $\int_A |f_n|\, d\mu \to \int_A |f|\, d\mu$ and the identity $\int_A |f|\, d\mu = \int_A |f_n|\, d\mu + \int_A [|f| - |f_n|]\, d\mu$ to complete the proof.]

5. (a) Show that if $|f_t| \leq g$, g is integrable and $f_t \to f_{t_0}$ as $t \to t_0$ along an arbitrary set T of real numbers in the extended real-number system, then $\int f_t\, d\mu \to \int f_{t_0}\, d\mu$. [*Hint:* The result follows from Theorem 3 by noting that $f_t \to f_{t_0}$ as $t \to t_0$ if and only if $f_{t_n} \to f_{t_0}$ for every sequence $t_n \in T$ converging to t_0.]

(*b*) Show that if f_t is continuous, $\int f_t\, d\mu$ is a continuous function of t.

6. Let $(X, \mathscr{S}, \mu)$ be the Borel line with the Lebesgue measure and suppose that $f(x) \geq 0$ is measurable on $[a, b]$. Furthermore, assume that $f(x)$ is integrable over $[a_n, b]$ where $\lim_{n\to\infty} a_n = a$ and $a_{n+1} > a_n$. Then f is integrable over $[a, b]$ if and only if $\lim_{n\to\infty} \int_{a_n}^{b} f(x)\, d\mu$ exists. What is the value of this limit?

7. (a) Show that if df_t/dt exists at $t_0 \in T$ and

$$\left|\frac{f_t - f_{t_0}}{t - t_0}\right| \leq g,$$

which is integrable, then

$$\frac{d}{dt}\left(\int f_t\, d\mu\right)_{t_0} = \int \left(\frac{df_t}{dt}\right)_{t_0} d\mu.$$

(b) Deduce from Exercise 7(a) that if on a finite interval $[a, b]$, df_t/dt exists and is bounded in absolute value by an integrable function, then

$$\frac{d}{dt}\int_a^b f_t\, d\mu = \int_a^b \frac{df_t}{dt}\, d\mu.$$

[*Hint:* $f_t = f_s = (t - s)(df_t/dt)_\theta$, where $t < \theta < s$].

8. (a) Show that if on a finite interval $[a, b]$ we have f_t continuous and bounded in absolute value by an integrable function, then

$$\int_a^x \left(\int f_t \, d\mu\right) dt = \int \left(\int_a^x f_t \, dt\right) d\mu$$

for every $x \in [a, b]$. The integrals with respect to t are Riemann integrals. [*Hint*: Differentiating the left-hand side of the above equation, we obtain $\int f_t \, d\mu$, since this is a continuous function of t. Apply Exercise 5(b) to the right-hand side.]

(b) Assume that $\int_{-\infty}^{\infty} |f_t| \, dt \leq h$, which is integrable, and that the hypotheses of Exercise 8(a) are fulfilled in every finite interval. Show that

$$\int_{-\infty}^{\infty} \left(\int f_t \, d\mu\right) dt = \int \left(\int_{-\infty}^{\infty} f_t \, dt\right) d\mu.$$

9. Find the expected value of the random variable described in Example 5.9.2.

10. For a finite measurable function f define

$$s_n = \sum_{i=-\infty}^{\infty} \frac{1}{2^{-n}} \mu \left[i2^{-n} \leq f < (i+1)2^{-n}\right]$$

for $n = 1, 2, 3, \ldots$. Suppose that μ is finite.

(a) Show that if f is integrable, then each s_n is absolutely convergent and $\int f \, d\mu = \lim_{n \to \infty} s_n$. [*Hint*: Prove these three problems for nonnegative functions.]

(b) Suppose that s_n converges absolutely for some n. Prove that s_n converges absolutely for all n and that $f(x)$ is integrable. [*Hint*: $f(x) \leq f_n(x) + 1$.]

(c) Prove that if f is integrable, then $\lim_{k \to \infty} k\mu(|f| \geq k) = 0$. [*Hint*: Consider s_0.]

7.6 LEBESGUE AND RIEMANN–STIELTJES INTEGRALS

As an example of the general theory of integration, we will study integrals on the Lebesgue–Stieltjes measure space $(X, \mathscr{L}, \mu)$. Here $X = (-\infty, \infty)$, $\mathscr{L}$ is the class of Lebesgue–Stieltjes sets, and μ is the Lebesgue–Stieltjes measure defined in Definitions 6.6.2 and 6.6.3. The integral $\int f \, d\mu$, where f is an integrable $\mathscr{L}$-measurable function, is called a *Lebesgue–Stieltjes integral*. This integral is also denoted by $\int f \, dF$, where F is a distribution function corresponding to μ (cf., Section 6.6). If $F(x) = x$, the integrals are called *Lebesgue integrals*.

We shall discuss the relation of the Lebesgue–Stieltjes integral to the Riemann–Stieltjes integral on the real line. This discussion can easily be extended to $N > 1$ dimensions.

THEOREM 1

If the Riemann–Stieltjes integral $I = \int_a^b f\,dF$ exists, then f is Lebesgue–Stieltjes-integrable on $(a, b]$, and $\int_{(a,b]} f\,d\mu = I$.

Proof

We first approximate I by upper and lower sums. Consider the partition of $(a, b]$ into n half-open subintervals by the points $a = x_0 < x_1 \cdots < x_{n-1} < x_n = b$. Then upper and lower sums are given by $\bar{S}_n = \sum_{k=0}^{n-1} M_k \,\Delta F_k$ and $\underline{S}_n = \sum_{k=0}^{n-1} m_k \,\Delta F_k$ respectively, where $M_k = \sup f(x)$ on $(x_k, x_{k+1}]$ and $m_k = \inf f(x)$ on the same interval, while $\Delta F_k = F(x_{k+1}) - F(x_k)$. We define simple functions $\bar{f}_n = \sum_{k=0}^{n-1} M_k I_k$ and $\underline{f}_n = \sum_{k=0}^{n-1} m_k I_k$ where I_k is the indicator function of $(x_k, x_{k+1}]$. It is easily verified that $\bar{S}_n = \int_{(a,b]} \bar{f}_n\,d\mu$ and $\underline{S}_n = \int_{(a,b]} \underline{f}_n\,d\mu$. By definition, the Riemann–Stieltjes integral is

$$I = \lim \bar{S}_n = \lim \underline{S}_n,$$

where the limit is taken in such a manner that $\max_{0 \le k \le n-1} (x_{k+1} - x_k) \to 0$. Now $\{\bar{f}_n\}$ is a decreasing sequence and $\{\underline{f}_n\}$ is an increasing sequence, and $\bar{f}_n \to \bar{f} \ge f$, $\underline{f}_n \to \underline{f} \le f$ a.e. Hence by Corollaries 1 and 4 of Theorem 5.1,

$$\int_{(a,b]} \bar{f}\,d\mu = \lim \bar{S}_n = I = \lim \underline{S}_n = \int_{(ab]} \underline{f}\,d\mu.$$

Similarly, the above relations hold for any $c \in (a, b]$ in place of b. Therefore, $\bar{f} = \underline{f}$ a.e. (cf., Exercise 4.14), which implies $\bar{f} = f = \underline{f}$ (a.e.), and consequently

$$\int_{(ab]} f\,d\mu = \int_{(ab]} \bar{f}\,d\mu = I,$$

which proves the theorem.

COROLLARY 1

If f is continuous on $[a, b]$, then the Lebesgue–Stieltjes integral becomes a Riemann–Stieltjes integral.

COROLLARY 2

Let $f(x)$ be bounded on $[a, b]$. Then $f(x)$ is Riemann-integrable if and only if it is continuous a.e.

The proofs of the corollaries follow from the fact that $f(x)$ is continuous at x if and only if $\bar{f}(x) = \underline{f}(x)$. Note that Corollary 2 would apply if f were discontinuous at an arbitrary countable set of points, since the Lebesgue measure of this set is zero. However, the Riemann–Stieltjes integral might not exist, since this same set of points might not have a measure zero with respect to a Lebesgue–Stieltjes measure.

An example of a bounded function that is Lebesgue-integrable but not Riemann-integrable appears in Example 5.9.2.

Even though f is continuous on $[a + \epsilon, b]$ for all $\epsilon > 0$, its improper Riemann–Stieltjes integral defined by

$$\int_a^b f\,dF = \lim_{\epsilon \to 0} \int_{a+\epsilon}^b f\,dF$$

does not necessarily coincide with its Lebesgue–Stieltjes integral. We might have

$$\int_a^b |f|\,dF = \lim_{\epsilon \to 0} \int_{a+\epsilon}^b |f|\,dF = \infty$$

so that $|f|$ is not Lebesgue–Stieltjes-integrable and consequently neither is f. An example appears below. However, if f is Lebesgue–Stieltjes-integrable, then both integrals coincide (see Exercise 5.6).

EXAMPLE 1

The function

$$f(x) = \frac{d}{dx}\left(x^2 \sin \frac{1}{x^2}\right) = 2x \sin \frac{1}{x^2} - \frac{2}{x} \cos \frac{1}{x^2}$$

is not Lebesgue-integrable over $[0, 1]$. Since $|f| \geq 2/x \cos(1/x^2) - 2x \geq x^{-1} - 2x$ on each of the intervals $[(2n + \frac{1}{3})\pi]^{-1/2} \leq x \leq [(2n - \frac{1}{3})\pi]^{-1/2}$ for $n = 1, 2, \ldots,$ then

$$\int_0^1 |f|\,d\mu \geq \sum_{n=1}^{\infty} \left[\frac{1}{2} \ln \frac{6n+1}{6n-1} - 2\left(\frac{3}{\pi}\right)\frac{1}{36n^2 - 1}\right].$$

Hence $\int_0^1 |f|\,d\mu$ does not exist since the first sum diverges. However, the improper Riemann integral $\lim_{\epsilon \to 0} \int_\epsilon^1 f(x)\,dx$ exists.

EXERCISES

1. Complete the proof of Corollaries 1 and 2 of Theorem 1.

2. Prove that a monotone function defined on an interval $[a, b]$ is Riemann-integrable on this interval.

3. Show that if $f(x)$ and $g(x)$ are Riemann-integrable on $[a, b]$, then their product is too.

4. Show that the Lebesgue integral of $(\sin x)/x$ over $(-\infty, \infty)$ does not exist, although the improper Riemann integral exists.

7.7 INDEFINITE INTEGRALS

The indefinite integral ν is a set function defined on the class of all measurable sets $\mathscr{S}$ by

$$\nu(A) = \int_A f\,d\mu = \int I_A f\,d\mu,$$

provided the indicated integrals exist. If f is a measurable function whose integral is defined, then the indefinite integral has the following properties:

1. $\int_A f\,d\mu$ exists for $A \in \mathscr{S}$; in fact ν is σ-finite if μ is σ-finite.
2. ν is a σ-additive set function on $\mathscr{S}$.
3. If $\mu A = 0$, then $\nu A = 0$.

Property 2 was proved earlier in Corollary 3 of Theorem 5.1, while property 3 follows directly from the observation that $I_A f = 0$ a.e. The existence of the indefinite integral is a consequence of the inequalities $I_A f^+ \leq f^+$ and $I_A f^- \leq f^-$ and the fact that at least one of the integrals $\int f^+\,d\mu$, $\int f^-\,d\mu$ is finite. In particular, if f is integrable then ν is finite. Finally, the indefinite integral is σ-finite, since by decomposing X into sets A_n of finite measure,

$$\int_E f\,d\mu = \sum_{m=-\infty}^{\infty} \sum_{n=1}^{\infty} \nu(EA_n(m \leq f < m + 1)),$$

where the terms involving the set $(f = \infty)$ or $(f = -\infty)$ have been omitted. These terms are of measure zero because f is integrable and so finite a.e. If follows that ν is σ-finite, since every term of the double sum is finite.

Property 3 is characteristic of other set functions; for example, $|\mu|\,A = 0$ implies that $\mu^+ A = 0$. Therefore, it is convenient to give it a name.

DEFINITION 1

If $(X, \mathscr{S})$ is a measurable space and μ and ν are signed measures on $\mathscr{S}$, then ν is said to be *absolutely μ-continuous*, in symbols $\nu \ll \mu$, if $\nu E = 0$ for every measurable set E for which $\mu E = 0$ (or $|\mu|E = 0$).

As an illustration, we note that $\mu \gg \mu^+$. Let $X = A + B$ be a Hahn decomposition of X; then $\mu E = 0$ implies that $\mu AE = \mu^+ E = 0$. The reader can verify that if $\mu \gg \mu_1$ and $\mu \gg \mu_2$, then $\mu \gg \mu_1 \pm \mu_2$. The proof that either μ or $|\mu|$ can be used in this definition should now be clear.

The question arises whether the above three properties characterize the indefinite integral. The answer is contained in the Radon–Nikodym theorem, whose proof requires the following lemma.

LEMMA 1

If μ and ν are finite measures, there exist measurable sets A and A_n $(n = 1, 2, 3 \ldots;\ A_n \supset A)$ such that $\nu(EA) = 0$, $\nu(EA') = \nu E$, and $\nu(EA'_n) \geq (1/n)\mu(EA'_n)$ for every measurable set E.

Proof

Let $A_n + B_n$ be a Hahn decomposition for the finite and σ-additive set function $\nu - (1/n)\mu$, for which A_n is the negative set. Writing $A = \bigcap_{n=1}^{\infty} A_n$,

$$0 \leq \nu(EA) = \nu(EAA_n) \leq \frac{1}{n}\,\mu(EAA_n) \leq \frac{1}{n}\,\mu(EA), \qquad (n = 1, 2, \ldots)$$

since $A \subset A_n$. Upon letting $n \to \infty$, it follows that $\nu EA = 0$ and so $\nu(EA') = \nu E$.

THEOREM 1

If $(X, \mathscr{S}, \mu)$ is a σ-finite measure space and if a σ-finite signed measure ν on $\mathscr{S}$ is absolutely μ-continuous, then there exists a finite-valued measurable function f on X, uniquely determined up to an equivalence, such that $\nu E = \int_E f\,d\mu$ for every measurable set E.

Proof

Since μ and ν are σ-finite, X can be written as a countable disjoint union of measurable sets on which both μ and ν are finite. Hence there is no loss in generality, because of the σ-additivity of the integral, in assuming that the signed measures are finite. Furthermore, the assumption $\nu \ll \mu$ is equivalent to the simultaneous validity of the conditions $\nu^+ \ll \mu$ and $\nu^- \ll \mu$.

Therefore, the problem reduces to proving the existence and uniqueness of f for finite measures μ and ν.

Let Φ be the class of all integrable $f \geq 0$ such that $\int_E f\,d\mu \leq \nu E$ for every measurable set E. Since Φ is not empty ($f \equiv 0$ belongs to it), there is a sequence $\{f_n\} \subset \Phi$ such that $\lim \int f_n\,d\mu = \alpha = \sup\ (\int f\,d\mu : f \in \Phi)$. Note that $\alpha \leq \nu X < \infty$. Let $f_0 = \sup f_n$; we will show that $f_0 \in \Phi$ and $\int f_0\,d\mu = \alpha$. To do this we express f_0 as the limit of an increasing sequence of functions $g_n = \max(f_1, \ldots, f_n)$. We can assume that there are an infinite number of distinct f_i or else the result follows trivially. Since any fixed measurable set E can be written as a finite disjoint union of measurable sets $E = E_1 + \cdots + E_n$, where $E_i = (f_i = g_n)E$ for $i = 1, 2, \ldots, n$, then

$$\int_E g_n\,d\mu = \sum_{j=1}^{n} \int_{E_j} f_j\,d\mu \leq \sum_{j=1}^{n} \nu E_j = \nu E,$$

for fixed n. Consequently, the desired result follows from Theorem 5.1. Since f_0 is integrable, there exists a finite-valued measurable function f such that $f_0 = f$ a.e. We will show that the measure

$$\nu_0(E) = \nu(E) - \int_E f\,d\mu \tag{1}$$

is identically zero.

Let $g = f + (1/n)\,I_n$, where I_n is the indicator of the set A'_n described in Lemma 1 for $\nu = \nu_0$. Then

$$\int_E g\,d\mu = \int_E f\,d\mu + \frac{1}{n}\,\mu(EA'_n) = -\nu_0(A'E) + \nu E + \frac{1}{n}\,\mu(EA'_n),$$

where we have replaced $\int_E f\,d\mu$ by using (1). However,

$$\nu_0(A'E) - \frac{1}{n}\,\mu(EA'_n) \geq \nu_0(A'_nE) - \frac{1}{n}\,\mu(EA'_n) \geq 0$$

and so $\int_E g\,d\mu \leq \nu E$, which implies that $g \in \Phi$. This conclusion is contradicted by

$$\int g\,d\mu = \alpha + \frac{1}{n}\,\mu A'_n > \alpha$$

unless $\mu A'_n = 0$. Therefore $\mu(\cup A'_n) = 0$, since $\mu A'_n = 0$ for all n. Hence $\mu(EA') = 0$ and, since $\nu_0 \ll \mu$, $\nu_0 E = \nu_0(EA') = 0$ for every measurable set E. This completes the proof of the existence of a finite measurable function f.

The uniqueness of f follows from the fact that if $\nu E = \int_E g\, d\mu$, then $\int_E (f - g)\, d\mu = 0$ for every measurable set E. This in turn implies that $f = g$ a.e. with respect to μ.

We emphasize the fact that f is not, in general, integrable although either the positive or the negative part of f is integrable. A necessary and sufficient condition that f be integrable is that ν be finite. The Radon–Nikodym theorem holds even if ν is not σ-finite, but in this case f is not necessarily finite valued (see Exercise 5). However, if μ is not σ-finite, then the theorem is not necessarily true even if ν is finite, as the following example shows.

EXAMPLE 1

Let X be an uncountable set and let a measurable set be any set that is countable or that has a countable complement. For every measurable set E, let $\mu(E)$ be the number of points in E and let νE be 0 or 1 according to whether or not E is countable. Suppose that $\nu E = \int_E f\, d\mu$ for all $E \in \mathscr{S}$. Then $E = (f > 0)$ is uncountable, since $1 = \int_E f\, d\mu + \int_{E'} f\, d\mu$. If $F \neq 0$ is a finite subset of E, then $\int_F f\, d\mu = 0$ is false.

For a Lebesgue integral $F(x) = \int_a^x f\, d\mu$, $F'(x) = f(x)$ at every point at which f is continuous (cf., Exercise 4.13). The notion that the derivative of an indefinite integral is the integrand can be generalized.

DEFINITION 2

If μ is a σ-finite measure and if $\nu E = \int_E f\, d\mu$ for every measurable set E, then we can write $f = d\nu/d\mu$ (or $d\nu = f\, d\mu$) and call such a function f a *Radon–Nikodym derivative*.

Note that the Radon–Nikodym derivative $d\nu/d\mu$ is unique only a.e. with respect to μ. These derivatives possess the same properties as ordinary derivatives; the proof of this statement is contained in the Exercises.

Theorem 1 can be generalized for σ-finite signed measures that are not absolutely μ-continuous. An examination of the theorem reveals that μ-continuity is not used until the last stage of the proof. If ν is not μ-continuous, then ν_0 is not μ-continuous; we cannot deduce that $\nu_0(A'E) = 0$ from $\mu(A'E) = 0$. In fact, ν_0 exhibits an extreme form of non-μ-continuity. Not only is it false that the vanishing of μ implies that of ν_0 but, in general, the only sets for which ν_0 does not vanish are the ones for which μ does (see Theorem 2).

DEFINITION 3

If $(X, \mathscr{S})$ is a measurable space, we say that two signed measures μ, ν on $\mathscr{S}$ are *singular*—that is, $\mu \perp \nu$—if there exists a set A such that for every measurable set E, AE and $A'E$ are measurable and $|\mu|(AE) = |\nu|\,(A'E) = 0$.

For example, for every signed measure μ, $\mu^+ \perp \mu^-$.

LEMMA 2

If $\nu \ll \mu$ and $\nu \perp \mu$, then $\nu = 0$.

Proof

We have $\nu(AE) = 0$ and $\mu(A'E) = 0$ for some A and for every measurable set E. Since $\nu \ll \mu$, this implies $\nu(A'E) = 0$.

The discussion above motivates the *Lebesgue decomposition* of a σ-finite signed measure into an absolutely continuous part and a singular part with respect to another σ-finite signed measure.

THEOREM 2

If $(X, \mathscr{S})$ is a measurable space and μ and ν are σ-finite signed measures on $\mathscr{S}$, then there exists a unique decomposition

$$\nu = \nu_0 + \nu_1$$

such that $\nu_0 \perp \mu$ and $\nu_1 \ll \mu$.

Proof

We may assume that μ and ν are finite. We may also assume that μ is a measure, since absolute continuity and singularity with respect to μ are equivalent to absolute continuity and singularity with respect to $|\mu|$. Finally we may consider ν^+ and ν^- separately, and so we may assume that ν is also a measure. Hence the proof is reduced to verifying that the decomposition theorem is valid for finite measures.

The remainder of the proof is similar to that of Theorem 1. It follows (see Exercise 7) that $\nu_1(E) = \int_E f\,d\mu$ and $\nu_0(E)$ defined by Equation (1) satisfy the requirements of the theorem.

If $\nu = \nu_0 + \nu_1 = \bar{\nu}_0 + \bar{\nu}_1$, then $\nu_0 - \bar{\nu}_0 = \bar{\nu}_1 - \nu_1$. Since $\nu_0 - \bar{\nu}_0$ is singular and $\bar{\nu}_1 - \nu_1$ is absolutely continuous with respect to μ, it follows from Lemma 2 that $\nu_0 = \bar{\nu}_0$ and $\nu_1 = \bar{\nu}_1$. This establishes the uniqueness of the decomposition and completes the proof of the theorem.

EXERCISES

1. Prove the following statements:

(a) If μ, ν are two measures on a σ-field $\mathscr{S}$, then $\nu \ll \mu + \nu$.
(b) For every signed measure μ, we have $\mu \ll |\mu|$ and $|\mu| \ll \mu$.
(c) If μ is a signed measure and E is a measurable set, then $|\mu|E = 0$ if and only if $\mu F = 0$ for every measurable subset $F \subset E$.

2. Show that $\nu \ll \mu$ is equivalent to the following:

(a) $\nu^+ \ll \mu$ and $\nu^- \ll \mu$.
(b) $|\nu| \ll |\mu|$.

3. If μ is a signed measure and f is a measurable function that is integrable with respect to $|\mu|$, we define

$$\nu E = \int_E f\,d\mu = \int_E f\,d\mu^+ - \int_E f\,d\mu^-.$$

Show that $\nu \ll \mu$.

4. (a) Prove that if μ is a signed measure, then for any finite signed measure $\nu \ll \mu$ there exists a $\delta = \delta(\epsilon)$ such that for any measurable set E, $|\mu|E < \delta$ implies $|\nu|E < \epsilon$ for any $\epsilon > 0$. [*Hint*: Find a sequence $\{E_n\}$ such that $|\mu|(E_n) < 1/2^n$ and $|\nu|E_n \geq \epsilon$. If $E = \limsup E_n$, then $|\mu|E = 0$ and $|\nu|E \geq \epsilon$.]
(b) The above result is not true if ν is not finite. [*Hint*: Let X be the set of all positive integers and let $\mu E = \sum_{n \in E} 2^{-n}$ and $\nu E = \sum_{n \in E} 2^n$ for any $E \subset X$.]

5. *Generalized Radon–Nikodym Theorem.* Prove Theorem 1 under the assumption that ν is only a signed measure. In this case f is not necessarily finite valued. [*Hint*: Let $\mathscr{C}$ be the class of all measurable sets C such that ν is σ-finite on C, and let α be the supremum of μ on $\mathscr{C}$. There exists a sequence $\{A_n\}$ such that $\alpha = \lim \mu A_n$; let $A = \bigcup A_n$. Show that ν on AE is σ-finite and that ν on $A'E$ can take the values 0 and ∞ only for any $E \in \mathscr{S}$. To find f on AE apply the Radon–Nikodym Theorem; show that $f = \infty$ on $A'E$.]

6. *Radon–Nikodym Derivatives.* If λ and μ are σ-finite measures and ν is a σ-

finite signed measure, prove the following theorems. (Here $[\mu]$ signifies that the statement is true a.e. with respect to μ.)

(a) $\dfrac{d(\nu_1 + \nu_2)}{d\mu} = \dfrac{d\nu_1}{d\mu} + \dfrac{d\nu_2}{d\mu}[\mu]$.
(b) $\nu[x: (d\nu/d\mu)(x) = 0] = 0$ if $\nu \ll \mu$.
(c) $d\nu/d\lambda = (d\nu/d\mu)(d\mu/d\lambda)[\lambda]$ if $\mu \ll \lambda$ and $\nu \ll \mu$.
[*Hint*: Assume that ν is a measure. Let $d\nu/d\mu = f$ and find $f_n \uparrow f$, where f_n are simple nonnegative functions. Let $g = d\mu/d\lambda$; show that $\int_E f_n g\, d\mu = \int_E f_n g\, d\lambda$ by first considering indicator functions.]
(d) $\int f\, d\mu = \int f (d\mu/d\lambda) d\lambda$ if $\mu \ll \lambda$.
[*Hint*: Apply Exercise 6(c).]
(e) $d\mu/d\lambda = 1/(d\lambda/d\mu)$ if $\lambda \ll \mu$ and $\mu \ll \lambda$.

7. Complete the proof of Theorem 2. [*Hint*: To show that $\nu_0 \perp \mu$, apply Lemma 1 to ν_0 and repeat the reasoning of Theorem 1.]

7.8 ITERATED INTEGRALS

In this section, we will assume that $(X_i, \mathscr{S}_i, \mu_i)$ are σ-finite measure spaces and that μ is the product measure (see Section 6.7) on $\times_{i=1}^{n} \mathscr{S}_i$.

If a function f on $\times_{i=1}^{n} X_i$ is such that its integral exists, then this integral is denoted by $\int f(x_1 \cdots x_n)\, d\mu\, (x_1 \cdots x_n)$ or $\int f\, d\mu$;it is called the *n-tuple integral*[8] of f. If f is such that the integrals

$$\int g_i(x_i, \ldots, x_n)\, d\mu_i = g_{i+1}(x_{i+1}, \ldots, x_n) \qquad (i = 1, \ldots, n)$$

are all defined, where $g_1 = f(x_1 \cdots x_n)$, we write

$$\int g_n\, d\mu_n = \iint \cdots \int f\, d\mu_1 d\mu_2 \cdots d\mu_n.$$

The symbol $\iint \cdots \int f\, d\mu_{i_1} d\mu_{i_2} \cdots d\mu_{i_n}$, where $i_1 \cdots i_n$ is a permutation of $(1, 2, \ldots, n)$, is defined similarly. These integrals are called the *iterated integrals* of f.

If we have a nonnegative measurable function, we know that the n-tuple integral exists. We shall show that the iterated integrals exist and are equal to the n-tuple integral. The existence of the iterated integral hinges upon the fact that if certain variables are kept constant in a measurable function, the resulting function is still measurable. This is made more precise in the following definitions and theorems.

[8]The double integral for $n = 2$ and the triple integral for $n = 3$.

DEFINITION 1

If f is any function defined on a subset E of the product space $\mathop{\times}\limits_{i=1}^{n} X_i$ and if $(x_1, \ldots, x_m) \in \mathop{\times}\limits_{i=1}^{m} X_i$, we will call the function $f_{x_1 x_2, \ldots, x_m}$, which is defined on the section $E(x_1, \ldots, x_m)$ by

$$f_{x_1 \ldots x_m}(x_{m+1}, \ldots, x_n) = f(x_1, x_2, \ldots, x_n),$$

the *section determined* by $(x_1, \ldots, x_m)$. The concept of a section of f determined by $x \in X_{i_1} \times \cdots \times X_{i_m}$ is defined similarly.

THEOREM 1

Every section of a measurable function is a measurable function.

Proof

Let f be a measurable function on $\mathop{\times}\limits_{i=1}^{n} X_i$, $x \in \mathop{\times}\limits_{i=1}^{m} X_i$, and let B be a Borel set of the real line. The relations

$$\begin{aligned} f^{-1}_{x_1 \cdots x_m}(B) &= ((x_{m+1}, \ldots, x_n) : f_{x_1 \ldots x_m}(x_{m+1}, \ldots, x_n) \in B) \\ &= ((x_{m+1}, \ldots, x_n) : f(x_1, \ldots, x_n) \in B) \\ &= ((x_{m+1}, \ldots, x_n) : (x_1, \ldots, x_n) \in f^{-1}(B)) \\ &= f^{-1}(B)(x_1, \ldots, x_m) \end{aligned}$$

and the results of Section 6.7 show that the above section of f is measurable. The proof of the measurablity of an arbitrary section of f is similar.

The following theorem is necessary for the proof that the value of the iterated integral equals the value of the n-tuple integral. It also provides us with an alternative method of defining product measures.

THEOREM 2

If $(X_i, \mathscr{S}_i, \mu_i)$ are σ-finite measure spaces, then the set function μ, defined for every set E in $\mathop{\times}\limits_{i=1}^{n} \mathscr{S}_i$ by

$$(1) \quad \mu E = \int \cdots \int \mu_i E(x_1 \cdots x_{i-1} x_{i+1} \cdots x_n)\, d\mu_1 \cdots d\mu_{i-1}\, d\mu_{i+1} \cdots d\mu_n, \quad (i = 1, \ldots, n)$$

is a σ-finite measure with the property that, for every measurable rectangle $\mathop{\times}\limits_{i=1}^{n} E_i$, $\mu\left(\mathop{\times}\limits_{i=1}^{n} E_i\right) = \prod\limits_{i=1}^{n} \mu_i E_i$. This last condition determines μ uniquely (in fact, μ is the product measure).

Proof

We will first prove that the set function μ is uniquely defined; that is, all n integrals in (1) are equal. Let $\mathscr{M}$ be the class of all those measurable sets for which the integrals are equal. It is easy to see that the class $\mathscr{M}$ is closed under the formation of countable sums.

Since the measures μ_i are σ-finite, the product space is decomposable into a countable sum of rectangles with sides of finite measure. It follows that without loss of generality, we can suppose that the measures μ_i are finite.

If $E = \underset{i=1}{\overset{n}{\times}} E_i$ is a measurable rectangle, then

$$\mu_i E(x_1 \cdots x_{i-1}x_{i+1} \cdots x_n) = (\mu_i E_i)(\prod_{j \neq i} I_j),$$

where I_j is the indicator of E_j. Thus the functions involved are measurable and the iterated integrals reduce to $\prod_{i=1}^{n} \mu_i E_i$. This proves the last asserted equality and shows that $\mathscr{M}$ contains all measurable rectangles. Consequently, $\mathscr{M}$ contains the field of finite disjoint unions of these rectangles.

We will show that $\mathscr{M}$ is a monotone class. The limit of an increasing sequence of sets belonging to $\mathscr{M}$ also belongs to $\mathscr{M}$ because of the monotone convergence theorem (5.1). $\mathscr{M}$ is also closed under decreasing passages to the limit because of the dominated convergence theorem (5.3) and the finiteness of the measures. Therefore, by Theorem 6.3.1, it contains the product σ-field $\underset{i=1}{\overset{n}{\times}} \mathscr{S}_i$, and the equality of the integrals is proved.

The fact that μ is a measure follows from Corollary 2 of Theorem 5.1. The σ-finiteness of μ is a consequence of the fact that every measurable subset $\underset{i=1}{\overset{n}{\times}} X_i$ may be covered by countably many measurable rectangles of finite measure; uniqueness is implied by Theorem 6.5.5.

COROLLARY 1

Under the hypothesis of Theorem 2, we have

$$\int \mu^{(m)}E(x_1, \ldots, x_m)\, d\mu_{(m)}$$

$$= \iint \cdots \int \mu^{(m)}E(x_1, \ldots, x_m)\, d\mu_1 d\mu_2 \cdots d\mu_m = \mu E,$$

where $\mu^{(m)} = \mu_{m+1} \times \mu_{m+2} \times \cdots \times \mu_n$ and $\mu_{(m)} = \mu_1 \times \mu_2 \times \cdots \times \mu_m$. This result also holds for an arbitrary section $E(x_{i_1}, x_{i_2}, \ldots, x_{i_m})$.

COROLLARY 2

A measurable subset E of $\times_{i=1}^{n} X_i$ has measure zero if and only if almost every section $E(x_i)$ (for a fixed i) has measure zero.

The proof of Corollary 1 is practically identical to that of the theorem; we leave the minor modifications in the proof to the reader. Corollary 2 follows from the relation $\mu E = \int \mu^{(i)} E(x_i)\, d\mu_i$ and from the fact that the integral of a nonnegative function vanishes if and only if the integrand vanishes a.e.

Theorem 2 is not true if the condition of σ-finiteness is omitted.

EXAMPLE 1

Let $X_1 = X_2 = [0, 1]$, $\mathscr{S}_1 = \mathscr{S}_2$ be the class of Borel sets, and let $\mu_1 E$ be the Lebesgue measure of E and $\mu_2 E$ the number of points in E. If $E = ((x_1, x_2)\colon x_1 = x_2)$, then E is a measurable subset of $X_1 \times X_2$, since E is the limit of a decreasing sum of squares. However, $\int \mu_2 E(x_1)\, d\mu_1 = 1$ and $\int \mu_1 E(x_2)\, d\mu_2 = 0$.

We are now in a position to study the relation between iterated and n-tuple integrals. The following result comes from Lebesgue and Fubini; Corollary 2 is usually called Fubini's Theorem.

THEOREM 3

If f is an integrable function on $\times_{j=1}^{n} X_j$, then

$$\int f\, d\mu = \int\int f\, d\mu_i\, dv = \int\int f\, dv\, d\mu_i, \qquad (i = 1, 2, \ldots, n)$$

where $v = \mu_1 \times \cdots \times \mu_{i-1} \times \mu_{i+1} \times \cdots \times \mu_n$.

Proof

It is sufficient to consider only nonnegative functions f, since a real-valued function is integrable if and only if its positive and negative parts are integrable.

If f is the indicator of a measurable set E, then $\int f\, d\mu_i = \mu_i E(x_1 \cdots x_{i-1} x_{i+1} \cdots x_n)$ and $\int f\, dv = vE(x_i)$; the desired result follows from Corollary 1 of Theorem 2. Since a simple function is a finite linear combination of indicators, the theorem holds for simple functions. For an arbitrary f, we may find an increasing sequence $\{f_n\}$ of nonnegative simple functions converging to f. By the monotone convergence theorem (5.1), $\int f\, d\mu = \lim \int f_n\, d\mu$. If $g_n = \int f_n\, d\mu_i$, then $\{g_n\}$ is an increasing sequence of nonnegative measurable functions, because of the properties of the sequence $\{f_n\}$. Since $\int g_n\, dv =$

$\int f_n\, d\mu$, the first equality in the theorem follows by two applications of the monotone convergence theorem. The other equality is proved in a similar manner.

COROLLARY 1

Under the hypothesis of Theorem 2 almost every section of f is integrable and $\int f\, d\mu = \int\int f\, d\alpha\, d\beta$, where α is the product measure on the product of an arbitrarily selected but fixed number of subspaces, while β is the product measure on the remainder.

COROLLARY 2

Under the hypothesis of Theorem 2,

$$\int f\, d\mu = \int \cdots \int f\, d\mu_{i_1}\, d\mu_{i_2} \cdots d\mu_{i_n},$$

where $i_1, \ldots, i_n$ is some permutation of the numbers $1, 2, \ldots, n$.

Both of the corollaries follow by repeatedly applying the theorem. Note that if f is a nonnegative measurable function, Theorem 3 is always valid.

EXERCISES

1. Verify the following statements:

(a) If I is the indicator of a rectangle $A \times B$, then $I(x, y) = I_A(x)I_B(y)$.
(b) Every section of a simple function is a simple function.

2. Prove Fubini's Theorem for $\bar{\mu}$, the completion of μ. [*Hint*: Every function which is measurable $\overline{\times \mathscr{S}_i}$ is equal a.e. with respect to $\bar{\lambda}$ to a function which is measurable $\times \mathscr{S}_i$.]

3. Show that

$$\int_E f\, d\mu = \int_{X_2} \int_{E(x_2)} f\, d\mu_1\, d\mu_2 = \int_{X_1} \int_{E(x_1)} f\, d\mu_2\, d\mu_1.$$

4. Let $E_1 = E_2 = [-1. 1]$ and $f(x, y) = xy/(x^2 + y^2)^2$. Show that $\int_{-1}^{1} f\, dx = \int_{-1}^{1} f\, dy = 0$ but that the Lebesgue double integral $\int_E f\, d\mu$ over the square $E = E_1 \times E_2$ does not exist.

5. Suppose that f and g are integrable over X_1 and X_2 respectively and $h = fg$.

Show that h is $\mu = \mu_1 \times \mu_2$-integrable over $X_1 \times X_2$ and $\int h\, d\mu = (\int f\, d\mu_1) \cdot (\int g\, d\mu_2)$.

6. *Cavalieri's Principle.* Show that if A and B are measurable subsets of $X_1 \times X_2$ such that $\mu_2 A(x_1) = \mu_2 B(x_1)$ for almost every $x_1 \in X_1$, then $\mu A = \mu B$.

7. (a) The αth fractional integral[9] of a Lebesgue integrable function f is defined by

$$I_\alpha(f)(x) = [\Gamma(\alpha)]^{-1} \int_0^x (x - t)^{\alpha-1} f(t)\, dt$$

for $a \leq x \leq b$, where $\Gamma(\alpha)$ is the gamma function. Prove that $I_\alpha(f)(x)$ is defined a.e. on $[0, b]$ by showing that it is integrable over $[0, a]$ for $0 \leq a \leq b$.
(b) If $\alpha > 0$, $\beta > 0$, show that

$$I_\beta(I_\alpha(f)(x)) = I_{\alpha+\beta}(f).$$

[*Hint*: For $p > 0$, $q > 0$, we have $\int_0^1 x^{p-1}(1 - x)^{q-1}\, dx = \Gamma(p)\Gamma(q)/\Gamma(p + q)$.]

8. Show that in order that $\sum_{n=1}^{\infty} \sum_{m=1}^{\infty} a_{mn} = \sum_{m=1}^{\infty} \sum_{n=1}^{\infty} a_{mn}$, it is sufficient for at least one of the following conditions to be satisfied:

(a) $a_{mn} \geq 0$ for all m, n.

(b) $\sum_{m=1}^{\infty} \sum_{n=1}^{\infty} |a_{mn}| < \infty$.

(c) $\sum_{n=1}^{\infty} \sum_{m=1}^{\infty} |a_{mn}| < \infty$.

7.9 MEASURABLE TRANSFORMATIONS

The natural generalization of change of variables for Riemann integration leads to the study of transformations of our more general integral. A *transformation* is merely a function T defined for every point of a set X, taking values in a set Y. Arbitrary transformations are too general for our purpose; therefore, we will consider the following special class.

DEFINITION 1

Let T be a transformation of X into Y and let $(X, \mathscr{S})$ and $(Y, \mathscr{F})$ be measurable spaces; then T is a *measurable transformation* if the inverse image of every measurable set is measurable.

The following lemma is an immediate consequence of this definition.

[9] If $\alpha = n$ an integer, then $I_n(f)$ is just the integral iterated n-times [cf., Exercise 7(b)].

LEMMA 1

If T is a measurable transformation from $(X, \mathscr{S})$ into $(Y, \mathscr{F})$, then $T^{-1}(\mathscr{F})$ is a σ-field contained in $\mathscr{S}$.

THEOREM 1

Let T be a measurable transformation from $(X, \mathscr{S})$ into $(Y, \mathscr{F})$. Every extended real-valued function g on Y determines a unique function f on X, denoted by $T^{-1}g$. If g is measurable, then f is measurable with respect to the σ-field $T^{-1}(\mathscr{F})$.

Proof

We define f by the equation $f = gT$. Let B be a Borel set; then $f^{-1}(B) = T^{-1}(g^{-1}B)$. Since g is measurable, $g^{-1}(B) \in \mathscr{F}$ and so $f^{-1}(B) \in T^{-1}(\mathscr{F})$.

LEMMA 2

If f is a measurable function with respect to $T^{-1}(\mathscr{F})$, then $T(x_1) = T(x_2)$ implies $f(x_1) = f(x_2)$.

Proof

Let $f(x_1) = c$; then $f^{-1}\{c\} = T^{-1}(F)$ for some $F \in \mathscr{F}$, since f is measurable with respect to $T^{-1}(\mathscr{F})$. Because $x_1 \in f^{-1}\{c\}$, $y = T(x_1) \in F$ and so $T^{-1}\{y\} \subset T^{-1}(F)$. Now $T(x_1) = T(x_2) = y$ implies that $x_2 \in T^{-1}(F) = f^{-1}\{c\}$; hence $f(x_2) = c$.

If T is onto Y (or the range R of T is a measurable set), it is simple to determine a measurable function g. We need only set $g(Tx) = f(x)$ on R and 0 otherwise; g is well defined by Lemma 2. The general situation is treated below.

THEOREM 2

Let T be a measurable transformation from $(X, \mathscr{S})$ into $(Y, \mathscr{F})$. Every extended real-valued function f on X measurable with respect to $T^{-1}(\mathscr{F})$ determines a measurable function g on Y, denoted by Tf. Moreover $f(x) = g(y)$ on the range of T, where $x \in T^{-1}\{y\}$.

Proof

Let $S_r = (x: f(x) \leq r)$ for rational r. By hypothesis, there exists a measurable set $M_r \in \mathscr{F}$ such that $S_r = T^{-1}(M_r)$. Although $S_r \uparrow$ as $r \uparrow$, the M_r, in general, do not. If we define $F_r = \bigcap_{s \geq r} M_s$, then F_r is measurable, $F_r \subset F_s$

for $r < s$, and $S_r = T^{-1}(F_r)$. Noting that $f(x) = (\inf r: x \in S_r)$, we define $g(y) = (\inf r: y \in F_r)$ for $y \in \bigcup_r F_r$ and $g(y) = 0$ for $y \notin \bigcup_r F_r$. Since for $s \leq 0, g(y) < s$ if and only if $y \in F_r$ for some $r < s$, we have $(y: g(y) < s) = \bigcup_{r<s} F_r$ for $s \leq 0$ and $(y: g(y) < s) = (\bigcup_{r<s} F_r) \cup (\bigcup_r F_r)'$ for $s > 0$; therefore, g is measurable.

On the range of T, we have $f(x) = g(y)$, where x is any one of the preimages of y (see Lemma 2). This follows from the fact that if $r_i \downarrow g(y)$, then $y \in \bigcap F_{r_i}$, which implies $x \in \bigcap S_{r_i}$. Hence $f(x) \leq g(y)$; a similar argument shows that $f(x) \geq g(y)$.

The function $g = Tf$ determined by f on X is not unique. For example, we could have defined g to be 1 on $(\bigcup_r F_r)'$ [cf., Exercise 5(c)]. On the other hand, the operator T^{-1} is single valued. Moreover, if g is a function on Y, then $TT^{-1}g = g$, while for a function f on X, $T^{-1}Tf = f$.

A measurable transformation T from $(X, \mathscr{S})$ into $(Y, \mathscr{F})$ also assigns a set function ν on $\mathscr{F}$ to every set function μ on $\mathscr{S}$ by $\nu = \mu T^{-1}$.

THEOREM 3

Let T be a measurable transformation from a measure space $(X, \mathscr{S}, \mu)$ into a measurable space $(Y, \mathscr{F})$. If g is an extended real-valued measurable function on Y, then $\int_Y g \, d\mu T^{-1} = \int_X (gT) \, d\mu$ if either of the integrals exist.

Proof

Since $g = g^+ - g^-$, it is sufficient to treat nonnegative functions g. If g is the indicator of a measurable set F in Y, then it is easy to see that gT is the indicator of $T^{-1}F$; therefore,

$$\int g \, d\mu T^{-1} = \mu T^{-1}(F) = \mu[T^{-1}(F)] = \int gT \, d\mu.$$

Hence the theorem is valid whenever g is a simple function. For arbitrary g, we find an increasing sequence of nonnegative simple functions g_n converging to g. Then $\{g_n T\}$ is an increasing sequence of simple functions converging to gT and the conclusion follows from the monotone convergence theorem (5.1).

COROLLARY

Under the hypothesis of the theorem and with $F \in \mathscr{F}$,

$$\int_F g(y) \, d\mu T^{-1}(y) = \int_{T^{-1}(F)} g[T(x)] \, d\mu(x).$$

The corollary follows immediately from the theorem by applying it to the function $I_F g$. Observe that the formal substitution $y = Tx$ reduces either side of the above equation to the other.

A further consequence of Theorem 3 is the rule corresponding to the law of transformation of variables in n-tuple Riemann integrals.

THEOREM 4

Let T be a measurable transformation from a measure space $(X, \mathscr{S}, \mu)$ into a σ-finite measure space $(Y, \mathscr{F}, \nu)$ and suppose that $\mu T^{-1} \ll \nu$. Then there exists a nonnegative measurable function Φ on Y such that

$$\int g[T(x)]\, d\mu(x) = \int g(y)\Phi(y)\, d\nu(y)$$

for every measurable function g, provided that one of these integrals exists.

Proof

It is sufficient to prove the theorem for nonnegative measurable functions g. By the Radon–Nikodym Theorem, $\mu T^{-1} = \int \Phi\, d\nu$ or $\Phi = d\mu T^{-1}/d\nu$. Let I be the indicator function of a measurable set F; then

$$\int I\, d\mu T^{-1} = \mu T^{-1}(F) = \int I\Phi\, d\nu.$$

Hence this equality between the integrals continues to hold for simple functions g_n. If $\{g_n\}$ is a sequence of simple measurable functions converging to g, then $\int g\, d\mu T^{-1} = \int g\Phi\, d\nu$, provided that either integral exists. The desired result follows from Theorem 3.

The function Φ corresponds to the Jacobian in the theory of the transformation of multiple integrals.

DEFINITION 2

A measurable transformation T from a measurable space $(X, \mathscr{S})$ into a measurable space $(Y, \mathscr{F})$ is *measurability-preserving*[10] if for every $S \in \mathscr{S}$ there exists an $F \in \mathscr{F}$ such that $S = T^{-1}(F)$.

DEFINITION 3

A measurability-preserving transformation T from a measure space $(X, \mathscr{S}, \mu)$ into a measure space $(Y, \mathscr{F}, \nu)$ is called *measure-preserving* if $\mu T^{-1} = \nu$.

[10]Some authors impose the more restrictive conditions of Exercise 3 in defining "measurability-preserving."

EXERCISES

1. Show that the product of two measurable transformations is measurable.

2. Show that if T is a measurable transformation from $(X, \mathcal{S})$ into $(Y, \mathcal{F})$, and if μ and ν are two measures on $\mathcal{S}$ such that $\nu \ll \mu$, then $\nu T^{-1} \ll \mu T^{-1}$.

3. Let T be a one-to-one transformation from a measurable space $(X, \mathcal{S})$ onto a measurable space $(Y, \mathcal{F})$ such that both T and T^{-1} are measurable. Prove that T is measurability-preserving.

4. Let T be a measure-preserving transformation from a measure space $(X, \mathcal{S}, \mu)$ into a measure space $(Y, \mathcal{F}, \nu)$. Prove that if g is an extended real-valued measurable function on Y, then $\int_Y g\, d\nu = \int_X g\, T\, d\mu$, provided that either integral exists.

5. With T as in Exercise 4; show that

(a) $T^{-1}Y = X$.
(b) If R is the range of T, then R has an outer measure νY in the sense that every measurable set containing R has a measure νY.
(c) If $g_1 = Tf$ and $g_2 = Tf$, then g_1 and g_2 have the same values except on a set of measure zero.

6. (a) Let $f_j = T^{-1}g_j$ be measurable functions (see Theorem 1) and let ϕ be a Baire function of n variables. Show that $\phi(f_1 \cdots f_n) = T^{-1}\phi(g_1 \cdots g_n)$.
(b) Show that $\lim f_n(x)$ exists for almost all x if and only if $\lim g_n(y)$ exists for almost all y. If these limits exist, they are transforms of each other under T, neglecting values on sets of measure zero.

7. Suppose that $X = Y = [0, 1]$, $\mathcal{S}$ is the class of Borel sets, and $\mathcal{F}$ is the class of all countable sets. Show that if $Tx = x$, then T is a one-to-one measurable transformation from X onto Y but T is not measurability-preserving.

8. *Linear Transformations in* R_n. Let T be a linear transformation defined by $T(x_1, \ldots, x_n) = (y_1, \ldots, y_n)$, where $y_i = \sum_{j=1}^{n} t_{ij}x_j + b_i$ $(i = 1, 2, \ldots, n)$. Show that for every $E \subset X$, $\mu^\circ[T(E)] = (\det T)\mu^\circ E$ in the following situations:

(a) $y_i = x_i + b_i$ for $i = 1, \ldots, n$. [*Hint:* It is sufficient to prove the assertions for intervals E.]
(b) $y_i = x_i$ for $i \neq j$ and $i \neq k$; $y_j = x_k$ and $y_k = x_j$.
(c) $y_i = x_i$ for $i \neq j$; $y_j = x_j \pm x_k$ for $k \neq j$.
(d) $y_i = x_i$ for $i \neq j$; $y_j = cx_j$.
(e) The general transformation T given above. [*Hint:* T can be written as the product of transformations of the form given in parts (a)–(d).]
(f) If $\det T = \pm 1$, T is measurability-preserving.

9. Let T be a nonsingular linear transformation in R_n, and let $J = |\det(T)|$. Then if f is Lebesgue-integrable over R_n, we have $\int f(x)\, d\mu = \int f(Tx) J\, d\mu$.

7.10 L_p SPACES

An arbitrary measure space $(X, \mathscr{S}, \mu)$ has some function spaces associated with it.

DEFINITION 1

If $1 \leq p < \infty$, a measurable function f is said to belong to $\mathscr{L}_p(\mu)$ if $|f|^p$ is integrable.

The space $\mathscr{L}_p(\mu)$ is a linear space, as the following lemma shows.

LEMMA 1

If $f, g \in \mathscr{L}_p(\mu)$, then $c_1 f + c_2 g \in \mathscr{L}_p(\mu)$ for any numbers c_1 and c_2.

The proof of the lemma is an immediate consequence of Minkowski's inequality [use Exercise 2(b) (Minkowski's inequality) or Exercise 2(c) (Schwarz's inequality) of Section 9.1].

If for $f \in \mathscr{L}_p(\mu)$ and $p \geq 1$ we define

$$||f||_p = \left(\int |f|^p \, d\mu\right)^{1/p},$$

the function $|| \; ||_p$ has the following properties.

LEMMA 2

1. $||f||_p \geq 0$.
2. $||cf||_p = |c| \, ||f||_p$, for any number c.
3. $||f + g||_p \leq ||f||_p + ||g||_p$.

Properties 1 and 2 follow directly from the definition of $|| \; ||_p$ and property 3 follows from Minkowski's inequality. The only property of a norm that is missing is that $||f||_p = 0$ does not imply $f = 0$, but only $f = 0$ a.e. with respect to μ. To overcome this difficulty, we partition $\mathscr{L}_p(\mu)$ into equivalence classes using the relation $f = g$ a.e. to identify functions belonging to the same equivalence class.

DEFINITION 2

The normed linear space (see [1]) whose elements are the classes of equivalent members of $\mathscr{L}_p(\mu)$ will be denoted by $L_p(\mu)$. The norm of an equivalence class is defined as equal to the norm of any one of its members.

We leave it to the reader to verify that this uniquely defines a norm on $L_p(\mu)$.

For convenience, we will write $\mathscr{L}_p$ and L_p instead of $\mathscr{L}_p(\mu)$ and $L_p(\mu)$; for the case where $p = 1$, we will drop the subscript altogether.

LEMMA 3

If $f \in \mathscr{L}_q$ and μ is finite, then $f \in \mathscr{L}_p$ for $1 \leq p \leq q < \infty$.

Proof

If $|f| \leq 1$, then $|f|^p \leq 1$, while if $|f| > 1$, then $|f|^p \leq |f|^q$. Hence in either situation, $|f|^p \leq 1 + |f|^q$. Therefore, $f \in \mathscr{L}_q$ implies $f \in \mathscr{L}_p$.

The introduction of a norm in L_p introduces a new type of convergence.

DEFINITION 3

If $f_n \in \mathscr{L}_p$, then f_n *mean converges* to f with an *index* p (that is, converges in norm) if $\|f_n - f\|_p \to 0$, and we write $f_n \to f(L_p)$.

LEMMA 4

If $f_n \in \mathscr{L}_p$ forms a fundamental sequence in norm, then $\{f_n\}$ is a fundamental sequence in measure.

Proof

Let $E_{mn} = (|f_m - f_n| \geq \epsilon)$; then

$$\int |f_m - f_n|^p \, d\mu \geq \epsilon^p \, \mu E_{mn}$$

and, since $\|f_m - f_n\|_p \to 0$, the result follows.

THEOREM 1

If $1 \leq p < \infty$, the space L^p is a Banach space.

Proof

Since we know that L_p is a normed linear space, we need only show that it is complete. This means that if $f_n \in \mathscr{L}_p$ and $\|f_m - f_n\|_p \to 0$ as $m, n \to \infty$, we must establish the existence of an $f \in \mathscr{L}_p$ such that $\|f_n - f\|_p \to 0$ as $n \to \infty$.

By Lemma 4 and Theorem 3.8, there is a measurable function f such that $f_n \to f(\mu)$. There exists a sub-sequence of $\{f_n\}$ which we denote by $\{g_n\}$, where $g_n \to f$ a.e.. Since $\{g_n\}$ is also fundamental in norm, we have $\int |g_m - g_n|^p \, d\mu < \epsilon$ for $m, n \geq N$. By Fatou's lemma (Theorem 5.2),

$$\int |g_m - f|^p \, d\mu \leq \liminf_n \int |g_m - g_n|^p \, d\mu \leq \epsilon$$

for fixed m. Hence $g_m - f \in \mathscr{L}_p$ if $m \geq N$; thus $f \in \mathscr{L}_p$, since $\mathscr{L}_p$ is a linear space. Moreover, $||g_m - f||_p \to 0$, which allows us to conclude from the inequality

$$||f_n - f||_p \leq ||f_n - g_m||_p + ||g_m - f||_p$$

that $||f_n - f||_p \to 0$.

COROLLARY

If a sequence $\{f_n\}$ of measurable functions converges in the mean of index p to a measurable function f, it contains a subsequence that converges to f a.e.

EXAMPLE 1

1. Mean convergence does not imply convergence a.e. The sequence of functions defined in Example 3.5 converges in the mean to $f = 0$ but does not converge a.e.
2. Convergence a.e. does not imply mean convergence. Let $f_n(x) = n$ for $x \in (0, 1/n]$ and $f_n(x) = 0$ for remaining values of x. The sequence $\{f_n\}$ converges to zero everywhere on $[0, 1]$, but $\int_0^1 f_n^p \, dx = n^{p-1} \to \infty$.

Further properties of L_p spaces are contained in the Exercises.

If the space $X = \{1, 2, 3, \ldots\}$ and each point has a measure 1, then each complex (or real) function on X is a sequence $f = \{f_n\}$ of complex (real) numbers, and each function is measurable (since each subset of X is measurable). The function space L_p for $1 \leq p < \infty$ then consists of all $\{f_n\}$ such that $\left(\sum_1^\infty |f_n|^p\right)^{1/p} < \infty$ and is usually denoted by l_p. Similarly, function spaces can be defined for $X = \{0, \pm 1, \pm 2, \ldots\}$.

EXERCISES

1. (a) Show that if $f \in \mathscr{L}_p$ and $g \in \mathscr{L}_q$, where $1/p + 1/q = 1$ and $1 < p < \infty$, then $fg \in \mathscr{L}$. [*Hint:* Use Hölder's inequality.]

(b) Show that if $f_n \in \mathscr{L}_p$ converges to f in the mean and $g \in \mathscr{L}_q$ for $p > 1$, then $\{f_n g\}$ converges to fg.

2. Prove that if $f_n \in \mathscr{L}_p$ and $||f_n - f||_p \to 0$, then $\lim \int |f_n|^p \, d\mu = \int |f|^p \, d\mu$.

3. (a) Is a simple function an element of $\mathscr{L}_p$ if it is integrable?
(b) Show that L is dense in L_p. [*Hint:* for $f \in \mathscr{L}_p$, define $A_n = (x\colon 1/n < |f| < n)$. Let $g_n = I_{A_n} f$; show that $|f - g_n| \downarrow 0$ and hence $||f - g_n||^p \to 0$.]
(c) Show that the integrable simple functions are dense in L_p.

4. Prove that if μ is finite and $f_n \to f(\mu)$, then $f \in \mathscr{L}_p$.

5. A measurable function f is said to be *essentially bounded* if $|f| \leq c$ a.e. on X. The number c is called an *essential upper bound* of f on R. Let $m = \inf (c : c$ is an essential upper bound). The number m is called the *essential supremum* of f (ess sup f). Let L_∞ be the collection of equivalence classes determined by the relation $f = g$ a.e., where f, g are essentially bounded functions.

(a) Show that L_∞ is a normed linear space, with $||f||_\infty = \text{ess sup } f$.
(b) Show that if μ is finite and $f \in \mathscr{L}_\infty$ then $f \in \mathscr{L}_p$ for all p and

$$||f||_\infty = \lim_{p\to\infty} \left(\int |f|^p \, d\mu \right)^{1/p}.$$

6. Show that if $X = [0, \infty)$ and μ is the Lebesgue measure, then the function $x^{-1/2}(1 + |\log x|)^{-1}$ is in $\mathscr{L}_2$ but not in $\mathscr{L}_p$ for any other value of p such that $1 \leq p < \infty$.

7. Show that if $X = [0, \frac{1}{2}]$, μ is the Lebesgue measure, and $1 \leq p < \infty$, then the function $x^{-1/p}(\log x^{-1})^{-2/p}$ is in $\mathscr{L}_p$ but not in $\mathscr{L}_r$ for any $r > p$. Show that $\log x^{-1} \in \mathscr{L}_p$ for $p \geq 1$ but $\log 1/x \in \mathscr{L}_\infty$ is false.

8. The metric space L_p on a measure space $(X, \mathscr{S}, \mu)$ is separable if and only if the space $\mathscr{M}$ of measurable sets of finite measure is separable. This means that there exists a countable collection $\mathscr{C}$ of measurable sets of finite measure such that if E is an arbitrary measurable set of finite measure, then there exists for each $\epsilon > 0$ a set $F \in \mathscr{C}$ such that $\mu(E \Delta F) < \epsilon$. [*Hint:* If a class of sets is dense in $\mathscr{M}$, then the set of all finite linear combinations with rational coefficients of the indicators of these sets is dense in $\mathscr{L}_p$.]

REFERENCE

1. Zaanen, A. C., *Linear Analysis*. Amsterdam: North-Holland, 1960.

CHAPTER 8

PROBABILITY AND MEASURE

8.1 INTRODUCTION

Probability theory arose from the study of random physical phenomena. Investigations of the common features of some of these processes led to the formulation of a set of axioms for a probability space. In this form it is not difficult to see that a probability space is a measure space. Although most mathematicians agree that probability theory is just a special branch of measure theory, the language of probability theory has been retained because it lends intuitive meaning to the subject and also makes it more accessible to applied workers in other fields. The correspondence between probability and measure theoretical terminology follows.

A probability space $(\Omega, \mathscr{S}, P)$ is just a normalized measure space; that is, $P\Omega = 1$. A point ω belonging to the space Ω is called an *elementary event*. The reader is cautioned not to confuse this term with the word *event* that denotes an element of the σ-field $\mathscr{S}$ or a measurable set. The whole space Ω is called the *sure event*, while the empty set is called the *impossible event*. The probability function P is just a normalized measure—that is, $P\Omega = 1$—defined on $\mathscr{S}$. Instead of convergence a.e., the terms *convergence almost surely* (convergence a.s.) or convergence with probability 1 are frequently used. Finite, real-valued, measurable functions are called *random variables* (r.v.'s); their integral is called the *expectation* and is denoted by E.

These definitions alone do not suffice to set up a theory of probability, as the reader will soon discover. However, many results in probability theory are merely translations of theorems in measure theory using the above correspondences. Some of these are stated in the exercises, which should be studied carefully to obtain facility with the terminology.

EXERCISES

Prove the following statements:

1. (a) If E_i are events (i.e., $E_i \in \mathscr{S}$), then E_i, $\bigcup_{i=1}^{\infty} E_i$ and $\bigcap_{i=1}^{\infty} E_i$ are events.
(b) $\liminf E_i$ and $\limsup E_i$ are events for $i \in I$, a countable index set.
(c) If $\lim E_i$ exists, then it is an event.

2. (a) $P\emptyset = 0$.
(b) If $E, F \in \mathscr{S}$ and $E \subset F$, then $PE \leq PF$.
(c) Under the hypothesis of Exercise 1 (a), we have $P(\liminf E_i) \leq \liminf PE_i \leq \limsup PE_i \leq P(\limsup E_i)$; if $\lim E_i$ exists, then $P(\lim E_i) = \lim PE_i$.

3. (a) Every r.v. is the uniform limit of a sequence of elementary r.v.'s; the converse is also true.
(b) Every r.v. is the limit of a sequence of simple r.v.'s; the converse is also true.
(c) Every nonnegative r.v. is the limit of an increasing sequence of nonnegative simple r.v.'s; the converse is also true.
(d) The usual operations of analysis applied to r.v.'s yield a r.v. provided that these operations yield a finite function.
(e) Every finite Borel function of a finite number of r.v.'s is a r.v.

4. (a) The expectation of a simple or elementary r.v.

$$f = \sum_{j=1}^{\infty} y_j I_j,$$

where I_j are the indicator functions of events A_j, is

$$Ef = \sum_{j=1}^{\infty} y_j PA_j.$$

(b) The expectation of a nonnegative r.v. $f \geq 0$ is the limit of nonnegative elementary r.v.'s that converge to f.
(c) The expectation of a r.v. $f = f^+ - f^-$ is given by

$$Ef = Ef^+ - Ef^-,$$

provided that the right-hand side is defined.

5. The r.v.'s f_n converge a.s. to f and we write $f_n \to f$ (a.s.) iff $P(\bigcup_{k \geq n} (|f_k - f| > \epsilon)) \to 0$ for every $\epsilon > 0$.

6. Let f_n, f denote r.v.'s:[1]

(a) $f_n \to f(P)$ iff $f_n - f_m \to 0(P)$.
$f_n \to f$ (a.s.) iff $f_n - f_m \to 0$ (a.s.).
(b) $f_n \to f$ (a.s.) implies that $f_n \to f(P)$.
(c) If $f_n \to f(P)$, then there exists a subsequence $f_{n_k} \to f$ (a.s.).

7. (a) A r.v. f is integrable iff $|f|$ is integrable.
(b) $E(c_1 f_1 + c_2 f_2) = c_1 E f_1 + c_2 E f_2$, where c_1 and c_2 are real numbers and f_1 and f_2 are integrable.
(c) If $f_1 \leq f_2$, then $Ef_1 \leq Ef_2$, provided that f_1 and f_2 are integrable.

8. (a) If $|f_1| \leq |f_2|$ and f_2 is integrable, then f_1 is integrable.
(b) Every bounded r.v. is integrable.
(c) If $f = c$ a.s., then $Ef = c$.

9. If $0 \leq f_n \uparrow f$, which may or may not be finite, then $Ef_n \uparrow Ef$. Note that the measurable function f need not be a r.v., since it can be infinite. (See the monotone convergence theorem.)

10. If $f_n \to f(P)$ and $|f_n| \leq g$, which is integrable, then f is integrable and $Ef_n \to Ef$. (See the dominated convergence theorem.)

11. If f and g are integrable r.v.'s, then

(a) $f \leq f_n$ implies that $E(\liminf f_n) \leq \liminf Ef_n$.
(b) $f_n \leq g$ implies that $\limsup Ef_n \leq E(\limsup f_n)$.

8.2 RANDOM VARIABLES AND DISTRIBUTION FUNCTIONS

We recall the definition of a r.v.

DEFINITION 1

A *random variable* (r.v.) r is a real-valued measurable function defined on a probability space $(\Omega, \mathscr{S}, P)$. If the r.v. assumes a countable (finite or infinite) number of values it is usually called *discrete*; otherwise it is called *continuous*.

This definition states that if $B \in \mathscr{B}$, the class of Borel sets in R, then $r^{-1}(B) \in \mathscr{S}$. This does not exclude the possibility that $r^{-1}(A) \in \mathscr{S}$, where A belongs to a larger field containing $\mathscr{B}$. In fact, the σ-field $\mathscr{C}$ in R defined by $\mathscr{C} = (A : r^{-1}(A) \in \mathscr{S},\ A \subset R)$ may contain $\mathscr{B}$ and may even properly contain the Lebesgue–Stieltjes sets.

[1]The same notational conventions are used as in the previous chapter; for example, $f_n \to f(P)$ means that f_n converges to f in measure where $\mu = P$.

A necessary and sufficient conditon that a function r on Ω be a r.v. is that $\bar{\mathscr{S}} = r^{-1}(\mathscr{B}) \subset \mathscr{S}$, where $\mathscr{B}$ is the class of Borel sets in R. The events in $\mathscr{S} - \bar{\mathscr{S}}$ are superfluous; that is, r could just as well be defined on $(\Omega, \bar{\mathscr{S}}, P)$.

DEFINITION 2

The real-valued function F defined by

$$F(x) = P(r \leq x) \tag{1}$$

is called the *distribution function* (d.f.) of the r.v. r.[2]

This definition is sensible in that F is defined for every real x, since the inequality $r(\omega) \leq x$ determines a measurable ω set. The next theorem states the properties of d.f.'s.

THEOREM 1

Let F be the d.f. of a r.v. r; then

1. It is defined for every real value of its argument.
2. It is increasing.
3. It has right-hand continuity.
4. $\lim_{x \to -\infty} F(x) = 0$.
5. $\lim_{x \to \infty} F(x) = 1$.

Proof

The proof of part 1 of the theorem follows from the definition.

The proof of part 2 is a consequence of the monotonicity of the measure.

To prove part 3, let $A = (r \leq x)$ and $A_i = (r \leq x + h_i)$, where $h_i > 0$ and $\lim_{i \to \infty} h_i = 0$; then $\lim_{i \to \infty} F(x + h_i) = \lim P(A_i) = P(\lim A_i)\, PA = F(x)$.

The properties in parts 4 and 5 of the theorem are a result of the finiteness of r.

Any function F satisfying the conditions of Theorem 1 will be called a *distribution function.* This name is appropriate since a given d.f. F is the d.f. of a r.v. on a probability space. For example, we can take Ω to be the real line, $\mathscr{S}$ to be the Borel class, and P to be the probability function determined by the correspondence theorem (6.6.1), while we define r by $r(x) = x$.

[2]Recall that $(r \leq x) \equiv (\omega : r(\omega) \leq x)$.

DEFINITION 3

If F is a d.f. and

$$F(x) = \int_{-\infty}^{x} f(t)\,dt \qquad (-\infty < x < \infty)$$

for some Lebesgue-measurable and integrable function f, then f is called the *density function* corresponding to F.

We have $F'(x) = f(x)$ (a.e.) Furthermore, it can be shown that a density function exists if and only if F is absolutely continuous[3] (see Reference [1]).

In Section 1.5, we saw that a number of different spaces could be used to describe a given phenomenon. For a r.v. defined on one of these spaces there is a corresponding r.v. defined on each of the other spaces. The exact correspondence will be discussed below (see also Example 1). It turns out that the d.f.'s of all these r.v.'s are the same. Hence the essential properties of a r.v. are those that can be expressed in terms of its d.f. F. This is one of the essential features of probability theory that distinguishes it from measure theory.

Let us now study the mathematical counterpart of the above statements. Suppose that T is a measurability-preserving transformation which is also measure-preserving (see Definitions 7.9.2 and 7.9.3) and which maps $(\Omega, \mathscr{S}, P)$ into $(\bar{\Omega}, \bar{\mathscr{S}}, \bar{P})$. We recall that if $\bar{A} \in \bar{\mathscr{S}}$, then $\bar{P}(\bar{A}) = P(A)$, where $A = T^{-1}\bar{A}$. First we let $\bar{r}$ be a r.v. on $\bar{\Omega}$. By Theorem 7.9.1, there exists a corresponding r on Ω defined by $r = \bar{r}T$. We will now show that r and $\bar{r}$ have the same d.f.'s.

$$\begin{aligned}\bar{F}(x) &= \bar{P}(\bar{r}(\bar{\omega}) \leq x) = \bar{P}(\bar{r}^{-1}(-\infty, x]) = P(T^{-1}\bar{r}^{-1}(-\infty, x]) \\ &= P(r^{-1}(-\infty, x]) = P(r(\omega) \leq x) = F(x).\end{aligned}$$

Next, suppose that r is a r.v. on Ω and let $\bar{r} = Tr$ be the corresponding r.v. defined in Theorem 7.9.2. Then r and $\bar{r}$ have the same d.f.'s. We use the notation of Theorem 7.9.2. with r playing the role of f and $\bar{r}$ the role of g. By the very definition of $\bar{r}$, we see that

$$\begin{aligned}(\bar{\omega} : \bar{r}(\bar{\omega}) < x) &= \bigcup_{i<x} F_i & (x \leq 0) \\ &= (\bigcup_{i<x} F_i) \cup (\bigcup_{i} F_i)'. & (x > 0)\end{aligned}$$

[3]Recall that a point function F is *absolutely continuous* in an interval (a, b) if, given $\epsilon > 0$, we can find a $\delta > 0$ such that

$$\sum_{i=1}^{n} |F(x_i + h_i) - F(x_i)| \leq \epsilon$$

for every set of disjoint intervals $(x_i, x_i + h_i)$ where $\sum h_i \leq \delta$. It is not difficult to show that F is absolutely continuous if and only if the corresponding Lebesgue-Stieltjes measure is absolutely continuous. The result then follows from the Radon-Nikodym theorem.

In either case, since $P((\bigcup_i F_i)') = 0$,

$$\bar{F}(x^-) = \bar{P}(\bar{r} < x) = \bar{P}(\bigcup_{i<x} F_i).$$

However,

$$\bar{P}(\bigcup_{i<x} F_i) = P(\bigcup_{i<x} T^{-1}(F_i)) = P(\bigcup_{i<x} S_i) = P(r < x) = F(x^-).$$

Since $F(x^-) = \bar{F}(x^-)$ for all x, it follows that $F = \bar{F}$.

EXAMPLE 1

Let Ω be the two-point set $\{-1, 1\}$, let $\mathscr{S}$ be all subsets of Ω, and let $P(-1) = \frac{1}{2} = P(1)$. Suppose that $\bar{\Omega}$ is the real line, $\bar{\mathscr{S}}$ are all subsets of $\bar{\Omega}$, and $\bar{P}(\bar{A}) = \frac{1}{2}$ (the number of points in $\{-1, 1\} \cap \bar{A}$). Let $T : \Omega \to \bar{\Omega}$ be the transformation defined by $T(1) = 1$ and $T(-1) = -1$.

Consider the random variable $\bar{r}(\bar{\omega}) = 5$ if $\bar{\omega} \neq 1$ or -1, $\bar{r}(-1) = 1$, and $\bar{r}(1) = 2$. Then $r = \bar{r}T$ is given by $r(-1) = 1$ and $r(1) = 2$; we leave it to the reader to verify that both of these r.v.'s have the same d.f.

Now suppose that r is a r.v. defined on Ω with $r(-1) = 1$ and $r(1) = 2$. Then an $\bar{r} = Tr$ such as the one defined in Theorem 7.9.2 is given by $\bar{r}(1) = 2$, $\bar{r}(-1) = 1$, and $\bar{r}(\bar{\omega}) = 0$ for $\bar{\omega} \neq \pm 1$. Once again both of these r.v.'s have the same d.f.

To further illustrate the role played by the space Ω and a r.v. r on this space, consider a function f that maps Ω into any arbitrary space $\bar{\Omega}$. Further assume that $A \in \mathscr{S}$ implies that there is an $\bar{A} \subset \bar{\Omega}$ such that $f^{-1}(\bar{A}) = A$. Then we can define a new probability space $(\bar{\Omega}, \bar{\mathscr{S}}, \bar{P})$, where $\bar{\mathscr{S}} = (\bar{A} : f^{-1}(\bar{A}) \in \mathscr{S}, \bar{A} \subset \bar{\Omega})$ and $\bar{P}$ is defined on $\bar{\mathscr{S}}$ by $\bar{P}(\bar{A}) = P(f^{-1}(\bar{A}))$. Under these definitions, f is a measurable transformation that is also measurability and measure preserving. Thus, by the preceding discussion, we can find a r.v. $\bar{r} = fr$ such that r and $\bar{r}$ have the same d.f. The space $(\bar{\Omega}, \bar{\mathscr{S}}, \bar{P})$ and the r.v. $\bar{r}$ are just as suitable as the original space $(\Omega, \mathscr{S}, P)$ and the r.v. r.

Conversely, suppose that $(\bar{\Omega}, \bar{\mathscr{S}}, \bar{P})$ is a given probability space and $\bar{r}$ is a r.v. on $\bar{\Omega}$. Let f map an arbitrary space Ω into $\bar{\Omega}$ in such a way that $f(\Omega) \in \bar{\mathscr{S}}$ and $\bar{P}(f(\Omega)) = 1$. Then we can form a new probability space $(\Omega, \mathscr{S}, P)$ by defining $\mathscr{S} = f^{-1}(\bar{\mathscr{S}})$ and $P(A) = \bar{P}(\bar{A})$ for $A \in \mathscr{S}$ and $A = f^{-1}(\bar{A})$. It is not difficult to see that f is a measurable transformation between $(\Omega, \mathscr{S}, P)$ and $(\bar{\Omega}, \bar{\mathscr{S}}, \bar{P})$ that is also measurability and measure preserving. Hence if $\bar{r}$ is a r.v. on $\bar{\Omega}$, we can find a corresponding r.v. on Ω denoted by $r = f^{-1}\bar{r}$, which has the same d.f. as $\bar{r}$.

In particular, a r.v. r on a probability space $(\Omega, \mathscr{S}, P)$ defines a nonnegative set function P_r in R, the space of real numbers by $P_r(A) = P(r^{-1}(A))$ for all $A \subset R$ such that $r^{-1}(A) \in \mathscr{S}$. For $A \in \mathscr{B}$, the class of Borel sets, $r^{-1}(A) \in \mathscr{S}$, since r is a r.v. Furthermore, since r is finite, $r^{-1}(R) = \Omega$, and since the inverse image of a sum of sets is the sum of their inverse images, it is not difficult to see that P_r is a probability on $\mathscr{B}$. The set function P_r is called the *probability distribution* of r and the new probability space $(R, \mathscr{B}, P_r)$ is the probability space *induced* by r on its range space, or the *sample probability space* of r. Note that it is possible for P_r to be defined on a wider class of sets than the Borel sets.

Distributions are set functions, and they are not easy to manipulate. Fortunately, there is a close connection between probability distributions and d.f.'s of r.v.'s. We recall that a class of d.f.'s corresponds to each finite measure P_r (see Theorem 6.6.1). The d.f. F of the r.v. r belongs to this class; selecting it yields $P_r(t : t \leq x) = F(x)$. More generally, for any Borel set B we have

$$P_r(B) = \int_B dF, \tag{2}$$

where the integral is the Lebesgue–Stieltjes integral. This follows from the uniqueness of the measure, since the right-hand side of Equation (2) defines a measure on the Borel sets that agrees with P_r on the intervals.

The distribution P_r of r determines the distributions of all gr, where g is a finite Borel function on R. Then $(gr)^{-1}(B) = r^{-1}g^{-1}(B) \in \mathscr{S}$, since $g^{-1}(B)$ is a Borel set for every Borel set B, and $P_{gr}(B) = P(gr \in B) = P(r^{-1}g^{-1}(B)) = P_r(g^{-1}(B))$. Observe that our composite function gr will still be measurable if we replace our class $\mathscr{S}$ by the class $r^{-1}(\mathscr{B})$, which will be denoted by $\mathscr{B}(r)$. In other words, gr is a r.v. on the space $(\Omega, \mathscr{B}(r), P)$.

Since we are more familiar with the manipulation of real-valued functions, it is useful to replace the original space Ω by the real line, with a measure P_r defined by Equation (2). An ω event $(r \in B)$ can be interpreted as the point set B of the real line. The measure P_r has been defined to make the probability the same under the two interpretations. Formally, this is equivalent to setting up a measure-preserving transformation from the space Ω to the line. If we define $x = T(\omega) = r(\omega)$ for each ω, then T is a measurable transformation between $(\Omega, \mathscr{S}, P)$ and $(R, \mathscr{B}, P_r)$. Hence the following result is an immediate consequence of Theorem 7.9.3 and of the definitions of expectation and P_r.

THEOREM 2

Let g be a finite Borel function on R. Then

$$E(gr) = \int_\Omega gr\, dP = \int_R g(x)\, dP_r = \int_{-\infty}^{\infty} g(x)\, dF(x). \tag{3}$$

The last equality is a result of the notation introduced in Section 6.6. If g is integrable and continuous on R, then the Lebesgue–Stieltjes integral becomes an improper Riemann–Stieltjes integral.

DEFINITION 4

A *random vector* r is a finite collection of r.v.'s r_i $(i = 1, 2, \ldots, n)$ called the *components* of the random vector, defined on a probability space $(\Omega, \mathscr{S}, P)$. The random vector r will be written as $r = (r_1, r_2, \ldots, r_n)$.

The range space of the random vector is the n-dimensional Euclidean space R^n with points $x = (x_1, x_2, \ldots, x_n)$. From Theorem 7.2.8, it follows that a random vector is a measurable function from $(\Omega, \mathscr{S}, P)$ to $(R^n, \mathscr{B}^n)$, where $\mathscr{B}^n$ denotes the class of n-dimensional Borel sets; that is, $\mathscr{B}^n = \times_{i=1}^{n} \mathscr{B}_i$, where each $\mathscr{B}_i$ is the class of one-dimensional Borel sets.

Just as in the one-dimensional case, the random vector r may still be measurable with respect to a smaller class of sets in Ω. Since each component r_i is a r.v., it induces a σ-field $r_i^{-1}(\mathscr{B}_i) \equiv \mathscr{B}(r_i)$ of events. Hence the minimal σ-field containing the union of all the $\mathscr{B}(r_i)$ is contained in $\mathscr{S}$. Conversely, if the minimal σ-field over all the $\mathscr{B}(r_i)$ is contained in $\mathscr{S}$, then each r_i is a r.v. and consequently r is a random vector. It follows that the function r will be measurable iff the sets of the minimal σ-field containing the union of all the $\mathscr{B}(r_i)$, where $(i = 1, 2, \ldots, n)$, are considered to be events. This σ-field, denoted by $\mathscr{B}(r) \equiv \mathscr{B}(r_1, r_2, \ldots, r_n)$, is called a *compound* or *union* σ-field with component σ-fields $\mathscr{B}(r_i)$. It is evident that $\mathscr{B}(r) \subset \mathscr{S}$ and that the function r is still a random vector on the probability space $(\Omega, \mathscr{B}(r), P)$.

The discussion on corresponding random vectors carries over to the n-dimensional space; we leave the details to the reader, since it is similar to the discussion of one-dimensional space.

We define Er to be $(Er_1, Er_2, \ldots, Er_n)$.

DEFINITION 5

The real-valued function $F(x)$ $(F(x_1, \ldots, x_n))$ defined by

$$F(x) = P(r \leq x)$$

where $F(x_1, \ldots, x_n) = P(r_1 \leq x_1, r_2 \leq x_2, \ldots, r_n \leq x_n)$ is called the *distribution function* of the random vector r (or the *multivariate* d.f. of the r.v.'s r_i).

THEOREM 3

If F is a d.f. of a random vector r, then

1. It is defined for every real value of its arguments.
2. All of its nth-order differences[4] are nonnegative; i.e.,

[4] This notation was introduced in Exercise 6.6.2.

$$\overset{b}{\underset{a}{\Delta}} F(x) \geq 0. \qquad (b \geq a)$$

3. It is continuous from the right in each variable.
4. $\lim\limits_{x_i \to -\infty} F(x_1, \ldots, x_n) = 0. \qquad (i = 1, 2, \ldots, n)$
5. $\lim\limits_{x_1 \cdots x_n \to \infty} F(x_1, \ldots, x_n) = 1,$

where every one of the variables must approach ∞.

Proof

The proof of the theorem, except for part 2, is analogous to the proof of Theorem 1 and we will omit the details. Part 2 follows from the fact that

$$\overset{b_1}{\underset{a_1}{\Delta}} \cdots \overset{b_n}{\underset{a_n}{\Delta}} F(x_1, \ldots, x_n) = P(a_i < r_i \leq b_i). \qquad (i = 1, 2, \ldots, n).$$

Any function with the properties given in Theorem 3 will be called a *distribution function.*

The d.f. is increasing in each variable; in fact, a more general result holds.

THEOREM 4

If F is a d.f., then all its ith-order differences are nonnegative.

Proof

Suppose that h_r $(r = 1, 2, \ldots, i)$ and k_s $(s = i + 1, i + 2, \ldots, n)$ are nonnegative; then, since the nth-order difference is nonnegative for every such k_s,

$$\text{(4)} \qquad \lim_{k_{i+1}, \ldots, k_n \to \infty} \overset{x_1+h_1}{\underset{x_1}{\Delta}} \cdots \overset{x_i+1}{\underset{x_{i+1}-k_{i+1}}{\Delta}} \cdots \overset{x_n}{\underset{x_n-k_n}{\Delta}} F(x_1, x_2, \ldots, x_n) \geq 0.$$

By property 4 of Theorem 3,

$$\text{(5)} \qquad \lim_{k_j \to \infty} \overset{x_j}{\underset{x_j-k_j}{\Delta}} F(x_1, \ldots, x_j, \ldots, x_n) =$$
$$\lim_{k_j \to \infty} F(x_1, \ldots, x_j, \ldots, x_n) - \lim_{k_j \to \infty} F(x_1, \ldots, x_j - k_j, \ldots, x_n)$$
$$= F(x_1, \ldots, x_j, \ldots, x_n).$$

By using the fact that the difference operators commute, we can show that the order of taking limits is immaterial. Observing that

$$\lim_{k_n \to \infty} \overset{x_1+h_1}{\underset{x_1}{\Delta}} \cdots \overset{x_n}{\underset{x_n-k_n}{\Delta}} F = \overset{x_1+h_1}{\underset{x_1}{\Delta}} \cdots \lim_{k_n \to \infty} \overset{x_n}{\underset{x_n-k_n}{\Delta}} F,$$

it follows inductively from (5) that (4) reduces to

$$\overset{x_i+h_1}{\underset{x_1}{\Delta}} \cdots \overset{x_i+h_i}{\underset{x_1}{\Delta}} F(x_1, \ldots, x_n) \geq 0.$$

The same result can be shown to hold for any other ith-order difference.

COROLLARY

If F is a d.f., it is increasing in each variable.

DEFINITION 6

Let F be a d.f. Suppose that for all $x_1, \ldots, x_n$,

$$F(x_1, \ldots, x_n) = \int_{-\infty}^{x_1} \cdots \int_{-\infty}^{x_n} f(s_1, \ldots, s_n)\, ds_1, \ldots, ds_n$$

for some Lebesgue-measurable and integrable function f; f is called the corresponding *density function.*

The fact that any real-valued function F with the properties enumerated in Theorem 3 is a d.f. is in agreement with Definition 5. Such a function can be considered to be the d.f. of the random vector r on R^n, where $r_i(x_1, \ldots, x_n) = x_i$ and the probability measure is the Lebesgue-Stieltjes measure defined on the events of $\mathscr{B}^n$, the Borel class, as in Section 6.6.

By analogy with the one-dimensional case, we define the distribution of the random vector $r = (r_1, r_2, \ldots, r_n)$ on the Borel field in R^n by $P_r(B) = P(r \in B)$, where $B \in \mathscr{B}^n$. The reader should have no difficulty in verifying that $(R^n, \mathscr{B}^n, P_r)$ is a probability space. For any Borel set B, we have

$$P_r(B) = \int_B dF, \tag{6}$$

where F is the d.f. of the random vector r.

The distribution P_r of r determines the distribution of all $g(r)$, where $g = (g_1, \ldots, g_m)$ is a finite Borel function on R^n to R^m; that is, each component g_i is a finite Borel function from R^n to R. This is equivalent to the statement that $(y : g(y) \in B)$ is a Borel set in R^n for $B \in \mathscr{B}^m$, the class of Borel sets in R^m. In other words, the inverse image of any Borel set is a Borel set. To prove this last statement, we consider the inverse image of a closed interval:

$$(g \le x) = \bigcap_{i=1}^{m} (g_i \le x_i).$$

This equality shows that $(g \le x)$ is a Borel set, since each g_i is a Borel function. However, the Borel class in R^m is the smallest σ-field over the class of all such intervals. The result follows from the fact that the inverse image of the σ-field generated by a class of sets equals the σ-field generated by the inverse images. Hence, if $A = (y : g(y) \in B)$, where $B \in \mathscr{B}^m$, then $A \in \mathscr{B}^n$. Therefore, $(gr \in B) = (r \in A)$; it is an event, since r is a random vector. This implies that g is a random vector and also that

$$P_g(B) = P(gr \in B) = P_r(A).$$

This statement shows that the distribution of g is determined by P_r.

For many applications, it is convenient to replace the original space Ω by the n-dimensional space R^n. Let $\bar{r}_i$ be the coordinate function on R^n defined by $\bar{r}_i(x) = x_i$, where $x = (x_1, \ldots, x_n)$. Then $\bar{r} = (\bar{r}_1 \cdots \bar{r}_n)$ is a random vector on $(R^n, \mathscr{B}^n, P_r)$ that maps the space R^n onto itself. It can be shown that the d.f. of the random vector $\bar{r}$ is the same as the d.f. of the random vector r by setting up a measurability-preserving transformation[5] T between $(\Omega, \mathscr{S}, P)$ and $(R^n, \mathscr{B}^n, P_r)$, where $T(\omega) = r(\omega) = x$; that is, $r_i(\omega) = x_i$. The next theorem is also a result of this transformation. The proof is similar to the proof of Theorem 2.

THEOREM 5

Let g be a finite Borel function on R^n to R. Then

$$E(gr) = \int_{\Omega} gr\, dP = \int_{R^n} g(x_1, \ldots, x_n)\, dP_r$$

$$= \int_{-\infty}^{\infty} \int_{-\infty}^{\infty} g(x_1, \ldots, x_n)\, dF(x_1, \ldots, x_n).$$

THEOREM 6

If F is a d.f. of n variables, then the function of k variables defined by

$$F_{i_1, \ldots, i_k}(x_{i_1}, \ldots, x_{i_k}) = \lim_{x_j \to \infty} F(x_1, \ldots, x_n),$$

where the limit is taken for all $j \neq i_r$ and $r = 1, \ldots, k$, is also a d.f. for $k = 1, 2, \ldots, n$.

Proof

The limit exists since F is an increasing function of each variable that is bounded above. The properties of a d.f. that are listed in Theorem 3 are easily verified (property 2 follows from Theorem 4).

There are $\binom{n}{k}$ possible d.f.'s of k variables; these are called the *marginal distributions* of the d.f. F. These d.f.'s are consistent in the sense that, if $m < n$, then

$$F_{i_1, \ldots, i_m}(x_{i_1}, \ldots, x_{i_m}) = \lim_{x_j \to \infty} F_{i_1, \ldots, i_n}(x_{i_1}, \ldots, x_{i_n}). \tag{7}$$

$$(j = i_{m+1}, \ldots, i_n)$$

Further suppose that $j_1, \ldots, j_m$ is some permutation of $i_1, i_2, \ldots, i_m$; then it is obvious that

[5] It is not difficult to find a direct proof of this result by using the definition of P_r.

$$(8) \qquad F_{i_1, \ldots, i_m}(x_{i_1}, \ldots, x_{i_m}) = F_{j_1, \ldots, j_m}(x_{j_1}, \ldots, x_{j_m}).$$

The probability distribution P_r of a random vector r also determines consistent marginal distributions that are the projections of P_r on the product subspaces of $R^n = \mathop{\times}\limits_{i=1}^{n} R_i$, where $R_i = R$, the real line for all i. Let $I_m = \{i_1, \ldots, i_m\}$ be a subset of the integers $I = (1, 2, \ldots, n)$, $J_m = I - I_m$, $R_{I_m} = \mathop{\times}\limits_{j=1}^{m} R_{i_j}$, and $\mathscr{B}_{I_m} = \mathop{\times}\limits_{j=1}^{m} \mathscr{B}_{i_j}$, where $\mathscr{B}_{i_j} = \mathscr{B}$, the one-dimensional Borel class. On the Borel class $\mathscr{B}_{I_m}$, we define the marginal probability P_{I_m} by

$$P_{I_m}(B_{I_m}) = P_r(B_{I_m} \times R_{J_m}).$$

Now let $\bar{R}$ and $\bar{\bar{R}}$ be product subspaces of R^n with marginal probabilities $\bar{P}$ and $\bar{\bar{P}}$ respectively. The reader should have no difficulty in verifying that the projections of $\bar{P}$ and $\bar{\bar{P}}$ on their common product subspace, which is assumed to be nonempty, are both equal to the projection of P_r on this subspace. This is the meaning of the term "consistent." The relation between the P_{I_m} and the d.f. appearing in Equation (7) is similar to the one in Equation (6), with $B \in \mathscr{B}_{I_m}$.

The preceding discussion can be generalized to an arbitrary family of r.v.'s.

DEFINITION 7

A *random function* (or a *stochastic process*) $r = (r_i, i \in I)$ is a family of r.v.'s r_i, where I is a fixed index set of arbitrary cardinality.

We define Er to be $(Er_i, i \in I)$.

In reality, a random function is a function of two variables i and ω; that is, $r = r(i, \omega)$. For a fixed value of ω it is just a function of i and is called a *sample function*. For example, if I is the set of real numbers, the sample function is merely an ordinary function of a real variable. On the other hand, for a fixed value of i, it is a r.v.

If $I = \{1, 2, \ldots, n\}$, then r is a random vector; for $I = \{1, 2, \ldots\}$, r is called a *random sequence* and is written as $r = (r_1, r_2, \ldots)$.

The range space of r is the space $R^I = \mathop{\times}\limits_{i \in I} R_i$ of points $x = (x_i, i \in I)$. By a semiclosed interval in R^I, we mean a set of the form $(x : x_{i_r} \leq b_r, r = 1, 2, \ldots, n)$ where the b's are real numbers. In other words, it is a set in which only a finite but arbitrary set of coordinates is restricted by inequalities. The Borel class $\mathscr{B}^I$ is the σ-field generated by these intervals. Alternatively, $\mathscr{B}^I$ is the minimal σ-field over the field of finite sums of Borel rectangles[6] whose sides are either one-dimensional Borel sets or the whole space

[6]A *rectangle* is a set of the form $\mathop{\times}\limits_{i} A_i$, where only a finite number of the A_i are not the whole space. The A_i are called *sides*.

R. A random function could also be defined as a measurable function on $(\Omega, \mathscr{S})$ to the Borel space $(R^I, \mathscr{B}^I)$.

The minimal σ-field $\mathscr{F}$ in Ω with respect to which r is measurable is obtained in exactly the same manner as in the finite-dimensional case. Let $\mathscr{B}(r_i) = r^{-1}(\mathscr{B}_i)$, where $\mathscr{B}_i$ is the one-dimensional Borel class in R_i; then $\mathscr{B}(r_i, i \in I)$, the compound or union σ-field, is the minimal σ-field containing the union of all the $\mathscr{B}(r_i)$.

Again by analogy with the finite-dimensional case, the random function r induces a probability function P_r on $\mathscr{B}^I$, defined by

$$P_r(B) = P(r \in B),$$

where $B \in \mathscr{B}^I$. We leave it to the reader to verify that $(R^I, \mathscr{B}^I, P_r)$ is a probability space called the *sample space* or the *induced probability space* of the random function r. The d.f. F of r is defined as $F(x) = P(r \leq x)$; that is, $r_i \leq x_i$ for $i \in I$.

If we define a Borel function g as a function on a Borel space $(R^I, \mathscr{B}^I)$ to $(R^J, \mathscr{B}^J)$ such that $g^{-1}(\mathscr{B}^J) \subset \mathscr{B}^I$, then it is not difficult to see that the distribution of g is determined by r; if r is a random function on $(\Omega, \mathscr{S})$ to $(R^I, \mathscr{B}^I)$, then $g(r)$ is a random function on $(\Omega, \mathscr{S})$ to $(R^J, \mathscr{B}^J)$. If $B \in \mathscr{B}^J$ and $A = g^{-1}(B) \in \mathscr{B}^I$, then

$$P_{gr}(B) = P(gr \in B) = P(r \in A) = P_r(A),$$

which verifies our assertion.

By using the notion of a measure-preserving transformation, it is possible to define a random function $\bar{r}$ on $(R^I, \mathscr{B}^I)$ that has the same probability distribution as r. If $x \in R^I$ and $x = (x_i, i \in I)$, we can define $\bar{r}_i$ on R^I by $\bar{r}_i(x) = x_i$; then $\bar{r} = (\bar{r}_i, i \in I)$ is a function on R^I to R^I. The $\bar{r}_i$ are coordinate variables in a space R^I whose dimension is the cardinal number of I. Defining a transformation T from Ω to R^I by $T(\omega) = r(\omega) = x$, it follows that T is a measure-preserving and measurable transformation from $(\Omega, \mathscr{S}, P)$ to $(R^I, \mathscr{B}^I, P_r)$. Since $\bar{r}(x) = x$, the event $(r \in B)$ occurs if and only if $\bar{r} \in B$ [that is, $T^{-1}(B) = (r \in B)$], and so $P_r(B) = P(r \in B) = P(\bar{r} \in B) = P_{\bar{r}}(B)$, which shows that r and $\bar{r}$ have the same distribution. In particular, they have the same d.f.'s, since $P(r \leq x) = P(\bar{r} \leq x)$. By considering the mappings $T: \Omega \to R^I$, $\bar{r}_i: R^I \to R$, and $r_i: \Omega \to R$, we obtain $T^{-1}\bar{r}_i = r_i$ in terms of the notation introduced in Section 7.9. More generally, if $i_1, \ldots, i_n$ is any finite I set and if g is a Borel function of n variables, then

$$g(r_{i_1}, \ldots, r_{i_n}) = T^{-1}g(\bar{r}_{i_1}, \ldots, \bar{r}_{i_n}).$$

This leads to the following result, which is a consequence of Theorem 7.9.3.

THEOREM 7

Let g be a finite Borel function of n variables on R^n, and let r be a random function on $(\Omega, \mathscr{S}, P)$. Then

$$Eg(r_{i_1}, \ldots, r_{i_n}) = \int_{\Omega} g(r_{i_1}, \ldots, r_{i_n})\, dP = \int_{R^I} g(\bar{r}_{i_1}, \ldots, \bar{r}_{i_n})\, dP_r;$$

that is, if one of the integrals is defined, both are defined and equal.

Now let P be a probability function defined on $\mathscr{B}^I$. Let r_i be the coordinate r.v. defined by $r_i(x) = x_i$, where $x = \{x_j, j \in I\}$. Every finite random vector $(r_{i_1}, \ldots, r_{i_n})$ has an n-dimensional probability function $P_{i_1, i_2, \ldots, i_n}$ and a corresponding d.f. $F_{i_1, \ldots, i_n}$. Letting $I_n = \{i_1, i_2, \ldots, i_n\}$ and $I'_n = I - I_n$, form the Borel space $(R^{I_n}, \mathscr{B}^{I_n})$, where $R^{I_n} = \underset{i \in I_n}{\times} R_i$; here each R_i is a real line and $\mathscr{B}^{I_n}$ denotes the corresponding Borel class. A *Borel cylinder* B of R^I is by definition a set of the form $B = B_{I_n} \times R^{I'_n}$, where $B_{I_n} \in \mathscr{B}^{I_n}$; B is called the *base*. The n-dimensional probability functions have been defined in such a way that

$$(9) \qquad P(B) = P(B_{I_n} \times R^{I'_n}) = P_{i_1, \ldots, i_n}(B)$$

for any Borel cylinder B. Accordingly, the probability function P is uniquely determined on the field $\mathscr{F}$ of all Borel cylinder sets by (9). However, for Borel sets, the values of $P_{i_1, \ldots, i_n}$ are uniquely determined by means of the corresponding d.f. as in (6). Therefore, if P is defined on $\mathscr{F}$, then by the extension theorem (6.5.5), it is uniquely determined on $\mathscr{S}(\mathscr{F}) = B^I$ by the values of the d.f.'s $F_{i_1, \ldots, i_n}$.

We now consider the converse problem. Suppose that an index set I of arbitrary cardinality is given. For every finite subset I_n of I, the multivariate d.f. $F_{i_1, \ldots, i_n}$ is given. The problem is to determine a probability space $(\Omega, \mathscr{S}, P)$ on which a random function $r = (r_t, t \in I)$ is defined and such that the multivariate d.f. of the random vector $(r_{i_1}, r_{i_2}, \ldots, r_{i_n})$ coincides with $F_{i_1, \ldots, i_n}$.

Such a probability space cannot be found for an arbitrary collection of d.f.'s. By Equations (7) and (8), we see that it is necessary that the following relations hold:

$$(10) \qquad F_{j_1, j_2, \ldots, j_n}(x_{j_1}, x_{j_2}, \ldots, x_{j_n}) = F_{i_1, i_2, \ldots, i_n}(x_{i_1}, x_{i_2}, \ldots, x_{i_n})$$

$$(11) \qquad F_{i_1, i_2, \ldots, i_m}(x_{i_1}, x_{i_2}, \ldots, x_{i_m}) = F_{i_1, \ldots, i_m, \ldots, i_n}(x_{i_1}, \ldots, x_{i_m}, \infty, \ldots, \infty),$$

where $m < n$ and $j_1, \ldots, j_n$ is an arbitrary permutation of $i_1 \cdots i_n$. We will see that these necessary conditions are also sufficient.

Our space Ω consists of the points $x = (x_t, t \in I)$, where each x_i is a real number; in other words, $\Omega = \underset{i \in I}{\times} R_i$. The r.v.'s r_i are taken to be coordinate r.v.'s in this space; that is, $r_i(x) = x_i$. The measure of the set $(r_{i_j}(x) \leq x_{i_j}, j = 1, \ldots, n)$ is defined as $F_{i_1, \ldots, i_n}(x_{i_1} \cdots x_{i_n})$. More generally, the probability function P is defined on the field $\mathscr{F}$ of all Borel cylinders by

$$(12) \qquad P((r_{i_1}(x), r_{i_2}(x), \ldots, r_{i_n}(x) \in B_{I_n}) = P_{i_1, i_2, \ldots, i_n}(B_{I_n})$$

$$= \int \cdots \int_{B_{I_n}} dF_{i_1, i_2, \ldots i_n}(x_{i_1}, \ldots, x_{i}).$$

Finally, the σ-field $\mathscr{S}$ is defined to be $\mathscr{B}^I = \mathscr{S}(\mathscr{F})$. Kolmogorov [2] proved the following fundamental theorem, which justifies the foregoing assignments.

THEOREM 8

Every system of d.f.'s $F_{i_1,\ldots,i_n}$ that satisfies the conditions of Equations (10) and (11) defines a probability function P on $\mathscr{F}$ by Equation (12). This probability function can be extended to $\mathscr{B}^I$ (by the extension theorem).

Proof

Every d.f. $F_{i_1,\ldots,i_n}$ uniquely defines a corresponding probability function $P_{i_1,\ldots,i_n}$ for all Borel sets in R^{I_n} by (12). However, since the same cylinder can be defined by various B_{I_n}, we must show that (12) always yields the same value. Suppose that the cylinder B is defined simultaneously by $B = B_{J_m} \times R^{J'_m}$ and $B = B_{I_n} \times R^{I'_n}$. Without loss of generality, we may suppose that the corresponding r.v.'s r_{i_k} $(k = 1, \ldots, n)$ and r_{j_l} $(l = 1, \ldots, m)$ belong to the system $(r_{i_1}, \ldots, r_{i_p})$, where $i_1, i_2, \ldots, i_p$ is some reordering of the parameter values $\{i_1, \ldots, i_n\} \cup \{j_1, \ldots, j_m\}$. Then the sets $(r_{i_1}, \ldots, r_{i_n}) \in B_{I_n}$ and $(r_{j_1} \cdots r_{j_m}) \in B_{J_m}$ are both equal to $(r_{i_1} \cdots r_{i_p}) \in B_{I_p}$ for some $B_{I_p} \in \mathscr{B}^{I_p}$. Therefore,

$$\begin{aligned} P_{i_1,\ldots,i_n}(B_{I_n}) &= P_{i_1,\ldots,i_p}((r_{i_1} \cdots r_{i_n}) \in B_{I_n}) = P_{i_1,\ldots,i_p}((r_{i_1} \cdots r_{i_p}) \in B_{I_p}) \\ &= P_{i_1,\ldots,i_p}((r_{j_1} \cdots r_{j_m}) \in B_{j_m}) = P_{j_1,\ldots,j_m}(B_{j_m}), \end{aligned}$$

by using (10) and (11). This proves our assertion about the uniqueness of the definition of P.

We will now prove that P is a probability function on the field $\mathscr{F}$. Obviously, $P(R^I) = P_1(R_1) = 1$ and P is nonnegative by definition. The additivity of P is slightly more complicated. Consider two cylinders $A = A_{I_m} \times R^{I'_m}$ and $B = B_{J_n} \times R^{J'_n}$ and assume that the r.v.'s r_{i_k} and r_{j_l} belong to one inclusive finite system $(r_{i_1}, \ldots, r_{i_2}, \ldots, r_{i_n})$. If $A \cap B = \emptyset$, then

$$((r_{i_1}, \ldots, r_{i_m}) \in A_{I_m}) \cap ((r_{j_1}, \ldots, r_{j_n}) \in B_{J_n}) = \emptyset.$$

Hence

$$\begin{aligned} P(A + B) &= P_{i_1,\ldots,i_p}((r_{i_1} \cdots r_{i_m}) \in A_{I_m}) + ((r_{j_1} \cdots r_{j_n}) \in B_{J_n})) \\ &= P_{i_1,\ldots,i_m}((r_{i_1} \cdots r_{i_m}) \in A_{I_m}) + P_{j_1,\ldots,j_n}((r_{j_1} \cdots r_{j_n}) \in B_{J_n}) \\ &= PA + PB, \end{aligned}$$

which shows that P is additive.

To show that P is σ-additive, we need only prove that P is continuous at $\emptyset$. However, it is easier to prove the contrapositive statement; that is, if B_n is a decreasing sequence of cylinders whose bases are finite sums of intervals[7]

[7]The Borel field B^I is the minimal σ-field over the class of all Borel cylinders or, equivalently, over the class of all cylinders whose bases are product Borel sets (or whose bases are finite sums of intervals).

and which satisfies the condition $\lim P(B_n) = k > 0$, then $\bigcap_{n=1}^{\infty} B_n \neq 0$. The set of indexes involved in defining the sequence B_n is countable, since each cylinder depends upon a finite number of the indexes. We can consider only the product space $\times_{n=1}^{\infty} R_n$, interchanging the indexes if necessary, without essentially restricting the problem. Therefore, we can assume that $B_n = A_n \times R^{(n)}$, where the base A_n is a finite sum of intervals in $R_1 \times \cdots \times R_n$ and $R^{(n)} = \times_{i=n+1}^{\infty} R_i$. For brevity, we will set $P_{1,\ldots,n}(A_n) = P_n(A_n)$. Since the B_n are decreasing,

$$P_n(A_n) = P(B) \geq k > 0.$$

In every interval of $\times_{i=1}^{n} R_i$, we can find a closed bounded interval whose P_n measure is arbitrarily close to the original interval, since P_n is bounded and continuous from below. Thus there exists a finite sum of closed bounded intervals $\bar{A}_n \subset A_n$ such that $P_n(A_n - \bar{A}_n) \leq \epsilon/2^n$. Accordingly, if $\bar{B}_n$ is the Borel cylinder with the base $\bar{A}_n$, then

$$P(B_n - \bar{B}_n) = P_n(A_n - \bar{A}_n) \leq \epsilon/2^n.$$

Letting $C_n = \bigcap_{i=1}^{n} \bar{B}_i$, it follows that $P(B_n - C_n) \leq \epsilon$. Since $C_n \subset \bar{B}_n \subset B_n$, we deduce that $P(C_n) \geq PB_n - \epsilon \geq k - \epsilon$. For sufficiently small ϵ we deduce that $P(C_n) > 0$ and so C_n is not empty. We select a point $x^{(n)} = (x_1^{(n)}, x_2^{(n)}, \ldots)$ in each C_n. Since the C_n form a decreasing sequence, $x^{(n+i)} \in C_n \subset \bar{B}_n$ for every $i = 0, 1, 2, \ldots$. Hence $(x_1^{(n+i)}, \ldots, x_n^{(n+i)}) \in \bar{A}_n$. By the Bolzano–Weierstrass theorem, the sequence $\{x_1^{(1+i)}\}$ contained in the bounded set $\bar{A}_1$ has a cluster point x_1. Therefore, there is a subsequence $\{x_1^{n_{1k}}\}$ whose limit is x_1. In a similar manner, using the boundedness of $\bar{A}_2$, a point x_2 and a subsequence $\{n_{2k}\}$ of the sequence $\{n_{1k}\}$ exist such that $x_2^{(n_{2k})} \to x_2$ and so on. Because of the method of construction, it follows that the diagonal subsequence of points $x^{(n_{kk})} = (x_1^{(n_{kk})}, x_2^{(n_{kk})}, \ldots)$ converges to the point $x = (x_1, x_2, \ldots)$. Hence $x \in \bar{B}_n$ for all n and so it belongs to B_n, since $B_n \supset \bar{B}_n$. Then $\bigcap_{n=1}^{\infty} B_n \neq \emptyset$, which completes the proof of the theorem.

The reader might wonder why the above result does not follow from the product probability theorem. Since the probabilities are consistent, we can obtain marginal probabilities on each of the factor spaces and then apply the product probability theorem to obtain the product probability $\bar{P}$ on the product space. In general, $P \neq \bar{P}$ and thus this method is incorrect. The reason for this can be seen by considering a probability measure P on a finite product space with n factors. Projecting on the factor spaces, we obtain marginal probability functions P_i, although in general, $P(A_1 \times \cdots \times A_n) \neq P_1(A_1)P_2(A_2) \cdots P_n(A_n)$ or, in terms of d.f.'s, $F(x_1 \cdots x_n) \neq F_1(x_1) \cdots F_n(x_n)$.

However, equality holds in the case of independence. Accordingly, the product probability function $\bar{P} \neq P$, since they do not coincide on rectangles. Nevertheless, assuming that the product probability theorem is known for a finite number of factors, we can deduce the more general result from Theorem 8.

COROLLARY

Suppose that F_i is a sequence of d.f.'s. Then there exists a probability space $(\Omega, \mathscr{S}, P)$ with independent random variables r_i defined on it such that the d.f. of r_n is F_n.

Proof

Given a finite subset of indexes $(i_1, \ldots, i_n)$, we choose the joint distribution $P_{i_1,\ldots,i_n}$ on R^n to be the product of the n measures on R' determined by the d.f.'s $F_{i_1}, \ldots, F_{i_n}$. It is clear that the conditions in Equations (10) and (11) are satisfied; thus the result follows by applying Theorem 8.

The following example shows that not every family of d.f.'s satisfies the hypothesis of Theorem 8.

EXAMPLE 1

Instead of using d.f.'s, we will work with density functions for convenience. For contrast we first give an example of a consistent family of density functions.

1. Let S_n denote the n-dimensional rectangle $(x : 0 \leq x_i \leq i, i = 1, \ldots, n)$, where $x = (x_1, \ldots, x_n)$. We define a set of density functions by $f_{1,\ldots,n}(x_1, \ldots, x_n) = 1$ if $x \in S_n$ and 0 if $x \notin S_n$. The reader can verify that the density function of any r-tuple, say $x_1 \cdots x_r$, is $f_{1,\ldots,r}(x_1 \cdots x_r) = 1$ if $x \in S_r$ and 0 otherwise.
2. Now suppose that T_n is the region defined by $(x : x_i \geq 0,\ x_1 + x_2 + \cdots + x_n \leq 1)$ and that a typical density function of the given family is defined as $g_{1,\ldots,n}(x_1 \cdots x_n) = n!$ for $x \in T_n$ and 0 otherwise. The marginal density function $g_{1,\ldots,n-1}(x_1, \ldots, x_{n-1}) = n!(1 - (x_1 + x_2 + \cdots + x_{n-1}))$ for $x \in T_{n-1}$ and 0 otherwise. It is evident that this density function does not coincide with the member of the given family that has the value $(n - 1)!$ in T_{n-1} and 0 otherwise.

Random variables were defined as finite measurable functions. However, this definition is frequently extended. One reason is that if two functions agree except on a set of measure zero (a null event) they have the same integral, which exists provided that one of them is integrable. Moreover, the concepts of convergence in probability and a.s. convergence are defined for equivalence classes; that is, the given sequence $\{f_n\}$ can be replaced by another sequence $\{g_n\}$ provided that $P(f_n \neq g_n) = 0$. Hence we extend the definition of an r.v. to include a function that is a.s. defined, a.s. finite, and a.s. measurable. To illustrate the convenience of this new definition, suppose that $g \leq f_n$, where g is an integrable function and $\liminf Ef_n < \infty$, then $\liminf f_n$ is a r.v. by the Fatou–Lebesgue theorem. In general, this limit would not have been an r.v. previously.

The definitions and theorems in this section can also be extended to nonfinite measurable functions by making some minor modifications. We replace R^n by $\bar{R}^n = \times_{i=1}^{n} \bar{R}_i$, where $\bar{R}_i$ is the extended real line $[-\infty, \infty]$. Similarly, $\mathscr{B}^n$ is replaced by $\overline{\mathscr{B}}^n$. For example, since $\overline{\mathscr{B}}$ is determined by $\mathscr{B}$ and the sets $\{\infty\}$ and $\{-\infty\}$, P_r on $\mathscr{B}$ determines P_r on $\overline{\mathscr{B}}$ except for sets containing the points $\pm\infty$. Hence we must now specify the values

$$P_r(+\infty) = P(r = +\infty) \qquad P_r(-\infty) = P(r = -\infty).$$

Note that in general $\lim_{x \to -\infty} F(x) \neq 0$, since $F(x) = P(r \leq x) = P_r[-\infty, x]$; therefore, $\lim_{x \to -\infty} F(x) = P(r = -\infty) \geq 0$. Similarly, $\lim_{x \to \infty} F(x) = 1 - P(r = \infty) \leq 1$.

EXERCISES

1. (a) Find a few probability spaces that describe the tossing of a die once.
(b) Define some r.v.'s on one of these spaces and find the corresponding r.v.'s by setting up a measurable transformation between these spaces.
(c) Find the d.f.'s of the above r.v.'s.

2. (a) Let f be a continuous function that maps R into R. Is f an r.v. on the probability space $(R, \mathscr{B}, P)$, where $\mathscr{B}$ is the class of Borel sets?
(b) Let $(\Omega, \mathscr{S}, P)$ be an arbitrary probability space. Suppose that Ω is also a topological space and f is a real-valued continuous function on Ω. When is f a r.v.?

3. Give an example of a real-valued function that is not a r.v.

4. (a) Let F be an arbitrary d.f. For every such F, find a single probability space and a r.v. on this space which has F as its d.f. [*Hint:* Let $\Omega = (0, 1)$, let $\mathscr{S}$ be the Borel sets of this interval, and let P be the Lebesgue measure. Finally, let the r.v. be the inverse function of F.]
(b) Generalize to n-dimensions.

5. Show that there is no single probability space for all possible r.v.'s on all possible probability spaces.

6. A fair coin is tossed an infinite number of times.

(a) Describe the probability space.
(b) Define the probability function P on Ω, indicating the σ-field.
(c) Let r_i be a function such that $r_i = 1$ if heads occurs on the ith toss and $r_i = 0$ otherwise. Show that r_i is a r.v.
(d) Describe a typical sample function.

7. Suppose that $F_n = 0$ for $x < 0$, $F_n = \frac{1}{2}$ for $0 \leq x < 1$, and $F_n = 1$ for $x \geq 1$. Show that the space $(\Omega, \mathscr{S}, P)$, which is constructed by means of the corollary to Theorem 8, can be made to correspond to the unit interval, with P corresponding to the Lebesgue measure.

8. Verify that the following are density functions over the indicated range of values (the function is assumed to be zero otherwise):

(a) *Cauchy*: $\frac{1}{\pi}\left(\frac{1}{1+x^2}\right)$. $(-\infty < x < \infty)$

(b) *Uniform (Rectangular)*: $\frac{1}{b-a}$. $(a \leq x \leq b)$

(c) *Triangular*: $\frac{2a - |x|}{4a^2}$. $(-2a \leq x \leq 2a)$

(d) *Normal*: $\frac{1}{\sqrt{2\pi}\sigma} \exp\left(-\frac{1}{2}\left(\frac{x-\mu}{\sigma}\right)^2\right)$. $(-\infty < x < \infty)$

(e) *Gamma*: $\frac{1}{\Gamma(s)} x^{s-1}e^{-x}$. $(x \geq 0)$

(f) *Beta*: $\frac{1}{B(r, s)} x^{r-1}(1-x)^{s-1}$. $(0 \leq x \leq 1)$

(g) *Student*: $\frac{1}{\sqrt{r}\,B\left(\frac{r}{2}, \frac{1}{2}\right)}\left(1 + \frac{x^2}{r}\right)^{-(r+1)/2}$. $(-\infty < x < \infty)$

(h) *Fisher*: $\frac{r^{r/2}s^{s/2}}{B\left(\frac{r}{2}, \frac{s}{2}\right)} \frac{x^{r/2}}{(rx+s)^{(r+s)/2}}$. $(x \geq 0)$

(i) *Exponential*: $\frac{1}{\lambda} e^{-x/\lambda}$. $(x > 0)$

(j) *Poisson*: $e^{-\mu}\frac{\mu^n}{n!}$. $(\mu > 0,\ n = 0, 1, 2, \ldots)$

(k) *Geometric*: pq^{n-1}. $(1 > p > 0,\ p + q = 1,\ n = 1, 2, 3, \ldots)$

(l) *Binomial*: $\binom{n}{k} p^k q^{n-k}$. $(1 > p > 0,\ p + q = 1,\ k = 0, 1, 2, \ldots, n)$

(m) *Negative Binomial*: $\binom{r+k-1}{k} p^r q$.
(r is a positive integer, $0 < p < 1$, $k = 0, 1, 2, \ldots$)

(n) *Hypergeometric*: $\frac{\binom{n}{k}\binom{m-n}{r-k}}{\binom{n}{r}}$.
[m, n, and r are positive integers, $k = 0, 1, 2, \ldots, \min(n, r)$]

9. The Pearson system of density functions f are those satisfying a differential equation of the form

$$\frac{df}{dx} = \frac{(x - a)f(x)}{b_0 + b_1 x + b_2 x^2}.$$

Verify that the following are density functions that belong to the Pearson system.

(a) The beta distribution.

(b) $K\left(1 + \frac{x^2}{a^2}\right)^{-m} \exp\left(\frac{x}{a}\theta \arctan \frac{x}{a}\right).$ $(-\infty < x < \infty)$

(c) $Kx^{-P}e^{-\gamma/x}.$ $(\gamma > 0, \ -\infty < x < \infty)$

(d) The gamma distribution.

(e) $Kx^{-r}(x - a)^s.$ $(a < 0 \text{ for every } x > 0 \text{ or } a > 0 \text{ for every } x \geq a)$

(f) $\frac{K}{(1 + x^2)^m}.$ $(m \geq 1, \ -\infty < x < \infty)$

8.3 CONVERGENCE

In probability theory, convergence a.e. and convergence in measure are called convergence a.s. and convergence in probability respectively. We recall that a sequence $\{r_n\}$ of measurable functions (r.v.'s) converges in the qth mean to a function $r \in L_q$ if and only if $\lim_{n\to\infty} (\int |r_n - r|^q \, dP)^{1/q} = 0$. In our new terminology we would write this last limit as $\lim_{n\to\infty} (E|r_n - r|^q)^{1/q} = 0$.

Using the new terminology and our previous results from measure theory it is not difficult to verify that the relations given in Figure 5 will

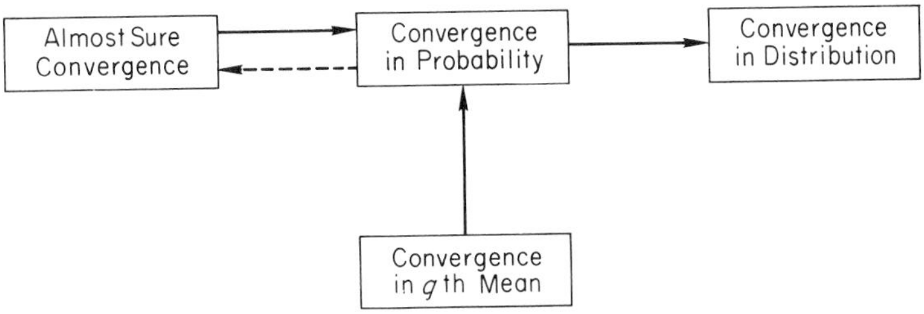

Figure 5

hold among these various convergence concepts. For completeness we have added convergence in distribution, which will be defined below. A dotted arrow means that a subsequence has the indicated properties. We recall that if f_n converges to f in the rth mean ($f_n \to f(L_r)$), then $f_n \to f(L_q)$ for $q < r$.

We will now define the concept of convergence for sequences of d.f.'s. The term "general d.f." refers to Definition 6.6.3 whereas "d.f." denotes a probability d.f.

DEFINITION 1

The sequence of general d.f.'s $\{F_n\}$ *converges weakly* to the general d.f. F if and only if $\lim_{n\to\infty} F_n(x) = F(x)$ at every continuity point $x \in R$ of F. Weak convergence will be denoted by $F_n \to F(w)$.

Part 2 of Example 3 will show that if a sequence F_n of probability d.f.'s converges weakly to a general d.f. F, then F need not be a probability d.f. However, if we assume that $F_n(\pm\infty) \to F(\pm\infty)$,[8] then F will be a probability d.f. Weak convergence with the additional hypothesis that $F_n(\pm\infty) \to F(\pm\infty)$ will be called *complete convergence* and will be denoted by $F_n \to F(c)$. Example 3 illustrates these points.

EXAMPLE 1

1. The function $F_n = 0$ for $x < -1/n$; $F_n = (x + 1/n)/(2/n)$ for $-1/n \le x < 1/n$, and $F_n = 1$ for $1/n \le x$ is a d.f. for $n = 1, 2, 3, \ldots$. The d.f.'s F_n converge weakly to the function $F(x)$, where

$$F(x) = \begin{cases} 0 & (x < 0) \\ 1 & (x \ge 0) \end{cases}.$$

Note that $\lim F_n(0) = \frac{1}{2} \neq F(0)$.

2. Using the same notation, $\lim_{n\to\infty} F_{1/n}(x) = \frac{1}{2}$. The function $F(x) = \frac{1}{2}$ is a general d.f. but not a probability d.f.

There are some obvious relations between weak and complete convergence for probability d.f.'s. If $F_n \to F(w)$ and F is also a d.f., then $F_n \to F(c)$. Conversely, if $F_n \to F(c)$ and the F_n are d.f.'s, then F is a d.f. Even though F, the weak limit of the d.f.'s F_n, is not generally a d.f., we still have $F(\infty) \le 1$. This follows from the inequality

$$F(x) \le F_n(x) + \epsilon \le 1 + \epsilon.$$

If the d.f.'s $F_n \to F(c)$, then for any $\epsilon > 0$ there exists a $b = b(\epsilon)$, independent of n, such that for every n,

$$F_n(x) > 1 - \epsilon \qquad F(x) > 1 - \epsilon,$$

provided that $x \ge b$. To prove this last assertion, choose a continuity point c of F such that $F(c) > 1 - \epsilon/2$. There exists an $N(\epsilon)$ such that $n > N$ implies that $|F(c) - F_n(c)| < \epsilon/2$; it follows that $F_n(c) > 1 - \epsilon$. For $n = 1, 2, \ldots, N$, there exist points c_n such that $F_n(c_n) > 1 - \epsilon$. It follows that the inequalities are valid for $b = \max(c, c_1, \ldots, c_n)$. Analogously, it can be shown that if $F_n \to F(c)$, then

[8] $\lim_{n\to\infty} \lim_{x\to\infty} F_n(x) = \lim_{x\to\infty} \lim_{n\to\infty} F_n(x)$.

$$F_n(x) < \epsilon \qquad F(x) < \epsilon$$

for $x \leq a = a(\epsilon)$. Combining these last two results, it follows that if $F_n \to F(c)$, then for every $\epsilon > 0$ there exist points a and b $(a < b)$ such that

$$F_n(b) - F_n(a) > 1 - \epsilon \qquad F(b) - F(a) > 1 - \epsilon.$$

This is called *equicontinuity at infinity*.

Conversely, if $F_n \to F(w)$ and the functions are equicontinuous at infinity, then $F_n \to F(c)$ (see Exercise 18).

LEMMA 1

If $\{F_n\}$ is a sequence of general d.f.'s that converges to F, then F is an increasing function.

Proof

If $x_1 < x_2$, then $F_n(x_1) \leq F_n(x_2)$; upon taking limits, we obtain $F(x_1) \leq F(x_2)$.

THEOREM 1 (POLYA'S THEOREM)

If the d.f.'s F_n converge to a continuous d.f. F, then the convergence is uniform.

Proof

Since $\lim_{x\to\infty} F(x) = 1$ and $\lim_{x\to-\infty} F(x) = 0$, there exist x_1 and x_2 such that $F(x_1) < \frac{1}{2}\epsilon$ and $F(x_2) > 1 - \frac{1}{2}\epsilon$ for an arbitrary $\epsilon > 0$. Now choose N_1 so large that

$$(1) \qquad |F_n(x_1) - F(x_1)| < \epsilon/2. \qquad (n \geq N_1)$$

We will show that

$$(2) \qquad |F_n(x) - F(x)| < \epsilon. \qquad (n \geq N_1,\ x \leq x_1)$$

From the choice of x_1 and (1), it follows that $0 \leq F_n(x_1) < \epsilon$. Since F_n and F are d.f.'s, then $0 \leq F_n(x) \leq F_n(x_1)$ and $0 \leq F(x) - F(x_1)$ if $x \leq x_1$. Therefore, $0 \leq F_n(x) < \epsilon$ and $0 \leq F(x) < \epsilon/2$ for any $x \leq x_1$ and $n \geq N_1$, which proves inequality (2). Similarly, it can be shown that

$$(3) \qquad |F_n(x) - F(x)| < \epsilon. \qquad (n \geq N_2,\ x \geq x_2)$$

By using the well-known fact that convergence on a closed bounded interval is uniform, we obtain

$$(4) \qquad |F_n(x) - F(x)| < \epsilon. \qquad (n \leq N_3,\ x_1 \leq x \leq x_2)$$

From (2), (3) and (4) it follows that $|F_n(x) - F(x)| < \epsilon$ for $n \geq \max(N_1, N_2, N_3)$.

If the d.f. F is not continuous, we can only state that the convergence is uniform for every finite set of closed continuity intervals of the d.f. F.

Lemma 1 can be generalized by showing that every set of d.f.'s is weakly compact; that is, every infinite sequence of d.f.'s of the set contains a weakly convergent subsequence. It suffices to prove the following theorem.

THEOREM 2 (HELLY'S THEOREM)

Let $\{F_n\}$ be a sequence of d.f.'s. Then there exists a subsequence that converges weakly to a general d.f. F.

Proof

Let $\{x_n\}$ be the set of rationals. For fixed x_1, the sequence $\{F_n(x_1)\}$ is uniformly bounded, since $0 \leq F_n(x) \leq 1$ for every $x \in R$. Therefore, by the Bolzano–Weierstrass theorem, this sequence contains a convergent subsequence $F_{n1}(x_1)$. Similarly, the numerical sequence $F_{n1}(x_2)$ contains a convergent subsequence $F_{n2}(x_2)$ and the sequence $F_{n2}(x_1)$ converges, since it is a subsequence of $F_{n1}(x_1)$. Repeating this procedure, we obtain a set of sequences F_{nk} $(k = 1, 2, 3, \ldots)$ such that for a fixed i, F_{ni} converges at the points $x_1, x_2, \ldots, x_i$. It follows that the diagonal sequence $\{F_{nn}\}$, that is contained in all the subsequences $\{F_{n1}\}, \{F_{n2}\}, \ldots$ converges on the rationals. Let $\bar{F}$ be the function defined on the rationals by $\bar{F}(x_i) = \lim_{n\to\infty} F_{nn}(x_i)$. From Lemma 1 it follows that $\bar{F}$ is an increasing function. We will use this property to show that $\bar{F}$ can be extended to a function F defined on all the real numbers. Let

$$F(x) = \inf_{x_n > x} \bar{F}(x_n) = \lim_{x_n \to x} \bar{F}(x_n),$$

where x_n is a sequence of rationals decreasing to x $(x_n > x)$. It is not difficult to see that if $x \leq y$, then $F(x) \leq F(y)$, and that F has right-hand continuity. Since F is also bounded, it follows that F is a d.f.

It remains to be proved that at every x at which F is continuous, $\lim_{n\to\infty} F_{nn}(x) = F(x)$. Since F is continuous at x, there exist rational numbers $x_i < x < x_j$ such that

$$\text{(5)} \qquad \bar{F}(x_j) - F(x) < \epsilon/2, \qquad \bar{F}(x_i) - F(x) > -\epsilon/2.$$

Moreover, the convergence of the sequence $\{F_{nn}\}$ at these points implies the existence of an N such that for $n \geq N$,

$$\text{(6)} \qquad |\bar{F}(x_j) - F_{nn}(x_j)| < \epsilon/2 \qquad |F_{nn}(x_i) - \bar{F}(x_i)| < \epsilon/2.$$

Hence for $n \geq N$,

$$F_{nn}(x) - F(x) < \bar{F}(x_j) + \epsilon/2 - F(x) < \epsilon.$$

The first inequality follows from (6), since $F_{nn}(x) \leq F_{nn}(x_j)$, while the second is a consequence of (5). Similarly, it can be shown that $F_{nn}(x)$

$-F(x) > -\epsilon$, by using the point x_i. This assures the convergence of $F_{nn}(x)$ to $F(x)$.

The reason that $\bar{F}$ must be used in the above proof is that $\bar{F}(x_i)$ can be different from $F(x_i)$. The conditions that are necessary and sufficient in order that the convergence to F be complete appear in Exercise 18.

The proof holds for general d.f.'s if we assume that $\{F_n\}$ is *uniformly bounded*—that is, $|F_n(x)| \leq M$ for all n and x, where M is a fixed constant.

During the course of the proof we showed that the values of a d.f. are completely determined by its values on the rationals. It is not difficult to see that a d.f. is determined by its values on any set dense in R.

The above definitions, theorems, and proofs can easily be generalized to sets of d.f.'s $F(x\,;\underline{t})$ that depend upon a vector $\underline{t}$ (see Exercise 5).

DEFINITION 2

The r.v.'s r_n *converge in distribution* to the r.v. r and we write $r_n \rightarrow r(d)$ if the d.f.'s F_{r_n} of r_n converge weakly to the d.f. F_r of r.

THEOREM 3

If $r_n \rightarrow r(P)$, then $r_n \rightarrow r(d)$.

Proof

We start with the following relation:

$$(r_n \leq x) = (r_n \leq x,\ r \leq x_2) + (r_n \leq x,\ r > x_2) \subset (r \leq x_2) + (r_n \leq x,\ r > x_2).$$

Hence $F_{r_n}(x) \leq F_r(x_2) + P(r_n \leq x, r > x_2)$. If $x_2 > x$, we have $P(r_n \leq x, r > x_2) \leq P(|r_n - r| \geq x_2 - x)$. Therefore, $F_{r_n}(x) - F_r(x_2) \leq P(|r_n - r| \geq x_2 - x) \rightarrow 0$, and so

$$\limsup F_{r_n}(x) \leq F_r(x_2). \tag{7}$$

A similar argument for $x_1 < x$ yields

$$\liminf F_{r_n}(x) \geq F_r(x_1). \tag{8}$$

If x is a continuity point of F_r, it follows from (7) and (8), upon letting $x_2 \downarrow x$ and $x_1 \uparrow x$, that $F_{r_n} \rightarrow F_r(w)$.

THEOREM 4

If $r_n - s_n \rightarrow 0(P)$ and $r_n \rightarrow r(d)$, then $s_n \rightarrow r(d)$.

The proof follows from Theorem 3 by replacing r_n by s_n and r by r_n in the various inequalities.

In general, if a sequence of r.v.'s converges in distribution, it does not converge in any other way. For instance, whenever the d.f.'s are identical and the r.v.'s are arbitrary otherwise, the sequence of r.v.'s converges in distribution. However, there is an important case in which convergence in distribution and convergence in probability are equivalent.

THEOREM 5

$r_n \to c\,(d)$ if and only if $r_n \to c\,(P)$.

Proof

Let F be the d.f. defined by $F(x) = 1$ if $x \geq c$ and $F(x) = 0$ otherwise. We need only show that $F_{r_n} \to F\,(w)$ implies that $r_n \to c\,(P)$. For an arbitrary $\epsilon > 0$,

$$P(|r_n - c| \leq \epsilon) = F_{r_n}(c + \epsilon) - F_{r_n}(c - \epsilon). \tag{9}$$

Since $c \pm \epsilon$ are continuity points of F, there exists an N such that $n \geq N$ implies that $F_{r_n}(c + \epsilon) > 1 - \eta/2$ and $F_{r_n}(c - \epsilon) < \eta/2$ for any $\eta > 0$. From (9) and these last two inequalities, it follows that $P(|r_n - c| \leq \epsilon) > 1 - \eta$; that is, $r_n \to c\,(P)$.

Observe that for the various types of convergence introduced above, the r.v.'s need not be consistent. We do not have to know their joint distribution. If $r_n \to r\,(P)$ or $r_n \to r\,(L_q)$, then we need only know the d.f. of (r_n, r), while for the limit in distribution, only the d.f.'s F_{r_n} and F_r are required.

THEOREM 6

$r_n \to r\,(P)$ if and only if

$$F_{r_n, r}(x_2, x_1) \to \min\,(F_r(x_1), F_r(x_2)) = F_r(\min\,(x_1, x_2))\ (w).$$

Proof

Assume that $x_2 > x_1$ and $r_n \to r(P)$. Since the event

$$(r \leq x_1) = (r \leq x_1,\ r_n \leq x_2) + (r \leq x_1,\ r_n > x_2),$$

it follows that

$$P(r \leq x_1,\ r_n > x_2) + F_{r_n, r}(x_2, x_1) = F_r(x_1).$$

From the inequality

$$P(r_n > x_2,\ r \leq x_1) \leq P(|r_n - r| \geq x_2 - x_1),$$

it follows that if x_1 is a continuity point of $F_{r_n, r}(x_2, x_1)$, then $\lim\limits_{n\to\infty} F_{r_n, r}(x_2, x_1) \geq F_r(x_1)$, since $\lim P(|r_n - r| \geq x_2 - x_1) = 0$. Obviously $\lim F_{r_n, r}(x_2, x_1) \leq F_r(x_1)$, and so $\lim F_{r_n, r}(x_2, x_1) = F_r(x_1)$. The proof for $x_1 \geq x_2$ is similar.

Now assume that $\lim_{n\to\infty} F_{r_n,r}(x_2, x_1) = F_r(\min(x_1, x_2))$. We will show that $\lim P(|r_n - r| < \epsilon) = 1$ for arbitrary $\epsilon > 0$. Let $A_k = (k\epsilon, (k+1)\epsilon] \times (k\epsilon, (k+1)\epsilon]$; then $(|r_n - r| < \epsilon) \subset \sum_{k=-\infty}^{\infty} ((r_n, r) \in A_k)$. However,

$$P((r_n, r) \in A_k) = F_{r_n,r}((k+1)\epsilon, (k+1)\epsilon) - F_r(k\epsilon, (k+1)\epsilon)$$
$$- F_{r_n,r}((k+1)\epsilon, k\epsilon) + F_{r_n,r}(k\epsilon, k\epsilon).$$

Therefore, $\lim_{n\to\infty} P((r_n, r) \in A_k) = F_r((k+1)\epsilon) - F_r(k\epsilon)$ by hypothesis. Hence

$\lim P(|r_n - r| < \epsilon) \geq \sum_{k=-\infty}^{\infty} F_r((k+1)\epsilon) - F_r(k\epsilon) = 1.$

We can introduce similar convergence concepts for random vectors. The following results indicate that we obtain nothing new. Convergence of random vectors is equivalent to component-wise convergence in the same sense.

DEFINITION 3

The random vectors r_k *converge in the qth mean* to the random vector r if $\lim_{k\to\infty} (E\|r_k - r\|^q)^{1/q} = 0$, where

$$\|x - y\| = \left(\sum_{i=1}^{n} (x_i - y_i)^2\right)^{1/2}.$$

We denote this by $r_k \to r\ (L_q)$.

THEOREM 7

The random vectors $r_k \to r\ (L_q)$ if and only if $r_{ik} \to r_i\ (L_q)$ for $i = 1, 2, \ldots, n$, where r_i denotes the components of r.

Proof

It is obvious that $(E\|r_k - r\|^q)^{1/q} \geq (E|r_{ik} - r_i|^q)^{1/q}$; from $(\sum |a_k|)^{1/2} \leq \sum |a_k|$ and Minkowski's inequality, it follows that

$$(E\|r_k - r\|^q)^{1/q} \leq \sum_{i=1}^{n} (E|r_{ki} - r_i|^q)^{1/q}.$$

The result follows from these two inequalities, upon taking limits.

DEFINITION 4

The random vectors r_k *converge almost surely* to the random vector r if $P(\lim_{k\to\infty} \|r_k - r\| = 0) = 1$.

THEOREM 8

$r_k \to r$ (a.s.) if and only if $r_{ik} \to r_i$ (a.s.) for $i = 1, 2, \ldots, n$.

Proof

This result follows from the following inequalities:

$$|x_i - y_i| \leq \|x - y\| \leq \sum_{i=1}^{n} |x_i - y_i|.$$

DEFINITION 5

The random vectors r_k *converge in probability* to the random vector r if $\lim_{k\to\infty} P(\|r_k - r\| < \epsilon) = 1$ for an arbitrary $\epsilon > 0$. This is denoted by $r_k \to r\ (P)$.

THEOREM 9

$r_k \to r\ (P)$ if and only if $r_{ik} \to r_i\ (P)$ for $i = 1, 2, \ldots, n$.

Proof

The theorem is a consequence of the following relations between the indicated events:

$$(|r_{ik} - r_i| < \epsilon) \supset (\|r_k - r\| < \epsilon) \supset \bigcap_{i=1}^{n} (|r_{ik} - r| < \epsilon/n).$$

THEOREM 10

If $r_k \to r\ (P)$, then $\|r_k\| \to \|r\|\ (P)$.

Proof

Since $\|r_k - r\| \geq |\ \|r_k\| - \|r\|\ |$, then $P(|\ \|r_k\| - \|r\|\ | < \epsilon) \geq P(\|r_k - r\| < \epsilon)$. The result follows upon taking limits, since by hypothesis $\lim_{k\to\infty} P(\|r_k - r\| < \epsilon) = 1$.

THEOREM 11

If $r_k \to r\ (P)$, then for every $\epsilon > 0$ there exists a number $s > 0$ such that $P(\|r_k\| \leq s) > 1 - \epsilon$ for $k = 1, 2, 3, \ldots$, and $P(\|r\| \leq s) > 1 - \epsilon$.

Proof

By choosing s_0 sufficiently large, we have $P(\|r\| \leq s_0) > 1 - \epsilon/2$. We can assume that s_0 is a continuity point of the d.f. $F(x) = P(\|r\| \leq x)$.

Since $r_k \to r\ (P)$, it follows from Theorem 10 that $||r_k|| \to ||r||\ (P)$; therefore, $\lim_{k\to\infty} P(||r_k|| \leq s_0) = F(s_0)$. There exists a positive integer N such that for all $k \geq N$, $P(||r_k|| \leq s_0) - F(s_0) > -\epsilon/2$; thus $P(||r_k|| \leq s_0) > 1 - \epsilon$. However, we can find positive numbers s_i such that $P(||r_i|| \leq s_i) > 1 - \epsilon$ for $i = 1, 2, \ldots, N - 1$. The theorem follows upon setting $s = \max(s_0, \ldots, s_N)$.

THEOREM 12 (FRÉCHET'S THEOREM)

If $r_k \to r\ (P)$ and $f(x_1, \ldots, x_n)$ is a continuous function, then as $k \to \infty$,

$$f(r_{1k}, \ldots, r_{nk}) \to f(r_1, \ldots, r_n)\ (P).$$

Proof

By Theorem 11, there exists an s such that $P(||r_k|| \leq s) > 1 - \epsilon$ and $P(||r|| \leq s) > 1 - \epsilon$. The continuous function f is uniformly continuous on the closed and bounded set $(||x|| \leq s)$. Hence there exists a $\delta = \delta(\eta)$ where $||x - y|| < \delta$ implies that $|f(x) - f(y)| < \eta$. Since $r_k \to r\ (P)$, there exists an N such that for $k \geq N$, $P(||r_k - r|| < \delta) > 1 - \epsilon$.

Therefore, for $k \geq N$, we have $|f(r_k) - f(r)| < \eta$ for all values of ω except those lying in a set of probability measure less than 3ϵ—that is, those ω for which $||r_k|| > s$, $||r|| > s$, or $||r_k - r|| \geq \delta$. In other words, $P(|f(r_k) - f(r)| < \eta) > 1 - 3\epsilon$ or $f(r_k) \to f(r)\ (P)$.

The theorem remains valid if the random vectors lie in some region in which the function is continuous—for example, if $a_i \leq r_{ik}$, $r_i \leq b_i$ for $i = 1, 2, \ldots, n$ and sufficiently large k, and f is continuous for $a_i \leq x_i \leq b_i$.

Theorems 1 and 2 are valid for n-dimensional d.f.'s. Their formulations and proofs are left to the reader.

EXERCISES

1. Show directly that $r_n \to r\ (L_q)$ implies that there is a subsequence $\{r_{nk}\}$ such that $r_{nk} \to r$ (a.s.).

2. Prove that $r_n \to r\ (P)$ if and only if every subsequence of the r_n's contains a further subsequence that converges to r with a probability of 1.

3. (a) Prove Markov's inequality: $P(|r_n - r| \geq \epsilon) \leq \epsilon^{-q} E|r_n - r|^q$.
(b) Deduce that $r_n \to r\ (L_q)$ implies that $r_n \to r\ (P)$.

4. Suppose that $\Omega = \sum_{i=1}^{\infty} A_i$ and $PA_i = 1/2^i$. Let $r_n = 2^{n/q} I_n$, where I_n is the indicator of the event A_n. Show that $\lim_{n\to\infty} P(|r_n| < \epsilon) = 1$ [i.e., $r_n \to 0\ (P)$] and

$(E|r_n|^q)^{1/q} = 1$. This illustrates that convergence in probability does not imply convergence in the qth mean.

5. (a) Define convergence concepts for r.v.'s that depend upon a vector; i.e., $r(\underline{t}) \to r$ as $\underline{t} \to \underline{t}_0$.
(b) Prove Theorems 1 and 2 for d.f.'s that depend upon a vector.

6. (a) Let $A(n, \epsilon) = \bigcap_{i=n}^{\infty} (|r_i - r| < \epsilon) = \inf_{i \geq r} (|r_i - r| < \epsilon)$. Show that $A(n, \epsilon) \subset A(n + 1, \epsilon)$.

(b) Let $A(\epsilon) = \bigcup_{n=0}^{\infty} A(n, \epsilon) = \liminf (|r_i - r| < \epsilon)$. Prove that $A(\epsilon_1) \supset A(\epsilon_2)$ if $\epsilon_1 > \epsilon_2 > 0$.
(c) Show that $P(A(\epsilon)) = \lim_{n \to \infty} P(A(n, \epsilon))$. We will denote this quantity by $P(\epsilon)$.
(d) Verify that $P(\bar{\epsilon})$ is the probability that for any $\epsilon \geq \bar{\epsilon}$ there exists an $N(\epsilon)$ such that for $n \geq N(\epsilon)$, $|r_n - r| < \epsilon$.
(e) Let $P = \lim_{\epsilon \to 0} P(\epsilon)$; then $P = 1$ if and only if $P(\epsilon) = 1$ for every $\epsilon > 0$. If $P = 1$, then $r_n \to r$ (a.s.) and the converse is also true. In general, P is the probability that $\lim r_n = r$.

7. Does $r_n \to r$ (a.s.) if and only if $\lim_{n \to \infty} P(\sup_{m \geq n} |r_m - r_n| \geq \epsilon) = 0$?

8. Let r_n be a sequence of independent r.v.'s (i.e., any n-dimensional d.f. is the product of marginal d.f.'s) that have the d.f.

$$F(x) = \begin{cases} 0 & (x < 0) \\ 1 - \dfrac{1}{1 + x} & (x \geq 0) \end{cases}.$$

(a) Let $s_n = (1/n)r_n$. Show that $F_{s_n}(x) = 1 - 1/(nx + 1)$ for $x \geq 0$ and deduce that $\lim s_n = 0\,(d)$ and therefore that $\lim s_n = 0\,(P)$.
(b) Show that $P(\lim s_n = 0) = 0$. This shows that convergence in probability does not imply convergence a.s.

(a) Give an example of two r.v.'s r and s such that $r \neq s$ but $P(r = s) = 1$. Prove or disprove parts (b) and (c).
(b) $r_n \to r\,(P)$ and $r_n \to s\,(P)$; then $r = s$ (a.s.).
(c) $r_n \to r\,(P)$ and $r = s$ (a.s.); then $r_n \to s\,(P)$.

10. State and prove results similar to those in Exercise 6 for random vectors. (*Hint:* Replace $|r_i - r|$ by $||r_i - r||$.)

11. Formulate a definition of convergence in distribution for random vectors. Does convergence in probability still imply convergence in distribution?

12. Show that if for the sequence $\{(r_n, s_n)\}$ of pairs of random vectors there exists an N such that for all $n \geq N$, $P(|r_n| \geq |s_n|) = 1$, then $r_n \to 0\,(P)$ implies that $s_n \to 0\,(P)$.

13. If $r_1 \geq r_2 \cdots$ are nonnegative r.v.'s and $r_n \to 0\,(P)$, show that $r_n \to 0$ (a.s.).

14. The *associated* r.v. s_p of a sequence $\{r_n\}$ is defined as $s_p(\omega) = \sup_{n \geq p} r_n(\omega)$ for each fixed ω. Note that $s_1 \geq s_2 \geq s_3 \ldots$. Show that $r_n \to 0$ (a.s.) if and only if $s_n \to 0\,(P)$.

15. Prove that if $\{s_n\}$ is the sequence associated with the r.v.'s $r_n - r$, then $r_n \to r$ (a.s.) if and only if $s_n \to 0\,(P)$.

16. Show that if $r_n \to r\,(P)$ and $0 < a \leq s_n$, then $r_n/s_n \to (r/s)\,(P)$.

17. The following example illustrates that the convergence of a sequence of r.v.'s in probability does not imply that their Cesàro sum converges. Let $\{r_i\}$ be a sequence of independent r.v.'s such that $P(r_n \leq x) = F(x) = 1 - 1/(n + x)$ for $x > 0$ and 0 otherwise.

(a) Show that $r_n \to 0\,(P)$.
(b) Let $s_n = (1/n)(r_1 + r_2 + \cdots + r_n)$. Show that s_n does not converge to $0\,(P)$. [*Hint:* Let $z_n = \max(r_1, \ldots, r_n)$. Find $P(z_n \leq x)$ and use the fact that $-z_n/n \leq s_n$ implies that $P(|z_n/n| < \epsilon) \geq P(|s_n| < \epsilon)$.]

18. The sequence $\{r_n\}$ (or F_n) is *stochastically bounded* if for every $\epsilon > 0$ there exists a b such that $P(|r_n| > b) < \epsilon$ for all sufficiently large n. Show that $F_n \to F(c)$ if and only if $F_n \to F(w)$ and the F_n are stochastically bounded.

19. Let Var F denote the total variation of the general d.f. F.

(a) If $F_n \to F$ weakly, then $\text{Var}\, F \leq \liminf \text{Var}\, F_n$.
(b) $F_n(\pm\infty) \to F(\pm\infty)$ if and only if $\text{Var}\, F_n \to \text{Var}\, F$.

20. Let S be the space of all r.v.'s (equivalent r.v.'s are identified). We define the following functions on $S \times S \to R$:

(a) $\rho_0(r, s) = \inf(P(|r - s| \geq \epsilon) + \epsilon$ for $\epsilon > 0$.
(b) $\rho_1(r, s) = \inf \epsilon$ such that $P(|r - s| \geq \epsilon)$.
(c) $\rho_2(r, s) = \inf E(f(|r - s|))$, where F is a continuous increasing function on $[0, \infty]$ with $f(0) = 0$.

Show that these functions are metrics on S and that convergence in probability is equivalent to ordinary convergence in any of these metric spaces.

21. Suppose that the graph of a d.f. F is completed at points of discontinuity by a vertical line segment; call this function $\bar{F}$. Let S be the space of these new functions, and define a function on $S \times S \to R$ by $\rho(\bar{F}, \bar{G})$, which is equal to the maximum distance between the points of the two completed graphs measured along the lines of the slope -1. Show that S is a complete metric space and that $\rho(\bar{F}_n, \bar{F}) \to 0$ if and only if $F_n \to F$ at all points of continuity of F.

8.4 TYPES OF DISTRIBUTION

Suppose that the solution of a problem requires the evaluation of the normal distribution

$$F(x) = \frac{1}{\sigma\sqrt{2\pi}} \int_{-\infty}^{x} e^{-(t-\mu)^2/2\sigma^2}\, dt.$$

For numerical computations it is not economical to prepare tables of this function for all possible values of x, μ, and σ. Instead we note that if we set $z = (t - \mu)/\sigma$, the above integral is of the form

$$N(x) = \frac{1}{\sqrt{2\pi}} \int_{-\infty}^{x} e^{-z^2/2}\, dz.$$

Tables of $N(x)$ are prepared and the value of $F(x)$ is obtained from the fact that $F(x) = N[(x - \mu)/\sigma]$.

More generally, suppose that $s = ar + b$ ($a > 0$, b is arbitrary), where r and s are r.v.'s. It is easily seen that the d.f.'s of r and s are connected by the equation $F_r(x) = F_s(ax + b)$. This motivates the following concept.

DEFINITION 1

The d.f.'s F_1 and F_2 *belong to the same type* if

$$F_2(x) = F_1(ax + b)$$

for some constants $a > 0$ and b.

The property of belonging to the same type is a symmetrical and transitive relation. Hence the totality of d.f.'s can be partitioned into mutually disjoint classes called *types*. All normal distributions form one type, the *normal* type. All improper (or singular[9]) d.f.'s form the *improper* type. The types of d.f.'s other than the improper type are called *proper* (or nonsingular).

The following question arises: Given a sequence of d.f.'s F_n, do all the limit d.f.'s of the form $F_n(a_n x + b_n)$ belong to the same type? The answer, by Khintchine, is contained in the next theorem.

THEOREM 1

If the d.f.'s F_n converge weakly to a proper d.f. F as $n \to \infty$, then for any constants $a_n > 0$ and b_n the d.f.'s $F_n(a_n x + b_n)$ can converge weakly to a proper d.f. G only if $G(x) = F(ax + b)$ with $a_n \to a$ and $b_n \to b$.

Proof

Choose a sequence of integers $n_1 < n_2 < \cdots < n_k < \cdots$ such that $\lim_{k\to\infty} a_{n_k} = a$ $(0 \leq a \leq \infty)$ and $\lim_{k\to\infty} b_{n_k} = b$ $(-\infty \leq b \leq \infty)$ exist. To simplify the notation, we will assume for the moment that $n_k = k$.

We will prove that $0 < a < \infty$. Suppose that $a = \infty$. Let m be the supremum of the numbers x for which $\overline{\lim}_{k\to\infty} (a_k x + b_k) < \infty$. If $m = -\infty$,

[9]Another synonym for singular is "degenerate"; that is, $F(x) = 0$ for $x < c$ and $F(x) = 1$ for $x \geq c$.

then $G \equiv 1$, and if $m = \infty$, then $G \equiv 0$ (see Exercise 1). Therefore, m is finite. For $x < y < m$,

$$\overline{\lim_{k\to\infty}}\,(a_k x + b_k) \leq \overline{\lim_{k\to\infty}}\,(x - y)a_k + \overline{\lim_{k\to\infty}}\,(a_k y + b_k).$$

By definition of m we have $\overline{\lim\limits_{k\to\infty}}\,(a_k x + b_k) = -\infty$ for every $x < m$; therefore, $G(x) = 0$ for $x < m$. Since $\overline{\lim}\,(a_k x + b_k) = \infty$ for $x > m$, $G(x) = 1$. The assumption $a = \infty$ contradicts the fact that $G(x)$ is proper and so $a \neq \infty$.

It follows that b must also be finite. In fact, $\lim\,(a_k x + b_k) = \infty$ for all x implies that $G \equiv 1$, while $\lim\,(a_k x + b_k) = -\infty$ for all x implies that $G \equiv 0$.

Finally, suppose that $a = 0$. Then for every x and $\epsilon > 0$ we have

$$b - \epsilon \leq a_k x + b_k \leq b + \epsilon$$

for sufficiently large k; hence

$$F_k(b - \epsilon) \leq F_k(a_k x + b_k) \leq F_k(b + \epsilon).$$

If ϵ is chosen so that the function F is continuous at $b - \epsilon$ and $b + \epsilon$, then

$$F(b - \epsilon) \leq G(x) \leq G(b + \epsilon).$$

Letting $x \to \infty$, we obtain $F(b + \epsilon) = 1$; letting $x \to -\infty$, we obtain $F(b - \epsilon) = 0$. This shows that F is improper, which contradicts the hypothesis of the theorem.

We are now in a position to prove that

$$G(x) = \lim_{k\to\infty} F_k(a_k x + b_k) = F(ax + b). \tag{1}$$

Let x and $ax + b$ be continuity points of G and F respectively. Since $\lim\limits_{k\to\infty}\,(a_k x + b_k) = ax + b$, we have, for sufficiently large k,

$$ax + b - \epsilon \leq a_k x + b_k \leq ax + b + \epsilon,$$

where $\epsilon > 0$ is chosen so that F is continuous at the points $ax + b - \epsilon$ and $ax + b + \epsilon$. Since

$$F_k(ax + b - \epsilon) \leq F_k(a_k x + b_k) \leq F_k(ax + b + \epsilon),$$

it follows that

$$\begin{aligned} F(ax + b - \epsilon) &\leq \underline{\lim}_{k\to\infty} F_k(a_k x + b_k) \leq \overline{\lim_{k\to\infty}}\, F_k(a_k x + b_k) \\ &\leq F(ax + b + \epsilon). \end{aligned}$$

Since ϵ is arbitrarily small and $ax + b$ is a continuity point of F, it follows that (1) is valid.

Recalling our change of notation, we see that we have succeeded in proving that $G(x) = F(ax + b)$, where $a = \lim\limits_{k\to\infty} a_{n_k}$ and $b = \lim\limits_{k\to\infty} b_{n_k}$. The proof that a and b are independent of the chosen subsequence $\{n_k\}$ follows immediately from the following theorem (see Reference [10]).

THEOREM 2

The d.f.'s F_n satisfy the following (weak) convergence relation: If

(2) $$\lim_{n\to\infty} F_n(a_n x + b_n) = F(x);$$

then

(3) $$\lim_{n\to\infty} F_n(\alpha_n x + \beta_n) = F(x),$$

where $a_n > 0$, $\alpha_n > 0$, b_n and β_n are real constants and F is a proper d.f. if and only if

(4) $$\lim_{n\to\infty} \alpha_n/a_n = 1 \qquad \lim_{n\to\infty} (b_n - \beta_n)/a_n = 0.$$

Proof

We will prove that Equations (2) and (4) imply Equation (3). Let $x - \epsilon$, x, and $x + \epsilon$ for $\epsilon > 0$ be continuity points of F. Then for sufficiently large n,

$$x - \epsilon < \frac{\alpha_n}{a_n} x + \frac{\beta_n - b_n}{a_n} < x + \epsilon$$

by (4). Hence

$$a_n(x - \epsilon) + b_n < \alpha_n x + \beta_n < a_n(x + \epsilon) + b_n,$$

which implies that

$$F_n(a_n(x - \epsilon) + b_n) < F_n(\alpha_n x + \beta_n) < F_n(a_n(x + \epsilon) + b_n).$$

Taking limits, this last inequality yields

$$F(x - \epsilon) \le \underline{\lim}_{n\to\infty} F_n(\alpha_n x + \beta_n) \le \overline{\lim}_{n\to\infty} F_n(\alpha_n x + \beta_n) \le F(x + \epsilon).$$

Letting $\epsilon \to 0$ gives (3).

Conversely, (2) and (3) imply (4). To prove this, set $A_n = \alpha_n/a_n$, $B_n = (\beta_n - b_n)/a_n$, and $G_n(x) = F_n(a_n x + b_n)$. It follows that relations (2) and (3) are transformed into

$$\lim_{n\to\infty} G_n(x) = F(x)$$

$$\lim_{n\to\infty} G_n(A_n x + B_n) = F(x).$$

Applying the proof of the first part of Theorem 1,

(5) $$F(x) = F(Ax + B),$$

where $A > 0$ and B are finite numbers and

$$\lim_{k\to\infty} A_{n_k} = A \qquad \lim_{k\to\infty} B_{n_k} = B.$$

We will prove that $A = 1$ and $B = 0$.

Suppose that $A < 1$. From (5) we obtain $F(Ax + B) = F(A(Ax + B) + B)$ or $F(x) = F(A^2x + B(1 + A))$. Repeating this procedure $n - 1$ times yields

$$F(x) = F(A^n x + B(1 + A + \cdots + A^{n-1})).$$

Since $\lim_{n\to\infty} A^n = 0$, we can conclude from the last relation that $F(x) = F(B/(1 - A))$ for $x > B/(1 - A)$ and so $F(x) = 1$ for $x \geq B/(1 - A)$. From (5) it follows that $F(B/(1 - A) - \epsilon) = F(B/(1 - A) - A^n\epsilon)$. Letting $n \to \infty$, we deduce that $F(B/(1 - A) - \epsilon) = F(B/(1 - A)-)$. This cannot hold for a proper d.f. The case where $A > 1$ can be reduced to the preceding situation by rewriting (5) in the following form: $F(x) = F(A^{-1}x - BA^{-1})$. Therefore, $A = 1$.

Suppose that $B \neq 0$; then (5) yields $F(x) = F(x + nB)$, when n is an arbitrary positive or negative integer. Hence for every x,

$$F(\infty) = \lim_{nB\to\infty} F(x + nB) = F(x) = \lim_{nB\to-\infty} F(x + nB) = F(-\infty),$$

which is a contradiction. It follows that $B = 0$,

The proof that $A = 1$ and $B = 0$ does not depend in any way upon the chosen subsequence $\{n_k\}$. Hence $\lim A_n = 1$ and $\lim B_n = 0$ as $n \to \infty$.

EXERCISES

1. In Theorem 1, prove that $G \equiv 1$ $(m = -\infty)$ and $G \equiv 0$ $(m = \infty)$. [*Hint:* If $m = -\infty$, then $\overline{\lim}\,(a_k x + b_k) = \infty$ for all x and so $G(x) = \lim F_k(a_k x + b_k) = 1$ for a certain subsequence by using equicontinuity at infinity.]

2. Show that for every finite c and for every sequence F_n of d.f.'s, there exist numbers $a_n \neq 0$ and $b_n \neq 0$ such that $F_n(a_n x + b_n) \to F$, where $F = 0$ for $x < c$ and $F = 1$ for $x \geq c$.

3. Prove Theorem 1 for arbitrary a_n under the modification $|a_n| \to |a|$.

4. Prove Theorem 2 for arbitrary a_n with the additional hypothesis that $a_n\alpha_n > 0$ for every n.

5. Show that Theorem 2 is false if F is degenerate.

6. Replace F in (3) by a d.f. G, where $G(x) = F(Ax + B)$, and replace the limits 1 and 0, in (4) by A and $-B$. Prove this version of Theorem 2.

8.5 REMARKS

We recall that there are many probability spaces suitable for describing a particular physical experiment. Given r.v.'s on one of these spaces, it is possible to find corresponding r.v.'s on any other of the spaces having the same d.f.'s (see Section 8.2). Hence we see that probability theory is primarily concerned with the joint distribution of families of r.v.'s which describe the characteristics of the phenomena and not with either the probability space or particular r.v.'s. It is this feature that distinguishes probability theory from measure theory.

Our probability measure is not assumed to be complete. This assumption is frequently made to preserve one's physical intuition, since a subset of an impossible event should be impossible. We can always complete the measure as described in Theorem 6.5.6.

In Section 8.2, we defined the induced probability measure in R^n by

$$\bar{P}(A) = P_r(A) = P(r^{-1}(A)), \tag{1}$$

where r is a random vector that maps Ω into R^n, and $r^{-1}(A) \in \mathscr{S}$, the σ-field of events in Ω. The function $\bar{P}$ is defined on the induced σ-field $\bar{\mathscr{S}} = (A : r^{-1}(A) \in \mathscr{S})$. Now let F be the n-dimensional d.f. of the random vector r. Using this d.f., we can define a Lebesgue—Stieltjes measure μ as in Section 6.6. The measures μ and $\bar{P}$ agree on any n-dimensional interval, and therefore on all Borel sets. Assume that P is complete; it is not difficult to see that $\bar{P}$ is also complete. Under this hypothesis, by the method of constructing μ (using outer measure), it follows that $\bar{\mathscr{S}}$ contains the μ-measurable sets and that μ and $\bar{P}$ coincide on the μ-measurable sets. However, $\bar{\mathscr{S}}$ can contain a set that is not μ-measurable. To avoid this inconvenience, we must assume that the probability measure is *perfect*.

In discussing this concept, it is convenient to introduce the following property, which is meaningful only if the measure space is also a topological space:

If $E \in \mathscr{S}$, then $\mu E = \inf 0$, where the infimum is taken over all open sets 0 that belong to $\mathscr{S}$ and contain E.

DEFINITION 1

Let r be a Borel measurable vector function that maps a measure space X into R^n. The measure μ is called *perfect* if the induced measure μ defined as in (1) has the property given in the preceding paragraph.

THEOREM 1

The probability measure P is perfect if and only if for every random vector r, $\bar{P}$ and the Lebesgue—Stieltjes measure based on F_r are identical. The proof of this result appears in Appendix I of Reference [3]. Perfect measure is also discussed in [11].

Theorem 1 shows that the assumption of a perfect measure restricts membership in the induced σ-ring $\bar{\mathscr{S}}$ to n-dimensional Lebesgue-Stieltjes sets. Hence problems involving random vectors can be reduced to those involving coordinate r.v.'s in Euclidean space using a natural and convenient measure.

To illustrate the convenience of the hypothesis of a perfect measure, we will cite two examples. In the first place, conditional probabilities and expectations can be defined more conveniently. However, it is still possible to treat these topics without this assumption. A detailed presentation appears in [4]. Second, let r_1 and r_2 be r.v.'s defined on the same probability space. We will say that r_1 and r_2 are *independent* (see Section 11.3) if

$$(2) \qquad P(r_1 \in B_1,\ r_2 \in B_2) = P(r_1 \in B_1)P(r_2 \in B_2)$$

for every pair of Borel sets B_1 and B_2. Some authors define the r.v's to be independent if (2) holds whenever its right-hand side is defined. This more natural definition is not equivalent to the first unless the measure is perfect, since B_1 and B_2 need not be Borel sets. For a further discussion of this point see Section 11.3.

The existence of a d.f. with prescribed properties is not equivalent to the existence of a r.v. on the given probability space. This point will be elucidated when we discuss infinitely divisible r.v.'s in Chapter 13.

In the study of stochastic processes, it is often necessary to find the limit superior (or inferior) of a nondenumerable collection of r.v.'s. For example, consider the family $(r_t, t \in T)$; we have

$$(3) \qquad (\sup_{t \in T} |r_t(\omega)| > x) = \bigcup_{t \in T} (|r_t(\omega)| > x).$$

This equality does not imply the measurability of the set on the left. Although each set on the right is measurable, we do not know whether a nondenumerable union of measurable sets is an event. Hence some method must be found for interpreting such probabilities. The study of such processes is beyond the scope of this volume; we refer the interested reader to [4].

There is another set of axioms for probability theory based upon frequency theory. The events do not form a σ-field; however, the probability is still σ-additive. This is the approach of R. von Mises [5].

A generalized probability theory has arisen from attempts to formulate a statistical model for quantum mechanics. The classical probability theory is inadequate mainly because the quantum mechanical events fail to form a σ-field but have a much less richly endowed algebraic structure. G. Bodiou (References [6] and [7]) was one of the first pioneers in this area. The reader interested in exploring this topic can also consult [8] and [9].

REFERENCES

1. Titchmarsh, E. C., *The Theory of Functions.* London: Oxford University Press, 1939.
2. Kolmogorov, A. N., *Foundations of the Theory of Probability.* New York: Chelsea, 1956. This is a translation of the German original, first published in 1933.
3. Gnedenko, B. V., and A. N. Kolmogorov. *Limit Distributions for Sums of Independent Random Variables.* Reading, Mass.: Addison-Wesley, 1954.
4. Doob, J. L., *Stochastic Processes.* New York: John Wiley, 1953.
5. Von Mises, R., *Mathematical Theory of Probability and Statistics.* New York: Academic Press, 1964.
6. Bodiou, G., *Probabilité sur un treillis non modulair.* Paris: Publ. de l'Inst. de Stat. de l'Univ. de Paris, VI–I, 1957.
7. Bodiou, G., *Théorie Dialectique des Probabilités.* Paris: Gauthier–Villars, 1964.
8. Varadarajan, V. S., "Probability in physics and a theorem on simultaneous observability," *Comm. Pure and Appl. Math,* **15** (1962), 189–217.
9. Gudder, S. P., "A Generalized Probability Model for Quantum Mechanics," Ph. D. dissertation, Univ. of Illinois, 1964.
10. Gnedenko, B. V., "Sur la distribution limitée du terme maximum d'une série aléatoire," *Ann. of Math.,* **44** (1943), 423–453.
11. Parthasarathy, K. R., *Probability Measures on Metric Spaces.* New York: Academic Press, 1967.

CHAPTER 9

DISTRIBUTIONS AND MOMENTS

9.1 MOMENTS

Several different kinds of moments are used quite frequently.

DEFINITION 1

The k*th moment about the point a* of a r.v. r is defined as $E(r - a)^k$. For $a = 0$ this moment will be called the k*th moment* and will be denoted by μ'_k, while if $a = \mu$, the mean, it will be called the k*th central moment* and will be denoted by μ_k.

Corresponding moments can be defined by taking absolute values; for example, the k*th absolute moment* is defined by $E|r|^k$. Since integrability is equivalent to absolute integrability, if the kth moment exists and is finite, then the same is true for the kth absolute moment; the converse is also true. For the same reason the even moments always exist, although they may be infinite; however, the odd moments do not necessarily exist. The remarks are illustrated in the following example.

EXAMPLE 1

1. Let r be a Cauchy r.v. with a density function $(1/\pi)[1/(1 + x^2)]$. Then Er does not exist.

2. Suppose that r is a discrete r.v. such that

$$P(r = (-1)^n 2^n) = \frac{1}{2^n}. \qquad (n = 1, 2, 3, \ldots)$$

Then $Er = -1 + 1 - 1 + \cdots$ does not exist, but $Er^2 = 2 + 2^2 + 2^3 + \cdots$ exists, although it is infinite.

DEFINITION 2

The *factorial moment* $\mu_{(k)}$ of the order k is defined as

$$\mu_{(k)} = E(r(r-1)\cdots(r-k+1)).$$

Since a r.v. is just a finite, real-valued, measurable function, all the results about L_p spaces derived in Section 7.10 hold. In particular, we have the following theorem (c.f., Lemma 7.10.3).

THEOREM 1

If $E|r|^k < \infty$, then $Er^l < \infty$ for $k \geq l$.

Proof

If $|r| > 1$, then $|r|^k \geq |r|^l$ for $k \geq l \geq 0$, and if $|r| \leq 1$, then $|r|^l \leq 1$. Thus in either situation, $|r|^l \leq 1 + |r|^k$, from which the result follows, since integrability is equivalent to absolute integrability.

We have seen that $r_n \to r\ (L_q)$ implies that $r_n \to r\ (P)$. However, convergence in probability does not imply convergence in the qth mean unless we add some extra conditions.

THEOREM 2

If the r.v.'s r_n are uniformly bounded a.s., then $r_n \to r(P)$ implies that $r_n \to r(L_q)$.

Proof

The uniform boundedness of the r_n means that $|r_n| \leq M$ (a constant) a.s. for all n. This implies that $|r| \leq M$ (a.s.) and so $(r_n - r)$ is uniformly bounded. Let $A_n = (|r_n - r| \geq \epsilon)$; then

$$(1) \qquad E|r_n - r|^q = \int_{A_n} |r_n - r|^q \, dP + \int_{A'_n} |r_n - r|^q \, dP \leq (2M)^q P(|r_n - r| \geq \epsilon) + \epsilon^q,$$

since on the set A'_n, $|r_n - r| < \epsilon$ and $\int_{A'_n} dP \leq \int dP = 1$. Since $\lim_{n\to\infty} P(|r_n - r| \geq \epsilon) = 0$, it follows from the inequality (1) by taking limits that $r_n \to r\,(L_q)$.

If we examine the proof, we see that the second integral appearing on the right-hand side of (1) is always small. If the first integral were small the result would follow. We might be tempted to say that since the indefinite integral is μ-continuous (see Section 7.7) and A_n is small, the result follows. However this is incorrect; we need *uniform* continuity.

DEFINITION 3

The integrals of the r.v.'s r_n are *uniformly* (P-*absolutely*) *continuous* if for every $\epsilon > 0$ there exists a $\delta = \delta(\epsilon)$ independent of n such that $PA < \delta$ implies that $\int_A r_n dP < \epsilon$.

With this concept in mind the reader should have no difficulty in proving the next lemma.

LEMMA 1

If $r_n \to r\,(P)$ and $\int |r_n - r|^q\, dP$ are uniformly continuous, then $r_n \to r\,(L_q)$.

Instead of considering the sets A_n in (1), we could have used the sets $B_n = (|r_n - r| \geq c)$ for large values of c and assumed that $\lim \int_{B_n} |r_n - r|^q\, dP = 0$ uniformly in c. If we choose a large but fixed value of c, the first integral in (1) can be made arbitrarily small, while the second integral is small by Theorem 2 since the r.v.'s $|r_n - r|I_n$, where I_n is the indicator function of B'_n, are uniformly bounded by c. To phrase our new result succinctly, we will formally define this new concept.

DEFINITION 4

Let $B_n = (|r_n| \geq c)$; the r.v.'s r_n are *uniformly integrable* if for every $\epsilon > 0$ there exists a $C = C(\epsilon)$ independent of n such that $c \geq C$ implies that $\int_{B_n} |r_n|\, dP < \epsilon$.

In view of this definition, the foregoing result can be stated as in Lemma 2.

LEMMA 2

If $r_n \to r\ (P)$ and the integrals $\int |r_n - r|^q\, dP$ are uniformly integrable, then $r_n \to r\ (L_q)$.

In Lemma 1 it would be preferable to have a criterion in terms of $\int |r_n|^q\, dP$ instead of $\int |r_n - r|^q\, dP$. We will show that this is possible by proving that the integrals $\int |r_n|^q\, dP$ are uniformly continuous if and only if the integrals $\int |r_n - r|^q\, dP$ are uniformly continuous. Since $r \in L_q$, it follows that $\int_A |r|^q\, dP$ is arbitrarily small for a sufficiently small PA by the absolute continuity of the indefinite integral. The equivalence of the two conditions is immediate upon applying the inequality in Exercise 2(d), which yields

$$\int_A |r_n - r|^q\, dP \leq k\left(\int_A |r|^q\, dP + \int_A |r_n|^q\, dP\right)$$

$$\int_A |r_n|^q\, dP \leq k\left(\int_A |r_n - r|^q\, dP + \int_A |r|^q\, dP\right).$$

Accordingly we can restate Lemma 1 as in Theorem 3.

THEOREM 3

If $r_n \to r\ (P)$ and the integrals $\int |r_n|^q\, dP$ are uniformly continuous, then $r_n \to r\ (L_q)$.

A similar restatement is possible for Lemma 2. This follows from Exercises 9 and 10, which also offer an alternate way of proving Lemma 2.

EXERCISES

1. Let $g = g_1 g_2$ be a finite Borel function on R^2 to R and let $F(x, y) = G(x)H(y)$ be the d.f. of the r.v. (r_1, r_2). Show (see Theorem 8.2.2) that

$$Eg(r_1, r_2) = \left(\int_{-\infty}^{\infty} g_1(x)\, dG\right)\left(\int_{-\infty}^{\infty} g_2(y)\, dH\right).$$

2. Prove the following inequalities:

(a) *Hölder's inequality.* $E|rs| \leq (E|r|^p)^{1/p}(E|s|^q)^{1/q}$, where $p > 1$ and $1/p + 1/q = 1$.

(b) *Minkowski's inequality.* $(E|r + s|^q)^{1/q} \leq (E|r|^q)^{1/q} + (E|s|^q)^{1/q}$.

(c) *Schwarz's inequality.* $(E|rs|)^2 \leq (E|r|^2)(E|s|^2)$.

(d) $\int_A |r+s|^q\,dP \le k(\int_A |r|^q\,dP + \int_A |s|^q\,dP)$. [*Hint:* Use the inequality $|x+y|^q \le k(|x|^q + |y|^q)$, where $k = \max(1, 2^{q-1})$.]

(e) $\dfrac{Er^q - a^q}{\text{a.s.}\sup r^q} \le P(|r| \ge a) \le \dfrac{E|r|^q}{a^q}$.

[*Hint:* See the proof of Chebyshev's inequality (Theorem 2.3). The right-hand inequality of the above equation is called the *Markov inequality*.]

3. (a) Show that $\log Er^q$ is a convex function of q.
(b) Deduce that $(E|r|^q)$ is a nondecreasing function of q, and hence $L_0 \supset L_p \supset L_q \supset L_\infty$ for $0 \le p \le q < \infty$, where L_0 is the space of all r.v.'s and L_∞ is the space of all a.s.-bounded r.v.'s.

4. Prove that $\lim_{q\to\infty} (E|r|^q)^{1/q} = \text{a.s.}\sup |r|$.

5. Show that if $r_n \to r\ (L_q)$, then $E|r_n|^q \to E|r|^q$. [*Hint:* Apply Exercise 2(d) if $r \le 1$ and Exercise 2(b) if $r > 1$.]

6. If $r_n \in L_q$ and $E|r_n - r|^q \to 0$, then $r \in L_q$. [*Hint:* Apply Exercise 2(d).]

7. Show that if $\int |r_n|dP \le c$—that is, the integrals are uniformly bounded—then $PB_n \to 0$. [*Hint:* Use Exercise 2(e).]

8. Prove that the r.v.'s r_n are uniformly integrable if and only if their integrals are uniformly continuous and uniformly bounded. [*Hint:* $\int_A |r_n|\,dP \le \int_{B_n} |r_n|\,dP + cPA$, where B_n is defined above.]

9. Prove that $r_n \to r\ (L_q)$ implies that $\int |r_n - r|^q\,dP$ is uniformly continuous. [*Hint:* $E|r_n - r|^q$ is small for sufficiently large $n = n(\epsilon)$ and $\int_A |r_n - r|\,dP$ is small for $i \le n$ if PA is small.]

10. Show that $r_n \to r\ (L_q)$ implies that the $|r_n|^q$ are uniformly integrable. [*Hint:* $r \in L_q$ by Exercise 6; use Exercise 2(d) and Exercise 8.]

11. Find the kth factorial moment of a r.v. that is binomially distributed.

9.2 DECOMPOSITION OF DISTRIBUTION FUNCTIONS

In order to decompose a given d.f. into a sum of d.f.'s, we need to know something about its discontinuities. We begin by studying the one-dimensional d.f. F that corresponds to a r.v. r. The function F is always continuous from the right and increasing. Furthermore, for every value of x such that $P(r = x) > 0$, F has a discontinuity at x with a jump of $P(r = x)$; the converse is also true. Hence for every value of x such that $P(r = x) = 0$, F is continuous at x. Discontinuity points of F are also called discontinuity points of the set function P_r.

THEOREM 1

The discontinuity points of F form a countable set D.

Proof

We classify the discontinuity points according to the magnitude of the jumps $F(x) - F(x-)$; let

$$A_n = \left(x : \frac{1}{n+1} < F(x) - F(x-)\right). \qquad (n = 1, 2, 3, \ldots)$$

Then the class A_n can have at most n distinct points; otherwise the sum of the jumps would be greater than 1, which is impossible. Since $D = \bigcup_{n=1}^{\infty} A_n$ and since each A_n has a finite number of points, the result follows.

The discontinuity set D can be empty, finite, or infinite. The limit points of this set can be continuity or discontinuity points of the d.f.; however, if the limit point is ∞ or $-\infty$ it must be a continuity point for finite r.v.'s.

EXAMPLE 1

1. For the normal d.f., $D = \phi$.
2. If $F(x) = 0$ for $x < 0$ and $F(x) = 1$ for $x \geq 0$, then $D = \{0\}$.
3. The discontinuity set D is infinite for the following d.f.:

$$F(x) = \begin{cases} 0 & (x < 0) \\ \sum_{1}^{n} \dfrac{1}{2^i} & (n - 1 \leq x < n, n = 1, 2, 3, \ldots) \end{cases}$$

From the nature of the set D, it follows that we can decompose F into a continuous d.f. and a step d.f. by using the function $G(x) = \sum_{d \leq x} F(d)$, where $d \in D$. The details will be given in Theorem 6.

A point x is a discontinuity point of F if and only if there exists $h_i \geq 0$ such that $|F(x_1, \ldots, x_n) - F(x_1 - h_1, \ldots, x_n - h_n)| \equiv |F(x) - F(x - h)| > 0$. In $n > 1$ dimensions, the discontinuity points of a d.f. can lie on hypersurfaces; however, we shall see that at such points it is not necessarily true that $P(r = x) > 0$, where $r = (r_1, \ldots, r_n)$ and x is interpreted in the usual manner.

THEOREM 2

If $d = (d_1 \cdots d_n)$ is a discontinuity point of the n-dimensional d.f. $F(x_1, \ldots, x_n)$, then at least one d_i $(i = 1, 2, \ldots, n)$ is a discontinuity point of the one-dimensional marginal d.f. F_i.

Proof

The result follows from the inequality

$$(1) \qquad |F(x) - F(x - h)| \le \sum_{i=1}^{n} |F_i(x_i) - F_i(x_i - h_i)|,$$

where $h = (h_1, \ldots, h_n)$, since if all the marginal d.f.'s are continuous, then so is F. The proof of (1) follows.

Let $A_i = (r_1 \le x_1, \ldots, r_{i-1} \le x_{i-1}, x_i - h_i < r_i \le x_i, r_{i+1} \le x_{i+1}, \ldots, r_n \le x_n)$ for $i = 1, 2, \ldots, n$, and $B_i = (x_i - h_i < r_i \le x_i)$. Since $B_i \supset A_i$, we have

$$(2) \qquad P(A_1 \cup A_2 \cup \cdots \cup A_n) \le \sum_{i=1}^{n} PA_i \le PB_1 + PB_2 + \cdots + PB_n,$$

which is just a restatement of (1).

Let $\{d_{ij}\}$ be the set of discontinuity points of the marginal distribution $F_i(x_i)$ for $i = 1, \ldots, n$. Then the discontinuity points of the n-dimensional d.f. F are contained in sets whose points x have at least one of the d_{ij}'s as their ith coordinate for some i. For example, if $n = 2$, d_0 and d_1 are discontinuity points for $F_1(x)$, and d_2 is a discontinuity point for $F_2(y)$, then the discontinuity points are contained in the sets $\{(d_0, y)\}$, $\{(d_1, y)\}$, $\{(d_0, d_2)\}$, $\{(d_1, d_2)\}$, and $\{(x, d_2)\}$. Sets of this form will be called *underdimensional* sets.

The following example shows that not every point of an underdimensional set is a discontinuity point for F.

EXAMPLE 2

Let

$$F(x, y) = \begin{cases} 0 & (x < 0, y < 0) \\ \dfrac{1}{2} \cdot & (0 \le x < 1, y \ge 0; 0 \le x, 0 \le y < 1) \\ 1 & (x \ge 1, y \ge 1) \end{cases}$$

Then $F_1(x) = 0$ for $x < 0$, $F_1(x) = \frac{1}{2}$ for $0 \le x < 1$, $F_1(x) = 1$ for $x \ge 1$, and $F_2(y) = F_1(y)$. The points 0 and 1 are discontinuity points of F_1 and F_2. However, $(1, y)$ is not a discontinuity point for $F(x, y)$ if $0 < y < 1$. The following result is a partial converse of Theorem 2.

THEOREM 3

Let d_i be a discontinuity point of F_i, the one-dimensional marginal d.f. of F. Then the discontinuity points of F are contained in the sets $(x : x_i = d_i)$ for $i = 1, 2, \ldots, n$. There are at most a countable number of such sets.

Proof

Let d_1 be a discontinuity point of F_1; then $(d_1, x_2, \ldots, x_n)$ must be a discontinuity point for some values of $x_2, \ldots, x_n$. If $P(r_1 = d_1, r_2 \leq x_2, \ldots, r_n \leq x_n) = 0$ for all possible values of $(x_2, \ldots, x_n)$, we would have

$$P(r_1 = d_1) = \lim_{x_2,\ldots,x_n\to\infty} P(r_1 = d_1, r_2 \leq x_2, \ldots, r_n \leq x_n) = 0.$$

The fact that $P(r_1 = d_1, r_2 \leq x_2, \ldots, r_n \leq x_n)$ is positive for some $(x_2, \ldots, x_n)$ leads to the result, since

$$P(r_1 = d_1, r_2 \leq x_2, \ldots, r_n \leq x_n) = \lim_{h_1\to 0} [F(d_1, \ldots, x_n) - F(d_1 - h_1, \ldots, x_n)].$$

The last assertion of the theorem is a direct consequence of Theorem 4.

We will call the limits appearing on the right-hand side of the last equation *differentials*. It is convenient to introduce the following notation:

$$\delta_i F = \lim_{h_i\to 0} [F(x_1, \ldots, x_{i-1}, x_i, x_{i+1}, \ldots, x_n) - F(x_1, \ldots, x_{i-1}, x_i - h_i, x_{i+1}, \ldots, x_n)];$$

then $\delta_{i_1} \cdots \delta_{i_k} F$ is called a *differential of the* k*th order*. We leave it to the reader to verify that

(3) $\delta_1 \cdots \delta_k F(x) = P(r_1 = x_1, \ldots, r_k = x_k, r_{k+1} \leq x_{k+1}, \ldots, r_n \leq x_n),$

The next result is a criterion for determining whether a point $x = (x_1, \ldots, x_n)$ is a discontinuity point.

THEOREM 4

The point x is a discontinuity point of the d.f. F if and only if some differential of the kth order ($k = 1, 2, \ldots, n$) evaluated at x is positive.

Proof

Suppose that x is a discontinuity point of F. It follows from Equation (2) that

$$0 < \lim_{h_i\to 0} |F(x) - F(x - h)| \leq \sum_{i=1}^{n} \delta_i F(x).$$

Hence at least one $\delta_i F(x) > 0$.

Conversely, suppose that $\delta_{i_1} \cdots \delta_{i_k} F(x) > 0$. For notational convenience we will assume that $i_1, \ldots, i_k$ are the first k integers. After some set-theoretic calculations, Equation (3) yields

$$0 < \delta_1 \cdots \delta_k F(x) \leq \lim_{h_1,\ldots,h_k\to 0} |F(x) - F(x - h)|,$$

where $h_{k+1} = h_{k+2} = \cdots = h_n = 0$. Therefore, x is a discontinuity point of F.

A set of points with at least one coordinate fixed is an underdimensional set. From Theorems 2 and 3 we see that a discontinuity point $(d_1, d_2, \ldots, d_n)$ is contained in one of the underdimensional sets $(x : x_i = d_i)$ for $i = 1, 2, \ldots, n$. These sets are not disjoint and not all points are discontinuity points. However, the discontinuity points can be decomposed into disjoint underdimensional sets. The Lebesgue measure of an underdimensional set is zero.

Theorem 4 gives us some insight into the nature of discontinuity points. For a one-dimensional d.f. we have $P(r = x) > 0$ if and only if x is a discontinuity point. However, for an n-dimensional d.f., a point is a discontinuity point of the d.f. if and only if a kth-order differential is positive, where $k = 1, 2, \ldots, n$. Accordingly, a point can be a discontinuity point without having $P(r = x) > 0$, where $r = (r_1, \ldots, r_n)$ and $x = (x_1 \cdots x_n)$. For instance, in Example 2, $P(r_1 = 0, r_2 = 1) = 0$ even though $(0, 1)$ is a discontinuity point; however, $P(1, 1) = \frac{1}{2}$. This necessitates singling out such special discontinuity points.

DEFINITION 1

A discontinuity point d such that $P(r = d) > 0$ will be called an *atomic point* (or *atom*).

THEOREM 5

The set S of atomic points is at most countable.

Since the proof of this result is similar to the proof of Theorem 1, we will omit the details.

DEFINITION 2

If the set S of atomic points is empty, then the d.f. is called *paracontinuous*.

In general, for $n > 1$, a paracontinuous d.f. is not continuous; however, it can be decomposed into an absolutely continuous d.f. and a singular d.f. (see Theorem 7).

EXAMPLE 3

The following is a paracontinuous d.f. that is not continuous.

$$F(x, y, z) = \begin{cases} 0 & [(x < 0) \cup (y < 0) \cup (z < 0)] \\ \frac{1}{2} z & [(0 \le x) \cap (y \ge 0) \cap (0 \le z < 1)] \\ \frac{1}{2} + \frac{1}{2} xy & [(0 \le x < 1) \cap (0 \le y < 1) \cap (1 \le z)] \\ \frac{1}{2} + \frac{1}{2} x & [(0 \le x < 1) \cap (1 \le y) \cap (1 \le z)] \\ \frac{1}{2} + \frac{1}{2} y & [(1 \le x) \cap (0 \le y < 1) \cap (1 \le z)] \\ 1 & [(1 \le x) \cap (1 \le y) \cap (1 \le z)] \end{cases}$$

Its one-dimensional marginal d.f.'s are: $F_1(x) = 0$ for $x < 0$, $F_1(x) = \frac{1}{2} + \frac{1}{2}x$ for $0 \le x < 1$, and $F_1(x) = 1$ for $x \ge 1$; $F_2(y) = F_1(y)$; and $F_3(z) = 0$ for $z < 0$, $F_3(z) = \frac{1}{2}z$ for $0 \le z < 1$, and $F_3(z) = 1$ for $z \ge 1$. Its discontinuity set is formed by the underdimensional sets $(0, 0, z)$ for $z \ge 0$, and $(x, y, 1)$ for $x \ge 0$ and $y \ge 0$.

DEFINITION 3

If a d.f. F can be written in the form

$$(4) \qquad F(x_1, \ldots, x_n) = \begin{cases} \sum_{d \in S} \delta_1 \cdots \delta_n F(d_1, \ldots, d_n) & (\text{if } d_1 \le x_1, d_2 \le x_2, \ldots, d_n \le x_n) \\ 0 & (\text{otherwise}) \end{cases}$$

then F is called a *step d.f.*

Examples of step d.f.'s are provided in parts 2 and 3 of Example 1.

THEOREM 6

Every d.f. F can be written in the form

$$F = aF_j + lF_p,$$

where F_j is a step d.f., F_p is a paracontinuous d.f., and a and l are nonnegative constants such that $a + l = 1$.

Proof

Let $S \ne \phi$ be the set of atoms of F and define $\bar{F}_j$ by Equation (4). We will verify that $\bar{F}_j$ satisfies conditions 1–4 in Theorem 8.2.3.

Obviously, $\bar{F}_j$ is nonnegative; in fact, $0 \le F_j \le 1$. The fact that all its kth-order differences are nonnegative follows from the identities

$$(5)\qquad \underset{h_1}{\Delta}\underset{h_2}{\Delta}\cdots\underset{h_k}{\Delta}\bar{F}(x_1, x_2, \ldots, x_n)$$
$$= P(x_1 < r_1 \leq x_1 + h_1, \ldots, x_k < r_k \leq x_k + h_k,$$
$$r_{k+1} \leq x_{k+1}, \ldots, r_n \leq x_n)$$
$$= \sum_d \delta_1 \cdots \delta_n F \geq 0,$$

where the last summation is over those d for which $x_i < d_i \leq x_i + h_i$ $(i = 1, 2, \ldots, k)$ and $d_i \leq x_i$ $(i = k + 1, k + 2, \ldots, n)$. From Equation (5), we see that $\bar{F}_j$ is continuous on the right in each variable. Hence $\bar{F}_j$ is a d.f. in the general sense. In addition, since $F \geq \bar{F}_j$, it follows that $\lim_{x\to-\infty} \bar{F}_j(x) = 0$.

The function $\bar{F}_j$ need not be a d.f. since $\lim_{x\to\infty} \bar{F}_j = \bar{a} \leq 1$. If $\bar{a} = 1$, the proof is complete and the original d.f. was a step d.f. Now suppose that $\bar{a} < 1$; then set $\bar{F}_p = F - \bar{F}_j$. The only nontrivial point in proving that F_p is a general d.f. is to show that the kth-order differences Δ^k of $\bar{F}_p$ are non-negative. This is true since $\Delta^k\bar{F}_p = \Delta^k F - \Delta^k\bar{F}_j$ and $\Delta^k F \geq \Delta^k \bar{F}_j$ from (5). Again we have $\lim_{x\to\infty} \bar{F}_p = 1 - \bar{a} \neq 1$.

The desired decomposition is obtained by setting $F_j = (1/\bar{a})\bar{F}_j$, $F_p = [1/(1 - \bar{a})]\bar{F}_p$, $a = \bar{a}$, and $l = 1 - \bar{a}$.

Observe that for one dimension, $\bar{F}_p$ is continuous. If $\bar{a} = 0$, then $S = \phi$ and the original d.f. F must have been paracontinuous.

EXAMPLE 4

Let $F(x) = 1 - \frac{1}{2}e^{-x}$ for $x \geq 0$ and 0 for $x < 0$; then $F_j = 0$ for $x < 0$, $F_j = 1$ for $x \geq 0$, $F_p = 1 - e^{-x}$ for $x \geq 0$, $F_p = 0$ for $x < 0$, and $a = l = \frac{1}{2}$.

A paracontinuous d.f. can be further decomposed into the d.f.'s described in the following definition.

DEFINITION 4

Suppose that the d.f. F corresponds to the Lebesgue–Stieltjes measure ν. Then F is *absolutely continuous* (*singular*) with respect to μ if and only if ν is absolutely continuous (singular) with respect to μ.

The phrase "with respect to μ" will be omitted, since we will always use μ as the Lebesgue measure in the following discussion. Equivalent definitions appear in Exercise 9.

THEOREM 7

Let F be a paracontinuous d.f.; then F can be written in the form

$$F = bF_s + cF_c, \tag{6}$$

where F_s is a paracontinuous singular d.f., F_c is an absolutely continuous d.f., and b and c are nonnegative constants such that $b + c = 1$.

Proof

The d.f. F defines a Lebesgue–Stieltjes measure μ according to Theorem 6.6.1. The measure μ can be decomposed by applying the Lebesgue decomposition theorem (7.7.2) with respect to the Lebesgue measure on $\mathscr{B}^n$, the Borel field in R^n. We obtain $\mu = \mu_c + \mu_s$, where μ_s is zero except on some set N_s of Lebesgue measure zero and μ_c is absolutely continuous with respect to the Lebesgue measure. By the Radon–Nikodym theorem (7.7.1) $\mu_c(E) = \int_E \bar{g}(x)\, dx$ for $E \in \mathscr{B}^n$, where $\bar{g} \geq 0$ is a Borel function of n variables. Finally, by Theorem 6.6.1, there are general d.f.'s $\bar{F}_c$ and $\bar{F}_s$ corresponding to the measures μ_c and μ_s, respectively, such that $F = \bar{F}_c + \bar{F}_s$. Observe that the points of increase of $\bar{F}_s$ all lie in N. In general, $\lim_{x\to\infty} F_c = c \neq 1$ and $\lim_{x\to\infty} \bar{F}_s = 1 - c \neq 0$. The required decomposition is obtained by setting $F_s = [1/(1-c)]\bar{F}_s$, $F_c = (1/c)\bar{F}_c$, and $b = 1 - c$. We obtain (6) where

$$F_c(x_1 \cdots x_n) = \int_{-\infty}^{x_n} \cdots \int_{-\infty}^{x_1} g(x_1 \cdots x_n)\, dx_1 \cdots dx_n;$$

$g = (1/c)\bar{g}$ is the density function of F_c.

Combining Theorems 6 and 7, we obtain the following result.

THEOREM 8

Every d.f. F can be written as

$$F = aF_j + bF_c + cF_s, \qquad (a + b + c = 1, a \geq 0, b \geq 0, c \geq 0)$$

where F_j is a step function, F_c is an absolutely continuous d.f. such that

$$F_c = \int_{-\infty}^{x} g(x)\, dx, \qquad (g \geq 0, x \in R^n)$$

and F_s is a paracontinuous singular d.f.; that is, there exists a set N of Lebesgue null measure such that $\int_N dF_s = 1$ (c.f., Exercise 2).

EXAMPLE 5

1. The d.f. $F(x) = 0$ for $x \leq 0, F = x$ for $0 < x \leq 1$, and $F = 1$ for $1 < x$ is absolutely continuous; $F(x) = \int_{-\infty}^{x} f(t)\,dt$ where $f = 0$ except for $0 < x < 1$, where $f = 1$.
2. The d.f. defined in the last example of Section 5.9 by means of the Cantor ternary set is a singular d.f. Observe that its derivative is equal to zero a.e. Exercise 5 shows that this is true for any singular d.f.; in fact this is how they are sometimes defined.

EXERCISES

1. (a) Verify the last statement of Theorem 3.
(b) Verify equation (5) and the statement following it.

2. (a) The d.f. F is defined by

$$F(x, y) = \begin{cases} 0 & [(x < 0) \cup (y < 0)] \\ \frac{1}{2} \min(x, y) & [(0 \leq x < \frac{1}{2}) \cup (0 \leq y)] \\ \frac{1}{2} \min(x, y) + \frac{1}{2} y. & [(\frac{1}{2} \leq x) \cap (0 \leq y < 1)] \\ \frac{1}{2} + \frac{1}{2} x & [(\frac{1}{2} \leq x < 1) \cap (1 \leq y)] \\ 1 & [(1 \leq x) \cap (1 \leq y)] \end{cases}$$

Show that its discontinuity set D is the underdimensional set of pairs $(\frac{1}{2}, y)$ with $y > 0$. Find the corresponding singular d.f. and show that the set N of Theorem 8 is formed by the intervals on $x = \frac{1}{2}$ for $0 \leq y \leq 1$ and on $x = y$ for $0 \leq y \leq 1$.
(b) Verify that the function

$$F(x, y) = \begin{cases} 0 & [(x < 0) \cup (y < 0)] \\ \min(x, y) & [(0 \leq x < 1) \cap (0 \leq y < 1)] \\ x & [(0 \leq x < 1) \cap (1 \leq y) \\ y & [(1 \leq x) \cap (0 \leq y < 1)] \\ 1 & [(1 \leq x) \cap (y \leq 1)] \end{cases}$$

is a continuous singular d.f., where N is the set $x = y$ for $0 \leq y \leq 1$.

3. Decompose the following d.f.:

$$F(x, y) = \begin{cases} 0 & [(x < 0) \cup (y < 0)] \\ xy & [(0 \leq x < 1) \cap (0 \leq y < 1)] \\ x. & [(0 \leq x < 1) \cap (1 \leq y)] \\ y & [(1 \leq x) \cap (0 \leq y < 1)] \\ 1 & [(1 \leq x) \cap (1 \leq y)] \end{cases}$$

4. Is the decomposition of Theorem 7 unique?

5. Give an alternate proof of Theorem 7 along the following lines:

(a) Use the fact that a d.f. F has a derivative $\partial^n F/(\partial x_1 \cdots \partial x_n)$ (a.e.).

(b) Let $\bar{F}_c = \int_{-\infty}^{x_1} \cdots \int_{-\infty}^{x_n} [\partial^n F/(\partial x_1 \cdots \partial x_n)]\, dx_1 \cdots dx_n$ and define $\bar{F}_s = F - \bar{F}_c$. Observe that this implies that $\partial^n F_s/(\partial x_1 \cdots \partial x_n) \equiv 0$ for a singular d.f.

6. By the use of Theorem 6, give an alternative proof of the correspondence between Lebesgue–Stieltjes measures and d.f.'s.

7. Is the decomposition given in Theorem 8 unique?

8. Show that if a sequence F_n of d.f.'s converges weakly to a d.f. F, then this weak limit is unique. [*Hint:* Use continuity from the right for the countable set of discontinuities of both functions.]

9. (a) Show that F is absolutely continuous if and only if $F = \int g\, dx$, where g is a Borel measurable function.

(b) Show that F is absolutely continuous if and only if for every $\epsilon > 0$ there exists a $\delta > 0$ such that for any collection of disjoint, right-semiclosed intervals $I_1, \ldots, I_k$, $\sum_1^k \mu I_j < \delta$ implies that $\sum_1^k \Delta^n F < \epsilon$, where the nth-order difference is evaluated on each interval.

(c) The d.f. F is singular if and only if F is concentrated on a set N (i.e., $\nu N' = 0$) such that $\mu N = 0$.

9.3 SEQUENCES OF INTEGRALS

In Section 8.3 we defined weak convergence for a sequence of d.f.'s F_n; we will denote this type of convergence by $F_n \to F(w)$. Recall that, in general, if $F_n \to F(w)$, then $\lim_{n\to\infty} F_n(\pm\infty) \neq F(\pm\infty)$. The type of convergence in which the equality sign always holds is complete (or strong) convergence; it will be denoted by $F_n \to F(c)$. Any sequence of d.f.'s gives rise to a sequence of integrals obtained by integrating a continuous function with respect to these distributions. We will study the properties of such sequences of integrals under the hypothesis of weak convergence of the d.f.'s.

For convenience, we will only state and prove these theorems in R^1. However, they can readily be extended to R^n; we leave their formulation and proof to the reader. Note that $\int g\,dF = \int g\,d\bar{F}$ if $\bar{F} = F + c$, where c is an arbitrary constant. From this observation it follows that the hypothesis "F_n converges to F in some manner" can be replaced by "F_n converges to F up to an arbitrary constant" in the statement of the theorems below.

The basic result for convergence of sequences of integrals is Helly's theorem.

THEOREM 1

Let g be a continuous function on a finite interval $[a, b]$ and let the d.f.'s $F_n \to F(w)$ on $[a, b]$, where a and b are continuity points of F; then

$$\lim_{n\to\infty} \int_a^b g\,dF_n = \int_a^b g\,dF.$$

Proof

Since g is continuous, we can find a subdivision of $[a, b]$ by means of the points $a = x_0 < x_1 < \cdots < x_m = b$ into the subintervals $(x_k, x_{k+1}]$ such that $|g(x_k) - g(x)| < \epsilon$ in each interval for a fixed $\epsilon > 0$. The points of subdivision can be chosen to be continuity points of F; therefore since $F_n \to F(w)$,

$$|F_n(x_k) - F(x_k)| < \frac{\epsilon}{mM} \tag{1}$$

for sufficiently large n, where $M = \sup g(x) < \infty$. Hence

$$\left|\int_a^b g\,dF - \int_a^b g\,dF_n\right| \le \left|\int_a^b g\,dF - \int_a^b g_\epsilon\,dF\right| + \left|\int_a^b g_\epsilon\,dF - \int_a^b g_\epsilon\,dF_n\right| + \left|\int_a^b g_\epsilon\,dF_n - \int_a^b g\,dF_n\right|, \tag{2}$$

where

$$g_\epsilon(x) = \sum_{k=0}^{m-1} g(x_k) I_{(x_k, x_{k+1}]}.$$

By the above remarks, the first and third terms on the right-hand side of Equation (2) are less than $\epsilon[F(b) - F(a)]$ and $\epsilon[F_n(b) - F_n(a)]$ respectively. The second term is the Riemann–Stieltjes sum approximating the integral and is equal to

$$\left|\sum_{k=0}^{m-1} g(x_k)\,\Delta F(x_k) - \sum_{k=0}^{m-1} g(x_k)\,\Delta F_n(x_k)\right|$$

$$= \left|\sum_{k=0}^{m-1} g(x_k)[F(x_{k+1}) - F_n(x_{k+1})] - \sum_{k=0}^{m-1} g(x_k)[F(x_k) - F_n(x_k)]\right|;$$

therefore it does not exceed 2ϵ because of inequality (1). Since the functions $F_n(x)$ are uniformly bounded by 1, the sum $\epsilon[F(b) - F(a)] + \epsilon[F_n(b) - F_n(a)] + 2\epsilon$ can be made as small as we wish.

Theorem 1 remains valid for general d.f.'s if we add the hypothesis that the F_n are uniformly bounded; in fact, the remaining theorems are valid for general d.f.'s under this added assumption.

THEOREM 2

If the real-valued function g is continuous and bounded over the entire real line, and if the d.f.'s $F_n \to F(c)$, then

$$\lim_{n\to\infty} \int g(x)\, dF_n(x) = \int g(x)\, dF(x).$$

Proof

The proof of the theorem is based upon the inequality

$$(3) \qquad \left|\int g\, dF_n - \int g\, dF\right| \le \left|\int_{-\infty}^{a} g\, dF_n - \int_{-\infty}^{a} g\, dF\right| + \left|\int_{a}^{b} g\, dF_n - \int_{a}^{b} g\, dF\right| + \left|\int_{b}^{\infty} g\, dF_n - \int_{b}^{\infty} g\, dF\right|,$$

where a and b are continuity points of F. The first and third terms appearing on the right-hand side of Equation (3) can be made arbitrarily small by choosing $|a|$, b, and n sufficiently large. If $M = \sup |g|$, then the first term is not greater than $M[F^n(a) + F(a)]$, while the third term is not greater than

$$M[F_n(\infty) - F_n(b)] + M[F(\infty) - F(b)].$$

Since a and b are continuity points of F, $\lim_{a\to-\infty} F_n(a) = 0$ and $\lim_{b\to\infty} [F_n(\infty) - F_n(b)] = 0$ uniformly in n, by reasoning similar to that used in Theorem 8.3.1. Moreover, from the assumption that $\lim_{n\to\infty} F_n(\pm\infty) = F(\pm\infty)$, it follows that $\lim_{b\to\infty} F(b) = F(\infty)$ and $\lim_{a\to-\infty} F(a) = 0$. The smallness of the first and third terms follows from these estimates. The second term appearing on the right-hand side of (3) can be made arbitrarily small for sufficiently large n by using Theorem 1.

THEOREM 3

If g is continuous, $g(\pm\infty) = 0$, and $F_n \to F(w)$, then $\int g\, dF_n \to \int g\, dF$.

Proof

Since g is continuous and $g(\pm\infty) = 0$, it follows that g is bounded on the whole real line. Therefore, the integrals $\int g\, dF$ and $\int g\, dF_n$ exist and are finite. The theorem follows by using inequality (3); the proof is similar to the proof of Theorem 2. The first and third terms in the right-hand side are not greater than $c_1(\sup |g(x)| : x \in (-\infty, a])$ and $c_3(\sup |g(x)| : x \in [b, \infty))$, where c_1 and c_3 are some fixed constants. Since $\sup |g(x)| \to 0$ as $a \to -\infty$ and $b \to \infty$, these terms can be made arbitrarily small. The remaining term is also small for sufficiently large n by Theorem 1.

Theorem 3 is a generalization of Theorem 2 in that we only require $F_n \to F(w)$. However, if we drop the assumption that $g(\pm\infty) = 0$, then, in general, $\int g\, dF_n$ does not approach $\int g\, dF$ and we can only prove the following theorem.

THEOREM 4

If g is continuous and $F_n \to F(w)$, then $\liminf \int |g|\, dF_n \geq \int |g|\, dF$.

The proof of this result is left as an exercise.

Further extensions of the Helly–Bray theorems (1–4) are possible; they depend upon the notion of uniform integrability.

DEFINITION 1

The function g is *uniformly integrable* in F_n if, given an $\epsilon > 0$,

$$\int |g|\, dF_n - \int_a^b |g|\, dF_n < \epsilon, \qquad [a \leq A(\epsilon), b \geq B(\epsilon)]$$

where $A(\epsilon)$ and $B(\epsilon)$ are independent of n.

THEOREM 5

Let g be continuous on R, let g be uniformly integrable in F_n, and let $F_n \to F(w)$; then $\int g\, dF_n \to \int g\, dF$.

THEOREM 6

If $F_n \to F(w)$, then $\int |g|\, dF_n \to \int |g|\, dF < \infty$ if and only if g is uniformly integrable in F_n.

The proofs of these theorems are outlined in the Exercises.

We will use these results to solve the Fréchet–Shohat moment convergence problem [3]. As a preliminary step, we will prove the following theorem.

THEOREM 7

If x^s is uniformly integrable in F_n for a fixed $s > 0$, then

1. For every subsequence $F_{n_i} \equiv G_i$ for $n_1 < n_2 < n_3 \cdots$ such that $G_i \to F(w)$, we have

$$\int x^l \, dG_i \to \int x^l \, dF < \infty \qquad \int |x|^l \, dG_i \to \int |x|^l \, dF < \infty$$

for every $l \leq s$.
2. There exists a subsequence $\{G_i\}$ such that $G_i \to F(c)$.

Proof

By Theorem 8.3.2 we can find a subsequence G_i and a d.f. F such that $G_i \to F(w)$. For every $l \leq s$,

$$\int_{|x|\geq c} |x|^l \, dG_i \leq c^{l-s} \int_{|x|\geq c} |x|^s \, dG_i. \tag{4}$$

The last integral approaches zero by the uniform integrability of $|x|^s$; hence we see that $|x|^l$ is uniformly integrable for every $l \leq s$. Accordingly, part (1) of this theorem follows from Theorems 5 and 6.

From inequality (4), for $l = 0$,

$$\int_b^\infty dG_i \leq \int_{x\geq b} |x|^s \, dG_i/b^s \qquad \int_{-\infty}^a dG_i \leq \int_{-\infty}^a |x|^s \, dG_i/|a|^s,$$

which can be rewritten as

$$0 \leq 1 - G_i(b) \leq \int |x|^s \, dG_i/b^s \qquad 0 \leq G_i(a) \leq \int |x|^s \, dG_i/|a|^s$$

by increasing the estimates of the right-hand sides. From these last inequalities it follows, by letting $i \to \infty$, $b \to \infty$, and $a \to -\infty$, that $F(\infty) = 1$ and $F(-\infty) = 0$, since the right-hand sides approach zero by the boundedness of $\int |x|^s \, dF$.

COROLLARY

If the sequence of integrals $\int |x|^{s+\epsilon} \, dF_n < M$ for all n and some $\epsilon > 0$, then the theorem holds.

Proof

The uniform integrability of $|x|^s$ follows from the inequality

$$\int_{|x|\geq c} |x|^s \, dF_n \leq c^{-\epsilon} \int_{|x|\geq c} |x|^{s+\epsilon} \, dF_n \leq c^{-\epsilon} M.$$

We are now in a position to consider the moment convergence problem. The kth moments of F_n will be denoted by μ'_{kn}.

THEOREM 8

If $\lim_{n\to\infty} \mu'_{kn} = \mu'_k < \infty$ for every $k \geq k_0$, then there exists at least one fixed d.f. F whose kth moment ($k = 0, 1, 2, 3, \ldots$) is μ'_k. A subsequence $G_i \equiv F_{n_i}$ can be extracted from the sequence $\{F_n\}$ such that $G_i \to F(c)$.

Proof

To prove this result for $k < k_0$, we choose an even number $l \geq k_0$ and apply the corollary to Theorem 7. Its hypothesis is satisfied since $\int x^l \, dF_n = \int |x|^l \, dF_n < \mu'_l + 1$ for sufficiently large values of n, as $\mu'_{ln} \to \mu'_l < \infty$.

This result can be strengthened if the moments determine the d.f. F uniquely—that is, if there can be at most one d.f. with the given moments μ'_k (there may be none).

COROLLARY

Under the hypothesis of the theorem, suppose that the moments μ'_k are such that there exists at most one d.f. having these moments; then there exists a unique d.f. F such that $F_n \to F(w)$.

Proof

We need only prove convergence, since the existence of F is a consequence of Theorem 8. Assume that $F_n(x_0)$ does not converge to $F(x_0)$ for some continuity point x_0 of F. Hence, a subsequence $G_i \equiv F_{n_i}$ can be extracted such that $G_i(x_0)$ converges to a number $y \neq F(x_0)$ as $i \to \infty$. However, we have seen that the sequence $\{G_i\}$ gives rise to a subsequence $\{g_i\}$ that converges completely to a d.f. g as $i \to \infty$. Since g has the same moments as F, then $F = g$ and, according to our hypothesis, $\lim_{i\to\infty} g_i(x_0) = g(x_0) = F(x_0)$. This is impossible since $\{g_i(x_0)\}$, being a subsequence of $\{G_i(x_0)\}$, must converge to y.

We will postpone until the next section the investigation of the conditions necessary for a d.f. to be determined by its moments.

A more detailed investigation of the concept of complete convergence reveals that it is both necessary and sufficient for the validity of the Helly–Bray theorem 2. The proof of this assertion is contained in the next theorem, which also lists other equivalent results. A concise statement of this theorem requires the following notation:

1. $C \equiv$ the set of all real-valued, bounded continuous functions defined on R.
2. $C_0^\infty \equiv$ the set of all real-valued, infinitely differentiable functions defined on R which vanish outside a bounded closed interval (i.e., of compact support).

THEOREM 9

Let F and F_n be probability d.f.'s and let μ and μ_n be the corresponding Lebesgue measures. The following statements are equivalent.

1. $\int g\, d\mu_n \to \int g\, d\mu$ as $n \to \infty$ for any $g \in C$.[1]
2. $\int g\, d\mu_n \to \int g\, d\mu$ as $n \to \infty$ for any $g \in C_0^\infty$.
3. $\mu_n(B) \to \mu(B)$ as $n \to \infty$ for any bounded Borel set $B \in \mathscr{B}$ such that $\mu(\partial B) = 0$.[2]
4. $F_n \to F(c)$ as $n \to \infty$.

Proof

The fact that part 1 implies part 2 (denoted by $1 \to 2$) is trivial since $C_0^\infty \subset C$. We will now show that $2 \to 3$.

Let $B \in \mathscr{B}$ be bounded and let $\mu(\bar{B} - \mathring{B}) = 0$, where $\bar{B} \equiv$ closure of B and $\mathring{B} \equiv$ interior of B. Since $\bar{B} \supset B \supset \mathring{B}$, it follows that

$$\mu(\bar{B}) = \mu(B) = \mu(\mathring{B}).$$

For any $\epsilon > 0$, there exists an open set $O \supset B$ such that $\mu(O) < \mu(\bar{B}) + \epsilon$.[3] Moreover, there exists an $f \in C_0^\infty$ such that $f = 1$ on $\bar{B}$, $f = 0$ on O', and $0 \le f \le 1$ on R (see [4]). The fact that $I_0 \ge f \ge I_{\bar{B}}$ implies that

$$\mu_n(\bar{B}) \le \int f\, d\mu_n \to \int f\, d\mu \le \mu(O) < \mu(\bar{B}) + \epsilon.$$

Since $\mu_n(B) \le \mu_n(\bar{B})$, it follows that

$$\varlimsup_{n\to\infty} \mu_n(B) \le \mu(\bar{B}) + \epsilon = \mu(B) + \epsilon. \tag{5}$$

Next, for every $\epsilon > 0$ there exists a closed set $F \subset \mathring{B}$ such that $\mu(\mathring{B}) <$

[1] We say that μ_n converges weakly to μ; this is known more precisely as convergence in the C-weak topology.

[2] $\partial B \equiv$ boundary of B = closure of B — interior of B. If B is an interval, ∂B = end points and so only intervals whose end points are not atoms are considered.

[3] See P. A. Halmos, *Measure Theory*. Princeton, N. J.: Van Nostrand, 1950.

$\mu(F) + \epsilon$.[4] Choosing a $g \in C_0^\infty$ such that $g = 1$ on F, $g = 0$ on $R - \mathring{B}$, and $0 \leq g \leq 1$ yields

$$\mu_n(\mathring{B}) \geq \int g\, d\mu_n \to \int g\, d\mu \geq \mu F > \mu(\mathring{B}) - \epsilon.$$

Since $\mu_n(B) \geq \mu_n(\mathring{B})$,

(6) $$\varliminf_{n\to\infty} \mu_n(B) \geq \mu(\mathring{B}) - \epsilon = \mu(B) - \epsilon.$$

Combining equations (5) and (6), we can deduce that $\lim \mu_n(B) = \mu(B)$.

To prove that $3 \to 4$, let $x, a \in R$ and $x > a$, and assume that F is continuous at x and a. If $B = (a, x)$, then $\mu(\partial B) = 0$. Hence $\mu_n(B) \to \mu(B)$, which implies that

$$F_n(x) - F_n(a) \to F(x) - F(a).$$

Choose a sufficiently small so that $F(a) < \epsilon$ for $\epsilon > 0$. Since $F_n(x) \geq F_n(x) - F_n(a)$, it follows that $\varliminf F_n(x) \geq F(x) - \epsilon$ and so

(7) $$\varliminf F_n(x) \geq F(x).$$

By choosing a continuity point $b > x$ large enough so that $F(b) > 1 - \epsilon$, it follows that

$$1 - F_n(x) \geq F_n(b) - F_n(x) \to F(b) - F(x) > (1 - \epsilon) - F(x).$$

Hence $F_n(x) < F(x) + \epsilon$ and so

(8) $$\varlimsup F_n(x) \leq F(x).$$

Inequalities (7) and (8) imply that $F_n \to F$ at each continuity point of F. It is obvious that $F_n(\pm\infty) \to F(\pm\infty)$ and so $F_n \to F(c)$.

The implication that part $4 \to 1$ is an immediate consequence of Theorem 2.

Since $1 \to 2 \to 3 \to 4 \to 1$, all the statements are equivalent.

As an exercise, the reader can try to extend Theorem 9 to R^n with complex-valued functions g.[5]

The example in Exercise 12 shows that if $F_n \to F(w)$, it does not follow that μ^n converges weakly to μ; that is, part 1 of Theorem 9 may not hold.

EXERCISES

1. Extend and prove the theorems of this section for R^n.
2. Prove Theorem 4. [*Hint:* Let $\pm c$ be continuity points of F. Then

[4] *Ibid.*

[5] More generally, the result remains valid in locally compact spaces.

$$\int |g|\, dF_n \geq \int_{-c}^{c} |g|\, dF_n \to \int_{-c}^{c} |g|\, dF$$

by Helly's theorem.]

3. Show that if $\int_{|x| \geq c_m} |g|\, dF_n \to 0$ uniformly in n as $c_m \to \infty$ with increasing m, then g is uniformly integrable in F_n.

4. Prove Theorem 5. [*Hint:* First show that $\int |g|\, dF < \infty$ by using uniform integrability and Helly's theorem. Then use the inequality $|\int g\, dF_n - \int g dF| \leq$

$$\int_{|x| \geq c} |g|\, dF_n + \left| \int_{-c}^{c} g\, dF_n - \int_{-c}^{c} g\, dF \right| + \int_{|x| \geq c} |g|\, dF.]$$

5. Prove Theorem 6. [*Hint:* The uniform integrability of $|g|$ follows from the inequality

$$\int_{|x| \geq c} |g|\, dF_n \leq \left| \int |g|\, dF_n - \int |g|\, dF \right| + \int_{|x| \geq c} |g|\, dF$$
$$+ \left| \int_{-c}^{c} |g|\, dF - \int_{-c}^{c} |g|\, dF_n \right|$$

by choosing $c = c_n$ so that $\int_{|x| \geq c} |g|\, dF < \epsilon/3$ and by choosing $n \geq N(\epsilon)$ so that the remaining terms on the right-hand side are both less than $\epsilon/3$. Then let $c(\epsilon) = \max(c_1, \ldots, c_n)$, where c_i are chosen such that $\int_{|x| \geq c_i} |g|\, dF_i < \epsilon$ for $(i = 1, 2, \ldots, N - 1)$. The reverse implication is proved in the same manner as in Theorem 5.]

6. Show that Theorem 8 remains valid if we only assume that $\lim \mu'_{kn} = \mu'_k$ for arbitrarily large even values of k.

7. (a) Verify that the following example shows that Theorem 5 is not true if $|g|$ is not uniformly integrable: Let $g(x) = x$ and let $F_n(x) = 0$ for $x < 0$, $F_n(x) = 1 - (\frac{1}{2})^n$ for $0 \leq x < 4^n$, and $F_n(x) = 1$ for $x \geq 4^n$.
(b) Part (2) of Theorem 7 is not obvious. Consider that $F_n(x) = 0$ for $x < -n$, $F_n(x) = \frac{1}{2}$ for $|x| \leq n$, $F_n(x) = 1$ for $x > n$, and $F_n \to \frac{1}{2}$.

8. Generalize Theorem 8.3.2 and Theorem 1 to functions of bounded variation and give their proofs. [*Hint:* A function of bounded variation can be written as the difference of two nondecreasing functions.]

9. Suppose that we say that g is uniformly integrable with respect to F_n if $\left| \int_b^{\infty} g\, dF_n \right| < \epsilon$ and $\left| \int_{-\infty}^{-a} g\, dF_n \right| < \epsilon$ for $b \geq b(\epsilon)$ and $a \geq a(\epsilon)$ independent of n. Are Theorems 5 and 6 valid under this new definition with $|g|$ replaced by g?

10. Show that if $\{F_n\}$ is a sequence of d.f.'s admitting moments of all orders and if

$$\lim_{n\to\infty}\int_{-\infty}^{\infty} x^k\, dF_n(x) = \pi^{-1/2}\int_{-\infty}^{\infty} x^k e^{-x^2}\, dx, \qquad (k = 0, 1, 2, \ldots)$$

then for all x, $\lim_{n\to\infty}\int_{-\infty}^{x} dF_n(x) = \pi^{-1/2}\int_{-\infty}^{x} e^{-x^2}\, dx.$[6]

11. Show that under the hypothesis of Theorem 8,

$$\lim_{x\to\infty} x^k(1 - F_n(x)) = 0 \qquad (n = 1, 2, 3, \ldots)$$

$$\lim_{x\to-\infty} |x|^k F_n(x) = 0. \qquad (k > 0)$$

12. Let $F_n = 0$ for $x < n$ and $F_n = 1$ for $x \geq n$. Then $F_n \to 0(w)$, where 0 is the d.f. of the zero measure μ_0. Show that μ_n does not converge weakly to μ_0.

13. If condition (1) of Theorem 9 holds, we can write $\mu_n \to \mu$. Show that if $\mu_n \to \mu$ and $\mu_n \to \nu$, then $\mu = \nu$.

14. (a) Let $\mu_n\{x_n\} = 1$. Show that $\mu_n \to \mu$, where $\mu\{x\} = 1$, if and only if $\lim x_n = x$, where $x_n, x \in R$.
(b) Let $\mu_n\{i/n\} = 1/(n + 1)$ for $i = 0, 1, 2, \ldots, n$. Show that $\mu_n \to \mu$, the Lebesgue measure on $[0, 1]$.

15. Show that Theorem 9 remains valid if the functions g in condition 2 are only differentiable r times with bounded and uniformly continuous derivatives of the orders $0, 1, \ldots, r$.

16. Prove the converse of Theorem 3.

17. (a) Define the norm of a measure μ by

$$||\mu|| = \sup \left(\left|\int f\, d\mu\right| : \sup |f(x)| \leq 1\right),$$

where f is a measurable function on $(R, \mathscr{B})$. Show that if $\lim ||\mu_n - \mu|| = 0$ as $n \to \infty$, then $\int f\, d\mu_n \to \int f\, d\mu$ for $f \in L_\infty$, but the converse is not true.
(b) Show that if $\int f_n\, d\mu_n \to \int f\, d\mu$ for $f \in L_\infty$, then condition 1 of Theorem 9 holds but the converse is not true.

9.4 THE MOMENTS PROBLEM

Suppose that an infinite sequence $\{c_k\}$ of real constants and an interval (a, b), finite or infinite, are given. The problem of moments is finding a real increasing function F that is continuous from the right and defined on (a, b) such that

[6] The original proof, derived by A. Markov, was quite complicated and was based on the properties of Hermite polynomials and Chebyshev inequalities in the theory of algebraic continued fractions. It served as a motivation for a paper by Fréchet and Shohat ([3]).

(1) $$\int_a^b x^k \, dF(x) = c_k. \qquad (k = 0, 1, 2, \ldots)$$

If F is a solution to the problem of moments that is bounded at b for the data c_k on (a, b), and if c_0 is finite,[7] then $G(x) = 0$ for $x < a$, $G(x) = (1/c_0)(F(x) - F(a))$ for $a \leq x < b$, and $G(x) = 1$ for $x \geq b$, where $G(x)$ is a d.f. that is a solution to the problem of moments with the modified data $1, c_1/c_0, c_2/c_0, c_3/c_0, \ldots$. Accordingly, we will consider F to be a d.f. in (1).

In general, for arbitrary data $\{c_k\}$, the problem of moments has no solution. This follows from the fact that $(E|r|^k)^{1/k}$ is an increasing function of k; hence the moments must obey certain inequalities. Further inequalities can be obtained from the fact that the quadratic form

(2) $$E\left(\sum_{i=1}^{n} x_i r^{k_i}\right)^2 = \sum_{j=1}^{n} \sum_{i=1}^{n} a_{ij} x_i x_j \qquad (a_{ij} = E(r^{k_i + k_j}))$$

is nonnegative and so $\det(a_{ij}) \geq 0$. This last condition yields a relation between the moments.

Next suppose that the moments problem has a solution; is it unique? Phrased in this manner, the answer is no, since if F is a solution, then so is $F + c$, where c is a constant. However, if we look for solutions on (a, b) such that $F(a) = 0$ (or any other convenient number) and agree to identity functions that coincide at all of their continuity points, then the moments problem has a unique solution (if it has any) for a finite interval.

THEOREM 1

If (a, b) is finite, then the problem of moments cannot have more than one solution.[8]

Proof

The existence of two solutions F_1 and F_2 implies that

(3) $$\int_a^b x^k \, dF = 0, \qquad (k = 0, 1, 2, \ldots)$$

where $F = F_1 - F_2$. Since $F_1(a) = F_2(a) = 0$ and $c_0 = \int_a^b dF_1 = \int_a^b dF_2$, then $F_1(b) = F_2(b)$; hence $F(b) = 0$. When Equation (3) is integrated by parts, this last fact yields

[7]We assume that $c_0 \neq 0$; otherwise the problem has a trivial solution.

[8]The proof comes from Stieltjes. See *Correspondance d'Hermite et Stieltjes* (Paris, 1905), pp. 337–338.

$$\int_a^b x^l F(x)\,dx = 0. \qquad (l = 0, 1, 2, \ldots) \tag{4}$$

Now assume that there exists a continuity point z $(a < z < b)$ of F such that $F(z) \neq 0$. The function $1 - (x - z)^2/M > 0$ $(a \leq x \leq b)$ for sufficiently large M, and $[1 - (x - z)^2/M]^n$ is close to zero except in a neighborhood of z for sufficiently large n. Hence it follows that $\int_a^b F(x)[1 - (x - z)^2/M]^n\,dx \neq 0$. This is impossible, since this integral is a linear combination of integrals of the type in Equation (4), which vanish. Hence $F = 0$ at every continuity point and so $F_1 \equiv F_2$, since they are d.f.'s.

On the other hand, if the interval (a, b) is infinite, then the problem of moments does not, in general, have a unique solution. This is illustrated by the following example.

EXAMPLE 1

1. The infinite family of increasing functions

$$F(t) = \int_0^t e^{-kx^\lambda}(1 + h \sin(kx^\lambda \tan \lambda\pi))\,dx,$$

where $k > 0$, $0 < \lambda < \frac{1}{2}$, and $-1 \leq h \leq 1$, are all solutions of the same problem of moments, since

$$\int_0^\infty x^n e^{-kx^\lambda} \sin(kx^\lambda \tan \lambda\pi)\,dx = 0. \tag{5}$$

To prove this last equality, set

$$b = k + di \qquad a = \frac{n+1}{\lambda} \qquad \frac{d}{k} = \tan \lambda\pi \qquad y = x^\lambda$$

in the formula

$$\int_0^\infty y^{a-1} e^{-by}\,dy = \frac{\Gamma(a)}{b^a}. \tag{6}$$

The imaginary part of $\Gamma(a)/b^a$ is zero, while the imaginary part of the transformed integral in (6) is proportional to the integral in (5).

2. The increasing functions

$$\int_{-\infty}^x e^{-kx^\mu}\left(1 + h \cos\left(kx^\mu \tan \frac{\mu\pi}{2}\right)\right) dx,$$

where $0 < \mu = 2s/(2s + 1) < 1$ and s is a positive integer, have the same moments. To verify this, set

$$a = \frac{2n+1}{\lambda} \qquad \frac{d}{k} = \tan\frac{\mu\pi}{2} \qquad y = x^{\mu}$$

in (6), obtaining

$$\int_{-\infty}^{\infty} x^n e^{-kx^\mu} \cos\left(kx^\mu \tan\frac{\mu\pi}{2}\right) dx = 0.$$

Further examples appear in the Exercises.

We will now state some conditions necessary in order that the problem of moments has a unique solution for an infinite interval. The first condition is that

(7) $$\sum_{n=1}^{\infty} c_{2n}^{-1/2n} \text{ diverges;}$$

it was derived by T. Carleman [5]. The second condition, derived by Stieltjes [6], is

$$dF = f(x)\, dx, \qquad [f(x) \geq 0 \text{ on } (a, b)]$$

with $f(x) < (M|x|^{\alpha-1})(e^{-k|x|^\lambda})$ for sufficiently large $|x| \geq x_0$, where N, α, and k are positive constants. Here $\lambda \geq \frac{1}{2}$ for $(a, b) = (0, \infty)$ and $\lambda \geq 1$ for $(a, b) = (-\infty, \infty)$. The proofs of these statements can be found in the references cited. Further conditions equivalent to condition (7) are given in the Exercises.

On the other hand, the problem of moments is indeterminate if $dF = f(x)\, dx$, $|x|$ is sufficiently large, and

$$f(k) > e^{-k|x|^\lambda}, \qquad (k > 0)$$

with $\lambda < \frac{1}{2}$ for $(0, \infty)$ and $\lambda < 1$ for $(-\infty, \infty)$ (see Example 1).

We will be content with proving a theorem of F. Hausdorff that gives a necessary and sufficient condition for the existence and uniqueness of a solution of the moments problem on the interval [0, 1].

THEOREM 2

There exists a unique d.f. F such that for the closed interval of integration,

(8) $$\int_0^1 x^n\, dF = c_n \qquad (c_0 = 1, n = 0, 1, 2, \ldots)$$

if and only if

(9) $$(-1)^n \Delta^n c_k \geq 0.$$

Proof

Suppose that there exists a unique d.f. F such that $F(0) = 0, F(1) = 1$, and (8) holds. Let r be a r.v. such that $r(x) = x$ on $[0, 1]$. Taking differences, we have $-\Delta c_k = E(r^k(1 - r))$; by induction,

$$(-1)^n \Delta^n c_k = E(r^k(1 - r)^n).$$

It follows that (9) is valid.

Conversely, suppose that (9) holds. Define $p_n(k)$ by

$$p_n(k) = \binom{n}{k}(-1)^{n-k} \Delta^{n-k} \mu_k. \qquad (k = 0, 1, \ldots, n)$$

Inverting the above formula by applying Equation 5.4.6 yields

$$\sum_{k=0}^{n} \binom{k}{s} p_n(k) = \binom{n}{s} \mu_s. \tag{10}$$

[The summation is really from s to n; however, $\binom{k}{s} = 0$ if $s > k$.] Multiplying (10) by $s!n^{-s}$, it follows that

$$\lim_{n\to\infty} \sum_{k=0}^{n} \frac{k}{n}\left(\frac{k}{n} - \frac{1}{n}\right) \cdots \left(\frac{k}{n} - \frac{(s-1)}{n}\right) p_n(k) = c_s.$$

Setting $s = 1, 2, 3, \ldots$ successively in this equation leads to

$$\lim_{n\to\infty} \sum_{k=0}^{n} \left(\frac{k}{n}\right)^s p_n(k) = c_s. \tag{11}$$

For $s = 0$, we conclude from (10) that $\sum_{k=0}^{n} p_n(k) = 1$. By hypothesis, $p_n(k) \geq 0$ and so for a fixed value of n, the $p_n(k)$ can be used to define a probability distribution. To be specific, set $P(r_n = k/n) = p_n(k)$ for $k = 0, 1, \ldots, n$. By (11) the sequence $\{c_s\}$ appears as a limit of moment sequences $\{Er_n^s\}$, and so by the corollary to Theorem 3.8 there exists a unique d.f. having these moments.

A limit formula for F appears in Exercise 2.

The problem of moments has been generalized to function spaces. The reader interested in pursuing this topic further can consult [7]–[10].

EXERCISES

1. (a) Let F be a d.f. such that $F(0) = 0$ and $F(1) = 1$. Show that the expectation of the Bernstein polynomial

$$B_n = \sum_{i=0}^{n} g\left(\frac{i}{n}\right)\binom{n}{i} x^i(1-x)^{n-i}$$

is given by

$$EB_n = \sum_{i=0}^{n} g\left(\frac{i}{n}\right) p_n(i)$$

using the notation of Theorem 2.

(b) Verify that if g is continuous on [0, 1], then as $n \to \infty$,

$$EB_n \to \int g\, dF.$$

[*Hint:* Recall that $B_n \to g$ uniformly on [0, 1].]

(c) Deduce that

$$F_n(t) = \sum_{i \le nt} \binom{n}{i}(-1)^{n-i}\, \Delta^{n-i}\, c_i \to F(t)\ (w).$$

[*Hint:* Define $g_t(x) = 1$ for $x < t$ and $g_t(x) = 0$ for $x > t$. Use Exercise 1(b).]

2. Show that the F_n defined in Exercise 1(c) are d.f.'s that converge weakly to the d.f. in Theorem 2.

3. Show that the following conditions for the uniqueness of the solution of the problem of moments for an infinite interval are special cases of Carleman's condition (7):

(a) *Von Mises:* $$c_{2n} \le K\left(\frac{n}{k^2 e}\right)^n. \qquad (K \text{ and } k \text{ are constants})$$

(b) *Polya:* $$\lim_{n\to\infty} \frac{c_{2n}^{1/2n}}{n} \text{ is finite.}$$

(c) *Cantelli:* $$c_{2n} \frac{[\psi(2n)]^{2n}}{2n!} < 1, \qquad (n > n_1)$$

where $\psi(n) \to \infty$ as $n \to \infty$.

4. Show that the d.f. F that is nonzero for $x > 0$ and that is given by

$$F(x;\lambda) = \left(\int_0^\infty x^{-\log x}\, dx\right)^{-1} \int_0^x x^{-\log x}[1 - \lambda \sin(4\pi \log x)]\, dx$$

has the same moments for all values of λ—that is,

$$E(r^k) = \left(\int_0^\infty x^{-\log x}\, dx\right)^{-1} \int_0^\infty x^k x^{-\log x}\, dx.$$

9.5 EXISTENCE OF MOMENTS

In this section we will present some theorems about moments. We begin by giving a sufficient condition for the existence of moments.

THEOREM 1

If for some $s > 0$ the d.f. F satisfies the conditions

$$F(x) < M_1|x|^{-s} \qquad (x \leq -d)$$
$$1 - F(x) < M_2 x^{-s}, \qquad (x \geq c)$$

where $c > 0$ and $d > 0$, then any absolute moment of order $r < s$ exists.

Proof

By definition of the improper integral, it is sufficient to show that $\int_a^b |x|^r \, dF < k$, where k is some constant independent of a and b, if $|a|$ and b are sufficiently large. Without loss of generality we can assume that $M_1 = M_2$, $c = d = 2^l$, $|a| > 2^l$, and $b > 2^l$. Then (for $i \geq l$),

$$\int_{2^i}^{2^{i+1}} |x|^r \, dF \leq 2^{r(i+1)}(F(2^{i+1}) - F(2^i)) \leq 2^{r(i+1)}(1 - F(2^i)) < \frac{M}{2^{i(s-r)}},$$

where M is a constant independent of i. A similar relation holds for the interval $(-2^{i+1}, -2^i)$. Summing over $i = l, l+1, l+2, \ldots$ and adding the integral over $(-c, c)$, which is a fixed finite constant, yields

$$\int_a^b |x|^r \, dF < \int_{-c}^{c} |x|^r \, dF + \frac{2M}{2^{l(s-r)}(1 - 2^{s-r})}.$$

THEOREM 2

For $r \geq 1$, $\int_{-\infty}^{\infty} |x|^r \, dF < \infty$ if and only if $\sum_{n=1}^{\infty} n^{r-1}(1 - F(n) + F(-n))$ $< \infty$, where F is a d.f.

Proof

$$(1) \qquad \int_{-\infty}^{\infty} |x|^r \, dF \geq \sum_{i=1}^{\infty} \int_i^{i+1} |x|^r \, dF + \sum_1^{\infty} \int_{-i-1}^{-i} |x|^r \, dF$$
$$\geq \sum_1^{\infty} i^r(F(i+1) - F(i) + F(-i) - F(-i-1)).$$

If $r \geq 1$ and $i \geq 1$,

$$(2) \qquad i^r \geq 1^{r-1} + 2^{r-1} + \cdots + i^{r-1} \geq \frac{1}{r} i^r;$$

therefore, the sum appearing on the right-hand side of (1) is not smaller than

$$\sum_{i=1}^{\infty}\left(\sum_{n=1}^{i} n^{r-1}\right)(F(i+1)-F(i)+F(-i)-F(-i-1))$$

$$=\sum_{n=1}^{\infty} n^{r-1}\sum_{i=n}^{\infty}(F(i+1)-F(i)+F(-i)-F(-i-1))$$

$$=\sum_{n=1}^{\infty} n^{r-1}(1-F(n)+F(-n)).$$

It follows immediately from (1) that if $\int_{-\infty}^{\infty} |x|^r\,dF$ is finite, then so is $\sum_{n=1}^{\infty} n^{r-1}(1-F(n)+F(-n))$.

Now assume that the series $\sum_{n=1}^{\infty} n^{r-1}(1-F(n)+F(-n))$ is finite; then the series $\sum_{n=0}^{\infty}(n+1)^{r-1}(1-F(n)+F(-n))$ is also finite, since for $n \geq 1$,

(3) $$(n+1)^{r-1} \leq 2^{r-1}\,n^{r-1}.$$

From this result it follows that the integral is finite, since

$$r\sum_{n=0}^{\infty}(n+1)^{r-1}(1-F(n)+F(-n))$$

$$=r\sum_{n=0}^{\infty}(n+1)^{r-1}\sum_{i=n}^{\infty}(F(i+1)-F(i)+F(-i)-F(-i-1))$$

$$=\sum_{i=0}^{\infty}\left(r\sum_{n=0}^{i}(n+1)^{r-1}\right)(F(i+1)-F(i)+F(-i)-F(-i-1))$$

$$\geq \sum_{i=0}^{\infty}(i+1)^r(F(i+1)-F(i)+F(-i)-F(-i-1)) \qquad \text{[by (2)]}$$

$$\geq \sum_{i=0}^{\infty}\left(\int_{i}^{i+1} x^r\,dF+\int_{-i-1}^{-i}|x|^r\,dF\right)=\int_{-\infty}^{\infty}|x|^r\,dF.$$

THEOREM 3

If $r \geq 0$ and $\alpha > 0$, then

$$\lim_{n\to\infty} n^r(1-F(n)+F(-n))=0$$

implies that for any d.f. F,

$$\lim_{n\to\infty}\frac{1}{n^\alpha}\int_{-n}^{n}|x|^{r+\alpha}\,dF=0,$$

where the integration is over the *open* interval $(-n, n)$.

Proof

$$\frac{1}{n^\alpha}\int_{-n}^{n}|x|^{r+\alpha}\,dF=\frac{1}{n^\alpha}\sum_{i=0}^{n-1}\left[\int_{i}^{i+1}|x|^{r+\alpha}\,dF+\int_{-i-1}^{-i}|x|^{r+\alpha}\,dF\right]$$

$$\le\sum_{i=0}^{n-1}\frac{(i+1)^{r+\alpha}}{n^\alpha}[F(i+1)-F(i)+F(-i)-F(-i-1)]$$

$$=\sum_{i=0}^{n-1}\frac{(i+1)^{r+\alpha}}{n^\alpha}[1-F(i)+F(-i)]$$

$$-\sum_{i=0}^{n-1}\frac{(i+1)^{r+\alpha}}{n^\alpha}[1-F(i+1)+F(-i-1)].$$

Setting $i+1=j$ in the last sum yields, after some simplification,

$$\frac{1}{n^\alpha}\int_{-n}^{n}|x|^{r+\alpha}\,dF\le\sum_{i=0}^{n-1}\frac{(i+1)^{r+\alpha}-i^{r+\alpha}}{n^\alpha}[1-F(i)+F(-i)],\tag{4}$$

where the term $-n^r(1-F(n)+F(-n))$ is omitted. In order to estimate the right-hand side of Equation (4) the following inequalities are needed:

$$(i+1)^{r+\alpha}-i^{r+\alpha}\le(r+\alpha)(i+1)^{r+\alpha-1}\qquad(r+\alpha\ge1, i\ge0)\tag{5}$$

$$(i+1)^{r+\alpha}-i^{r+\alpha}\le(r+\alpha)i^{r+\alpha-1}.\qquad(r+\alpha<1, i>0)\tag{6}$$

Assuming that Equation (5) holds,

$$\frac{1}{n^\alpha}\int_{-n}^{n}|x|^{r+\alpha}\,dF\le(r+\alpha)\sum_{i=0}^{n-1}\frac{(i+1)^{r+\alpha-1}}{n^\alpha}(1-F(i)+F(-i)).\tag{7}$$

By hypothesis, there exists an integer $N=N(\epsilon)$ such that

$$n^r(1-F(n)+F(-n))<\frac{\epsilon}{2}\tag{8}$$

for $n\ge N$. Using this result, (7) can be rewritten as

$$\frac{1}{n^\alpha}\int_{-n}^{n}|x|^{r+\alpha}\,dF\le(r+\alpha)\frac{1}{n^\alpha}\sum_{i=0}^{N-1}(i+1)^{r+\alpha-1}[1-F(i)+F(-i)]\tag{9}$$

$$+\frac{1}{2}\epsilon\,\frac{(N+1)^{\alpha-1}+(N+2)^{\alpha-1}+\cdots+n^{\alpha-1}}{n^\alpha}.$$

With fixed N, choose n so large that the first sum in Equation (9) is less than $\epsilon/2$. Hence $(1/n^\alpha)\int_{-n}^{n}|x|^{r+\alpha}\,dF<\epsilon$, since

$$\frac{(N+1)^{\alpha-1}+(N+2)^{\alpha-1}+\cdots+n^{\alpha-1}}{n^\alpha}<1.$$

The other situation described in (6) can be handled in an analogous fashion.

THEOREM 4

If $r \geq 0, \alpha > 0$, and F is a d.f., then $\int_{-\infty}^{\infty} |x|^r \, dF < \infty$ if and only if $\sum_{n=1}^{\infty} (1/n^{\alpha+1}) \int_{-n}^{n} |x|^{r+\alpha} \, dF < \infty$, where the integration is over the open interval $(-n, n)$.

Proof

$$\sum_{n=1}^{\infty} \frac{1}{n^{\alpha+1}} \int_{-n}^{n} |x|^{r+\alpha} \, dF = \sum_{n=1}^{\infty} \frac{1}{n^{\alpha+1}} \sum_{i=0}^{n-1} \left(\int_{i}^{i+1} x^{r+\alpha} \, dF + \int_{-i-1}^{-i} |x|^{r+\alpha} \, dF \right) \tag{10}$$

$$= \sum_{i=0}^{\infty} \left(\int_{i}^{i+1} x^{r+\alpha} \, dF + \int_{-i-1}^{-i} |x|^{r+\alpha} \, dF \right) \sum_{n=i+1}^{\infty} \frac{1}{n^{\alpha+1}} .$$

To simplify the last sum, the inequality

$$\frac{1}{\alpha} \frac{1}{(i+1)^{\alpha}} < \sum_{n=i+1}^{\infty} \frac{1}{n^{\alpha+1}} \frac{1}{\alpha} \frac{1}{i^{\alpha}} \tag{11}$$

will be used; furthermore, if i is large enough,

$$\frac{1}{2i^{\alpha}} < \frac{1}{(1+i)^{\alpha}} . \tag{12}$$

For sufficiently large values of i, say $i \geq L$, the last sum in Equation (10) can be written as

$$\sum_{i=L}^{\infty} \left(\int_{i}^{i+1} x^{r+\alpha} \, dF + \int_{-i-1}^{-i} |x|^{r+\alpha} \, dF \right) \sum_{n=i+1}^{\infty} \frac{1}{n^{\alpha+1}}$$

$$> \frac{1}{2^{\alpha}} \sum_{i=L}^{\infty} \frac{1}{i^{\alpha}} \left(\int_{i}^{i+1} x^{r+\alpha} \, dF + \int_{-i-1}^{-i} |x|^{r+\alpha} \, dF \right)$$

$$> \frac{1}{2^{\alpha}} \sum_{i=L}^{\infty} \left(\int_{i}^{i+1} x^{r} \, dF + \int_{-i-1}^{-i} |x|^{r} \, dF \right)$$

$$= \frac{1}{2^{\alpha}} \left(\int_{L}^{\infty} x^{r} \, dF + \int_{-\infty}^{-L} |x|^{r} \, dF \right)$$

by using inequalities (11) and (12). Similarly, using the right-hand side of inequality (11),

$$\sum_{i=L}^{\infty}\left(\int_{i}^{i+1} x^{r+\alpha}\,dF + \int_{-i-1}^{-i} |x|^{r+\alpha}\,dF\right)\sum_{n=i+1}^{\infty}\frac{1}{n^{\alpha+1}}$$

$$< \frac{2}{\alpha}\sum_{i=L}^{\infty}\left(\int_{i}^{i+1} x^{r}\,dF + \int_{-i-1}^{-i} |x|^{r}\,dF\right),$$

which completes the proof of the theorem.

EXERCISES

1. Prove inequalities (2), (3), (5), (6), (11), and (12).

2. If F is a d.f., is it true that $\sum_{n=1}^{\infty} n^{-1}[1 - F(n) + F(-n)] < \infty$?

3. Complete the proofs of Theorems 3 and 4.

4. Show that for $\alpha > 0$, $\lim_{n\to\infty} (1/n^{\alpha}) \int_{-n}^{n} |x|^{\alpha}\,dF = 0$, where the integration is over the open interval.

REFERENCES

1. Helly, E., "Über lineare Funktionaloperationen," *Sitz. Nat. Kais. Akad. Wiss.*, **121** (1912).
2. Bray, H. E., "Elementary properties of the Stieltjes integral," Ann. Math., **20** (1919).
3. Fréchet, M., and J. A. Shohat, "A proof of the generalized second limit theorem in the theory of probability," Trans. Am. Math. Soc. **33** (1931).
4. Yosida, K., *Functional Analysis*. Berlin: Academic Press, 1965.
5. Carleman, T., *Sur les équations intégrales singulières*. Uppsala, 1923.
6. Stieltjes, T. J., "Recherches sur les fractions continuées," *Oeuvres*, II.
7. Royden, H. L., "Bounds on a distribution function when its first n moments are given," Ann. Math. Stat. (1953).
8. Akheizer, N., and M. Krein, *Some Questions in the Theory of Moments*. Providence, R. I.: Am. Math. Soc., 1962.

9. Shohat, J. A., and J. D. Tamarkin, *The Problem of Moments.* New York: Am. Math. Soc., 1943.

10. Karlin, S., and W. Studden, *Tchebycheffian Systems and Applications to Analysis.* New York: Interscience, 1966.

CHAPTER 10

CHARACTERISTIC FUNCTIONS

10.1 INTRODUCTION

In Section 5.4 we introduced the notion of generating functions for discrete random variables—that is, r.v.'s whose range is an at most countable set. To apply these concepts to arbitrary r.v.'s we might try, by analogy with Definition 5.4.1, to define a function $G(x) = E(x^r)$. Since it is more convenient to write arbitrary powers in exponential form, we will replace the variable x by e^λ, where λ is a real number. Hence if F is the d.f. of r, then

(1) $$G(\lambda) = E(e^{\lambda r}) = \int_{-\infty}^{\infty} e^{\lambda y}\, dF.$$

In probability theory this function is known as the *moment generating function*. It can be shown that this function has properties similar to those of the generating function; these will be listed in the Exercises. Unfortunately, the integral appearing in (1) can diverge and the definition breaks down. This difficulty can easily be overcome by introducing complex-valued r.v.'s—complex valued functions whose real and imaginary parts are r.v.'s. Then Equation (1) becomes the Fourier–Stieltjes transform of F.

DEFINITION 1

The function χ defined for all real λ by

(2) $$\chi(\lambda) = E(e^{i\lambda r}) = \int_{-\infty}^{\infty} (\cos \lambda x + i \sin \lambda x)\, dF$$

is called the *characteristic function* (c.f.) of the r.v. r (or of its d.f. F). Since $|e^{i\lambda x}| = 1$ for all real values of λ, it follows that a c.f. can be defined for every r.v. (see Exercise 1).

Note that the c.f. $\chi = U + iV$ is completely determined by its real part

$$U = \int_{-\infty}^{\infty} \cos(\lambda x)\, dF(x) \tag{3}$$

and its imaginary part

$$V = \int_{-\infty}^{\infty} \sin \lambda x\, dF(x). \tag{4}$$

The importance of c.f.'s in probabilistic analysis is due to the one-to-one correspondence between d.f.'s and c.f.'s[1] which will be established in Section 10.5. This fact enables us to use uniformly bounded, continuous c.f.'s (this will be established in Section 10.2) instead of discontinuous d.f.'s in many arguments. Moreover, sums of independent r.v.'s correspond to products of their c.f.'s (c.f., Theorem 5.4.2). This fact along with the fact that complete (weak) convergence of d.f.'s corresponds to ordinary convergence (convergence of integrals) of c.f.'s, leads to a considerable simplification in the proofs of limit theorems (c.f., Theorem 5.4.3).

Characteristic functions for random vectors are also defined by (2) under the convention that a single letter denotes a vector.

To be specific, suppose that $(\lambda_1, \ldots, \lambda_n)$ is a row vector, $x = (x_1, \ldots, x_n)'$ is a column vector, and λx denotes their scalar product $\lambda_1 x_1 + \lambda_2 x_2 + \cdots + \lambda_n x_n$; then

$$\chi(\lambda) = \chi(\lambda_1, \lambda_2, \ldots, \lambda_n) = E(e^{i\lambda r}). \tag{5}$$

In general, we will restrict our attention to c.f.'s of r.v.'s. The properties of c.f.'s of random vectors are left as exercises for the reader.

The first person to use a technique similar to the moment generating function was Lagrange [1]. Equivalent functions were systematically employed by Laplace [2]. However, the first complete and thorough study of the properties of c.f.'s was made by P. Lévy [3].

EXAMPLE 1

1. Suppose that a r.v. r is normally distributed, with expectation 0 and variance 1. The c.f. of r is

$$\chi(\lambda) = \frac{1}{\sqrt{2\pi}} \int_{-\infty}^{\infty} e^{i\lambda x - x^2/2}\, dx.$$

[1] Here we consider only normalized d.f.'s—i.e., $F(-\infty) = 0$, $F(\infty) = 1$. Much of the theory remains valid for general bounded d.f.'s. However, there is no longer a one-to-one correspondence, since F and $F + c$ have the same c.f.

Setting $z = x - i\lambda$ reduces $\chi(\lambda)$ to the form

$$\chi(\lambda) = e^{-\lambda^2/2} \frac{1}{\sqrt{2\pi}} \int_{-\infty - i\lambda}^{\infty - i\lambda} e^{-z^2/2}\, dz = e^{-\lambda^2/2}.$$

Similarly, it can be shown that if the r.v. has a mean μ and a variance σ^2, then $\chi(\lambda) = \exp(i\mu\lambda - \sigma^2\lambda^2/2)$.

2. If the r.v. r is distributed according to the Poisson law—i.e., $P(r = k) = \mu^k e^{-\mu}/k!$ $(k = 0, 1, 2, \ldots)$—then its c.f. is

$$\chi(\lambda) = Ee^{i\lambda r} = \sum_{k=0}^{\infty} e^{i\lambda k}(\mu^k/k!)e^{-\mu} = e^{\mu(e^{i\lambda}-1)}.$$

More generally, consider a discrete r.v. r with a *lattice distribution*;[2] i.e., there exist numbers a and $b > 0$ such that all possible values of r can be written in the form $a + kb$ $(k = 0, \pm 1, \pm 2, \ldots)$. If $p_k = P(r = a + kb)$, then

$$\chi(\lambda) = e^{ia\lambda} \sum_{k=-\infty}^{\infty} p_k e^{i\lambda kb}.$$

3. The *triangular density* $f(x)$ is defined by $f(x) = (1 - |x|/a)a^{-1}$ for $|x| < a$ and $f(x) = 0$ otherwise. Its c.f. is

$$\chi(\lambda) = \int_0^a e^{i\lambda x}\left(1 - \frac{x}{a}\right)\frac{1}{a}\, dx + \int_{-a}^0 e^{i\lambda x}\left(1 + \frac{x}{a}\right)\frac{1}{a}\, dx.$$

Replacing x by $-t$ in the second integral yields, after some slight simplification, $\chi(\lambda) = (2/a)\int_0^a (1 - x/a)\cos \lambda x\, dx$, and an integration by parts leads to

$$\chi(\lambda) = 2\,\frac{1 - \cos a\lambda}{a^2\lambda^2}.$$

From Corollary 1 of Theorem 5.3, it follows that the c.f. of the density $(1/\pi)[(1 - \cos ax)/ax^2]$ is $1 - a^{-1}|\lambda|$ for $|\lambda| \leq a$ and 0 for $|\lambda| > a$.

4. Let the vector $r = (r_1, \ldots, r_n)$ have a nondegenerate normal distribution (see Exercise 11) and suppose that $Er = 0$. Then $\lambda r' = \lambda_1 r_1 + \lambda_2 r_2 + \cdots + \lambda_n r_n$ is normally distributed with a mean 0 and a variance $\lambda' C\lambda$, where C is the covariance matrix defined by $E(rr')$. Hence from the result in part 1 of the example it follows that the c.f. of r is $\chi = e^{-1/2\lambda' C\lambda}$.

5. If r is a Cauchy r.v. with density $(1/\pi)[1/(1 + x^2)]$, its c.f. is

$$\chi(\lambda) = \frac{1}{\pi}\int_{-\infty}^{\infty} \frac{e^{i\lambda x}}{1 + x^2}\, dx.$$

To evaluate this integral we use contour integration. The contour is formed

[2]The r.v. r is often called *arithmetic* in this case.

by joining the points $(-R, 0)$ and $(R, 0)$ by a line segment and also by a semicircle C lying in the upper half-plane. Since the residue of $e^{i\alpha z}/(1 + z^2)$ at $z = i$ is $e^{-\alpha}/2i$, then

(6)
$$\pi e^{-\alpha} = \int_C \frac{e^{i\alpha z}}{1 + z^2} dz + \int_{-R}^{R} \frac{e^{i\alpha x}}{1 + x^2} dx.$$

If $\lambda > 0$, then

(7)
$$\left| \int_C \frac{e^{i\lambda z}}{1 + x^2} dz \right| \leq \int_C \frac{|e^{i\lambda z}|}{|1 - |z|^2|} dz \leq e \left(\frac{\pi R}{|1 - R^2|} \right),$$

since $e^{-\lambda y} < e$ and $|1 + z^2| \geq |1 - |z|^2|$. As $R \to \infty$, it follows from (6) and (7) that

$$\int_{-\infty}^{\infty} \frac{e^{i\lambda x}}{1 + x^2} dx = \pi e^{-\lambda}.$$

If $\lambda < 0$, a similar argument using a contour in the lower half-plane shows that

$$\int_{-\infty}^{\infty} \frac{e^{i\lambda x}}{1 + x^2} = \pi e^{\lambda}.$$

Hence $\chi(\lambda) = e^{-|\lambda|}$.

EXERCISES

1. Let r, s be real-valued r.v.'s; then $E(r + is) = Er + iEs$ by definition. Show that

$$|E(r + is)| \leq E|r + is|.$$

In Exercises 2–6, assume that all of the series converge. If the expansions are not valid, the coefficients of the series are still used to define the pertinent parameters.

2. The function (for real α_i)

$$F(\alpha_1, \alpha_2, \ldots, \alpha_n) = E(e^{\alpha_1 r_1 + \alpha_2 r_2 + \cdots + \alpha_n r_n})$$

is called the *moment generating function* (m.g.f.) of the r.v. $(r_1, r_2, \ldots, r_n)$. By rewriting the function in a series, show that formally the coefficient of $\alpha_1^i \alpha_2^j \cdots \alpha_n^k$ is some multiple of

$$\mu_{i,j,\ldots,k} = E(r_1^i r_2^j \cdots r_n^k).$$

3. The *factorial moment generating function* (f.m.g.f) is defined by

$$N(\beta^1, \beta_2, \ldots, \beta_n) = E(1 + \beta_1)^{r_1}(1 + \beta_2)^{r_2} \cdots (1 + \beta_n)^{r_n}.$$

Show that the factorial moment

$$\mu_{(i,j,\ldots,k)} = E(r_1)(r_1 - 1)\cdots(r_1 - i + 1)(r_2)(r_2 - 1)\cdots (r_2 - j + 1)\cdots(r_n)(r_n - 1)\cdots(r_n - k + 1)$$

appears in the coefficient of $\beta_1^i\beta_2^j\cdots\beta_n^k$.

4. Let $\chi(\alpha_1,\ldots,\alpha_n)$ be a c.f.; then

$$\Lambda(\alpha_1,\ldots,\alpha_n) = \log \chi(\beta_1,\alpha_2,\ldots,\alpha_n)$$
$$= \sum_{j=0}^{\infty}\cdots\sum_{k=0}^{\infty}\sum_{l=0}^{\infty}\frac{(i\alpha_1)^j}{j!}\frac{(i\alpha_2)^k}{k!}\cdots\frac{(i\alpha_n)^l}{l!}\lambda_{j,k\ldots,l}$$

is called the *semi-invariant generating function* (s.i.g.f) and the $\lambda_{j,k,\ldots,l}$ are called *semi-invariants.*[3] For $n = 1$, verify that

$$\lambda_1 = \mu_1$$
$$\lambda_2 = \mu_2 - \mu_1^2$$
$$\lambda_3 = \mu_3 - 3\mu_1\mu_2 + 2\mu_1^3$$
$$\lambda_4 = \mu_4 - 3\mu_2^2 - 4\mu_1\mu_3 + 12\mu_1^2\mu_2 - 6\mu_1^4.$$

5. Verify the following relations between the moments and factorial moments:

(a) $\mu_{(1)} = \mu_1'$
$\mu_{(2)} = \mu_2' - \mu_1'$
$\mu_{(3)} = \mu_3' - 3\mu_2' + 2\mu_1'$
$\mu_{(4)} = \mu_4' - 6\mu_3' + 11\mu_2' - 6\mu_1'.$

(b) $\mu_1' = \mu_{(1)}$
$\mu_2' = \mu_{(2)} + \mu_{(1)}$
$\mu_3' = \mu_{(3)} + 3\mu_{(2)} + \mu_{(1)}$
$\mu_4' = \mu_{(4)} + 6\mu_{(3)} + 7\mu_{(2)} + \mu_{(1)}.$

6. Let $N(\beta)$ be a f.m.g.f.; then

$$D(\beta) = \log N(\beta) = \sum_{1}^{\infty}\frac{\beta r}{r!}\delta_r$$

is called the *factorial parameter generating function* (f.p.g.f.) and the δ_r are called the *factorial parameters.* Express the first four factorial parmeters in terms of the factorial moments.

7. If the d.f. F is defined on $[0, \infty)$,[4] the m.g.f. defined in Exercise 2 is frequently written as

[3]Semi-invariants and their generating functions were introduced by T. N. Thiele in *Theory of Observations* (London, 1903). Methods of finding the general relations between various types of moments can be found in [14].

[4]We assume that $F(0) = 0$. If not, $\phi(\lambda) = F(0) + \int_{0+}^{\infty} e^{-\lambda x}\,dF.$

$$\phi(\lambda) = \int_0^\infty e^{-\lambda x}\, dF$$

for $\lambda > 0$ and is called the *Laplace–Stieltjes transform* of F. It is also defined for unbounded d.f.'s as long as $\phi(\lambda)$ exists for all $\lambda > 0$. Show that if F is an unbounded d.f. and $\phi(\lambda)$ exists for all $\lambda > 0$, then $F^*(\lambda) = \int_0^x e^{-\lambda t}\, dF$ for $x > 0$ and $F^*(\lambda)$ is bounded; conversely, given F, $F(x) = \int_0^x e^{\lambda t}\, dF^*$.

8. Find the c.f.'s for the densities given in Exercise 8.2.8.

9. Find a decomposition theorem for c.f.'s analogous to the one for d.f.'s.

10. Show that if A is a set of positive Lebesgue measure, then A contains an interval of positive measure.

11. *Multivariate Normal Density.* Let q be a quadratic form defined by

$$q(x) = \sum_{j,k=1}^{n} c_{jk} x_j x_k = x'Cx,$$

where $C = (c_{ij})$ is a positive definite symmetric matrix and $x = (x_1, x_2, \ldots, x_n)$. A density f in n dimensions is called *normal* and is centered at the origin if

$$f(x) = \gamma^{-1} e^{-1/2q(x)},$$

where γ is a constant. A normal density centered at $\mu = (\mu_1, \ldots, \mu_n)$ is given by $f(x - \mu)$.

(a) Show that if r has the above density and A has an inverse, then $s = rA$ has a normal density with a matrix $\bar{C} = A^{-1}C(A^{-1})'$.

(b) Show that the substitution defined by

$$s_1 = r_1, \ldots, s_{n-1} = r_{n-1} \qquad s_n = c_{1n} r_1 + \cdots + c_{nn} r_n$$

is nonsingular and

$$q(x) = c_{nn}^{-1} y_n^2 + q'(y),$$

where q' is a quadratic form in $y_1, \ldots, y_{n-1}$.

(c) Under the transformation defined in part (b), show that the marginal densities for s_n and $(s_1, \ldots, s_{n-1})$ are normal. Prove that all marginal densities of a normal density are normal.

(d) Show that there exists a nonsingular linear transformation $s = rA$ such that s has mutually independent normal components s_k with $Es_k = 0$. Verify that s has a normal density with $\bar{C} = D$, diagonal matrix with elements σ_k^{-2}, and that $\gamma^2 = (2\pi)^n \det D^{-1}$.

(e) Define the covariance matrix $\operatorname{Cov} r$ by $\operatorname{Cov} r = E(rr')$. Show that $D^{-1} = A' \operatorname{Cov} r\, A$ and conclude that $C^{-1} = \operatorname{Cov} r$ and $\gamma^2 = (2\pi)^n \det C$.

(f) Verify the statements made in part 4 of Example 1. [*Hint:* See Theorem 2.3 and Exercise 5(h) of Section 10.2].

10.2 ELEMENTARY PROPERTIES

We recall that U denotes the real part and V denotes the imaginary part of a c.f. χ. Since $\cos\theta$ is an even function and $\sin\theta$ is an odd function, it follows from Equations 1.3 and 1.4 that

$$U(-\lambda) = U(\lambda) \qquad V(-\lambda) = -V(\lambda). \tag{1}$$

Consequently,

$$\chi(-\lambda) = \overline{\chi(\lambda)}. \tag{2}$$

THEOREM 1

A c.f. χ satisfies the following relations:

$$\chi(0) = 1 = \operatorname{Var} F \qquad |\chi(\lambda)| \leq 1. \qquad (-\infty < \lambda < \infty)$$

Proof

The first result follows from $E(1) = 1$ and the second from $|Ee^{i\lambda r}| \leq E|e^{i\lambda r}|$.

THEOREM 2

A c.f. is uniformly continuous over the whole real line.

Proof

The result follows immediately from the inequality

$$|\chi(\lambda) - \chi(\lambda + h)|^2 \leq 2(1 - U(h)), \tag{3}$$

which is valid for any real numbers λ and h. To prove (3), note that

$$|\chi(\lambda) - \chi(\lambda + h)|^2 = \left|\int_{-\infty}^{\infty} e^{i\lambda x}(1 - e^{ihx})\, dF\right|^2.$$

From Schwarz's inequality,

$$\left|\int_{-\infty}^{\infty} e^{i\lambda x}(1 - e^{ihx})\, dF\right|^2 \leq \int_{-\infty}^{\infty} dF \int_{-\infty}^{\infty} |1 - e^{ihx}|^2\, dF$$

$$= 2\chi(0) \int_{-\infty}^{\infty} (1 - \cos hx)\, dF = 2\chi(0)(\chi(0) - U(h)),$$

which yields the result.

COROLLARY 1

If the sequence $\{\chi_n\}$ of c.f.'s approaches a function χ that is continuous at $\lambda = 0$, then χ is uniformly continuous on R.

Proof

Inequality (3) holds for every χ_n; upon taking limits, it remains valid for χ.

COROLLARY 2

If the sequence $\{\chi_n\}$ of c.f.'s is equicontinuous at $\lambda = 0$, then it is equicontinuous at every $\lambda \in R$.

Proof

As $h \to 0$, $|\chi_n(\lambda + h) - \chi_n(\lambda)|^2 \leq 2(1 - U_n(h)) \to 0$ uniformly in n.

COROLLARY 3

Let $\{\chi_n\}$ be a sequence of c.f.'s such that $\chi_n \to 1$ on a set A of positive Lebesgue measure. Then $\chi_n \to 1$ on the whole real line.

Proof

The set A contains the origin since, for $\lambda \in A$,

$$1 = \chi_n(0) \geq |\chi_n(\lambda)| \to 1.$$

Furthermore, A is symmetrical with respect to the origin, since

$$\chi_n(-\lambda) = \overline{\chi_n(\lambda)} \to 1$$

for $\lambda \in A$. Because

$$|\chi_n(\lambda) - \chi_n(\lambda - \mu)|^2 \leq 2(1 - U(\mu)) \to 0$$

for $\lambda, \mu \in A$, it follows that $\chi_n(\lambda - \mu) \to 1$. Using the results of Exercise 1.10, the theorem is reduced to showing that if $\chi_n \to 1$ on $(-a, a)$, then $\chi_n \to 1$ on A. If $|\lambda| < a$, then $\lambda \in A$ and so $\lambda - (-\lambda) = 2\lambda \in A$. Hence $\chi_n(2\lambda) \to 1$ for $|\lambda| < a$; the result follows by induction.

THEOREM 3

If $s = ar + b$, where a and b are constants, then

$$\chi_s(\lambda) = \chi_r(a\lambda)e^{ib\lambda},$$

where χ_s and χ_r are the c.f.'s of the r.v.'s s and r.

Proof

$$\chi_s(\lambda) = E(e^{i\lambda(ar+b)}) = e^{i\lambda b} Ee^{i\lambda ar} = e^{i\lambda b}\chi_r(a\lambda).$$

COROLLARY

The c.f. of the r.v. $-r$ is $\overline{\chi_r(\lambda)}$.

EXAMPLE 1

Suppose that r is a r.v. that is normally distributed with mean μ and variance σ^2. Then $s = (r - \mu)/\sigma$ is normally distributed with mean 0 and variance 1. It is less troublesome to calculate χ_s than χ_r. In fact, from part 1 of Example 1, we have $\chi_s = e^{-\lambda^2/2}$. Theorem 3 yields

$$e^{-\lambda^2/2} = \chi_r(\sigma^{-1})e^{-i\mu\lambda/\sigma};$$

hence $\chi_r(t) = e^{i\mu t - \sigma^2 t^2/2}$, where $t = \lambda/\sigma$.

DEFINITION 1

A r.v. r is *symmetric* if r and $-r$ have the same d.f.; that is, $P(r \leq x) = P(r \geq -x)$.

The d.f. of $-r$ is given by $1 - F(-x)$ at all points of continuity. However, the value of the d.f. at its points of continuity determines it uniquely.

THEOREM 4

A c.f. χ is real if and only if r is symmetric.

Proof

If r is symmetric, $\chi(\lambda) = \chi(-\lambda) = \overline{\chi(\lambda)}$; therefore, χ is real. From Equation 1.3 it is not difficult to verify that the real part U of χ is the c.f. of the distribution $G = \frac{1}{2}(F(x) + 1 - F(-x))$ at the points of continuity of F. We shall see (Theorem 5.2) that no other distribution has the same c.f. Since G is symmetric, it follows that if χ is real, then r is symmetric.

Further insight into the structure of c.f.'s can be obtained by examining the c.f.'s of lattice r.v.'s. Suppose that $P(r = kb) = p_k$ and $\sum_{k=-\infty}^{\infty} p_k = 1$; then, as we saw in part 2 of Example 1.1, $\chi = \sum_{k=-\infty}^{\infty} p_k e^{i\lambda kb}$. This function obviously has a (true) period[5] $2\pi/b$ and $\chi(2\pi/b) = 1$. All of these factors are equivalent, as shown by the following theorem.

THEOREM 5

Let r be a r.v. with a d.f. F and a c.f. χ. If $c \neq 0$, the following statements are equivalent.

1. r is a lattice r.v. whose range is contained in the set $0, \pm b, \pm 2b, \ldots,$ where $b = 2\pi c^{-1}$.
2. $\chi(\lambda + nc) = \chi(\lambda)$ for $n = \pm 1, \pm 2, \ldots$; that is, χ has a period c.
3. $\chi(c) = 1$.

Proof

We have seen that part 1 implies part 2 (and part 3). Taking $\lambda = 0$ and $n = 1$ in part 2, we see that part 2 implies part 3. It still must be shown that part 3 implies part 1. If $\chi(c) = 1$, then $\int_{-\infty}^{\infty} (1 - \cos cx)\, dF = 0$. Since the integrand is nonnegative, this is only possible if $1 - \cos cx$ vanishes at each point of increase of F. Hence the latter are of the form $n2\pi c^{-1}$, where $n = 0, \pm 1, \pm 2, \ldots$.

COROLLARY 1

If $|\chi(c)| = 1$ and $c \neq 0$, then there exists a constant a such that $\chi(\lambda)e^{-ia\lambda}$ has a period c and $r - a$ is a lattice r.v. with a range in the set $n2\pi c^{-1}$, where $n = 0, \pm 1, \pm 2, \ldots$.

Proof

Since $|\chi(c)| = 1$, there exists a real number θ such that $\chi(c) = e^{i\theta}$. The function $\chi_1(\lambda) = \chi(\lambda)e^{-ia\lambda}$ (where $a = \theta/c$) is a c.f.; in fact, it is the c.f. of the r.v. $r - a$ by Theorem 3. Since $\chi_1(c) = 1$, statements 1 and 2 of Theorem 5 yield the desired result.

A d.f. F with exactly one point of increase b—that is, $F(b) - F(b^-) = 1$—is called *degenerate* (or *unitary*). The corresponding degenerate c.f. is $e^{ib\lambda}$; its modulus is 1 for all λ. Conversely, if $|\chi(\lambda)| = 1$ for all λ, then by the last corollary, $\chi(\lambda)e^{-ia\lambda}$ has a period c for every real number c. In other words, $\chi(\lambda)e^{-ia\lambda} = k$, where k is a constant. Since $\chi(0) = 1$, it follows that $k = 1$ and $\chi(\lambda) = e^{ia\lambda}$.

[5] If $f(x + np) = f(x)$ for $n = 0, 1, 2, \ldots$, then p is a *period* of f. Obviously, p is not unique. The least period $p > 0$ is called the *true period*.

COROLLARY 2

A c.f. χ is degenerate if and only if $|\chi(\lambda)| = 1$ for all values of λ.

This corollary can be strengthened by noting that the number c appearing in Corollary 1 is also a period of $|\chi(\lambda)|$. Unless this function is a constant, it has a smallest period of $p > 0$. *Hence if* F *is not degenerate,* $|\chi(\lambda)| < 1$ *for all* $\lambda \neq 0$ *or else* $|\chi|$ *has a true period* $\mathrm{p} > 0$ *and* $|\chi(\lambda)| < 1$ *for* $0 < |\lambda| < \mathrm{p}$. In other words, a c.f. is degenerate if and only if $|\chi| = 1$ for two values $b \neq 0$ and $rb \neq 0$ whose ratio r is irrational. In particular, a c.f. χ is degenerate if $|\chi(\lambda)| = 1$ in a nondegenerate interval.

However, two c.f.'s can be distinct and yet coincide on a nondegenerate interval. To construct such examples, we require the fact that if χ_i $(i = 1, \ldots, n)$ are c.f.'s and $p_i \geq 0$ with $\sum_{i=1}^{n} p_i = 1, \chi = \sum_{i=1}^{n} p_i \chi_i$ is a c.f. The last statement follows from the observation that if F_i is the d.f. corresponding to χ_i, then $F = \sum_{i=1}^{n} p_i F_i$ is a d.f. with a c.f. χ. This result can easily be generalized to infinite sums and integrals of c.f.'s.

EXAMPLE 2

The c.f. χ of the density $(1/\pi)[(1 - \cos x)/x^2]$ is $\chi(\lambda) = 1 - |\lambda|$ for $|\lambda| \leq 1$ and $\chi(\lambda) = 0$ for $|\lambda| > 1$ (cf., part 3 of Example 1). Let $0 < a_1 < a_2 < \cdots < a_n$, and let $p_i > 0$ with $\sum_{i=1}^{n} p_i = 1$; then $\phi = \sum_{i=1}^{n} p_i \chi(\lambda/a_i)$ is a c.f. and a polygonal concave arc (see Figure 6). The vertexes of the polygon

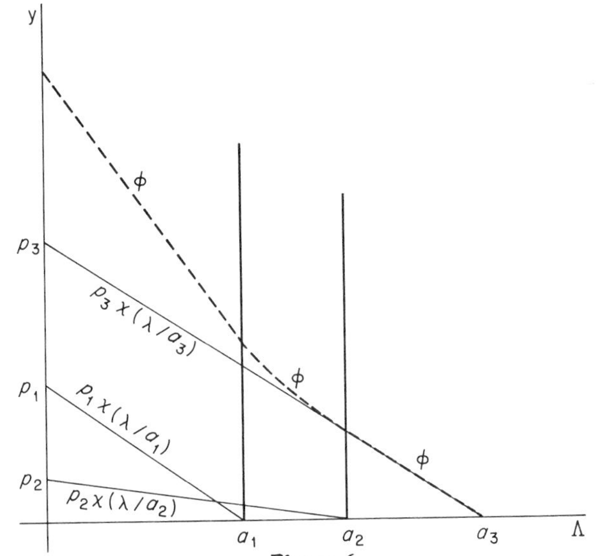

Figure 6

lie on the lines $\lambda = \pm a_i$, where $i = 1, \ldots, n$, and $\lambda = 0$. The proofs of these statements follow from the fact that the ordinate of the curve between a_i and a_{i+1} is given by $\sum_{j=i+1}^{n} p_j \chi(\lambda/a_j)$. The function $p_j\chi(\lambda/a_j)$ is the straight line $y_j/p_j + \lambda/a_j = 1$ and so the ordinate is

$$y = \sum_{j=i+1}^{n} y_j = -\left(\sum_{j=i+1}^{n} \frac{p_j}{a_j}\right)\lambda + \sum_{j=i+1}^{n} p_j.$$

Accordingly, the curve is a straight line which we will call the $(i+1)$th side. The slopes of the sides $n, n-1, n-2, \ldots, 1$ are

$$-\frac{p_n}{a_n}, \quad -\left(\frac{p_n}{a_n} + \frac{p_{n-1}}{a_{n-1}}\right), \quad -\left(\frac{p_n}{a_n} + \frac{p_{n-1}}{a_{n-1}} + \frac{p_{n-2}}{a_{n-2}}\right), \ldots, -\left(\sum_{i=1}^{n} \frac{p_i}{a_i}\right)$$

respectively. Therefore, the slopes of the sides increase as the number of sides increases and so the curve is concave. Since $\phi(-\lambda) = \phi(\lambda)$, the same conclusions are valid in the negative half-plane.

It is not difficult to see that two polygons can be different yet have some of their sides coinciding. This shows that two c.f.'s can be distinct yet coincide on a nondegenerate interval.

Conversely, given an arbitrary symmetrical polygon of the form described above, with $\phi(a_n) = 0$ and $\phi(0) = 1$, we can express ϕ as a linear combination of triangular functions. The λ coordinates of the vertexes give the a_i, and the intersection of the successive sides of the polygon with the y-axis yield the p_i. The equation for the $(i+1)$th sides shows that if the sides $n, n-1, n-2, \ldots, 1$ are prolonged, they intersect the y-axis at $p_n, p_n + p_{n-1}, p_n + p_{n-1} + p_{n-2}, \ldots, 1$ respectively and so successive subtractions will yield the p_i.

EXERCISES

1. Show that the set of all differences of numbers belonging to a set of positive Lebesgue measure contains a nondegenerate interval $(-a, a)$.

2. Prove that the following inequality holds for the real part U of χ:

$$0 \leq 1 - U(2\lambda) \leq 4(1 - U(\lambda)).$$

[*Hint:* $1 - U(2\lambda) = \int_{-\infty}^{\infty} (1 - \cos^2 \lambda x)\, dF = 2\int_{-\infty}^{\infty} (1 - \cos^2 \lambda x)\, dF$. Use the inequality $0 \leq 1 + \cos \lambda x \leq 2$.]

3. Verify that for $v > 0$ there exist functions $0 < m(v) < M(v) < \infty$ such that

$$m(\nu)\int_0^\nu (1-U(\lambda))\,d\lambda \leq \int_{-\infty}^{\infty}\frac{x^2}{1+x^2}\,dF \leq M(\nu)\int_0^\nu (1-U(\lambda))\,d\lambda.$$

[*Hint:* Use the inequality

$$0 < M^{-1}(\nu) \leq |\nu|\left(1-\frac{\sin \nu x}{\nu x}\right)\frac{1+x^2}{x^2} \leq m^{-1}(\nu) < \infty$$

and the fact that

$$\int_0^\nu d\lambda \int_{-\infty}^{\infty}(1-\cos\lambda x)\,dF = \nu\int_{-\infty}^{\infty}\left(1-\frac{\sin\nu x}{\nu x}\right)\left(\frac{1+x^2}{x^2}\right)\left(\frac{x^2}{1+x^2}\right)dF.\Big]$$

4. Show that

$$\int_{|x|<\nu^{-1}} x^2\,dF \leq \frac{3}{2}(1-U(\nu))$$

$$\int_{|x|\geq\nu^{-1}} dF \leq \frac{7}{\nu}\int_0^\nu (1-U(\lambda))\,d\lambda.$$

[*Hint*:

$$1-U(\nu) \geq \int_{|x|<\nu^{-1}}\frac{\nu^2x^2}{2}\left(1-\frac{\nu^2x^2}{12}\right)dF \geq \frac{11\nu^2}{24}\int_{|x|<\nu^{-1}} x^2\,dF$$

$$\frac{1}{\nu}\int_0^\nu (1-U(\lambda))\,d\lambda = \int_{-\infty}^{\infty}\left(1-\frac{\sin\nu x}{\nu x}\right)dF \geq (1-\sin 1)\int_{|x|\geq\nu^{-1}} dF.\Big]$$

5. *Properties of Multivariate Characteristic Functions.* Prove the following:

(a) $\chi(\lambda)$ exists and $|\chi(\lambda)| \leq 1$.
(b) $\chi(0, \ldots, 0) = 1$.
(c) $\chi(\lambda) = U(\lambda) + iV(\lambda)$, where

$$U(\lambda_1, \ldots, \lambda_n) = E[\cos(\lambda_1 r_1 + \lambda_2 r_2 + \cdots + \lambda_n r_n)]$$
$$V(\lambda_1, \ldots, \lambda_n) = E[\sin(\lambda_1 r_1 + \lambda_2 r_2 + \cdots + \lambda_n r_n)].$$

(d) $\chi(-\lambda_1, -\lambda_2, \ldots, -\lambda_n) = \overline{\chi(\lambda)}$.
(e) The c.f. $\chi(\lambda_1, \ldots, \lambda_n)$ is continuous.
(g) If $\chi_1(\lambda_1)$ is the c.f. of r_1, then $\chi_1(\lambda_1) = \chi(\lambda_1, 0, 0, \ldots, 0)$. If $\phi(\lambda_1, \ldots, \lambda_k)$ is the c.f. of $(r_1, r_2, \ldots, r_k)$, then

$$\phi(\lambda_1, \ldots, \lambda_k) = \chi(\lambda_1, \lambda_2, \ldots, \lambda_k, 0, \ldots, 0). \qquad (k < n)$$

(h) If ϕ is the c.f. of the r.v. $r = r_1 + r_2 + \cdots + r_n$, then

$$\phi(t) = \chi(t, t, \ldots, t).$$

6. If a r.v. r has a f.m.g.f. $N(\beta)$, show that the r.v. $ar + b$ for $a \neq 0$ has a f.m.g.f. $(1+\beta)^b N((1+\beta)^a - 1)$ (cf., Exercise 1.3).

7. (a) Show that if λ_i $(i = 1, 2, \ldots)$ are semi-invariants[6] (cf., Exercise 1.4) of an

[6]Note that if $a = 1$, $\lambda_i' = \lambda_i$ $(i = 2, 3, \ldots)$; this justifies the name "semi-invariant."

r.v. r, then the semi-invariants λ_i' of the r.v. $ar+b$ for $a \neq 0$ are $\lambda_1' = a\lambda_1 + b$ and $\lambda_i' = a^i\lambda_i$ $(i = 2, 3, \ldots)$.

(b) Show that if an r.v. r is not constant and all semi-invariants λ_i with $i > k$ are zero, then $k = 2$.

8. *Laplace Transform.* Verify the following (cf., Exercise 1.7), where ϕ is the transform of F:

(a) *Integration by Parts:*

$$\int_0^\infty e^{-\lambda x} F(x)\,dx = \lambda^{-1}\phi(\lambda) \qquad (\lambda > 0)$$

$$\int_0^\infty e^{-\lambda x}(1 - F)\,dx = \frac{1 - \phi(\lambda)}{\lambda}. \qquad (\lambda > 0)$$

(b) *Change of Scale:*

$$\int_0^\infty e^{-\lambda x}\,dF(x/a) = \phi(a\lambda).$$

10.3 DERIVATIVES

In this section we will gather together some results on derivatives of c.f.'s. As usual, we denote the real part of χ by U, the imaginary part by V, and the corresponding d.f. by F.

Observe that if U' exists, then $U'(0) = 0$, since $U(\lambda) = U(-\lambda)$.

THEOREM 1

Let F be the d.f. associated with the c.f. χ. Then $U'(0)$ exists if and only if

(*) $$\lim_{x\to\infty} x(1 - F(x) + F(-x)) = 0.$$

Proof

Since $U(0) = 1$ and $U'(0) = 0$, it follows from the definition of a derivative that for every $\epsilon > 0$ there exists a $\delta = \delta(\epsilon) > 0$ such that

(1) $$0 \leq 1 - U(\lambda) < \epsilon\lambda$$

if $|\lambda| < \delta$. To show that $\lim x(1 - F(x) + F(-x)) = 0$, we obtain a bound for this quantity using the integrals

(2) $$x^2 \int_0^{1/x} (1 - U(\lambda))\,d\lambda = x^2 \int_0^{1/x} d\lambda \int_{-\infty}^{\infty} (1 - \cos\lambda y)\,dF(y)$$

$$= x^2 \int_{-\infty}^{\infty} dF \int_0^{1/x} (1 - \cos\lambda y)\,d\lambda = x \int_{-\infty}^{\infty} \left(1 - \frac{\sin y/x}{y/x}\right) dF(y).$$

Since $(\sin y/x)/(y/x) \leq 1$ if $-x \leq y \leq x$ and $(\sin y/x)/(y/x) \leq \sin 1$ otherwise, then

$$(3) \qquad x \int_{-\infty}^{\infty} \left(1 - \frac{\sin y/x}{y/x}\right) dF \leq x(1 - \sin 1)(1 - F(x) + F(-x)).$$

Using the estimate in (1), it follows from (2) and (3) that

$$\epsilon/2 \geq x^2 \int_0^{1/x} (1 - U(\lambda))\, d\lambda \geq x(1 - \sin 1)(1 - F(x) + F(-x))$$

for $1/x < \delta$. Hence $x > 1/\delta$ implies that $x(1 - F(x) + F(-x)) \leq \epsilon/2(1 - \sin 1)$, which proves one half of the theorem.

Conversely, suppose that $(*)$ holds. If $h > 0$ and $x = 1/h$, then

$$(4) \qquad \frac{U(0) - U(h)}{h} = x\left(1 - U\left(\frac{1}{x}\right)\right) = x \int_{-\infty}^{\infty} \left(1 - \cos\frac{y}{x}\right) dF(y).$$

However, for $x > 0$, $\cos y/x \geq 1 - \frac{1}{2}y^2/x^2$ if $-x \leq y < x$ and $\cos y/x \geq -1$ otherwise. Using this inequality to estimate the integrand in (4) it follows that

$$(5) \qquad 0 \leq \frac{U(0) - U(h)}{h} \leq \frac{1}{2}\frac{1}{x} \int_{-x}^{x} y^2\, dF(y) + 2x(1 - F(x) + F(-x)).$$

However, Theorem 9.5.3 shows that if $(*)$ holds, then the integral appearing in (5) approaches zero as $x \to \infty$, and so the right-hand derivative is zero. If $h < 0$ and $x = 1/|h|$, then

$$0 \geq \frac{U(0) - U(h)}{h} \geq -\frac{1}{2} x \int_{-x}^{x} y^2\, dF(y) - 2x(1 - F(x) + F(-x))$$

and it follows as before that the left-hand derivative is also zero, which completes the proof.

THEOREM 2

If $\lim_{x \to \infty} x(1 - F(x) + F(-x)) = 0$, then $V'(0)$ exists if and only if

$$(6) \qquad \lim_{n \to \infty} \int_{-n}^{n} x\, dF(x)$$

exists.

Proof

We first see that

$$\frac{V(h) - V(0)}{h} = \frac{1}{h} \int_{-\infty}^{\infty} \sin hx \, dF(x) \tag{7}$$

$$\left| \int_{-\infty}^{-1/|h|} \sin hx \, dF \right| \leq \int_{-\infty}^{-1/|h|} dF = F(-1/|h|); \tag{8}$$

$$\left| \int_{1/|h|}^{\infty} \sin hx \, dF \right| \leq 1 - F(1/|h|). \tag{9}$$

Utilizing our hypothesis that $\lim_{x \to \infty} x(1 - F(x) + F(-x)) = 0$, it follows from Equations (7)–(9) that $V'(0)$ exists, which implies the existence of

$$\lim_{h \to 0} \frac{1}{h} \int_{-1/|h|}^{1/|h|} \sin hx \, dF. \tag{10}$$

Conversely, if this limit exists, it is obvious that $V'(0)$ exists.

We will show that the existence of the limits in (6) and (10) are equivalent and thus complete the proof. Setting $n = 1/h$ in (10) and subtracting the two integrals,

$$\left| \int_{-|n|}^{|n|} x \, dF - n \int_{-|n|}^{|n|} \sin \frac{x}{n} \, dF \right| \leq \int_{-|n|}^{|n|} x - n \sin \frac{x}{n} \, dF \tag{11}$$

$$\leq \frac{1}{3!} \frac{1}{n^2} \int_{-|n|}^{|n|} |x|^3 \, dF$$

by utilizing the inequality $\sin x/n \geq x/n - (1/3!)(x^3/n^3)$ if $0 < x < n$ and the reverse inequality if $x < 0$. From our hypothesis, it follows from Theorem 9.5.3 that the right-hand side of inequality (11) approaches zero as $n \to \infty$.

Since $\mathcal{X}'(0)$ exists if and only if $U'(0)$ and $V'(0)$ exist, we immediately obtain the following theorem.

THEOREM 3

If $\mathcal{X}$ is a c.f., then $\mathcal{X}'(0)$ exists if and only if $\lim_{x \to \infty} x(1 - F(x) + F(-x)) = 0$ and $\lim_{n \to \infty} \int_{-n}^{n} x \, dF$ exist.

The fact that the derivative does not exist at the origin has no bearing on derivatives at other points. The Cauchy density has a c.f. $e^{-|\lambda|}$ that is differentiable everywhere but at the origin (see part 5 of Example 1.1).

The conditions imposed by the existence of $\chi'(0)$ are not as stringent as those required for the existence of the first moment. From Theorem 9.5.2, we see that Er exists if and only if $\sum_{n=1}^{\infty} (1 - F(n) + F(-n)) < \infty$. The convergence of this series implies that

$$\lim_{n\to\infty} \sum_{k=n}^{2n-1} (1 - F(-k) + F(k)) = 0. \tag{12}$$

Since a d.f. is increasing, it follows that

$$\sum_{k=n}^{2n-1} (1 - F(-k) + F(k)) \geq n - nF(-n) + nF(n);$$

therefore, by Equation (12), $\lim_{n\to\infty} n(1 - F(-n) + F(n)) = 0$. Since the existence of Er implies the existence of $\lim_{n\to\infty} \int_{-n}^{n} x\, dF$, we see that the conditions of the theorem are satisfied. The next theorem shows that this result can be generalized.

THEOREM 4

If $\mu_k = Er^k$ exists, then $\mu_k = (1/i^k)\chi^{(k)}(0)$.

Proof

Differentiating the c.f. χ j times,

$$\chi^{(j)}(\lambda) = \frac{d^j}{d\lambda^j} \int_{-\infty}^{\infty} e^{i\lambda x}\, dF = \int_{-\infty}^{\infty} \frac{d^j}{d\lambda^j} (e^{i\lambda x})\, dF = i^j \int_{-\infty}^{\infty} x^j e^{i\lambda x}\, dF. \tag{13}$$

The interchange of integration and differentiation is justified by the fact that the last integral is absolutely convergent for $j \leq k$, which is implied by the existence of μ_k. Setting $\lambda = 0$ and $j = k$ in (13) yields the result.

Note that Equation (13) shows that $\chi^{(k)}(\lambda)$ is continuous and bounded by μ_k.

The direct converse of this theorem is false. If k is odd, $\chi^{(k)}(\lambda)$ can exist while μ_k does not exist.

EXAMPLE 1

Utilizing the technique of Example 2.2, we will construct a c.f. χ such that χ' exists and is finite on R while $\int_{-\infty}^{\infty} |x|\, dF$ is infinite. Let F_n be a d.f.

with jumps of magnitude $\frac{1}{2}$ at the points $\pm n$, and let $p_n = k/(n^2 \log n)$ for $n = 2, 3, \ldots,$ where k is chosen so that $\sum_{n=2}^{\infty} p_n = 1$. Then $F = \sum_{n=2}^{\infty} p_n F_n$ is a d.f. whose c.f.

$$\chi(\lambda) = k \sum_{n=2}^{\infty} \frac{\cos n\lambda}{n^2 \log n}.$$

Now

$$\chi'(\lambda) = -k \sum_{n=2}^{\infty} \frac{\sin n\lambda}{n \log n}$$

is uniformly convergent,[7] while

$$\int_{-\infty}^{\infty} |x| \, dF = k \sum_{n=2}^{\infty} \frac{1}{n \log n}$$

diverges.

We will see presently that the existence of $\chi^{(2k)}(0)$ implies the existence of μ_{2k}. To this end we require the following background. The *symmetric derivative* of a function $f(x)$ at x is defined as

$$\lim_{h \to 0+} \frac{f(x+h) - f(x-h)}{2h} = \lim_{h \to 0+} \left[\left(\frac{E - B}{2h}\right) f(x)\right],$$

where the operators E and B are defined by

$$Ef(x) = f(x + h) \qquad Bf(x) = f(x - h).$$

In operator notation, the nth symmetric is

$$\lim_{h \to 0+} \left[\left(\frac{E - B}{2h}\right)^n f(x)\right]. \tag{14}$$

It is not difficult to show that if the nth derivative exists, then it coincides with the nth symmetric derivative. For example, for the function a^x we have $Ea^x = a^{x+h} = a^x a^h$ and $Ba^x = a^x a^{-h}$. From these two results it follows that $B^r E^s a^x = a^{-rh} a^{sh} a^x$; that is, the effect of operating on a^x by $B^r E^s$ is equivalent to multiplying by $a^{-rh} a^{sh}$. Consequently, $(E - B)^n a^x = (a^h - a^{-h})^n a^x$ and therefore it follows from (14) that if the nth derivative of a^x at the origin exists, then

$$\left[\frac{d^n}{dx^n}(a^x)\right]_{x=0} = \lim_{h \to 0+} \left(\frac{a^h - a^{-h}}{2h}\right)^n. \tag{15}$$

[7]See T. J. Bromwich, *An Introduction to the Theory of Infinite Series* (London: Macmillan, 1908), pp. 31, 50.

LEMMA 1

If $\chi^{(n)}(0)$ exists, then

$$\chi^{(n)}(0) = \lim_{h\to 0+} \int_{-\infty}^{\infty} \left(\frac{e^{ihx} - e^{-ihx}}{2h}\right)^n dF(x).$$

Proof

The result follows by using the nth symmetrical derivative (14).

THEOREM 5

If $\chi^{(2n)}(0)$ exists, then

$$\mu_{2n} = (-1)^n \chi^{(2n)}(0).$$

Proof

From the lemma,

$$\chi^{(2n)}(0) = \lim_{h\to 0+} \int_{-\infty}^{\infty} \left(\frac{e^{ihx} - e^{-ihx}}{2n}\right)^{2n} dF$$

$$= \lim_{h\to 0+} (-1)^n \int_{-\infty}^{\infty} \left(\frac{\sin hx}{hx}\right)^{2n} x^{2n}\, dF.$$

By the Fatou–Lebesgue theorem (7.5.2),

$$\lim_{h\to 0+} \int_{-\infty}^{\infty} \left(\frac{\sin hx}{hx}\right)^{2n} x^{2n}\, dF \geq \int_{-\infty}^{\infty} x^{2n}\, dF$$

and so μ_{2n} exists, which implies the result by Theorem 4.

An analog of this result does not hold for odd derivatives, as shown by Example 1.

THEOREM 6

$U''(0)$ exists if and only if μ_2 is finite.

Proof

According to Theorem 4, $\mu_2 < \infty$ implies that $U''(0)$ exists and so we must only prove that the converse is true. For $h > 0$,

$$(16) \qquad \frac{1 - U(h)}{h^2} = \int_{-\infty}^{\infty} \frac{1 - \cos hx}{h^2}\, dF \geq \int_{-1/h}^{1/h} \frac{1 - \cos hx}{h^2}\, dF$$

$$\geq \int_{-1/h}^{1/h} \left(\frac{1}{2}x^2 - \frac{1}{4!}h^2 x^4\right) dF$$

by using the series expansion of cos hx. Since

$$\int_{-1/h}^{1/h} x^4\, dF \leq \frac{1}{h^2} \int_{-1/h}^{1/h} x^2\, dF,$$

it follows from Equation (16) that

$$\frac{1 - U(h)}{h^2} \geq \frac{11}{24} \int_{-1/h}^{1/h} x^2\, dF. \tag{17}$$

The existence of $U''(0)$ implies the existence of the symmetric second derivative, and

$$U''(0) = \lim_{h\to 0+} \frac{U(2h) - 2U(0) + U(-2h)}{(2h)^2} = \lim_{h\to 0+} \frac{2(U(2h) - 1)}{(2h)^2};$$

that is, the existence of $U''(0)$ implies that $\lim_{h\to 0+} [1 - U(h)]/h^2$ exists. It follows from (17) that μ_2 exists, since

$$\mu_2 = \lim_{h\to 0+} \int_{-1/h}^{1/h} x^2\, dF \leq \frac{24}{11} \lim_{h\to 0+} \frac{1 - U(h)}{h^2}.$$

EXERCISES

1. Show that if μ_k exists, then $\mu_l = (1/i^l)\,\chi^{(l)}(0)$ for every $l \leq k$.

2. Prove the following results:

(a) If the moment μ_r exists and the f.m.g.f. $F(\alpha)$ is defined in the open interval $(-a, a)$ for $a > 0$, then $\mu_r = F^{(r)}(0)$.
(b) If both the factorial moment $\mu_{(k_1,\ldots,k_n)}$ and $[(\partial^{k_1+\cdots+k_n}N)/(\partial\beta_1^{k_1}\cdots\partial\beta_n^{k_n})]\,\beta = 0$ exist (where N is the f.m.g.f.), then they are equal.
(c) If the semi-invariant λ_r exists, then $\lambda_r = (i)^{-r}\Lambda^{(r)}(0)$, where Λ is the s.i.g.f.

3. (a) Let $\phi(\lambda)$ denote the Laplace transform of F. Show that

$$(-1)^n \phi^{(n)}(\lambda) = \int_0^\infty e^{-\lambda x} x^n\, dF.$$

(b) Verify that F possesses a finite moment if and only if $\phi^{(n)}(\lambda)$ tends to a finite limit as $\lambda \to 0+$.

4. (a) Let χ be the c.f. of the r.v. $(r_1, \ldots, r_n)$. Show that if the moment $\mu_{k_1,\ldots,k_n}$ exists, then

$$i^{k_1+\cdots+k_n}\mu_{k_1,\ldots,k_n} = \left[\frac{\partial^{k_1+\cdots+k_n}\chi}{\partial\lambda_1^{k_1}\cdots\partial\lambda_n^{k_m}}\right]_{\lambda=0}.$$

(b) Prove the converse result in the case where all the integers $k_1, \ldots, k_n$ are even.

10.4 REGULARITY

Theorem 3.4 leads to the conjecture that if all the moments exist, then the c.f. can be expanded in a Taylor series. To verify this conjecture, we require the following lemma.

LEMMA 1

For any real x,

$$\left|e^{ix} - \sum_{k=0}^{n} \frac{(ix)^k}{k!}\right| \leq \frac{|x|^{n+1}}{(n+1)!}. \qquad (n = 0, 1, 2, \ldots) \tag{1}$$

Proof

We will prove the result by induction. Assume that $x > 0$. Since

$$\int_0^x e^{is}\, ds = -i(e^{ix} - 1) \quad \text{and} \quad \int_0^x e^{is}\, ds = xe^{i\xi}, \qquad (0 < \xi < x)$$

by the mean-value theorem, the assertion is true for $n = 0$. Let

$$r_n(x) = e^{ix} - \sum_{k=0}^{n} \frac{(ix)^k}{k!};$$

then $i \int_0^x r_n(s)\, ds = r_{n+1}(x)$. If the theorem is true for n,

$$|r_{n+1}(x)| \leq \int_0^x |r_n(s)|\, ds \leq \int_0^x \frac{s^{n+1}}{(n+1)!}\, ds = \frac{x^{n+2}}{(n+2)!};$$

therefore, it is true for $n + 1$ and so for all n by induction. A similar proof holds for $x < 0$; for $x = 0$, the assertion is trivial.

THEOREM 1

If $m_{n+1} = \int_{-\infty}^{\infty} |x|^{n+1}\, dF < \infty$, then

$$|\chi(\lambda) - \sum_{k=0}^{n} \frac{\mu_k}{k!} (i\lambda)^k| \leq \frac{m_{n+1}}{(n+1)!} |\lambda|^{n+1}. \tag{2}$$

Proof

The result follows from (1) if we replace x by λx and integrate with respect to F.

Other estimates for the remainder appear in the Exercises.

COROLLARY

If the moments of all orders exist, then $\chi^{(k)}(0) = i^k \mu_k$ for every k and

$$\chi(\lambda) = \sum_{k=0}^{\infty} \frac{\mu_k}{k!}(i\lambda)^k$$

in any interval $|\lambda| < a$ where the series converges.

Proof

The first statement follows from Theorem 3.4. Let $s_n = \sum_{k=0}^{n} (\mu_k/k!)(i\lambda)^k$. If the series converges at $\lambda_0 > 0$, then $\mu_{2k}\lambda_0 2k/2k! = m_{2k}\lambda_0 2k/2k! \to 0$ as $k \to \infty$. Hence $s_{2n-1} \to \chi(\lambda)$ uniformly in $|\lambda| < \lambda_0$. Since $|s_{2n} - s_{2n-1}| = m_{2n}\lambda 2n/2n!$, It follows that $\lim_{n\to\infty} s_{2n} = \chi(\lambda)$.

The above corollary leads to Hamburger's sufficient condition for a d.f. F to be uniquely determined by its moments (cf., Section 9.4). We will see in Section 10.5 that the d.f. F is uniquely determined by its c.f., which in turn is determined by a knowledge of the μ_n ($n = 0, 1, 2, \ldots$) whenever the series in Equation (3) converges for some $\lambda \neq 0$. The series in Equation (3) converges if $\sum_{k=0}^{\infty} (m_k/k!)(i\lambda)^k$ converges. By the root test, this will be the case if $\overline{\lim}\,(m_k/k!)^{1/k} < \infty$, which by Stirling's formula is equivalent to

$$\overline{\lim}\,\frac{1}{k}\, m_k^{1/k} < \infty.$$

Actually, if λ is within the circle of convergence of Equation (3), then $\sum_{k=0}^{\infty} (m_k/k!)(i\lambda)^k < \infty$ (see Lemma 3).

EXAMPLE 1

Consider the normal density function $f = e^{-x^2/2}/\sqrt{2\pi}$. Since $f(-x) = f(x)$, it follows that all odd moments vanish. By integrating by parts, we obtain $\mu_{2n} = (2n)!/(2^n n!)$ Therefore, from (3) we obtain

$$\chi(\lambda) = \sum_{n=0}^{\infty} \frac{(-\lambda^2/2)^n}{n!} = e^{-\lambda^2/2},$$

which agrees with our earlier result (see Example 1.1).

A function that is regular in a region can sometimes be defined in a larger region by analytic continuation; this extension is unique. This leads to the question of whether it is possible to extend uniquely a c.f. defined on an interval to the whole real line R. For example, if $\chi = 1$ on $(-a, a)$, the χ

is uniquely determined on R; in fact, $\chi = 1$. The c.f. $\chi(\lambda) = \int_{-\infty}^{\infty} e^{it}\, dF(t)$ is regarded as a complex-valued function of $\lambda = x + iy \equiv \mathscr{R}\lambda + i\mathscr{I}\lambda$.

The next lemma gives a criterion for determining when χ is regular.

LEMMA 2

The c.f. $\chi(\lambda)$ is regular in the open disk $|\lambda| < R$ if and only if for every positive $r < R$, $\int_{-\infty}^{\infty} e^{r|t|}\, dF(t) < \infty$.

Proof

If $\int_{-\infty}^{\infty} e^{r|t|}\, dF < \infty$, then we can expand $e^{r|t|}$ in powers of r and integrate term by term. This shows that all of the absolute moments are finite and so Equation (3) is valid.

Now suppose that $\chi(\lambda)$ is regular for $|\lambda| < R$; then for $0 < r < R$,

$$\sum \frac{1}{n!} |\mu_n| r^n < \infty$$

and, in particular,

$$\sum \frac{1}{2n!} m_{2n} r^{2n} < \infty.$$

Since $m_{2n-1}^{1/(2n-1)} \leq m_{2n}^{1/2n}$,[8] it follows from inequality (4) that

$$\sum \frac{1}{(2n-1)} m_{2n-1} r^{2n-1} < \infty$$

and so $\sum (1/n!)\, m^n r^n = \int_{-\infty}^{\infty} e^{r|t|}\, dF < \infty$.

Actually, if χ is regular in $|\lambda| < R$, it is regular in $|\lambda - a| < R$ for any real number a, since the criterion for regularity only involves the distance from λ to the center of the disk. We give an alternative proof below.

LEMMA 3

If $\chi(\lambda)$ is regular in $|\lambda| < R$, then it is regular in $|\lambda - a| < R$ for every real number a.

[8]See W. Feller, *An Introduction to Probability Theory and Its Applications*, Vol. 2 (New York: John Wiley & Sons, Inc., 1966), p. 153.

Proof

Let $\lambda - a = w$; then $\chi(\lambda) = \chi(w + a) = \chi(w)\chi(a)$ is the product of two regular functions, if $|w| < R$, and so it is regular in $|\lambda - a| < R$.

COROLLARY 1

If $\chi(\lambda)$ is regular in $|\lambda| < R$, then $\chi(\lambda)$ is regular in the strip $|\mathscr{I}\lambda| < R$.

This result follows directly from the lemma by using analytic continuation. It can be formulated in a slightly different manner.

COROLLARY 2

If $\chi(\lambda)$ is regular in the rectangle $|\mathscr{R}\lambda| < S$, $|\mathscr{I}\lambda| < R$, then $\chi(\lambda)$ is regular in the strip $|\mathscr{I}\lambda| < R$.

Proof

If $S \geq R$, then $\chi(\lambda)$ is regular in $|\lambda| < R$ and the result follows from Corollary 1. Next suppose that $S < R$; then $\chi(\lambda)$ is regular in a circle $|\lambda| < S$ and therefore it is regular in the strip $|\mathscr{I}\lambda| < S$. It follows that $\chi(\lambda)$ is regular in the circle $|\lambda| < \min(-\sqrt{2}\, S, R)$, since the interior of such a circle is either in the rectangle R or $|\mathscr{I}\lambda| < S$. Hence $\chi(\lambda)$ is regular in $|\mathscr{I}\lambda| < \sqrt{2}\,S$ if $\sqrt{2}\,S \geq R$ and in $|\mathscr{I}\lambda| < R$ if $\sqrt{2}\,S \geq R$. If $\sqrt{2}\,S < R$, the above process can be repeated and we see that $\chi(\lambda)$ is regular in the circle $|\lambda| < \min(\sqrt{3}\,S, R)$. Since $\sqrt{n}\,S \geq R$ for some n, the process eventually terminates.

The first of the two extension theorems, proposed by Marcienkiewicz, follows directly from the corollaries above because of the uniqueness of analytic continuation.

THEOREM 2

If the restriction χ_a of a c.f. χ to an interval $(-a, a)$ is regular, then χ_a determines χ uniquely.

The theorem is not true for arbitrary c.f.'s, as shown by Example 2.2. An alternate condition on χ can be obtained by writing $\chi = \chi_+ + \chi_-$, where

$$\chi_+(\lambda) = \int_0^\infty e^{i\lambda t}\, dF(t) \qquad \chi_-(\lambda) = \int_{-\infty}^0 e^{i\lambda t}\, dF(t).$$

It is not difficult to see that χ_+ is regular for $\mathscr{I}\lambda > 0$ and χ_- for $\mathscr{I}\lambda < 0$. Therefore, if χ_+ is regular for $0 > \mathscr{I}\lambda > -R$, then χ is regular for $0 >$

$\mathscr{I}\lambda > -R$, since χ_- is always regular in this strip. In such a situation we say that the c.f. $\chi(\lambda)$, where λ is real, is the *boundary function* of a regular function. Actually, if χ_+ is only regular in a semicircle $|\lambda| < R, 0 > \mathscr{I}\lambda > -R$ or in a rectangle, then it is regular in the whole strip $0 > \mathscr{I}\lambda > -R$. The proof of this result is similar to those of Lemmas 2 and 3. In fact, if we define a function $G = (F - F(0))/k$, where $k = \int_0^\infty dF$, then $k^{-1}\chi_+$ is the c.f. of G and the previous results hold for $k^{-1}\chi_+$ and so for χ_+. For example, Lemma 2 assumes the following form.

LEMMA 4

$\chi_+(\lambda)$ is regular for $0 > \mathscr{I}\lambda > -R$ if and only if for every positive $r < R$, $\int_0^\infty e^{rt}\, dF(t) < \infty$.

We leave it to the reader to fill in the details of the proof of the following extension.

THEOREM 3

If the restriction χ_a of a c.f. χ to an interval $(-a, a)$ is the boundary function of a regular function, then χ_a determines χ uniquely.

EXERCISES

1. (a) Show that if $\chi(\lambda)$ is regular in $|\mathscr{I}\lambda| < R$, then the m.g.f $F(\alpha)$ exists in the open interval $(-R, R)$.

(b) Prove that if $F(\bar{\alpha}_2)$ exists and is finite, then

$$\max_{\alpha_1} \chi(\alpha_1 - i\bar{\alpha}_2) = F(\bar{\alpha}_2),$$

where $\bar{\alpha}_2$ denotes the complex conjugate of α_2.

2. Verify the following statements for c.f.'s χ:

(a) If $\chi = e^{i\lambda b}$ on $(-a, a)$, then $\chi = e^{i\lambda b}$ for every $\lambda \in R$.
(b) If $\chi = e^{-\lambda^2/2}$ on $(-a, a)$, then $\chi = e^{-\lambda^2/2}$ for every $\lambda \in R$.
(c) If $\chi = f$ on $(-a, a)$ and f is a c.f. of a r.v. bounded above (below), then $\chi = f$ on R.

3. (a) *Riemann–Lebesgue.* Show that if f is an integrable function, and $\chi(\lambda)$

$= \int_{-\infty}^{\infty} e^{i\lambda x} f(x)\, dx$, then $\chi(\lambda) \to 0$ as $\lambda \to \pm\infty$. [*Hint:* Show that the result is true for any step function vanishing at $\pm\infty$, and so the result is true for the limit of such step functions. Finally, use the fact that an arbitrary integrable function f can be approximated by continuous functions f_n with bounded support.]
(b) Deduce that χ vanishes at infinity whenever the c.f. F has a density. Give an example to show that the result is not true for an arbitrary F.
(c) Prove that if the nth derivative $\chi^{(n)}$ exists and is integrable, then $\chi(\lambda) = O(|\lambda|^{-n})$. [*Hint:* Use integration by parts.]

4. Suppose that the m.g.f. $F(\alpha)$ can be expanded in a Taylor series for $|\alpha| < r$. Prove that the moments detemine the d.f. uniquely. (*Hint:* Use a one-to-one correspondence between c.f.'s and d.f.'s.)

5. Show that the following expressions for the remainder $r_n(\lambda)$ in the limited expansions of $\chi(\lambda) = \sum_{k=0}^{n-1} \mu_k\, [(i\lambda)^k/k!] + r_n(\lambda)$ are valid:

(a) $r_n(\lambda) = \lambda^n \int_0^1 \frac{(1-t)^{n-1}}{(n-1)!} \chi^{(n)}(t\lambda)\, dt.$

[*Hint:* Integrate the expansion of $e^{i\lambda t}$ with the various remainders in parts (a) and (b).]

(b) $r_n(\lambda) = \mu_n \frac{(i\lambda)^n}{n!} + o(\lambda^n) = \xi\, m_n \frac{|\lambda|^n}{n!}.$ $\qquad (|\xi| \leq 1)$

(c) $r_n(\lambda) = \mu_n \frac{(i\lambda)^n}{n!} + 2^{1-\delta}\theta m_{n+\delta} \frac{|\lambda|^{n+\delta}}{(1+\delta)(2+\delta)\cdots(n+\delta)},$

where $0 < \delta \leq 1$ and $|\theta| \leq 1$. [*Hint:*

$$\left| \int_0^1 \frac{(1-t)^{n-1}}{(n-1)!} (e^{it\lambda x} - 1)\, dt \right| \leq 2^{1-\delta} |\lambda x|^\delta \int_0^1 \frac{(1-t)^{n-1}}{(n-1)!} t^\delta\, dt$$

$$= \frac{2^{1-\delta}|\lambda x|^\delta}{(1+\delta)(2+\delta)\cdots(n+\delta)}.\Big]$$

10.5 INVERSION

In Theorem 2.4 and in the solution of the moment problem (cf., Section 10.4), we used the fact that every c.f. χ uniquely determines a d.f. F such that

$$\chi(\lambda) = \int_{-\infty}^{\infty} e^{i\lambda x}\, dF. \tag{1}$$

We will justify this claim by presenting an analytic expression for finding F given χ. This inversion formula for c.f.'s was discovered by P. Lévy. There are many other suitable inversion formulas (see Exercise 7 of this section and Exercise 1 of Section 10.7). The proof of the inversion theorem requires the well-known Dirichlet formula.

LEMMA 1

(2)
$$\frac{1}{\pi}\int_{-\infty}^{\infty}\frac{\sin x}{x}\,dx = 1.$$

The proof is left as an exercise for the reader.[9]

THEOREM 1

If $a < b$, then

(3)
$$\frac{F(b-)+F(b)}{2} - \frac{F(a-)+F(a)}{2} = \lim_{v\to\infty}\frac{1}{2\pi}\int_{-v}^{v}\frac{e^{-i\lambda a}-e^{-i\lambda b}}{i\lambda}\chi(\lambda)\,d\lambda.$$

Proof

Replace $\chi(\lambda)$ by its defining integral in (1) in $I(v)$, where

$$I(v) = \frac{1}{2\pi}\int_{-v}^{v}\frac{e^{-i\lambda a}-e^{-i\lambda b}}{i\lambda}\chi(\lambda)d\lambda.$$

Now $I(v)$ is an ordinary Riemann integral, since the integrand is continuous, if we define its value at zero to be $b - a$. Moreover, the integrand is bounded on the real line by $b - a$ and so (see Exercise 7.5.8) we can interchange the order of integration. We obtain

(4)
$$I(v) = \int_{-\infty}^{\infty}\left(\int_{v(x-b)}^{v(x-a)}\frac{\sin t}{t}\,dt\right)dF(x)$$

after some elementary computations that involve showing that

$$\int_{-v}^{v}\frac{e^{i\lambda(x-a)}}{i\lambda}\,d\lambda = 2\int_{0}^{v(x-a)}\frac{\sin t}{t}\,dt.$$

From Dirichlet's formula (2), it follows that the integrand appearing in (4) is uniformly bounded as $v \to \infty$. Hence the integration and passage to the limit as $v \to \infty$ can be interchanged in (4). Since

$$\lim_{v\to\infty}\frac{1}{\pi}\int_{v(x-b)}^{v(x-a)}\frac{\sin t}{t}\,dt = \frac{1}{2}(sg(x-a) - sg(x-b)),$$

[9]Recall that an improper Riemann integral $\int_{-\infty}^{\infty} f(x)\,dx$ is defined as $\lim \int_{a}^{b} f(x)\,dx$ as $a \to -\infty$ and $b \to \infty$. If this integral exists and is finite, then $\lim_{a\to\infty}\int_{-a}^{a} f\,dx = \int_{-\infty}^{\infty} f\,dx$. However, the left-hand side can be well defined while the right-hand side does not exist.

where $sgx = -1$ for $x < 0$, $sgx = 0$ for $x = 0$, and $sgx = 1$ for $x > 0$, then

$$\lim_{v\to\infty} I(v) = \frac{1}{2}(F(a+) - F(a-)) + (F(b-) - F(a+))$$
$$+ \frac{1}{2}(F(b+) - F(b-)) = \frac{F(b-) + F(b)}{2} - \frac{F(a-) + F(a)}{2}.$$

Note that if a and b are continuity points of F, the inversion formula reduces to

$$F(b) - F(a) = \lim_{v\to\infty} \frac{1}{2\pi} \int_{-v}^{v} \frac{e^{-i\lambda a} - e^{-i\lambda b}}{i\lambda} \chi(\lambda)d\lambda. \tag{5}$$

However, if a (b) is a discontinuity point, we get the average of the right and left-hand limits of F at a (b) instead of $F(a)$ ($F(b)$).

For convenience in writing Equation (3), we sometimes define a normalized d.f. F^* by $F^*(x) = [F(x-) + F(x)]/2$. Strictly speaking, F^* is not a d.f., since it is no longer continuous from the right; however, it determines F uniquely—in fact, $F(x) = F^*(x+)$.

In practice, it is expedient to have alternative forms for (3). The integrand can be rewritten as

$$\left(\frac{e^{-i\lambda a} - e^{-i\lambda b}}{i\lambda}\right)(U + iV) = \lambda^{-1}(\cos \lambda a - \cos \lambda b)V(\lambda)$$
$$+ \lambda^{-1}(\sin \lambda b - \sin \lambda a)U(\lambda) + i\lambda^{-1}(\cos \lambda b - \cos \lambda a)U(\lambda)$$
$$+ i\lambda^{-1}(\sin \lambda b - \sin \lambda a)V(\lambda).$$

Since the imaginary part of this expression is an even function, its integral $\int_{-v}^{v} = 0$. Hence we obtain the following corollary.

COROLLARY 1

If $a < b$ and $\chi = U + iV$, then

$$F^*(b) - F^*(a) \tag{6}$$
$$= \lim_{h\to\infty} \frac{1}{2\pi} \int_{-h}^{h} \frac{(\cos \lambda a - \cos \lambda b)V(\lambda) + (\sin \lambda b - \sin \lambda a)U(\lambda)}{\lambda} d\lambda.$$

Since the integrand in (6) is an odd function, then by writing $\int_{-h}^{h} = \int_{-h}^{0} + \int_{0}^{h}$ and replacing h by $-h$ in the first integral on the right, we obtain the following corollary.

COROLLARY 2

If $a < b$ and $\chi = U + iV$, then

$$F^*(b) - F^*(a) = \frac{1}{\pi}\int_0^\infty \lambda^{-1}[(\cos \lambda a - \cos \lambda b)V(\lambda) + (\sin \lambda b - \sin \lambda a)U(\lambda)]\, d\lambda. \tag{7}$$

THEOREM 2

If two d.f.'s F and G have the same c.f.

$$\int_{-\infty}^{\infty} e^{i\lambda x}\, dF(x) = \int_{-\infty}^{\infty} e^{i\lambda x}\, dG(x),$$

then $F(x) = G(x)$.

Proof

If follows immediately from Lévy's theorem that $F(b) - F(a) = G(b) - G(a)$ for almost all values of a and b, since F and G are both increasing and thus have at most a countable number of discontinuities. Hence $G(x) = F(x) + C$ (a.e.) where C is a constant, but $G(-\infty) = F(-\infty) = 0$, which implies that $C = 0$. However, $F = G$ at all continuity points implies that $F \equiv G$.

The following examples illustrate the method of using these formulas.

EXAMPLE 1

1. Suppose that $\chi(\lambda) = e^{i\lambda c} = \cos \lambda c + i \sin \lambda c$; the corresponding d.f. F is obtained by using Equation (6). Then

$$F^*(b) - F^*(a) = \frac{1}{2\pi}\lim_{h\to\infty}\int_{-h}^{h} [\lambda^{-1}(\cos \lambda a \sin \lambda c - \sin \lambda a \cos \lambda c) + \lambda^{-1}(\sin \lambda b \cos \lambda c - \cos \lambda b \sin \lambda c)]\, d\lambda$$

$$= \frac{1}{2\pi}\lim_{h\to\infty}\int_{-h}^{h} \left[\frac{\sin \lambda(c-a)}{\lambda} + \frac{\sin \lambda(b-c)}{\lambda}\right] d\lambda$$

$$= \frac{1}{2}[sg(c-a) + sg(b-c)],$$

where $sgx = 1$ for $x > 0$, $sgx = -1$ for $x < 0$, and $sgx = 0$ for $x = 0$. If $a > c$, then $F^*(b) - F^*(a) = \frac{1}{2}[1 + sg(b-c)]$; that is, the difference $F^*(b) - F^*(a)$ is independent of a. Hence $F^*(b) - F^*(a_1) = F^*(b) - F^*(a_2)$ or

$F^*(a_1) = F^*(a_2)$, and so the function F^* is constant if $a < c$. Since F is a d.f., $\lim_{x\to-\infty} F(x) = 0$; therefore, $F(x) = 0$ if $x < c$. Choosing $a < c$ and replacing b by x yields

$$F^*(x) = \frac{1}{2}(1 + sg(x - c))$$

and so $F^*(x) = 0$ for $x < c$, $F^*(x) = \frac{1}{2}$ for $x = c$, and $F^*(x) = 1$ for $x > c$. Since $F(x) = F^*(x+)$, it follows that $F(x) = 0$ for $x < c$ and $F(x) = 1$ for $x \geq c$.

2. Let $X(\lambda) = \frac{1}{3} + \frac{2}{3}e^{i\lambda} = (\frac{1}{3} + \frac{2}{3}\cos\lambda) + i(\frac{2}{3}\sin\lambda)$. Using Equation (6),

$F^*(b) - F^*(a) =$

$$\lim_{h\to\infty}\frac{1}{2\pi}\int_{-h}^{h}\left[\frac{2}{3}\frac{\sin\lambda(1-a)}{\lambda} - \frac{1}{3}\frac{\sin\lambda a}{\lambda} + \frac{1}{3}\frac{\sin\lambda b}{\lambda} + \frac{2}{3}\frac{\sin\lambda(b-1)}{\lambda}\right]dh$$

$$= \frac{1}{2}\left[\frac{2}{3}sg(1-a) - \frac{1}{3}sga + \frac{1}{3}sgb + \frac{2}{3}sg(b-1)\right].$$

If $a < 0$, then

$$F^*(b) - F^*(a) = \frac{1}{2}\left[1 + \frac{1}{3}sgb + \frac{2}{3}sg(b-1)\right];$$

that is, $F^*(x) = 0$ for $x < 0$, $F^*(x) = \frac{1}{6}$ for $x = 0$, $F^*(x) = \frac{1}{3}$ for $0 < x < 1$, and $F^*(x) = \frac{2}{3}$ for $x = 1$ and $1 < x$. Thus $F(x) = F^*(x+) = 0$ for $x < 0$, $\frac{1}{3}$ for $0 \leq x < 1$, and 1 for $1 \leq x$.

3. If $X(\lambda) = e^{-|\lambda|}$, then (7) yields

$$F^*(b) - F^*(a) = \frac{1}{\pi}\int_0^\infty \frac{\sin\lambda b - \sin\lambda a}{\lambda}e^{-\lambda}\,d\lambda.$$

Since

$$\frac{\sin\lambda b - \sin\lambda a}{\lambda} = \int_a^b \cos\lambda x\,dx,$$

it follows that, upon interchanging the order of integration,

$$F^*(b) - F^*(a) = \frac{1}{\pi}\int_a^b dx\int_0^\infty e^{-\lambda}\cos\lambda x\,d\lambda$$

$$= \frac{1}{\pi}\int_a^b \frac{1}{1+x^2}\,dx.$$

Therefore, $F(x) = (1/\pi)\int_{-\infty}^{x}[dx/(1+x^2)]$, the Cauchy distribution.

If F is differentiable at a Equation (3) can be used to deduce a formula for F'. Replacing b by $a+h$, dividing by h, and taking the limit as $h \to 0$ results in the following expression for $F'(a)$.

COROLLARY 3

If F is differentiable at a, then its derivative

$$F'(a) = \lim_{h\to 0} \lim_{v\to\infty} \frac{1}{2\pi} \int_{-v}^{v} \left(\frac{1 - e^{-i\lambda h}}{i\lambda h}\right) e^{-i\lambda a}\, \chi(\lambda)\, d\lambda. \tag{8}$$

Conversely, if the right side exists, then F is differentiable at a and $F'(a)$ is given by (8).

The preceding result leads to a criterion, in terms of the c.f. χ, for determining whether F has a density function and a method for finding this density function (Fourier inversion).

THEOREM 3

If χ is absolutely integrable on the whole real line R, then F has a bounded continuous density function F' given by

$$F'(x) = \frac{1}{2\pi} \int_{-\infty}^{\infty} e^{-i\lambda x}\, \chi(\lambda)\, d\lambda \tag{9}$$

for every $x \in R$.

Proof

Since ($\quad$)$e^{-i\lambda a}$ in (8) is bounded by 1 (see Lemma 4.1) and $\int_{-\infty}^{\infty} |\chi(\lambda)|\, d\lambda < \infty$, it follows from the definition of the improper integral that $\lim_{v\to\infty} \int_{v}^{v} = \int_{-\infty}^{\infty}$.

Therefore, (8) reduces to

$$F'(x) = \frac{1}{2\pi} \int_{-\infty}^{\infty} \lim_{h\to 0} \frac{1 - e^{-i\lambda h}}{i\lambda h}\, e^{-i\lambda x}\, \chi(\lambda)\, d\lambda = \frac{1}{2\pi} \int_{-\infty}^{\infty} e^{-i\lambda x} \chi(\lambda)\, d\lambda,$$

where the interchange of integration and the operation of taking a limit is justified by Theorem 7.5.3 (see Exercise 7.5.5).

If $\chi \geq 0$ and is integrable, (9) reduces to

$$F'(x) = \frac{1}{2\pi} \int_{-\infty}^{\infty} \cos \lambda x\, \chi(\lambda)\, d\lambda = \frac{1}{\pi} \int_{0}^{\infty} \cos \lambda x\, \chi(\lambda)\, d\lambda. \tag{10}$$

Since χ is a c.f., it is defined by

(11) $$\chi(\lambda) = \int_{-\infty}^{\infty} \cos \lambda x F'(x)\, dx.$$

Equations (10) and (11) are of the same form, except for a multiplicative constant. Therefore, if we normalize χ by multiplying by some constant c, we can regard it as a density function with c.f. $2\pi c F'$. The normalizing constant c is obtained from the condition $2\,\pi c F'(0) = 1$. Hence we obtain the following corollary.

COROLLARY 1

If a c.f. $\chi \geq 0$ is integrable, then $\chi/2\pi F'(0)$ is a density function whose c.f. is $F'(x)/F'(0)$.

This result was used in part 3 of Example 1.1. As another illustration, consider the c.f. $\chi(\lambda) = e^{-|\lambda|}$ that corresponds to the Cauchy density $F' = (1/\pi)[1/(1 + x^2)]$. We can regard it as the density $\frac{1}{2}\, e^{-|\lambda|}$ by reading the formula associating F' to χ backwards. Its c.f. is $[F'(x)/F'(0)] = [1/(1 + x^2)]$.

COROLLARY 2

If the c.f.'s χ_n of the d.f.'s F_n are uniformly Lebesgue-integrable on R, and if $\chi_n \to \chi$, the c.f. of F, then χ is Lebesgue-integrable on R and $F'_n \to F'$.

Proof

Since the c.f.'s χ_n are uniformly absolutely integrable, we can apply (9) for every n. The result follows from Theorem 7.5.3 by taking limits.

Given the c.f. χ of a r.v. r, we can find $P(r = a)$ from the c.f. by the following result.

THEOREM 4

For any real a,

(12) $$F(a + 0) - F(a - 0) = \lim_{v\to\infty} \frac{1}{2v} \int_{-v}^{v} e^{-i\lambda a}\chi(\lambda)\, d\lambda.$$

Proof

The proof that we can interchange the order of integration is justified as in Theorem 1; hence

$$\lim_{v\to\infty}\frac{1}{2v}\int_{-v}^{v} e^{-i\lambda a}\chi(\lambda)\,d\lambda = \lim_{v\to\infty}\frac{1}{2v}\int_{-v}^{v} e^{-i\lambda a}\left(\int_{-\infty}^{\infty} e^{i\lambda x}\,dF(x)\right)d\lambda$$

$$= \lim_{v\to\infty}\int_{-\infty}^{\infty}\frac{\sin v(x-a)}{v(x-a)}\,dF(x).$$

Since

$$\lim_{v\to\infty}\frac{\sin v(x-a)}{v(x-a)} = 0 \qquad (x \neq a)$$

$$= 1, \qquad (x = a)$$

Equation (12) is obtained by interchanging integration and passing to the limit in the last integral (see Theorem 7.5.3).

Finally we will give a brief outline of the inversion formula for a c.f. $\chi(\lambda)$ of n-variables. For an n-dimensional d.f. $F(x_1, \ldots, x_n)$, we define the "normalized" d.f.

$$F^*(x_1, \ldots, x_n) = \frac{1}{2^n}[F(x_1^-, x_2^-, \ldots, x_n^-) + \cdots + F(x_1, x_2, \ldots, x_n)],$$

where the sum in brackets is obtained by forming 2^n left-hand limits of $F(x_1, \ldots, x_n)$, by changing some of the coordinates of the point $(x_1, x_2, \ldots, x_n)$ and adding the resultant functions. For example, $F^*(x_1, x_2) = \frac{1}{4}[F(x_1^-, x_2^-) + F(x_1^-, x_2) + F(x_1, x_2^-) + F(x_1, x_2)]$. F^* differs from F only at discontinuity points and $F^*(x_1^+, x_2^+, \ldots, x_n^+) = F(x_1, x_2, \ldots, x_n)$.

THEOREM 5

If $a_i < b_i$ for $i = 1, 2, \ldots, n$, then

$$(13)\qquad \Delta_{a_1}^{b_1}\cdots\Delta_{a_n}^{b_n}F^*(x_1, \ldots, x_n)$$

$$= \frac{1}{(2\pi)^n}\lim_{\substack{v_1\to\infty\\ \cdots\\ v_n\to\infty}}\int_{-v_1}^{v_1}\cdots\int_{-v_n}^{v_n}\prod_{k=1}^{n}\left(\frac{e^{-i\lambda_k c_k} - e^{-i\lambda_k b_k}}{i\lambda_k}\right)\chi(\lambda_1, \ldots, \lambda_n)\,d\lambda_1\cdots d\lambda_n.$$

Proof

If we substitute the value of χ and interchange the order of integration, the integral on the right-hand-side of (13) becomes

$$\int_{-\infty}^{\infty}\cdots\int_{-\infty}^{\infty} dF(x_1, \ldots, x_n)\int_{-v_1}^{v_1}\cdots\int_{-v_n}^{v_n}\prod_{k=1}^{n}\left(\frac{e^{i\lambda_k(x_k-a_k)} - e^{i\lambda_k(x_k-b_k)}}{i\lambda_k}\right)d\lambda_1\cdots d\lambda_n$$

$$= \int_{-\infty}^{\infty}\cdots\int_{-\infty}^{\infty} dF\prod_{k=1}^{n}\left(\int_{-v_k}^{v_k}\left(\frac{e^{i\lambda_k(x_k-a_k)} - e^{i\lambda_k(x_k-b_k)}}{i\lambda_k}\right)d\lambda_k\right)$$

$$= \int_{-\infty}^{\infty}\cdots\int_{-\infty}^{\infty} dF\prod_{k=1}^{n}\left(\int_{v_k(x_k-b_k)}^{v_k(x_k-a_k)} 2\frac{\sin t}{t}\,dt\right).$$

Since

$$\lim_{v_k\to\infty}\int_{v_k(x_k-b_k)}^{v_k(x_k-a_k)}\frac{\sin t}{t}\,dt=\frac{\pi}{2}(sg(x_k-a_k)-sg(x_k-b_k)),$$

the right-hand side of (13) reduces to

$$\text{(14)}\qquad \frac{\pi^n}{(2\pi)^n}\int_{-\infty}^{\infty}\cdots\int_{-\infty}^{\infty}\prod_{k=1}^{n}(sg(x_k-a_k)-sg(x_k-b_k))\,dF.$$

A typical term of the sum that is formed by expanding the product appearing in the integrand of the integral (14) is $(-1)sg(x_1-c_1)sg(x_2-c_2)\cdots sg(x_n-c_n)$, where $c_i=a_i$ or b_i; the $(-1)^s$ arises from the s c's that are b's. To evaluate

$$\text{(15)}\qquad (-1)^s\int_{-\infty}^{\infty}\cdots\int_{-\infty}^{\infty}sg[(x_1-c_1)(x_2-c_2)\cdots(x_n-c_n)]\,dF(x_1\cdots x_n),$$

note that the integrand takes the value ± 1 depending upon which of the 2^n regions contains the r.v. $r=(r_1,\ldots,r_n)$. For example, suppose that the region is

$$R=(r_1>c_1,r_2>c_2,\ldots,r_k>c_k,r_{k+1}<c_{k+1},\ldots,r_n<c_n);$$

then $sg[(x_1-c_1)\cdots(x_n-c_n)]=(-1)^{n-k}$. Hence the value of the integral is just the sum of the probabilities that the r.v. lies in one of these regions, along with the appropriate sign. To find the probability that $r\in R$, let $A_i=(r_i\le c_i)$ for $i=1,2,\ldots,k$, and $B_i=(r_i<c_i)$ for $i=k+1,\ldots,n$. Then

$$\begin{aligned}R&=(\Omega-A_1)\cdots(\Omega-A_k)B_{k+1}\cdots B_n\\&=B_{k+1}\cdots B_n-A_1B_{k+1}\cdots B_n-A_2B_{k+1}\cdots B_n\\&\quad\cdots+(-1)^kA_1A_2\cdots A_kB_{k+1}\cdots B_n,\end{aligned}$$

and so

$$\begin{aligned}PR=F(&\infty,\ldots,\infty,c_{k+1}^-,c_{k+2}^-,\ldots,c_n^-)\\&-F(c_1,\infty,\ldots,\infty,c_{k+1}^-,\ldots,c_n^-)+\cdots\\&+(-1)^kF(c_1,c_2,\ldots,c_k,c_{k+1}^-,\ldots,c_n^-).\end{aligned}$$

Accordingly, this term contributes the quantity

$$(-1)^nF(c_1,c_2,\ldots,c_k,c_{k+1}^-,\ldots,c_n^-)+L$$

to the integral (15), where L denotes the remaining terms in each of which the d.f. appears with at least one $x_i=\infty$. Hence as the $(x_1,\ldots,x_n)$ range over the points in the 2^n regions, we obtain all possible combinations of left-hand limits of $F(x_1,\ldots,x_n)$ at $(c_1,\ldots,c_n)$; therefore, the integral (15) reduces to

$$\text{(16)}\qquad (-1)^{n+s}F^*(c_1,c_2,\ldots,c_n)2^n+L(c),$$

where $L(c)$ is a sum of d.f.'s with at least one $x_i = \infty$ and the remaining coordinates c's. Now

$$\Delta_{a_1}^{b_1} \Delta_{a_2}^{b_2} \cdots \Delta_{a_n}^{b_n} G(x_1, x_2, \ldots, x_n)$$

can be written as

$$(17) \qquad (S_{b_1} - S_{a_1})(S_{b_2} - S_{a_2}) \cdots (S_{b_n} - S_{a_n})G = (-1)^n \prod_{i=1}^{n} (S_{a_i} - S_{b_i})G,$$

where S_{b_i} is an operator that substitutes b_i for x_i. There is a one-to-one correspondence between terms appearing in Equation (17) and the terms of the product in the integrand of (14) if $S_{a_i} \longleftrightarrow sg(x - a_i)$ and $S_{b_i} \longleftrightarrow sg(x - b_i)$. For example, $(-1)^s sg((x_1 - c_1) \cdots (x_n - c_n))$ corresponds to $(-1)^s(S_{c_1}S_{c_2} \cdots S_{c_n})$. Using this correspondence and the fact that (15) is equivalent to (16), it is not difficult to see that (14) is just

$$\Delta_{a_1}^{b_1} \cdots \Delta_{a_n}^{b_n} (F^*(x) + L(x)).$$

Since $L(x)$ is a sum of functions with at most $n - 1$ of the x_i's present, its nth difference is zero and we obtain Equation (13).

If χ is the c.f. of the r.v. $(r_1, \ldots, r_n)$, then $\chi(\lambda a_1, \lambda a_2, \ldots, \lambda a_n)$ is the c.f. of the r.v. $r = a_1 r_1 + a_2 r_2 + \cdots + a_n r_n$, where the a_i are fixed numbers. Conversely, if we know the distributions of r as the a_i vary, we can calculate all the numbers $\chi(\lambda a_1, \lambda a_2, \ldots, \lambda a_n)$ and so the multivariate c.f. $\chi(\lambda_1, \lambda_2, \ldots, \lambda_n)$ is known. By the preceding theorem, this uniquely determines the d.f. Hence a probability distribution in R^n is uniquely determined by the probabilities of all semihyperplanes. This observation was made by Cramer and Wold.

EXERCISES

1. Verify Lemma 1.

2. Find the d.f. corresponding to the following c.f.'s: $(\cosh \lambda)^{-1}, \lambda(\sinh \lambda)^{-1}$, $(\cosh \lambda)^{-2}$, $\cos \lambda$, $\cos^2 \lambda$, and $(1 + i\lambda)^{-1}(\sin a\lambda)/a\lambda$.

3. Let r be a r.v. with a c.f. $\chi(\lambda) = \exp(\mu(e^{a(e^{i\lambda}-1)} - 1))$, where a and μ are positive constants. Find

(a) Er.
(b) Var r.
(c) $P(r = 0)$.

4. Let F be a d.f., χ its corresponding c.f., and x_n the abscissas of the jumps of F. Show that

$$\lim_{v \to \infty} \frac{1}{2v} \int_{-v}^{v} |\chi(\lambda)|^2 \, d\lambda = \sum_n (F(x_n^+) - F(x_n^-))^2.$$

5. *Plancherel.* Let $\chi_n(\lambda) = (1/\sqrt{2\pi}) \int_{-n}^{n} e^{i\lambda x} f(x)\, dx$, where $f \in L^2(-\infty, \infty)$. Then prove that $\chi_n \in L^2$, $\chi_n \to \chi$ in the mean square, and

$$\int_{-\infty}^{\infty} |\chi(\lambda)|^2\, d\lambda = \int_{-\infty}^{\infty} |f(x)|^2\, dx.$$

Conversely, show that if $f_n = (1/\sqrt{2\pi}) \int_{-n}^{n} e^{-i\lambda x}\chi(\lambda)\, d\lambda$, then $f_n \in L^2$ and $\lim_{n\to\infty} f_n = f$ exist in the mean square and

$$\int_{-\infty}^{\infty} |f(x)|^2\, dx = \int_{-\infty}^{\infty} |\chi(\lambda)|^2\, d\lambda.$$

[*Hint:* Define linear operators T and A by $Tf = \lim_{n\to\infty} \int_{-n}^{n} e^{i\lambda x} f(x)\, dx$ and $Ax = \int_{-\infty}^{\infty} e^{-i\lambda x}\chi(\lambda)\, dx$. Let $\mathscr{M}$ ($\mathscr{N}$) be the linear subspace consisting of all finite linear combinations of $f_{\mu\sigma}$ ($\chi_{\mu\sigma}$), where

$$f_{\mu\sigma} = (\pi\sigma)^{-1/4} e^{-(x-\mu)^2/2\sigma} \qquad \chi_{\mu\sigma} = \left(\frac{\sigma}{\pi}\right)^{-1/4} e^{i\lambda\mu - \sigma\lambda^2/2}.$$

Show that T maps $\mathscr{M}$ onto $\mathscr{N}$ and A maps $\mathscr{N}$ onto $\mathscr{M}$ by showing that A and T are inverses. Finally, show that A and T are continuous by showing that they are isometric; complete the proof by showing that $\mathscr{M}$ and $\mathscr{N}$ are dense in L^2.]

6. Suppose that $F(x)$ is represented by three components as in Theorem 9.2.8; then we have $\chi = a\chi_s + b\chi_c + c\chi_p$.

(a) Show that $\chi_c \to 0$ as $|\lambda| \to \infty$ and

$$\lim_{T\to\infty} \frac{1}{2T} \int_{-T}^{T} |\chi_c(\lambda)|^2\, d\lambda = 0.$$

(b) Show that χ_s is an almost periodic function and $\overline{\lim_{\lambda\to\infty}} |\chi_s| = 1$; moreover,

$$\lim_{T\to\infty} \frac{1}{2T} \int_{-T}^{T} |\chi_s(\lambda)|\, d\lambda.$$

[*Hint:* See Besicovitch[10], pages 6, 19.]
(c) Show that[11]

[10] A. S. Besicovitch, *Almost Periodic Functions.* (Cambridge: The University Press, 1932.)

[11] It can be shown that χ_p does not necessarily tend to zero as $\lambda \to \infty$ See B. Jessen, and A. Wintner, "Distribution Functions and the Riemann Zeta Function," *Trans. Am. Math. Soc.* **38** (1935), 48–88.

$$\lim_{T\to\infty} \frac{1}{2T} \int_{-T}^{T} |\chi_p|^2 \, d\lambda = 0.$$

[*Hint:* Use the fact that if χ is the c.f. of a continuous d.f., the same holds for $|\chi|^2$.]

7. Show that if $(a - h, a + h)$ is a continuity interval of the d.f. F, then

$$F(a + h) - F(a - h) = \lim_{T\to\infty} \frac{1}{\pi} \int_{-T}^{T} \frac{\sin h\lambda}{\lambda} e^{-i\lambda a} \chi(\lambda) \, d\lambda.$$

8. (a) Verify that for any real a and $h > 0$,

$$\int_0^h F(a + x) - F(a - x) \, dx = \frac{1}{\pi} \int_{-\infty}^{\infty} \frac{1 - \cos h\lambda}{\lambda^2} e^{-i\lambda a} \chi(\lambda) \, d\lambda.$$

(b) Deduce that if $\chi_n \to \phi$ is continuous at 0, then $F(\infty) - F(-\infty) = 1$.

9. Let F and G be bounded or unbounded d.f.'s with Laplace transforms ϕ and γ defined for all $\lambda > 0$. If $\phi(a + bn) = \gamma(a + bn)$ for $n = 1, 2, 3, \ldots$, where $a > 0$ and $b > 0$ are constants, then $F = G$ (a.e.).

10.6 CONVERGENCE

One of the most important uses of c.f.'s is in the derivation of asymptotic formulas in probability theory. To this end, we frequently use the the result that a sequence of c.f.'s χ_n converges to a c.f. χ if and only if the corresponding d.f.'s F_n converge completely to the d.f. F that corresponds to χ. In order to establish this result, we first study weak convergence of d.f.'s. We recall that if $F_n \to F(w)$, then F need not be a proper d.f. For example, if $F_n(x) = 0$ for $x < -n$, $F_n(x) = \frac{1}{2}$ for $-n \leq x < n$, and $F_n(x) = 1$ for $x \geq n$, then F_n converges weakly to $F = \frac{1}{2}$. Hence we would not expect the corresponding c.f.'s χ_n to approach a c.f. In fact, in the example above, the sequence $\chi_n = \cos n\lambda$ does not converge. Observe, however, that $\int_0^\lambda \cos nt \, dt$ $= (\sin n\lambda)/n \to 0$, which is $\int_0^\lambda \chi(t) \, dt$, where $\chi(\lambda) = \int_{-\infty}^{\infty} e^{i\lambda x} \, dF$.

This suggests that we should consider the *integral c.f.*

$$\hat{\chi}(\lambda) = \int_0^\lambda \chi(t) \, dt = \int_{-\infty}^{\infty} \frac{e^{i\lambda x} - 1}{ix} \, dF(x) \tag{1}$$

in studying weak convergence. The last integral appearing in (1) is obtained by replacing χ by its defining value and interchanging the order of integration (see Exercise 7.5.8). Since there is a one-to-one correspondence between χ and F defined up to an additive constant (since F need not be a probability d.f.) and a one-to-one correspondence between $\hat{\chi}$ and its continuous derivative χ, it follows that there is a one-to-one correspondence between $\hat{\chi}$ and F defined up to an additive constant. This necessitates the introduction of a new type of convergence. We will say that $F_n \to F(w)$ *up to additive constants* if, for every subsequence F_n, there exists a constant $\bar{c}$ such that $\bar{F}_n \to F + \bar{c}(w)$.

THEOREM 1

If $F_n \to F(w)$ up to additive constants, then $\hat{\chi}_n \to \hat{\chi}$.

Proof

Since $g(x) = (e^{i\lambda x} - 1)/ix$ for $x \neq 0$, $g(0) = \lambda$ is continuous, and $g(\pm\infty) = 0$, it follows from Theorem 9.3.3, by using the definition of $\hat{\chi}_n$ in Equation (1), that $\hat{\chi}_n \to \chi$.

COROLLARY

Every sequence of integral c.f.'s has a convergent subsequence.

Proof

The corresponding d.f.'s have a weakly convergent subsequence (by Theorem 8.3.2). By the theorem, the subsequence of c.f.'s corresponding to this subsequence converges.

THEOREM 2

If $\hat{\chi}_n$ converges to some function $\hat{g}$, then there exists a general d.f. F such that $F_n \to F(w)$ up to additive constants and $\hat{g} = \hat{\chi}$, the integral c.f. of F.

Proof

By Theorem 8.3.2, there is a general d.f. F and a subsequence F_m such that $F_m \to F(w)$ as $m \to \infty$. From Theorem 1, it follows that $\hat{\chi}_m \to \hat{\chi}$; however, that $\hat{\chi}_m$ is a subsequence of $\hat{\chi}_n$ implies that $\hat{\chi}_m \to \hat{g}$. Hence $\hat{g} = \hat{\chi}$ and since $\hat{\chi}$ determines F up to an additive constant, it follows that all weakly convergent subsequences of the sequence $\{F_n\}$ have the same limit F up to additive constants.

COROLLARY

If $\chi_n \to g$ (a.e.) with respect to the Lebesgue measure on R, then there exists a general d.f. F such that $F_n \to F(w)$ up to additive constants, and whose c.f. $\chi = g$ (a.e.).

Proof

Since the χ_n are continuous, $|\chi_n| \leq 1$, and $\chi_n \to g$ (a.e.), it follows that g is measurable and bounded a.e. Accordingly, $\hat{\chi}_n \to \hat{g}$ by the dominated convergence theorem and so $F_n \to F(w)$ up to additive constants and $\hat{\chi} = \hat{g}$ by the theorem above. The derivative of $\int_0^\lambda g(\lambda)\, d\lambda$ is equal to g a.e., while the derivative of $\hat{\chi}$ is χ and so $\chi = g$ (a.e.).

We are now in a position to prove the assertion about complete convergence made at the beginning of this section.

THEOREM 3

If $F_n \to F(c)$, then $\chi_n \to \chi$ uniformly with respect to λ in every finite interval.

Proof

If $F_n \to F(c)$, then by the generalized Helly theorem (9.3.2), $\chi_n \to \chi$.
To prove the second half of the theorem we use Ascoli's theorem, which states that a uniformly bounded, equicontinuous family of functions on R that converges, converges uniformly on compact subsets of R (closed and bounded subsets of R). Since the χ_n are c.f.'s, they are uniformly bounded and the theorem reduces to showing that the χ_n are equicontinuous on R, which in turn is equivalent to equicontinuity at 0 by Corollary 2 of Theorem 2.2. To prove equicontinuity at 0, we recall that if $F_n \to F(c)$, then $\{F_n, F\}$ are equicontinuous at $\pm\infty$. Hence there exist points a and b ($|a| < b$) such that for all n, $F_n(a) < \epsilon/8$ and $1 - \epsilon/8 < F_n(b) \leq 1$ for an arbitrary $\epsilon > 0$. Hence

$$|\chi_n(h) - \chi_n(0)| \leq \int_{-\infty}^{a} |1 - e^{ihx}|\, dF_n + \int_a^b |1 - e^{ihx}|\, dF_n$$

$$+ \int_b^\infty |1 - e^{ihx}|\, dF_n \leq \epsilon/2 + b|h|,$$

by using the estimates $|1 - e^{ihx}| \leq 2$ in the first and third integrals above and the estimate $|1 - e^{ihx}| < |hx|$ in the second integral. Choosing $|h| < \epsilon/2b$, it follows that $|\chi_n(h) - \chi_n(0)| < \epsilon$ for every n.

If the F_n are differentiable—that is, if there exist density functions—then the previous result can be strengthened.

THEOREM 4

If the d.f.'s F_n and F are differentiable and $F'_n \to F'$ on R, then $\chi_n \to \chi$ uniformly on the whole real line R.

Proof

Obviously

$$(*) \qquad \lim_{n\to\infty} \int_R F'_n \, dx = \int_R \lim_{n\to\infty} F'_n \, dx,$$

since both integrals equal 1. Since $0 \leq (F' - F'_n)^+ \leq F'$ and F' is integrable, it follows by the Lebesgue dominated convergence theorem that $\lim_{n\to\infty} \int_R (F' - F'_n)^+ \, dx = 0$. Hence $(F' - F'_n)^- = (F' - F'_n)^+ - (F' - F'_n)$ is integrable and so $\lim_{n\to\infty} \int_R (F' - F'_n)^- \, dx = 0$. Therefore, $\lim_{n\to\infty} \int_R |F' - F'_n| \, dx = 0$, which implies the desired result.

The converse of Theorem 3 is known as *Lévy's continuity theorem* for c.f.'s.

THEOREM 5

If $\chi_n \to \phi$, which is continuous at 0, then $F_n \to F(c)$ and $\phi = \chi$, the c.f. of F.

Proof

For every $\lambda \in R$,

$$\hat{\chi}_n(\lambda) = \int_0^\lambda \chi_n(t) \, dt \to \int_0^\lambda \phi(t) \, dt = \hat{\phi}(\lambda),$$

since $\chi_n \to \phi$. Hence by Theorem 2, there exists a d.f. F with a c.f. χ such that $F_n \to F(w)$ and $\hat{\chi} = \hat{\phi}$; therefore,

$$(2) \qquad \frac{1}{\lambda} \int_0^\lambda \chi(t) \, dt = \frac{1}{\lambda} \int_0^\lambda \phi(t) \, dt.$$

Since χ and ϕ are both continuous at the origin, it follows from (2), upon taking limits as $\lambda \to 0$, that $\chi(0) = \phi(0)$. Accordingly,

$$1 = \chi_n(0) \to \phi(0) = \chi(0) = F(\infty) - F(-\infty),$$

which shows that $F_n \to F(c)$. Applying Theorem 3, it follows that $\chi_n \to \chi$.

The theorem is not true if ϕ is not continuous at $\lambda = 0$ (Exercise 8). The following fact that is contained in the above theorem is quite useful.

COROLLARY

A function that is continuous at the origin and is the pointwise limit of a sequence of c.f.'s is also a c.f.

The following result, derived by Polya, can easily be proved with the aid of the last corollary.

EXAMPLE 1

If the graph in the first quadrant of a nonnegative even function ϕ is a continuous concave curve and $\phi(0) = 1$, then ϕ is a c.f. The result follows since every continuous curve is the limit of inscribed polygons, and the inscribed polygons are concave and therefore are c.f.'s by Example 2.2. We see immediately that $e^{-|\lambda|^\alpha}$ is a c.f. when $0 < \alpha \leq 1$ and that $e^{\mu(e^{-|\lambda|}-1)}$, where $\mu > 0$ is a constant, is a c.f., by noting that their second derivatives are nonnegative.

THEOREM 5′

If the c.f.'s χ_n and χ are integrable over R, $\chi_n \to \chi$, and $\int_{-\infty}^{\infty} |\chi_n(\lambda) - \chi(\lambda)|\, d\lambda \to 0$, then the corresponding densities $f_n \to f$ uniformly on R.

Proof

Under the hypothesis $\int_{-\infty}^{\infty} |\chi_n - \chi|\, d\lambda \to 0$, it follows immediately from the Fourier inversion formula (5.9) that $f_n \to f$ uniformly on R. The continuity theorem shows that χ corresponds to f.

The following example illustrates the use of the continuity theorem in deriving limit laws. The method will be exploited to its full extent in Chapter 13.

EXAMPLE 2

We give an alternate proof of the De Moivre–Laplace theorem using the same notation as in Section 4.3. We leave it to the reader to show that the c.f. of the r.v. $r_n = (s_n - np)/\sqrt{npq}$ is $\chi_n(\lambda) = (qe^{-i\lambda\sqrt{p/nq}} + pe^{i\lambda\sqrt{q/np}})^n$.

By utilizing the Maclaurin expansion for e^x,

$$\chi_n^{1/n} = 1 - (\lambda^2/2n)(1 + R_n),$$

where

$$R_n = 2\sum_{k=3}^{\infty} \frac{1}{k!}\left(\frac{i\lambda}{\sqrt{n}}\right)^{k-2} \frac{pq^k + q(-p)^k}{\sqrt{(pq)^k}}.$$

Letting $n \to \infty$, $\chi_n \to e^{-\lambda 2/2}$ since $R_n \to 0$. By the continuity theorem it follows that as $n \to \infty$,

$$P(r_n \leq x) \to \frac{1}{2\pi}\int_{-\infty}^{x} e^{-t2/2}\, dt.$$

Suppose that $\chi_n \to \infty$, which is continuous at $\lambda = 0$ on some finite interval $(-a, a)$. Generally, the relation $F_n \to F(c)$ does not follow, since we have seen that distinct c.f.'s can coincide on a finite interval. In fact, we are not really sure that ϕ, defined only on $(-a, a)$, can be extended to a c.f. It turns out that such an extension is possible; we require the following lemmas to show this.

LEMMA 1

If $\hat{\chi}$ is the integral c.f. that corresponds to the c.f. $\chi = U + iV$, then

(3) $$\left|\frac{\hat{\chi}(\lambda + k) - \hat{\chi}(\lambda - k)}{2k}\right|^2 \leq \frac{1}{2}(1 + U(k)).$$

Proof

By using the definition $\hat{\chi}(\lambda)$ given by the integral on the right-hand side of (1), it can be seen that

(4) $$\left|\frac{\hat{\chi}(\lambda + k) - \hat{\chi}(\lambda - k)}{2k}\right|^2 = \left|\int_{-\infty}^{\infty} e^{i\lambda x} \frac{\sin kx}{kx}\, dF\right|^2 \leq \int_{-\infty}^{\infty} \left|\frac{\sin kx}{kx}\right|^2 dF(x),$$

where the inequality is a consequence of Schwarz's inequality. To simplify the last integral, use the result

$$\left|\frac{\sin x}{x}\right| = \left|\frac{\sin x/2}{x/2}\right|\left|\cos\frac{x}{2}\right| \leq \left|\cos\frac{x}{2}\right|,$$

which implies that

$$\left|\frac{\sin x}{x}\right|^2 \leq \left|\cos^2\frac{x}{2}\right| = \frac{1+\cos x}{2}.$$

Applying this last estimate to the second integral on the right-hand side of (4) yields (3).

COROLLARY

If $\hat{\chi}$ is the integral c.f. corresponding to $\chi = U + iV$, then

$$\left|\frac{1}{2ks}\int_{-ks}^{ks} \chi(\lambda)\, d\lambda\right|^2 \leq \frac{1}{2}(1 + U(k)).$$

Proof

The above inequality is identical to

$$\left|\frac{\hat{\chi}(ks) - \hat{\chi}(-ks)}{2ks}\right|^2 \leq \frac{1}{2}(1+U(k)). \tag{5}$$

To prove (5), set $\lambda = jk$ in (3) and sum over $j = -s+1,\ -s+3,\ -s+5,\ \ldots, s-3, s-1$, obtaining

$$\left|\frac{\hat{\chi}((-s+2)k) - \hat{\chi}(-sk)}{2k}\right|^2 + \left|\frac{\hat{\chi}((-s+4)k) - \hat{\chi}((-s+2)k)}{2k}\right|^2 + \cdots$$
$$+ \left|\frac{\hat{\chi}(sk) - \hat{\chi}((s-2)k)}{2k}\right|^2 \leq \frac{s}{2}(1 + U(k)).$$

Simplifying the left-hand side of this expression by using the inequality

$$|e_1|^2 + \cdots + |e_s|^2 \geq \frac{1}{s}\,|e_1 + e_2 + \cdots + e_s|^2 \tag{6}$$

leads to Equation (5).

LEMMA 2

If $\chi_n \to \phi$ on $(-a, a)$ and ϕ is continuous at $\lambda = 0$, then the χ_n are equicontinuous and the convergence is uniform.

Proof

We will show that the χ_n are equicontinuous at $\lambda = 0$, using the method of contradiction. Suppose that the χ_n are not equicontinuous at zero; then there exist $\epsilon > 0$ and a sequence $\lambda_m \to 0$ as $m \to \infty$, so that for some subsequence $\{\chi_m\}$ of $\{\chi_n\}$ we have $|\chi_m(\lambda_m)| < 1 - \epsilon$ for all m. Given a positive $h \in (-a, a)$, we define $t_m = [h/\lambda_m]$. It is clear that $\lambda_m t_m \to h$ as $m \to \infty$. By the corollary to Lemma 1, with $k = \lambda_m$, $s = t_m$, and $\chi = \chi_m$,

$$\left|\frac{1}{2\lambda_m t_m}\int_{-\lambda_m t_m}^{\lambda_m t_m} \chi_m(v)\, dv\right|^2 \leq \frac{1}{2}(1 + U_m(\lambda_m)) < 1 - \epsilon/2,$$

since $|U_m(\lambda_m)| < 1 - \epsilon$. Letting $m \to \infty$,

$$\left|\frac{1}{2h}\int_{-h}^{h} \phi(v)\, dv\right|^2 \leq 1 - \epsilon/2.$$

The integral is well defined, since $\phi(v)$ exists on $(-a, a)$. Letting $h \to 0$, it follows that the integral approaches $\phi(0)$ since ϕ is continuous at the origin. Since $\phi(0) = \lim \chi_n(0) = 1$, it follows that $1 \leq -\epsilon/2 + 1$, a contradiction.

Therefore, the χ_n are equicontinuous at $\lambda = 0$ and so by Corollary 2 of Theorem 2.2 they are equicontinuous everywhere.

THEOREM 6

If $\chi_n \to \phi_a$ on $(-a, a)$ and ϕ_a is continuous at $\lambda = 0$, then, ϕ_a extends to a c.f. χ on R. Moreover, if the extension χ is unique, then $\chi_n \to \chi$ on R.

Proof

Since the χ_n are c.f.'s, they are uniformly bounded by 1; moreover, they are equicontinuous by Lemma 2. By Ascoli's theorem, the sequence χ_n is compact in the sense of uniform convergence; hence there is at least one subsequence that converges uniformly to a limit function χ on R. By Theorem 5, χ is a c.f., since the uniform limit of a sequence of continuous functions is continuous. Since χ coincides with ϕ_a on $(-a, a)$, it is an extension of ϕ_a. Moreover, if χ is unique, then obviously $\chi_n \to \chi$ on R.

Theorems 4.2 and 4.3 give criteria necessary to make the extension unique as well as offering a method for finding this extension. Krein [4] has given the necessary and sufficient conditions for two c.f.'s that coincide on some interval $(-a, a)$ to be identical.

EXERCISES

1. Verify inequality (6).

2. Prove the following statements:

(a) If $\chi_n(\lambda) \to e^{i\lambda b}$ on $(-a, a)$, then $\chi_n(\lambda) \to e^{i\lambda b}$ on R.

(b) If $\chi_n(\lambda) \to e^{-\lambda^2/2}$ on $(-a, a)$, then $\chi_n(\lambda) \to e^{-\lambda^2/2}$ on R.

(c) If $\chi_n \to \chi$ on $(-a, a)$ and if χ is the c.f. of a r.v. bounded either above or below, then $\chi_n \to \chi$ on R.

3. A r.v. r is distributed according to the Poisson law with $Er = \lambda$.[12] Prove that for $\lambda \to \infty$, the distribution of the variable $(r - \lambda)/\sqrt{\lambda}$ tends to the normal distribution with parameters $\mu = 0$ and $\sigma = 1$.

[12]This result enables us to use the normal distribution tables to compute the probabilities $P(a < r \leq b)$ for large values of λ (α). The accuracy is quite good for the χ^2 distribution when $\lambda \geq 30$. The tables can also be used in Exercise 4.

4. The r.v. r has the density function

$$f(x) = \begin{cases} 0 & (x \leq 0) \\ \dfrac{\beta^\alpha}{\Gamma(\alpha)} x^{\alpha-1} e^{-\beta x} & (x > 0) \end{cases}$$

Show that the distribution of the r.v. $(\beta r - \alpha)/\sqrt{\alpha}$ tends to the normal distribution with parameters $\mu = 0$ and $\sigma = 1$ as $\alpha \to \infty$.

5. Prove that if $\chi(\lambda)$ is a c.f. and if $\phi(\lambda)$ is a function such that for some sequence $h_n \to \infty$ the products $\chi(\lambda)\phi(h_n\lambda) \equiv f_n(\lambda)$ are also c.f.'s, then ϕ is a c.f.

6. If $\chi_n \to \chi$ and $\lambda_n \to \lambda$ is finite, then $\chi_n(\lambda_n) \to \chi(\lambda)$.

7. Is a set $\{F_t\}$ of d.f.'s completely compact if and only if the corresponding set $\{\chi_t\}$ of c.f.'s is equicontinuous at $\lambda = 0$?

8. Let $\chi_n(t) = e^{-nt^2}$; then $\phi(t) = 1$ for $t = 0$ and $\phi(t) = 0$ for $t \neq 0$. Show that F is not a d.f. (see Theorem 5).

10.7 CONVOLUTIONS

According to Theorem 5.4.2 the convolution of the density functions of two discrete-valued r.v.'s r_1 and r_2 is just the density function of the r.v. $r_1 + r_2$. Moreover, their probability generating function is just the product of the probability generating functions of r_1 and r_2. Here we formally study the generalization of the notion of convolution to arbitrary d.f.'s and its relation with c.f.'s. We will postpone the analogous interpretation in terms of r.v.'s until after we have defined independence in the following chapter.

DEFINITION 1

A function F on R is said to be the *convolution* of the d.f.'s F_1 and F_2; it is written as $F_1 * F_2$ if

$$F(x) = \int_{-\infty}^{\infty} F_1(x - y)\, dF_2(y)$$

for every $x \in R$.

THEOREM 1

If F_1 and F_2 are d.f.'s, then $F = F_1 * F_2$ is a d.f.

Proof

Since $|F_1(x - y)| \leq 1$, $F(x)$ is defined for all values of x. Moreover, since $F_1(x_1 - y) \geq F_1(x_2 - y)$ if $x_1 > x_2$, it follows that $F(x_1) \geq F(x_2)$ if $x_1 > x_2$. By the dominated convergence theorem, the process of taking limits with respect to x and integration can be interchanged. Hence, $F(x)$ is continuous from the right, and $F(-\infty) = 0$. Finally, $F(\infty) = 1$ since $\int_{-\infty}^{\infty} dF_2(y) = 1$.

A similar definition and theorem hold for general d.f.'s. For such functions, if $F = F_1 * F_2$, then $\text{Var } F = (\text{Var } F_1)(\text{Var } F_2)$.[13]

THEOREM 2

If F, F_1, and F_2 are d.f.'s and χ, χ_1, and χ_2 are their corresponding c.f.'s, then $F = F_1 * F_2$ if and only if $\chi = \chi_1\chi_2$.

Proof

Suppose that $F = F_1 * F_2$; we then will show that $\chi = \chi_1\chi_2$. Let $a = x_{n,1} < \cdots < x_{n,j(n)+1} = b$ be a partition of $[a, b]$ such that $\sup_j (x_{n,j+1} - x_{n,j}) \to 0$ as $n \to \infty$. Then by definition of the integral

$$\int_a^b e^{i\lambda x}\, dF(x) = \lim_{n\to\infty} \sum_j e^{i\lambda x_{nj}} \Delta_j F(x_{nj})$$

$$= \lim_{n\to\infty} \int_{-\infty}^{\infty} \sum_j e^{i\lambda(x_{nj}-y)}\, \Delta_j F_1(x_{nj} - y) e^{i\lambda y}\, dF_2(y).$$

Again, by definition of the integral, it follows that

$$\int_a^b e^{i\lambda x}\, dF = \int_{-\infty}^{\infty} \left(\left(\int_{a-y}^{b-y} e^{i\lambda x}\, dF_1(x) e^{i\lambda y}\, dF_2(y)\right)\right).$$

Letting $a \to -\infty$ and $b \to \infty$ in the above equality gives

$$\int_{-\infty}^{\infty} e^{i\lambda x}\, dF = \left(\int_{-\infty}^{\infty} e^{i\lambda x}\, dF_1\right)\left(\int_{-\infty}^{\infty} e^{i\lambda y}\, dF_2(y)\right),$$

so that $\chi = \chi_1\chi_2$.

Conversely, suppose that $\chi = \chi_1\chi_2$. By the first half of the theorem we know that $\chi = \chi_1\chi_2$ is the c.f. of $F_1 * F_2$. The one-to-one correspondence between c.f.'s and d.f.'s show that $F = F_1 * F_2$.

COROLLARY

A product of c.f.'s is a c.f. If χ is a c.f., then so is $|\chi|^2$.

The first statement is obviously true; the second follows from the first by using the corollary to Theorem 2.3.

[13] $\text{Var } F = F(\infty) - F(-\infty)$.

THEOREM 3

The convolution operation has the following properties:

1. $F_1 * F_2 = F_2 * F_1$ (commutativity).
2. $(F_1 * F_2) * F_3 = F_1 * (F_2 * F_3)$ (associativity).

Proof

The proof follows immediately from Theorem 2 upon observing that multiplication of c.f.'s is associative and commutative.

Note that Theorem 3 implies that $F_1 * F_2 * \cdots * F_n$ is well defined and exists for all n.

EXAMPLE 1

1. *Poisson distribution.* The r.v. r with a density function

$$P(r = b_n) = \frac{\mu^n}{n!} e^{-\mu}, \qquad (n = 0, 1, 2, \ldots)$$

where $b_n = b + nh$, $\mu > 0$, $h > 0$, and b is arbitrary, has a c.f.

$$\chi = \exp(i\lambda b + \mu(e^{i\lambda h} - 1)).$$

The convolution of two such d.f.'s with parameters μ_i, b_i, and h for $i = 1, 2$, is again a Poisson distribution with parameters $\mu_1 + \mu_2$, $b_1 + b_2$, and h, as can readily be seen by multiplying the corresponding c.f.'s.

2. *Normal Distribution.* Let $N(\mu, \sigma)$ denote the normal d.f.; that is,

$$N(\mu, \sigma) = \frac{1}{\sqrt{2\pi}\,\sigma} \int_{-\infty}^{x} e^{-(x-\mu)^2/2\sigma^2}\, dx.$$

Then its c.f. is given by

$$\chi = \exp(i\lambda\mu - \lambda^2\sigma^2/2).$$

It follows, by multiplying corresponding c.f.'s, that

$$N(\mu_1 + \mu_2, \sigma_1 + \sigma_2) = N(\mu_1, \sigma_1) * N(\mu_2, \sigma_2).$$

THEOREM 4

If $F_n \to F(c)$, then $F_n * G \to F * G(c)$.

Proof

Let χ_n, χ, and ϕ be the c.f.'s of F_n, F, and G respectively. Now $F_n \to F(c)$ implies that $\chi_n \to \chi$ by Theorem 6.3 and so $\chi_n\phi \to \chi\phi$. The result follows from Theorems 2 and 6.5.

THEOREM 5

Let the c.f.'s χ, χ_1, and χ_2 be related by $\chi = \chi_1\chi_2$. Then $\chi(z)$ is regular in the strip $|\mathscr{I} z| < R$ if and only if $\chi_1(z)$ and $\chi_2(z)$ are both regular in $|\mathscr{I} z| < R$.

Proof

If χ_1 and χ_2 are both regular in $|\mathscr{I} z| < R$, it is obvious that χ is regular in the same region.

Conversely, suppose that $\chi(z)$ is regular in $|\mathscr{I} z| < R$. Set $z = -i\lambda$ with $|\lambda| < R$; then $\chi(-i\lambda)$ exists and $|\chi(-i\lambda)| < \infty$, where

$$\chi(-i\lambda) = \int_{-\infty}^{\infty} e^{\lambda x}\, dF(x).$$

However, if $F = F_1 * F_2$; then

$$\int_{-\infty}^{\infty} e^{\lambda x}\, dF = \left(\int_{-\infty}^{\infty} e^{\lambda x}\, dF_1\right)\left(\int_{-\infty}^{\infty} e^{\lambda x}\, dF_2\right),$$

where the proof is the same as that employed in Theorem 2. Therefore, if $|\lambda| < R$, then

$$\chi_1(-i\lambda) = \int_{-\infty}^{\infty} e^{\lambda x}\, dF_1 < \infty \qquad \chi_2(-i\lambda) = \int_{-\infty}^{\infty} e^{\lambda x}\, dF_2 < \infty.$$

In particular, $\int_{0}^{\infty} e^{\lambda x}\, dF_1 < \infty$ and $\int_{-\infty}^{\infty} e^{\lambda x}\, dF_1 < \infty$ for $|\lambda| < R$. It follows that for every $0 < \lambda < R$,

$$\int_{-\infty}^{\infty} e^{\lambda |x|}\, dF_1 < \infty \qquad \int_{-\infty}^{\infty} e^{\lambda |x|}\, dF_2 < \infty,$$

and so by Lemma 4.2 the functions χ_1 and χ_2 are regular in the circle $|z| < R$. Corollary 1 of Lemma 4 shows that these functions are regular in the strip $|\mathscr{I} z| < R$.

EXERCISES

1. (a) Let $n_\sigma(\lambda)$ denote the normal density function, with $\mu = 0$. Show that

(1) $$\frac{1}{\sqrt{2\pi}} \int_{-\infty}^{\infty} e^{-i\lambda t} \chi(\lambda) e^{-\lambda^2/2\sigma^2}\, d\lambda = \int_{-\infty}^{\infty} n_{1/\sigma}(t - x)\sigma\, dF,$$

and so find $N_{1/\sigma} * F$. Use Theorem 4 to show that as $\sigma \to \infty$, $N_{1/\sigma} * F \to F$.

(b) Deduce Equation (5.10) by using (1).
(c) Let $t = 0$ in (1), and deduce in the continuity theorem (6.4) that F is proper—that is, $F(\infty) - F(-\infty) = 1$.

2. Show that if the m.g.f. $F(\alpha)$ exists for $|\alpha| < R$, then the function $\chi(z)$ is regular in $|\mathscr{I} z| < R$ and coincides with the c.f. $\chi(\lambda)$ on the real axis.

3. (a) Show that if $F = F_1 * F_2$, then for every λ,

$$\int_{-\infty}^{\infty} e^{\lambda x}\, dF = \left(\int_{-\infty}^{\infty} e^{\lambda x}\, dF_1\right)\left(\int_{-\infty}^{\infty} e^{\lambda x}\, dF_2\right).$$

(b) Prove that there exist finite numbers $\alpha_j > 0$, $\beta_j > 0$ such that

$$\int_{-\infty}^{\infty} e^{\lambda x}\, dF(x) \geq \alpha_j e^{-\beta_j|\lambda|} \int_{-\infty}^{\infty} e^{\lambda x}\, dF_j(x). \qquad (j = 1, 2)$$

[*Hint:* $\int_{-\infty}^{\infty} e^{\lambda x}\, dF_1(x) \geq e^{b\lambda}(1 - F_1(b))$ or $e^{b\lambda} F_1(b)$, depending on whether $\lambda \geq 0$ or $\lambda < 0$. Let $\alpha_2 = \max(F_1(b), 1 - F_1(b))$ and let $\beta_2 = \max(|b_1|, |b_2|)$.]

4. Prove Theorem 3 directly by integration without using c.f.'s.

5. Show that if $F * F_1 = F * F_2$ and F is a normal d.f., then $F_1 = F_2$. Is the result true if F is not normal?

10.8 DECOMPOSITION

If χ_1 and χ_2 are c.f.'s and $\chi = \chi_1\chi_2$, then we say that χ is *composed* of χ_1 and χ_2. We have seen that the convolution (composition) of two Poisson (normal) distributions is a Poisson (normal) distribution (see Example 7.1). In particular, the composition of two degenerate distributions is a degenerate distribution, since such distributions are special cases of both the normal ($\sigma = 0$) and the Poisson ($\mu = 0$) distributions.

Conversely, given a c.f. χ, we could ask if it is possible to decompose χ into two c.f.'s χ_1 and χ_2 such that $\chi = \chi_1\chi_2$, with both χ_1 and χ_2 corresponding to the same type of distribution as χ. For example, suppose that the c.f. χ of the degenerate type is given by $\chi(\lambda) = e^{i\lambda a}$. If for every $\lambda \in R$

$$\chi_1(\lambda)\chi_2(\lambda) = e^{i\lambda a},$$

then $|\chi_1|\,|\chi_2| = 1$, which implies that $|\chi_1| = |\chi_2| = 1$, since $|\chi_1| \leq 1$ and $|\chi_2| \leq 1$. By Corollary 2 of Theorem 2.5, we have $\chi_1 = e^{i\lambda a_1}$ and $\chi_2 = e^{i\lambda a_2}$, with $a_1 + a_2 = a$. Such a decomposition is also possible for the nondegenerate Poisson and normal distributions; however, the proofs are more difficult.

To prove the normal decomposition theorem, we first must prove the following interesting result. If a c.f. $\chi(\lambda) = e^{P(\lambda)}$, where $P(\lambda)$ is a polynomial, then $P(\lambda)$ is of the second degree. We require the following lemmas.

LEMMA 1

Suppose that r is a r.v. that takes positive as well as negative values with a probability greater than zero. Let $L(\alpha)$ exist for all $\alpha \in R$, where $L(\alpha) = \chi(-i\alpha)$ and χ is the c.f. of r; then

$$\lim_{\alpha\to\infty} L(\alpha) = \lim_{\alpha\to-\infty} L(\alpha) = \infty. \tag{1}$$

Proof

By its definition, $L(\alpha) = \int_{-\infty}^{\infty} e^{\alpha x}\, dF(x)$, where F is the d.f. of r. It is clear that $L(\alpha) \geq \int_{x_0}^{\infty} e^{x}\, dF$ for any $x_0 > 0$. Choosing x_0 so that $P(r > x_0) = c > 0$, it follows that for $\alpha > 0$,

$$L(\alpha) \geq ce^{\alpha x_0};$$

taking limits, we see that $\lim_{\alpha\to\infty} L(\alpha) = \infty$. A similar proof based on the inequality $L(\alpha) \geq \int_{-\infty}^{0} e^{\alpha x}\, dF$ shows that $\lim_{\alpha\to-\infty} L(\alpha) = \infty$.

COROLLARY

If r is a nonconstant r.v. whose m.g.f. $E(\alpha)$ exists for all $\alpha \in R$, then there exists a constant c such that

$$\lim_{\alpha\to\infty} e^{\alpha c}\, E(\alpha) = \lim_{\alpha\to-\infty} e^{\alpha c} E(\alpha) = \infty. \tag{2}$$

Proof

Under the above hypothesis on r, it is always possible to find a constant c such that $r + c$ takes positive and negative values with nonzero probabilities. The m.g.f. of $r + c$ is easily seen to be $e^{\alpha c} E(\alpha)$; the result follows by applying (1) to this new m.g.f.

LEMMA 2

If $\chi(i\alpha_2)$ exists and $|\chi(\alpha_2)| < \infty$, then $\max_{\alpha_1} |\chi(\alpha_1 + i\alpha_2)| = \chi(i\alpha_2)$.

Proof

$$\chi(\alpha_1 + i\alpha_2) = \int_{-\infty}^{\infty} e^{i\alpha_1 x} e^{-\alpha_2 x}\, dF(x);$$

hence $|\chi| \leq \int_{-\infty}^{\infty} |e^{-\alpha_2 x}|\, dF = \chi(i\alpha_2)$. To obtain the maximum value, we need only set $\alpha_1 = 0$.

The following theorem was proposed by Marcienkiewicz.

THEOREM 1

If a c.f. χ has the form $\chi(\lambda) = e^{P(\lambda)}$, where $P(\lambda)$ is a polynomial, then $P(\lambda)$ is of the second degree.

Proof

The function $\chi(z) = e^{P(z)}$ is an entire function; therefore, $L(\alpha) = \chi(-i\alpha)$ for real α is also an entire function. Since $L(\alpha)$ is a real function, the polynomial $P(-i\alpha)$ must be real; thus P has the form

$$P(\alpha) = \sum_{r=0}^{n} (i)^r a_r \alpha^r,$$

where the coefficients a_r are real numbers. From the fact that $\chi(0) = 1 = e^{a_0}$, it follows that $a_0 = 0$.

The degree n of our polynomial is not zero; we will show that n is an even integer. The corollary to Lemma 1 shows that $\lim_{\alpha\to\beta} [\alpha c + P(\alpha)] = \infty$, when $\beta = \pm\infty$, for some constant c. However, this is possible only if $n = 2s$ and $a_{2s} > 0$.

Next we will demonstrate that $s = 1$. Letting $z = \alpha_1 + i\alpha_2$, it follows from Lemma 2 that $\max_{\alpha_1} |\chi(z)| = \chi(i\alpha_2)$. Since $|\chi(z)| = e^{\mathscr{R}P(z)}$, where $\mathscr{R}$ denotes "the real part of" and $\chi(i\alpha_2) = e^{P(i\alpha_2)}$, then

$$\mathscr{R}P(\alpha_1 + i\alpha_2) \leq P(i\alpha_2).$$

If we substitute $\alpha_1 = \rho\cos\theta$, $\alpha_2 = \rho\sin\theta$, the above inequality reduces to

$$(-1)^s a_{2s}\rho^{2s}\cos 2s\,\theta + \cdots \leq a_{2s}\rho^{2s}\sin^{2s}\theta + \cdots$$

for every value of ρ and θ. Since ρ can be made arbitrarily large, the above inequality must hold for the first terms of each sum, and since $a_{2s} > 0$,

$$(-1)^s \cos 2s\theta \leq \sin^{2s}\theta. \tag{3}$$

When $\theta = 0$, $\cos\theta = 1$ and $\sin\theta = 0$; therefore, s must be odd in order for (3) to hold. If $s > 1$, then for $\theta = \pi/2s$, $\sin^{2s}\theta < 1$ and $(-1)^s\cos 2s\theta = 1$, and (3) is violated. However, if $s = 1$, then (3) always holds since $-\cos 2\theta = \sin^2\theta - \cos^2\theta \leq \sin^2\theta$.

If we set $a_1 = \mu$ and $a_2 = \frac{1}{3}\sigma^2$, then

$$P(\lambda) = i\mu\lambda - \frac{1}{2}\sigma^2\lambda^2$$

and χ becomes the c.f. of the normal distribution (cf., Example 7.1).

Considering the degenerate distribution as a special case of the normal distribution, we immediately obtain the following corollary from the theorem.

COROLLARY

If a c.f. χ has the form $\chi(\lambda) = e^{P(\lambda)}$, where $P(\lambda)$ is a polynomial, then χ is the c.f. of a normal distribution.

The following theorem was surmised by P. Lévy and proved by H. Cramer [5].

THEOREM 2

If F_1 and F_2 are both normal d.f.'s, then $F_1 * F_2$ is also a normal d.f. Conversely, if F is a normal d.f., then there exist two normal d.f.'s F_1 and F_2 such that $F = F_1 * F_2$.

Proof

The closure under composition is trivial; it was outlined in Example 7.1. The proof of the converse theorem is more involved. We can always make a linear change of variables so that the mean and variance of the d.f. are 0 and 1 respectively. In this situation, the c.f. $\chi(\lambda) = e^{-\lambda^2/2}$; by Theorem 7.2, the theorem reduces to finding two c.f.'s χ_1 and χ_2 that correspond to normal d.f.'s such that for every $\lambda \in R$, $\chi_1\chi_2 = e^{-\lambda^2/2}$. Hence $|\chi_1|\,|\chi_2| \leq e^{|\lambda|^2/2}$ or $\log|\chi_1| + \log|\chi_2| \leq |\lambda|^2/2$ and so $\log|\chi_1| \leq |\lambda|^2/2$. By an extension of Liouville's theorem,[14] $\log \chi_1$ is a polynomial in λ of at most the second degree. Hence $\chi_1 = e^{P(\lambda)}$, where $P(\lambda)$ is a polynomial, and so by the corollary to Theorem 1, χ_1 is the c.f. of a normal distribution. The same proof is valid for χ_2.

Theorem 2 can be proved directly without using Theorem 1 (c.f., Exercise 1).

Example 7. 1 showed that the convolution of two Poisson distributions

[14]See E. C. Titchmarsh, *The Theory of Functions* (New York: Oxford University Press, Inc., 1939), p. 85.

was a Poisson distribution. To state the decomposition theorem derived by Raikov [6], we denote the d.f. of the Poisson distribution described in Example 7. 1 by $F(h, b, \mu)$ to indicate the pertinent parameters.

THEOREM 3

If $F_1 * F_2 = F(h, b, \mu)$, then there exist $\mu_1, \mu_2 > 0, b_1, b_2$ such that $\mu = \mu_1 + \mu_2$, $b = b_1 + b_2$, $F_1 = F(h, b_1, \mu_1)$ and $F_2 = F(h, b_2, \mu_2)$.

Proof

By a change of variables, we can reduce the theorem to the situation in which $h = 1$ and $b = 0$. To prove this simpler result, we first study the relation between the points of increase of F_1 and F_2 and $F(h, b, \mu)$.

Let r_1 and r_2 be two independent r.v.'s with d.f.'s F_1 and F_2. We will use two facts that are generalizations of the corresponding notions for discrete-valued r.v.'s—first, that the d.f. F of their sum is $F = F_1 * F_2$ and second, that $P(a_1 < r_1 \leq b_1)P(a_2 < r_2 \leq b_2) = P(a_1 < r_1 \leq b_1, a_2 < r_2 \leq b_2)$.[15] Since

$$(a_1 < r_1 \leq b_1) \cap (a_2 < r_2 \leq b_2) \subset (a_1 + a_2 < r_1 + r_2 \leq b_1 + b_2)$$

and r_1 and r_2 are independent, it follows by taking the probability of both sides of the inequality and using the definition of the d.f. that

$$[F_1(b_1) - F_1(a_1)][F_2(b_2) - F_2(a_2)] \leq F(b_1 + b_2) - F(a_1 + a_2). \tag{4}$$

Suppose that λ_1 and λ_2 are points of increase of F_1 and F_2 respectively. If $\lambda_1 \in (a_1, b_1), \lambda_2 \in (a_2, b_2)$ and the left-hand side in (4) is positive; then $\lambda_1 + \lambda_2 \in (a_1 + a_2, b_1 + b_2)$ is a point of increase of F. Letting $b_1, b_2 \to \infty$ in (4) yields, after some simplification,

$$F(a_1 + a_2) \leq F_1(a_1) + F_2(a_2). \tag{5}$$

Now let λ_1 and λ_2 be the first points of increase of F_1 and F_2 respectively. If $a_1 < \lambda_1$ and $a_2 < \lambda_2$, then the right-hand side of (5) is zero. Hence $F(a_1 + a_2) = 0$ and therefore $\lambda_1 + \lambda_2$ is the first point of increase of F.

Let $F = F(1, 0, \mu)$; its points of increase are $k = 0, 1, 2, \ldots$. By the results of the preceding paragraph, all points of increase λ_1 and λ_2 of F_1 and F_2 are such that $\lambda_1 + \lambda_2 =$ some k. To find the first points of increase of F_1 and F_2, we set $k = 0$ and obtain $\lambda_1 = \lambda$ and $\lambda_2 = -\lambda$. Replacing $F_1(x)$ by $F_1(x - \lambda)$ and $F_2(x)$ by $F_2(x + \lambda)$, it follows that the new functions both have zero as a point of increase. Since this change of variables does not affect F, F still has the same points of increase. If λ_1 and λ_2 are the points of increase of the new d.f.'s, then we still have $\lambda_1 + \lambda_2 =$ some k. Suppose that λ_1 is not equal to a nonnegative integer; then $\lambda_1 + 0$ is a point of in-

[15] A justification of these statements appears in the following chapter.

crease of F. This is a contradiction; thus the new d.f.'s have $k = 0, 1, 2, \ldots$ as the only possible points of increase. The corresponding c.f.'s therefore have the form

$$\chi_1(\lambda) = \sum_{k=0}^{\infty} p_k e^{i\lambda k} \qquad \chi_2(\lambda) = \sum_{k=0}^{\infty} q_k e^{i\lambda k},$$

where

$$p_0, q_0 > 0, p_k, q_k \geq 0 \qquad (k = 1, 2, 3, \ldots)$$

$$\sum_{k=0}^{\infty} p_k = \sum_{k=0}^{\infty} q_k = 1.$$

Let $z = e^{i\lambda}$, $\phi_1(z) = \chi_1(\lambda)$, and $\phi_2(z) = \chi_2(\lambda)$. Since $\chi_1 \chi_2$ is the c.f. of F,

$$\phi_1(z)\phi_2(z) = \sum_{k,l=0}^{\infty} p_k q_l z^{k+l} = \sum_{k=0}^{\infty} \frac{\mu^k e^{-\mu}}{k!} z^k.$$

Therefore, equating coefficients,

$$p_0 q_k + p_1 q_{k-1} + \cdots + p_k q_0 = \frac{\mu^k e^{-\mu}}{k!}. \qquad (k = 0, 1, 2, \ldots)$$

It follows that

$$p_k \leq \frac{1}{q_0} \frac{\mu^k e^{-\mu}}{k!} \qquad |\phi_1(z)| \leq \frac{1}{q_0} e^{\mu(|z|-1)};$$

a similar result holds for ϕ_2. Since their product is an entire function without zeros, ϕ_1 and ϕ_2 are entire functions without zeros. Hence $\log |\phi_1(z)| \leq c_1 + c_2|z|$, which implies that $\log \phi_1(z)$ is a polynomial in z of at most the first degree by Liouville's theorem. Then $\chi_1(\lambda) = e^{\alpha + \beta e^{i\lambda}}$ and, since $\chi_1(0) = 1$, $\alpha = -\beta$. From the fact that $|\chi_1(\lambda)| = |e^{\beta(e^{i\lambda}-1)}| \leq 1$, it follows that $\beta \geq 0$. A similar result holds for χ_2. Therefore, χ_1 and χ_2 are c.f.'s of Poisson distributions.

EXERCISES

1. Prove Theorem 3 without using Theorem 1. [*Hint:* Use the facts that $\chi_1(0) = 1$, $|\chi_1(\lambda)| \leq 1$, and $\chi_1(\lambda) = \overline{\chi_1(-\lambda)}$.]

2. Show that if r_1 and r_2 are independent r.v.'s such that $r_1 + r_2$ and $r_1 - r_2$ are independent of each other, then r_1 and r_2 have normal distributions with the same variance. [*Hint:* Reduce the problem to the special case where r_1 and r_2 have a common symmetric distribution. Let χ_1, where χ_1 is real, be their common c.f. Show that χ satisfies the functional equation $\chi(\lambda_1 + \lambda_2)\chi(\lambda_1 - \lambda_2) = \chi^2(\lambda_1)\chi^2(\lambda_2)$ and solve it.]

10.9 NONNEGATIVE DEFINITENESS

The purpose of this section is to characterize c.f.'s; in other words, is there some criterion by which we can determine whether a given function ϕ is a c.f.? Obviously, ϕ must be continuous and bounded, and $\phi(0) = 1$.[16] If these conditions are satisfied, we can apply Lévy's inversion formula and see whether $\Delta F^* \geq 0$ for all pairs $a < b$ of finite numbers. A more aesthetically pleasing criterion is a consequence of the following property of c.f.'s.

THEOREM 1

Let ϕ be a c.f.; then

$$\sum_{r=1}^{n} \sum_{s=1}^{n} \phi(\lambda_r - \lambda_s)\xi_r\bar{\xi}_s \leq 0 \tag{1}$$

for any real numbers $\lambda_1, \ldots, \lambda_n$ and any complex numbers $\xi_1, \ldots, \xi_n$ and for any $n = 1, 2, 3, \ldots$.

Proof

Replacing ϕ in (1) by its definition and interchanging the order of summation and inegration gives

$$\int_{-\infty}^{\infty} \left(\sum_{r=1}^{n} \sum_{s=1}^{n} e^{i(\lambda_r - \lambda_s)x}\xi_r\bar{\xi}_s\right) dF = \int_{-\infty}^{\infty} \left|\sum_{r=1}^{n} e^{i\lambda_r x}\xi_r\right|^2 dF \geq 0.$$

DEFINITION 1

A real- or complex-valued function $\phi(\lambda)$ defined on R is said to be *nonnegative definite* if it satisfies condition (1).

Such functions are also called *positive semidefinite*. Recalling the definition of a positive semidefinite matrix, we see that condition (1) can be replaced by the statement that the matrix $(\phi(\lambda_r - \lambda_s))$ is positive semidefinite for $r, s = 1, 2, \ldots, n$. Other equivalent definitions of nonnegative definite functions appear in Exercise 1.

Positive semidefinite functions have the following properties.

LEMMA 1

If $\phi(\lambda)$ is nonnegative definite, then

1. $\phi(0) \geq 0$.

[16] This condition is not necessary if we consider general d.f.'s.

2. $\phi(-\lambda) = \overline{\phi(\lambda)}$.
3. $|\phi(\lambda)| \leq \phi(0)$.
4. $|\phi(\lambda) - \phi(\mu)| \leq \sqrt{2\phi(0)[\phi(0) - \mathscr{R}\phi(\lambda - \mu)]}$.

Proof

1. If we take $n=1, \lambda_1=0$, and $\xi_1=1$ in Equation (1), we obtain $\phi(0) \geq 0$.
2. Letting $n = 2, \lambda_1 = \lambda, \lambda_2 = 0, \xi_1 = 1$, and $\xi_2 = \xi$, Equation (1) reduces to

$$\phi(0)(1 + |\xi|^2) + \phi(\lambda)\bar{\xi} + \phi(-\lambda)\xi \geq 0. \tag{2}$$

Since $\phi(0)(1 + |\xi|^2)$ is real, we must have, for all real λ and complex ξ,

$$\mathscr{I}(\phi(\lambda)\bar{\xi} + \phi(-\lambda)\xi) = 0. \tag{3}$$

For $\xi = 1$ in (3),

$$\mathscr{I}(\phi(\lambda) + \phi(-\lambda)) = 0$$

or $\mathscr{I}\phi(\lambda) = -\mathscr{I}\phi(-\lambda)$, and for $\xi = i$ in (3),

$$\mathscr{I}(-i\phi(\lambda) + i\phi(-\lambda)) = \mathscr{R}(-\phi(\lambda) + \phi(-\lambda)) = 0$$

or $\mathscr{R}\phi(-\lambda) = \mathscr{R}\phi(\lambda)$; accordingly, $\phi(-\lambda) = \overline{\phi(\lambda)}$.

3. Let $\xi = -\eta/\overline{\phi(\lambda)}$, where $\eta > 0$, in (2); then (2) reduces to

$$\frac{\phi(0)}{|\phi(\lambda)|^2}\eta^2 - 2\eta + \phi(0) \geq 0.$$

For this inequality to hold, the discriminant must not be positive; that is, $4 - 4\phi^2(0)/|\phi(\lambda)|^2 \leq 0$, which yields the desired result after a slight simplification.

4. Equation (1), with $n = 3, \lambda_1 = 0, \lambda_2 = \lambda, \lambda_3 = \mu, \xi_1 = 1, \xi_2 = \xi$, and $\xi_3 = -\xi$, yields

$$\phi(0) + 2\mathscr{R}[\phi(\lambda)\xi - \phi(\mu)\xi] + [2\phi(0) - 2\mathscr{R}\phi(\lambda-\mu)]|\xi|^2 \geq 0. \tag{4}$$

Now let $\xi=re^{i\theta}$ in Equation (4); choose θ so that $\mathscr{R}[\phi(\lambda)\xi - \phi(\mu)\xi] = |\phi(\lambda) - \phi(\mu)|r$. Therefore, Equation (4) reduces to

$$\phi(0) + 2|\phi(\lambda) - \phi(\mu)|r + 2[\phi(0) - \mathscr{R}\phi(\lambda - \mu)]r^2 \geq 0.$$

To satisfy this inequality, the discriminant must not be positive or

$$|\phi(\lambda) - \phi(\mu)|^2 \leq 2\phi(0)[\phi(0) - \mathscr{R}\phi(\lambda - \mu)],$$

which is equivalent to property 4 of the lemma.

Property (3) implies that if a nonnegative function is zero at the origin, it is always zero, while property 4 implies that if ϕ is continuous at 0, then ϕ is uniformly continuous on R. This last statement follows from the observation that

$$|\phi(0) - \mathscr{R}\phi(\lambda)| = |\mathscr{R}(\phi(0) - \phi(\lambda))| \leq |\phi(0) - \phi(\lambda)|.$$

Since positive semidefinite functions have many of the properties of c.f.'s, we might conjecture the following theorem discovered by Bochner (c.f., Theorem 5.4.4).

THEOREM 2

In order for a continuous function χ which satisfies the condition $\chi(0) = 1$ to be a c.f., it is necessary and sufficient that it be nonnegative definite.

Proof

If χ is a c.f., then it is nonnegative definite by Theorem 1. Conversely, suppose that we have a nonnegative definite continuous function χ such that $\chi(0) = 1$. If χ were absolutely integrable, we could define a logical candidate for a density function by using the inversion formula. However, we do not know if $\chi(t)$ is absolutely integrable and so we modify the function by multiplying it by the c.f. $1 - t/T$ for $(|t| \leq T)$ and 0 for $(|t| > T)$. Accordingly, define

$$p_T(x) = \frac{1}{2\pi}\int_{-\infty}^{\infty} \phi_T(t)e^{-itx}\,dt = \frac{1}{2\pi}\int_{-T}^{T}\left(1 - \frac{|t|}{T}\right)\chi(t)e^{-itx}\,dt, \tag{5}$$

where

$$\phi_T(t) = \begin{cases} \left(1 - \dfrac{|t|}{T}\right)\chi(t) & (|t| \leq T) \\ 0 & (|t| > T) \end{cases} \tag{6}$$

is absolutely integrable over the real line. We will show that $p_T(x)$ is a density function. Now

$$p_T(x) = \frac{1}{2\pi T}\int_0^T\int_0^T \chi(\lambda - \mu)e^{-i(\lambda-\mu)x}\,d\lambda\,d\mu,$$

which can easily be seen by making the change of variables $\lambda = t + \mu$ in the last double integral and integrating with respect to μ. Since χ is nonnegative definite and continuous on R, the double integral can be written as a limit of nonnegative Riemann sums and so $p_T(x) \geq 0$. To show that $\int_{-\infty}^{\infty} p_T(x)\,dx = 1$, invert the integral in (5) by multiplying by $e^{i\lambda x}$ and integrating over $[-v, v]$; this yields

$$\int_{-v}^{v} p_T(x)e^{i\lambda x}\,dx = \frac{1}{\pi}\int_{-\infty}^{\infty}\frac{\sin(\lambda - t)}{(\lambda - t)}\phi_T(t)\,dt. \tag{7}$$

Taking limits as $v \to \infty$ in (7) gives

$$\int_{-\infty}^{\infty} p_T(x) e^{i\lambda x}\, dx = \phi_T(\lambda) \tag{8}$$

since

$$\lim_{x \to \infty} \int_{-\infty}^{\infty} \frac{\sin(\lambda - t)x}{(\lambda - t)} g(t)\, dt = g(\lambda) \tag{9}$$

(see Exercise 2). Since $1 = \phi_T(0) = \int_{-\infty}^{\infty} p_T(x)\, dx$, $p_T(x)$ is a density function and $\phi_T(\lambda)$ is its corresponding c.f. As $T \to \infty$, ϕ_T converges uniformly to the function $\chi(t)$ in every finite interval $-T \leq t \leq T$; this is apparent from the defining relation (6). It follows from this by the corollary of Theorem 6.5 that χ is a c.f.

Observe that (8) would follow directly from (5) by the inversion formula if we knew that $p_T(x)$ was integrable on R and hence was a general density function. However, this fact is not obvious; we will outline its proof in Exercise 3.

EXERCISES

1. Show that the following are equivalent definitions of nonnegative definiteness:

(a) $\sum_{r=1}^{n} \sum_{s=1}^{n} (\lambda_r - \lambda_s) h(\lambda_r) \overline{h(\lambda_s)} \geq 0$ for every real- or complex-valued function h, where $\lambda_1, \ldots, \lambda_n$ are real numbers and $n = 1, 2, 3, \ldots$.

(b) $\int_{-\infty}^{\infty} \int_{-\infty}^{\infty} \phi(u - v) h(u) \overline{h(v)}\, du\, dv > 0$ for every real- or complex-valued measurable function h, whenever the integral exists.

(c) $\int_{-\infty}^{\infty} \int_{-\infty}^{\infty} \phi(u) h(u + v) \overline{h(v)}\, du\, dv \geq 0$ for every real- or complex-valued measurable function $h(u)$, whenever the integral exists.

(d) $\int_{-\infty}^{\infty} \phi(u) \eta(u)\, du \geq 0$ and $\eta(u) = \int_{-\infty}^{\infty} h(u + v) \bar{h}(v)\, dv$ for every real- or complex-valued measurable function h, whenever the integral exists.

2. Verify Equation (9). [*Hint:* Show that $\lim_{x \to \infty} \int_{-\infty}^{\infty} \sin(\lambda - t) x g_\lambda(t)\, dt = 0$, where

$g_\lambda = [g(t) - g(\lambda)]/(\lambda - t)$, by integrating by parts, using the assumption that g is differentiable; if not, approximate it by an absolutely continuous function. This result is known as the Riemann–Lebesgue lemma.]

3. (a) *Linnik.* Show that if the function $f(t)$ is measurable, bounded, and summable in the interval $(-T, T)$, and

$$p(x) = \int_{-T}^{T} e^{-i\lambda x} f(t)\, dt \geq 0,$$

then the function $p(x)$ is integrable over the entire line. [*Hint:* Let $G(x) = \int_{-x}^{x} p(t)\, dt$. To show that $G(x)$ is bounded, consider $F(x) = (1/x) \int_{x}^{2x} G(u)\, du$. Show that $G(x) = 2 \int_{-T}^{T} [(\sin xt)/t] f(t)\, dt$, $F(x) \geq G(x)$, and $F(x)$ is bounded.]

(b) Complete the proof of Theorem 2 as outlined in the last paragraph of this section.

4. (a) Define $D_s = (x - y | x \in S, y \in S, S$ is a set of real numbers). Show that if $S = (0, a)$, then $D_s = (-a, a)$, and if S equals the set of all positive integers, then D_s equals all integers.

(b) We say that ϕ on D_s is nonnegative definite if (1) holds for ϕ, where $\lambda_i \in S$ for $i = 1, 2, \ldots, n$. Show that if ϕ is nonnegative definite on D_s, then for every $\lambda \in D_s$,

$$\phi(0) \geq 0 \qquad \phi(-\lambda) = \overline{\phi(\lambda)} \qquad |\phi(\lambda)| \leq \phi(0).$$

(c) Prove that that if $D_s \supset (-a, a)$ and ϕ is continuous at the origin, then ϕ is continuous at every limit point of D_s.

5. (a) *Herglotz Lemma.* A function ϕ on the set $D_s = \{\ldots, -2c, -c, 0, c, 2c, \ldots\}$ is nonnegative definite if and only if it coincides on this set with a c.f. $\chi(\lambda) = \int_{-\pi/c}^{\pi/c} e^{i\lambda x}\, dF(x)$. [*Hint:* Show that

$$f_n = \frac{1}{2\pi} \sum_{k=n+1}^{n-1} \left(1 - \frac{|k|}{n}\right) \phi(kc) e^{-ikx}$$

is a density on $(-\pi, \pi)$.]

(b) Show that the Herglotz lemma remains valid, with $D_s = \{-Nc, \ldots, -c, 0, c, \ldots, Nc\}$ for all fixed integers N.

6. Prove Theorem 2 by using the Herglotz lemma and verifying the following chain of reasoning:

(a) A nonnegative definite continuous function ϕ on R such that $\phi(0) = 1$, which coincides with a c.f. on R, does so on the set S of all rationals of the form $k/2^n$, where $k = 0, \pm 1, \pm 2, \ldots,$ and $n = 1, 2, 3, \ldots$.

(b) Show that there exist c.f.'s χ_n such that $\phi(k/2^n) = \chi_n(k/2^n)$ for all k and n, and $\chi_n \to \phi$ on S.

(c) Show that $1 - \mathscr{R}\chi_n(\theta/2^n) \leq 1 - \mathscr{R}\phi(1/2^n)$, where $0 \leq \theta \leq 1$; by using the inequality $|a+b|^2 \leq 2(|a|^2 + |b|)^2$ and part 4 of Lemma 1, verify that for $0 \leq \theta_n \leq 1$,

$$\left|1 - \chi_n\left(\frac{k_n + \theta_n}{2^n}\right)\right|^2 = \left|1 - \chi_n\left(\frac{k_n}{2^n}\right) + \chi_n\left(\frac{k_n}{2^n}\right)\left(1 - \chi_n\left(\frac{\theta_n}{2^n}\right)\right)\right|^2$$

$$\leq 2\left|1 - \phi\left(\frac{k_n}{2^n}\right)\right| + 4\left(1 - \mathscr{R}\phi\left(\frac{1}{2^n}\right)\right).$$

(d) Use Corollary 2 of Theorem 2.2 to show that $\chi_n \to \chi$, a c.f. on S, and so $\phi = \chi$.

7. *Krein.* A function ϕ on $(-a, a)$ is nonnegative definite and continuous, and $\phi(0) = 1$, if and only if it coincides on $(-a, a)$ with a c.f. [*Hint:* Use Exercise 5(b) and replace S by its intersection with $(-a, a)$ in (6).]

8. *Riesz.* A function ϕ on R is nonnegative definite and Lebesgue measurable if and only if it coincides a.e. with a c.f. [*Hint:* Use Exercise 1(b) with $h(u) = e^{iux}$; proceed as in the proof of Bochner's theorem but use Theorem 6.2 to complete the proof.]

9. A function ϕ defined on $(0, \infty)$ is *completely monotone* if it possesses derivatives of all orders and if $(-1)^n\phi^{(n)}(\lambda) \geq 0$ for $\lambda > 0$. Show that $1/\lambda$ and $1/(1+\lambda)$ are completely monotone.

10. Prove the following statements:

(a) Let ϕ be completely monotone and let f be absolutely monotone on $(\phi(0), \phi(\infty))$ (i.e., all derivatives $f^{(n)}$ exist and are positive); then $f(\phi)$ is completely monotone.
(b) If ϕ and ψ are completely monotone, then so is their product.
(c) If $g \geq 0$ and g' is completely monotone, then so is $\phi(g)$ for any completely monotone function ϕ. (*Hint:* Use an inductive argument.)

11. *Bernstein.* A function ϕ is completely monotone on $(0, \infty)$ if and only if there exists a d.f. F (possibly unbounded) such that for all $\lambda > 0$, $\phi(\lambda) = \int_\sigma^\infty e^{-\lambda t}\, dF(t)$.

(*Hint:* See [7].)

REFERENCES

1. Lagrange, J. L., "Mémoire sur l'utilité de la méthode de prendre le milieu entre les résultats de plusieurs observations," *Misc.*, Vol. 5 (1770–1773).
2. Laplace, P. S., *Théorie Analytique des Probabilités.* Paris, 1812.
3. Lévy, P., *Calcul des Probabilités.* Paris, 1925.
4. Krein, M. G., "On representing functions by means of Fourier–Stieltjes Integrals," *Uchën. Zap Kuibysh. Ped. In-ta.*, No. 7 (1943).

5. Cramer, H., "Uber eine Eigenschaft der normalen Verteilungsfunktion," *Math. Zeitschrift*, **41** (1936).
6. Raikov, D. A., "On the decomposition of the Gaussian and Poisson Laws," *Izvestiya Akad. Nauk SSSR,* Ser. Mat., (1923), 91–124.
7. Tamarkin, J. D., "On a theorem of S. Bernstein-Widder," *Trans. Am. Math. Soc.*, **33** (1931).
8. Bochner, S., "Monotone Funktionen, Stieltjes Integrals, und harmonische Analyse," *Math. Ann.*, **108** (1933).
9. Marcienkiewicz, "Sur les fonctions indépendantes," *Fund. Math.*, **31** (1939).
10. Marcienkiewicz, "Sur une propriété de la loi de Gauss," *Math. Zeit.*, **44** 1939).
11. Polya, G., "Remark on characteristic functions," *First Berkeley Symp. on Stat. and Prob.* (1949).
12. Zygmund, A., "A remark on characteristic functions," Ann. Math. Stat. **18** (1947).
13. Lukacz, E., *Characteristic Functions,* Griffin's Statistical Monographs, No. 5 (1960).
14. Riordan, J., *An Introduction to Combinatorial Analysis*, John Wiley, New York (1958).
15. Linnik Yu. V., *Decomposition of Probability Distributions,* Dover Publications, New York (1964).

CHAPTER 11

INDEPENDENCE

11.1 INTRODUCTION

In Section 3.3 we introduced the concept of independence for a finite number of events and for simple r.v.'s defined in finite probability spaces. In this chapter we will introduce a natural generalization of this notion to an arbitrary number of events and random functions defined in arbitrary probability spaces. As a byproduct of these definitions, we will obtain a simple and useful criterion for independence in terms of d.f.'s and c.f.'s. Next we will investigate the connection between independence and product spaces, generalizing the results of Section 3.4 for repeated independent trials. Finally, we will study some fundamental properties of sequences of independent r.v.'s.

11.2 INDEPENDENT CLASS

Throughout this chapter, events belong to an arbitrary probability space $(\Omega, \mathscr{E}, P)$; unless otherwise specified, the indexes i are used to distinguish events belonging to a fixed but otherwise arbitrary index set I. For an arbitrary index set I, Definition 3.3.2 generalizes to the following definition.

DEFINITION 1

An arbitrary number of events E_i are *independent* if and only if

$$P \bigcap_{k=1}^{n} E_{i_k} = \prod_{k=1}^{n} PE_{i_k} \tag{1}$$

for every finite subset $(i_1, \ldots, i_n)$ of I.

Sometimes, instead of referring to the index set explicitly, we will say that events in a class $\mathscr{C}$ are independent iff (1) holds for every finite set of events chosen from $\mathscr{C}$.

We recall that it is not sufficient for the independence of the sets E_i for any two distinct sets of $\{E_i\}$ (or some fixed number of the sets) to be independent, even if the index set is finite.

Similarly, Definition 3.3.3 generalizes to the following definition.

DEFINITION 2

Classes $\mathscr{C}_i$ of events are independent if events selected arbitrarily, one from each class, are independent.

In other words, equation (1) holds, where $E_{i_k} \in \mathscr{C}_{i_k}$. *Unless otherwise stated, we will continue to denote the events of a class by using the index of the class.*

LEMMA 1

Let $\mathscr{C}_i$ be independent classes and let $\overline{\mathscr{C}}_i \subset \mathscr{C}_i$ for $i \in J \subset I$. Then the classes $\overline{\mathscr{C}}_i$ are also independent.

The proof is a direct consequence of Definition 2.

We have seen, in Theorem 3.3.2, that all "Boolean functions" of independent events are independent. Thus there are some operations on independent classes that will enlarge these classes without destroying independence.

LEMMA 2

Independent classes $\mathscr{C}_i$ remain independent if to each class we adjoin

1. The almost sure and null events.
2. The complements of its elements.
3. The countable sums of its elements.
4. The limits of sequences of its elements.

Proof

1. If an event A is almost sure, then $P(AB) = P(B)$ for any event B. Now suppose that $E_{i_k}(k = 1, \ldots, r)$ are almost sure events; then

$$P\bigcap_{k=1}^{n} E_{i_k} = P\bigcap_{k=r+1}^{n} E_{i_k} = \prod_{k=r+1}^{n} PE_{i_k} = \prod_{k=1}^{n} PE_{i_k}.$$

On the other hand, if some of the events appearing in equation (1) are null, then both sides reduce to zero.

2. Since only a finite number of events are involved, the proof follows from Theorem 3.3.1.
3. For convenience of notation, assume that the subscripts in (1) are the first n natural numbers. The result follows from the definition and the identity

$$P\left(\left(\sum_j E_1^j\right)E_2 \cdots E_n\right) = P\left(\sum_j (E_1^j E_2 \cdots E_n)\right)$$
$$= \sum_j P(E_1^j E_2 \cdots E_n) = \sum_j (PE_1^j)PE_2 \cdots PE_n$$
$$= P\left(\sum_j E_1^j\right) PE_2 \cdots PE_n.$$

The details for more than one infinite sum are left to the reader.
4. Suppose that $E_1^m \to E_1$ as $m \to \infty$; then

$$P(E_1 E_2 \cdots E_n) = P(\lim_m E_1^m E_2 \cdots E_n) = \lim_m P(E_1^m E_2 \cdots E_n)$$
$$= \lim_m PE_1^m PE_2 \cdots PE_n = PE_1 PE_2 \cdots PE_n.$$

The proof is completed by using induction on the number of limit events when there are more than one.

COROLLARY

If the $\mathscr{C}_i$ are independent and closed under finite intersections, then countable unions of elements from each $\mathscr{C}_i$ can be adjoined to $\mathscr{C}_i$ and the resulting classes remain independent.

The above results enable us to prove the following extension theorem.

THEOREM 1

Let $\mathscr{C}_i$ be independent classes that are closed under finite intersections. Then the minimal σ-fields, $\mathscr{S}(\mathscr{C}_i)$, over the classes $\mathscr{C}_i$ are also independent.

Proof

The definition of independence for classes shows that we only must prove the theorem for a finite number of classes $\mathscr{C}_i$ $(i = 1, 2, \ldots, n)$. Let $\mathscr{E}_1$ be the largest class of events containing $\mathscr{C}_1$ that is closed under finite intersections such that $\mathscr{E}_1, \mathscr{C}_2, \ldots, \mathscr{C}_n$ are independent. Since finite intersections of countable unions of elements of $\mathscr{E}_1$ can be written as a countable union of elements of $\mathscr{E}_1$, it follows from the previous lemma and its corollary that $\mathscr{E}_1$ is a σ-field. Hence $\mathscr{E}_1 \supset \mathscr{S}(C_1)$ and by Lemma 1, $\mathscr{S}(C_1), C_2, \ldots, C_n$ are independent. A repetition of this procedure yields the result.

EXAMPLE 1

1. The minimal σ-field over the event A_i is $\mathscr{C}_i = \{A_i, A_i', \Omega, \emptyset\}$. Hence if the events A_i are independent, so are the σ-fields $\mathscr{C}_i$.
2. Let r_i be a set of r.v.'s and let $\mathscr{C}_i = (r_i^{-1}(-\infty, x] | x \in R)$. If the $\mathscr{C}_i$ are independent, then so are the classes $r_i^{-1}(\mathscr{B})$, where $\mathscr{B}$ is the Borel field in R. This is true because $\mathscr{B}$ is the minimal σ-field over the class of all intervals of the form $(-\infty, x]$, $\mathscr{S}(\mathscr{C}_i) = r_i^{-1}(\mathscr{B})$, and every $\mathscr{C}_i$ is closed under finite intersections.

DEFINITION 3

Let $\mathscr{S}_i$ be σ-fields. The *compound σ-field* $\mathscr{S}_I$ is the minimal σ-field over the class $\mathscr{C}_I$ of all finite intersections of events $E_i, i \in I$. The $\mathscr{S}_i$ are called *components* of the compound σ-field $\mathscr{S}_I$. If $J \subset I$, then we say that the compound σ-field $\mathscr{S}_J$, with components $\mathscr{S}_j, j \in J$, is a *compound sub-σ-field* of $\mathscr{S}_I$. In particular, if J is a finite index set, $\mathscr{S}_J$ is called a *finitely compound sub-σ-field*. Alternatively, we can define the compound σ-field as the minimal σ-field containing all the $\mathscr{S}_i$. The two definitions are equivalent.

LEMMA 3

Suppose that $\mathscr{S}_I$ is a compound σ-field with component σ-fields $\mathscr{S}_i$. Let $\mathscr{C}$ be the class of all events E that belong to some finitely compound sub-σ-field. Then the minimal σ-field over $\mathscr{C}$ is $\mathscr{S}_I$; moreover, $\mathscr{C}$ is closed under finite intersections.

The first assertion follows from the fact that $\mathscr{S}_I \supset \mathscr{C} \supset \mathscr{C}_I$ and so $\mathscr{S}_I \supset \mathscr{S}(\mathscr{C}) \supset \mathscr{S}(\mathscr{C}_I)$. The remainder of the proof is left as an exercise for the reader.

THEOREM 2

A necessary and sufficient condition for compound σ-fields to be independent is that their corresponding finitely compound sub-σ-fields must be independent.

Proof

The necessity of the condition follows immediately from Lemma 1. Now suppose that the finitely compound sub-σ-fields are independent. For each compound σ-field, form the class $\mathscr{C}$ described in Lemma 3. These classes are closed under finite intersections. Moreover, they are independent, since the elements in each $\mathscr{C}$ are just members of some finitely compound sub-σ-field.

Hence, by Theorem 1, the minimal σ-fields over these classes are independent; by Lemma 3, these minimal σ-fields are just the compound σ-fields.

EXERCISES

1. (a) Prove the corollary to Lemma 2. (*Hint*: Write the union as a sum.)
(b) Show that independence is preserved if, to every $\mathscr{C}_i$, we adjoin the proper difference of its elements.

2. (a) Show that the class $\mathscr{C}_I$ of Definition 3 is closed under finite intersections.
(b) Supply the missing details in the proof of Lemma 3.

3. Can Theorem 1 be proved by giving a "constructive" procedure for obtaining $\mathscr{S}(\mathscr{C}_i)$?

4. Let Ω be the unit square with Lebesgue measure, let $E_1 = ((x, y)|0 \le x \le 1, a \le y \le b)$, and let $E_2 = ((x, y)|c \le x \le d, 0 \le y \le 1)$, where a, b, c, and d are arbitrary numbers in the closed unit interval. Show that E_1 and E_2 are independent.

11.3 INDEPENDENT FUNCTIONS

We begin by defining independent random functions indirectly by using our definition of independent classes.

DEFINITION 1

Let r_i be random functions and let $\mathscr{S}(r_i)$ be the inverse image under r_i of the Borel field in the range space of r_i. The r_i are said to be *independent* if they induce independent σ-fields $\mathscr{S}(r_i)$.

Of course, independence for a set of r.v.'s or random vectors, which are random functions, is covered by this definition. More explicitly, the definition remains valid if the phrases "r.v.'s" or "random vectors" are substituted for the phrase "random functions."

Frequently a more direct definition is useful.

DEFINITION 2

Random variables r_i, $i \in I$ are independent if and only if for every finite class $\{B_{i_1}, \ldots, B_{i_n}\}$ of Borel sets in R,

$$P \bigcap_{k=1}^{n} (r_{i_k} \in B_{i_k}) = \prod_{k=1}^{n} P(r_{i_k} \in B_{i_k}). \tag{1}$$

If the r_i represent the random vectors $(r_{i1}, \ldots, r_{im_i})$ and if B_i is correspondingly an m_i-dimensional Borel set, the last statement furnishes the definition of independence of random vectors. Similarly, random functions r_i are said to be independent if and only if (1) holds, where B_i are Borel sets in the range space of r_i.

Equivalently, we can require (1) to hold for any class of sets $\mathscr{C}_i$ that are closed under finite intersections and that generate the Borel fields in R_i—for example, open sets, semi-infinite intervals $(-\infty, b]$, etc. The proof of this statement is similar to that used in part 2 of Example 2.1.

The equivalence of the two definitions for r.v.'s and random vectors follows directly from the definition of independence for classes. It is more difficult to show that Definitions 1 and 2 are equivalent for random functions; we require the following result.

THEOREM 1

The random functions r_i are independent, according to Definition 1, if and only if the finite random vectors v_i are independent, where each v_i is an arbitrary but fixed finite selection of components of r_i.

Proof

Let $r_i = \{r_{i,t}, t \in T_i\}$, where $r_{i,t}$ are r.v.'s. The random functions r_i induce σ-fields $\mathscr{S}(r_i)$ of events, where every $\mathscr{S}(r_i)$ is the minimal σ-field over the class $\mathscr{C}(r_i)$ of inverse images of all intervals in the range space R_i of r_i. By definition, the intervals in R_i, the range space of r_i are of the form

$$B = \mathop{\times}_{k=1}^{n} B_{t_k}\left(\mathop{\times}_{t \neq t_k} R_{i,t}\right),$$

where B_{t_k} is an interval (proper) in the range space R_{i,t_k} of r_{i,t_k}, and n is a finite number.

Now

$$r_i^{-1}(B) = \bigcap_{k=1}^{n} r_{i,t_k}^{-1}(B_{t_k})$$

and so $r_i^{-1}(B)$ is formed from the intersection of elements of $\mathscr{S}(r_{i,t_k})$, where $k = 1, \ldots, n$. Therefore, the sets in $\mathscr{C}(r_i)$ are just the finite intersections of sets of $\mathscr{S}(r_{i,t})$. Consequently, $\mathscr{S}(r_i)$ is a compound field with components $\mathscr{S}(r_{i,t})$ and the theorem follows by an application of Theorem 2.2.

THEOREM 2

Definition 1 and Definition 2 are equivalent.

Proof

We have seen that Definitions 1 and 2 are equivalent for random vectors; therefore, the result follows from Theorem 1.

THEOREM 3

Let g_i be Borel functions defined on the range of the random functions r_i. If the r_i are independent, then so are $g_i(r_i)$.

Proof

The result follows from Lemma 2.1 upon using Definition 1 of independence by noting that a Borel function g_i of r_i induces a sub-σ-field of events contained in $\mathscr{S}(r_i)$.

An interesting example of independent functions on $\Omega = ((x, y)|0 \leq x \leq 1,\ 0 \leq y \leq 1)$ with ordinary Lebesgue measure are the functions $r_1(x_1, x_2) = x_1$ and $r_2(x_1, x_2) = x_2$.

The preceding definitions of independence are applicable to complex r.v.'s or random vectors by replacing any m-dimensional Borel set appearing in (1) by a ($2m$-dimensional) Borel set of complex m-dimensional space. An alternative but equivalent definition is that the complex r.v.'s or random vectors $r_t = U_t + iV_t$ $(t \in T)$ are independent if and only if the corresponding real random vectors defined by $r_t = (U_t, V_t)$ are independent. Similar considerations apply to complex random functions.

Definition 2 of independence was formulated by Steinhaus and Kac [1] for the special case of a finite number of real functions defined on (0, 1) using Lebesgue measure for P. Functions independent according to Definition 2 will be called *S-independent*.

Earlier (Reference [2]), Kolmogorov had given the following definition of independence: If equation (1) holds for all B_i (measurable or not) for which the sets of Ω involved are measurable, then the functions are said to be *K-independent*.

Clearly, K-independence implies S-independence. Marczewski proposed the problem of determining whether S-independence always implies K-independence, Hartman [3] showed that S-independence implies K-independence if P is the Lebesgue measure. An outline of his proof is given in Exercise 1. However, in general, the above implication is not true, as shown by Doob [4] and Jessen [5]. The following counterexample was proposed by Jessen.

EXAMPLE 1

Let μ and μ^* denote the Lebesgue measure and the exterior Lebesgue measure on $I = (t : 0 \leq t \leq 1)$. The *inner measure* of a set Q is defined by

(2) $$\mu_* Q = \mu M - \mu^*(MQ'),$$

where M is any measurable set containing Q. It can be shown (see Exercise 2) that this definition is independent of the measurable set $M \supset Q$. It is known that there exists a nonmeasurable set Q, with $\mu^*(Q) = 1$ and $\mu_*(Q) = 0$ (see page 70 of Reference [6]). Let $\mathscr{S}$ be the system of all sets of the form

(3) $$A = BQ + CQ',$$

where $B, C \in \mathscr{L}$, the class of Lebesgue-measurable sets. It is not difficult to prove that $\mathscr{S} \supset \mathscr{L}$ and that $\mathscr{S}$ is a σ-field. We can define two functions on $\mathscr{S}$ by

$$\mu_1(A) = \mu B \qquad \mu_2(A) = \mu C.$$

The fact that these functions are measures and extensions of μ follows easily from their definition. However, it is more difficult to show that their values are independent of the sets B and C appearing in (1). We require the fact that for any measurable set B,

(4) $$B = \mu^*(BQ) + \mu_*(BQ'),$$

and the fact that if $L \supset M$, then $\mu_* L \geq \mu_* M$. The proofs of these results are outlined in Exercise 2. If Q is the set described above, it follows from (2), by letting $M = I$, that $\mu_* Q' = 0$. Hence $0 = \mu_* Q' \geq \mu_*(BQ') \geq 0$ and $\mu B = \mu^* BQ$ from (4). Suppose that, in addition to (3), $A = B_1 Q + C_1 Q'$; then $AQ = B_1 Q = BQ$ and so $\mu B = \mu^* BQ = \mu^* B_1 Q$. Accordingly, $\mu_1 A$ is independent of the set B and a similar result holds for μ_2. In particular, choosing $B = I$ and $C = \emptyset$,

$$\mu_1 Q = 1 \qquad \mu_2 Q = 0.$$

Now let $\mathscr{S} = ((x, y) : 0 \leq x \leq 1, 0 \leq y \leq 1)$. Then $\nu_1 = \mu_1 \times \mu_2$. $\nu_2 = \mu_2 \times \mu_1$ and $\lambda = \frac{1}{2}(\nu_1 + \nu_2)$ are extensions of the measure $\mu \times \mu$ on $\mathscr{S}$. However, λ is not a product measure, since $\lambda(I \times Q) = \lambda(Q \times I) = \frac{1}{2}$ and $\lambda(Q \times Q) = 0$.

Finally, the functions $r_1(x, y) = x$ and $r_2(x, y) = y$ are S-independent but not K-independent.

In Section 8.5, we introduced the concept of perfect measure. We recall that S-independence is equivalent to K-independence if the measure is perfect.

The word "independence" will mean S-independence in the remainder of the book unless otherwise stated.

EXERCISES

1. *Hartman*. Let $r_1 \cdots r_n$ be a system of Lebesgue-measurable real-valued functions defined on the interval $I = (0, 1)$ and let μ be the Lebesgue measure.

Show that S-independence implies K-independence by the following chain of reasoning.

(a) Let f and g be two S-independent L-measurable functions and let U and V be open sets. Show that $\mu(f^{-1}(U)g^{-1}(V)) = \mu(f^{-1}(U))\mu(g^{-1}(V))$. [*Hint:* Write U and V as a sum of intervals and use the definition of S-independence.]

(b) Prove that the above result remains valid if V and U are closed sets.

(c) Let P be a set of the y-axis such that $f^{-1}(P)$ is Lebesgue measurable. Show that for every $\epsilon > 0$ there exists a closed set $M \subset P$ such that,

$$(f^{-1}(P) - f^{-1}(M)) < \epsilon.$$

[*Hint:* Apply Lusin's theorem to show that there exist closed sets C and K such that $\mu(I - C) < \epsilon/2$ and $\mu(Cf^{-1}(P) - K) < \epsilon/2$. Let f restricted to C be denoted by ϕ (a continuous function) and let $M = f(K)$; then

$$K \subset \phi^{-1}(M) \subset f^{-1}(M) \subset f^{-1}(P).]$$

2. (a) If A is a measurable set, then $\mu^*(A + B) = \mu A + \mu^* B$. (*Hint:* Use $A + B$ as a test set in the definitions of μ^*-measurability.)

(b) Show that $\mu_* Q$ defined by (2) is independent of the set $M \supset Q$. [*Hint:* Let C be a measurable set containing Q. Write $M = MC + MC'$, substitute in (2), and show that $\mu_* Q = \mu(MC) - \mu^*(MCQ')$.

(c) Verify equation (4). (*Hint:* $B \supset BQ$.)

(d) If $L \supset M$, then $\mu_* L \geq \mu_* M$.

3. Let $(\Omega, \mathscr{E}, P)$ be a probability space, let P_r be the induced measure on the real line, and let F_r be the d.f. of r. Show that if P is not a perfect measure, there exists a set Q such that

(a) $\mu_r Q$ is well defined.

(b) Q is not a member of the σ-field of measurable sets constructed by the Carathéodory method from the outer measure determined by F_r.

4. *Doob.* Let S be the unit square $0 \leq x, y \leq 1$, and let Q be the set in the interval $[0, 1]$ described in Example 1. Let $F = ((x, y)|x \in Q, 0 \leq y \leq 1)$ and let $\mathscr{S}$ be the field of all sets of the form $A = L_1 F + L_2 F'$, where L_1 and L_2 are (two-dimensional) Lebesgue-measurable sets. Let $\phi(x, y)$ be a real, Lebesgue-measurable function defined in the unit square, with $0 \leq \phi \leq 1$, and define $m(A)$ by

$$m(A) = \iint_{L_1} \phi(x, y)\, dx\, dy + \iint_{L_2} (1 - \phi(x, y))\, dx\, dy.$$

(a) Show that m is uniquely defined and is an extension of the Lebesgue measure.

(b) The functions $f = x$ and $g = y$ are S-independent but not K-independent unless $\int_0^1 \phi(x, y)\, dx$ is constant for almost all y.

5. Show that the events of the class $\{r_i^{-1}(B_i)\}$, where $i \in I$ and B_i is an arbitrary (one-dimensional) Borel set, are independent if and only if the r_i are independent r.v.'s.

6. Prove Theorem 3.3.2 by using Theorem 3. [*Hint:* Use the indicator functions of events.]

11.4 MULTIPLICATION PROPERTIES

Independence of r.v.'s is equivalent to the fact that the probability space of the corresponding coordinate r.v.'s is just their product probability space (cf., Section 11.5). Let r and s be two independent r.v.'s on the probability space $(\Omega, \mathscr{S}, P)$ and let $\underline{r} = (r, s)$. Let T denote the measurable transformation from Ω into the plane, defined by $T = \underline{r}$; that is

$$T(\omega) = (r(\omega), s(\omega)). \tag{1}$$

T is a measurable transformation since $\underline{r}^{-1}(B)$, with B a Borel set, is an event. Since the functions r and s are r.v.'s, each of them is also a measurable transformation from Ω into the real line. To prove our first assertion, we need only show that the induced probability is a product probability—that is, $P_{\underline{r}} = P_r \times P_s$—or verify the following equivalent lemma.

LEMMA 1

The r.v.'s r and s are independent if and only if

$$PT^{-1} = Pr^{-1} \times Ps^{-1},$$

where the measurable transformation T is defined in (1).

Proof

The above relationship between the measures is equivalent to the definition of independence—that is, $P(\underline{r} \in B_1 \times B_2) = P(r \in B_1)P(s \in B_2)$, where B_1 and B_2 are (one-dimensional) Borel sets, since the Borel field in the plane is generated by sets of the form $B_1 \times B_2$.

COROLLARY

The r.v.'s r_1 and r_2 are independent if and only if

$$F_{\underline{r}} = F_{r_1}F_{r_2},$$

where the F's, in order of appearance, denote the d.f.'s of the r.v.'s $\underline{r} = (r_1, r_2)$, r_1, and r_2.

The necessity follows immediately from the lemma by choosing $B_1 = (-\infty, x_1]$ and $B_2 = (-\infty, x_2]$ and the sufficiency follows from part 2 of Example 2.1.

Similar results are true for arbitrary collections of r.v.'s; we leave the proof and formulation as an exercise for the reader.

Now let the functions g_1 and g_2 on the plane be defined by

$$g_1(x_1, x_2) = x_1 \qquad g_2(x_1, x_2) = x_2;$$

then $r = g_1 T$ and $s = g_2 T$.

THEOREM 1

If the r.v.'s $r_1, r_2, \ldots, r_n$ are independent and integrable, then so is their product, and

$$E(r_1 r_2 \cdots r_n) = (Er_1)(Er_2) \cdots (Er_n).$$

Conversely, if their product is integrable and no $r_i = 0$ (a.s.), then each r_i is integrable.

Proof

It suffices to prove the theorem for two independent r.v.'s r and s since $r = r_1$ and $s = r_2 r_3 \cdots r_n$ are independent by Theorem 3.3.

Suppose that r and s are integrable; then (by Theorem 7.9.3) g_1 and g_2 are integrable. Fubini's Theorem (7.8.3) and its first corollary show that $g_1 g_2$ is integrable and

$$\int g_1 g_2 \, dPT^{-1} = \left(\int x_1 \, d(Pr^{-1})\right)\left(\int x_2 \, d(Ps^{-1})\right). \tag{2}$$

Another application of Theorem 7.9.3 shows that $(Er)(Es) = E(rs)$.

Conversely, suppose that rs is integrable; then almost every section of $g_1 g_2$ is integrable. Each section is a constant multiple of either g_1 or g_2; it is not zero because r and s do not vanish a.s. It follows that g_1 and g_2 are themselves integrable by Corollary 1 of Theorem 7.8.3. Hence (2) is again valid and $(Er)(Es) = E(rs)$.

COROLLARY 1

Let r_i be independent r.v.'s and let $r = (r_1 \cdots r_n)$. Then the relation

$$\chi_r = \prod_{i=1}^{n} \chi_{r_i}$$

is valid for the corresponding c.f.'s.

The corollary follows directly from the theorem, which is valid for independent complex r.v.'s. Suppose that $r_k = u_k + iv_k$ for $k = 1, 2$; then $r_1 r_2 = u_1 u_2 + i(u_1 v_2 + u_2 v_1) - v_1 v_2$ and

$$\begin{aligned} E(r_1 r_2) &= E(u_1 u_2) + iE(u_1 v_2 + u_2 v_1) - E(v_1 v_2) \\ &= Eu_1 Eu_2 + i(Eu_1 Ev_2 + Eu_2 Ev_1) - Ev_1 Ev_2 \\ &= (Eu_1 + iEv_1)(Eu_2 + iEv_2) = Er_1 Er_2, \end{aligned}$$

because of the independence of the components of the random vectors (u_1, v_1) and (u_2, v_2) (cf., Theorem 3.1). This shows that the theorem remains valid for independent complex r.v.'s. We leave it to the reader to verify that the Borel Function Theorem (3.3) remains valid for complex Borel functions. Hence if the r_k are independent, then so are $e^{i\lambda_k r_k}$ and

$$E \exp\left(i \sum_{k=1}^{n} \lambda_k r_k\right) = E \prod_{k=1}^{n} \exp(i\lambda_k r_k) = \prod_{k=1}^{n} E \exp(i\lambda_k r_k),$$

which proves the corollary.

COROLLARY 2

Let r_i be independent r.v.'s and let $r = \sum_{k=1}^{n} r_k$; then

$$\chi_r = \prod_{i=1}^{n} \chi_{r_i}.$$

This follows from Corollary 1 by setting $\lambda = \lambda_1 = \lambda_2 = \cdots = \lambda_n$.

By using the one-to-one correspondence between c.f.'s and d.f.'s, the last corollary implies that the d.f. of a sum of independent r.v.'s is the convolution of their individual d.f.'s. However, it is interesting to prove this result directly. It suffices to prove the result for $n = 2$.

THEOREM 2

Let r_1 and r_2 be independent r.v.'s, let $r = r_1 + r_2$, and let F_1, F_2, and F be the corresponding d.f.'s, as indicated by the subscripts. Then

$$F = F_1 * F_2.$$

Proof

Let $g(x_1, x_2)$ be the indicator function of the set of points $(x_1 + x_2 \leq x)$ in R^2. Then by Lemma 1 and Theorem 7.9.3 with $T = (r_1, r_2)$,

$$\begin{aligned} (3) \qquad \int_{(r_1+r_2\leq x)} dP &= \int gT\, dP = \iint_{(x_1+x_2\leq x)} dF_1(x_1)\, dF_2(x_2) \\ &= \int_{-\infty}^{\infty} \left(\int_{-\infty}^{x-x_2} dF_1(x_1)\right) dF_2(x_2) \\ &= \int_{-\infty}^{\infty} F_1(x - x_2)\, dF_2(x_2) = F_1 * F_2. \end{aligned}$$

The corollaries to Theorem 2 show that independence of r.v.'s can also be defined by using c.f.'s or d.f.'s. For ease in reference, we will summarize these equivalent definitions in the following theorem.

THEOREM 3

Let F_i and χ_i, $F_{i_1\cdots i_n}$ and $\chi_{i_1\cdots i_n}$ be the d.f.'s and c.f.'s of the r.v. r_i and the random vector $(r_{i_1}, \ldots, r_{i_n})$, respectively. Then the following are equivalent conditions for the independence of the r.v.'s r_i.

1. $P \bigcap_{k=1}^{n} (r_{i_k} \in B_{i_k}) = \prod_{k=1}^{n} P(r_{i_k} \in B_{i_k})$.
2. $F_{i_1\cdots i_n}(x_{i_1} \cdots x_{i_n}) = F_{i_1}(x_{i_1}) \cdots F_{i_n}(x_{i_n})$.
3. $\chi_{i_1\cdots i_n}(\lambda_{i_1}, \ldots, \lambda_{i_n}) = \chi_{i_1}(\lambda_{i_1}) \cdots \chi_{i_n}(\lambda_{i_n})$

for *every* finite class of Borel sets B_1 and of points λ_i, $x_i \in R$.

Proof

The corollary of Lemma 1 shows that part 1 of the theorem is equivalent to part 2 and that part 1 implies part 3 by Corollary 2 of Theorem 1. To complete the proof, we need only show that part 3 implies part 2. However, by the inversion formula, part 3 implies that

$$\Delta_{a_{i_n}}^{b_{i_n}} \cdots \Delta_{a_{i_1}}^{b_{i_1}} F_{i_1\cdots i_n}(x_1 \cdots x_n) = \prod_{k=1}^{n} F(b_{i_n}) - F(a_{i_n})$$

for all continuity intervals. Part 2 follows by letting $a_i \to -\infty$ and $b_i \downarrow x_i$.

Note that the word "every" plays an important role in Theorem 3. For example, if the c.f. of a sum of two r.v.'s is the product of the c.f.'s of the summands, the summands may not be independent (see Exercise 3). Here $\lambda_1 = \lambda_2$. Theorem 3 can easily be generalized to random vectors.

EXERCISES

1. Show that a necessary and sufficient condition for the r.v.'s r_i $(i = 1, 2, \ldots)$ to be independent is that $PT^{-1} = (Pr_1^{-1})(Pr_2^{-1}) \cdots$, where

$$T(\omega) = (r_1(\omega), r_2(\omega), \ldots).$$

Show that if the functions $g_n(y_1, y_2, \ldots) = y_n$ are the coordinate functions on the range space of T, then $r_n = g_n T$. Generalize these results to uncountable sets of r.v.'s.

2. Prove Theorem 3.3 by using the notion of a measurable transformation.

3. Let r_1 be a Cauchy r.v. with a c.f. $e^{-|\lambda|}$, and let $r_2 = cr_1$, where $c \neq 0$. Then $\chi_{r_1+r_2} = \chi_{r_1}\chi_{r_2}$, even though r_1 and r_2 are not independent.

4. If $E(r_1 r_2) = (Er_1)(Er_2)$, are the r.v.'s r_1 and r_2 always independent?

5. Using the notation of Theorem 3, show that the r.v.'s r_i are independent if and only if

$$f_{i_1\cdots i_n}(x_{i_1}, \ldots, x_{i_n}) = f_{i_1}(x_{i_1}) \cdots f_{i_n}(x_{i_n}),$$

where the f's are the density functions of the r.v.'s r_i (assume that the density functions exist).

6. Prove Theorem 1 by using the following chain of reasoning:

(a) Verify the result to be true for two independent elementary r.v.'s.
(b) Generalize to nonnegative r.v.'s by approximating them by a sequence of increasing elementary r.v.'s and using the monotone convergence theorem.
(c) Show that if r and s are independent, then so are $\bar{r}$ and $\bar{s}$, where $\bar{r} = r^+$ or r^- or $|r|$, and $\bar{s} = s^+$ or s^- or $|s|$.

7. Show that if r_1 and r_2 are independent r.v.'s such that r_1, r_2, and $(r_1 + r_2)^2$ are integrable, then r_1^2 and r_2^2 are integrable.

8. Show that if r_i are independent r.v.'s and ϕ is any integrable function on R^n with respect to the induced probability measure, then

$$\int_\Omega \phi(r_1 \cdots r_n)\, dP = \int_{R^n} \phi(x_1 \cdots x_n)\, dF_1 \cdots dF_n,$$

where F_i is the d.f. of r_i.

11.5 PRODUCT SPACES

In the last section, we stated that the probability space induced by independent r.v.'s is a product probability space and we indicated the proof for a finite number of r.v.'s. This relationship will now be considered in greater detail. Let r_i be a set of independent r.v.'s on a probability space $(\Omega, \mathscr{S}, P)$. Each r_i induces a measure P_i on the Borel field B_i in its range space R_i. More generally, $r_{i_1\cdots i_n}$ induces a family of consistent probabilities, for arbitrary $i_1, \ldots, i_n$, on the product Borel sets $\times_{k=1}^{n} \mathscr{B}_{i_k}$. By Theorem 8.2.7, this family of probabilities determines a probability P_r on $\times\, \mathscr{B}_i$. P_r is uniquely determined by its values on Borel cylinders, which only depend on a finite number of coordinates. Therefore by Lemma 4.1 or directly from the definition of independence,

$$P_{i_1\cdots i_n} = P_{i_1} \times \cdots \times P_{i_n},$$

which implies that $P_r = \times P_i$. In other words, the probability space induced on the range space by a family of independent r.v.'s is just $(\times R_i, \times \mathscr{B}_i, \times P_i)$, the product probability space of $(R_i, \mathscr{B}_i, P_i)$. For a fixed i, $(R_i, \mathscr{B}_i, P_i)$ is just the probability space induced on R_i by r_i.

We frequently consider repeated independent trials—for example, tossing a coin n times. Suppose that we set $r_i = 1$ (0) if heads (tails) occurs on the ith toss. In many elementary textbooks, it is then stated that the r_i are independent r.v.'s. Intuitively this statement is true, but formally we must have a probability space on which the r_i are defined. A natural candidate for this space is $(\times_{i=1}^{n} R_i, \times_{i=1}^{n} \mathscr{S}_i, \times_{i=1}^{n} P_i)$, where $R_i = \{0, 1\}$, $\mathscr{S}_i = \{\phi, R_i, \{0\}, \{1\}\}$, and $P_i(0) = P_i(1) = \frac{1}{2}$. The r_i are just the coordinate r.v.'s—that is, $r_i(x_1, \ldots, x_n) = x_i$.

We now show that for a given product probability space $(\times R_i, \times \mathscr{S}_i, \times P_i)$ there exist a probability space $(\Omega, \mathscr{S}, P)$ and a family r_i of independent r.v.'s on this space which induce the given product probability space. Motivated by the preceding example, we will let $(\Omega, \mathscr{S}, P)$ be the given product space and we will define the r.v.'s r_i on this probability space by $r_i(x) = x_i$, where $x = (x_i, i \in I)$. The r_i's are independent and thus the induced space is $(\times R_i, \times \mathscr{S}_i, \times P_i)$.

However, note that if $(\Omega, \mathscr{S}, P)$ is fixed in advance, the construction above is invalid. In general, we cannot find r.v.'s on this space whose induced probability space coincides with $(\times R_i, \times \mathscr{S}_i, \times P_i)$, since even one r.v. with a specified probability distribution might not exist. For example, if $\mathscr{S} = (\Omega, \phi)$ the only r.v.'s on Ω are the constant functions.

Nevertheless, there exists a fixed probability space for each index set I provided that the d.f.'s F_i corresponding to P_i are continuous. To verify this statement, we first consider the situation in which I contains a single element. We will let $\Omega = [0, 1]$ and will let μ be the Lebesgue measure and $\mathscr{S}$ the class of Borel sets in Ω. Suppose that we are given the probability space $(R, \mathscr{S}, P)$. Then we must find a r.v. on Ω whose induced space coincides with the given space. Let $X(x) = x$ be the coordinate r.v. on R and let F be the corresponding d.f.; that is, $P(X \leq x) = F(x)$. Since F is continuous, there exists an inverse function, which we denote by F^{-1}, whose domain is Ω. Let $Y(y) = y$ be the coordinate r.v. on Ω and define a r.v. r on Ω by $r = F^{-1}Y$. Then

$$(1) \qquad F_r(x) = \mu(r \leq x) = \mu(F^{-1}(Y) \leq x) = \mu(Y \leq F(x)) = F(x),$$

and so the induced space of r is $(R, \mathscr{S}, P)$ as required. Now let the index set I be arbitrary. Consider the space $(\times \Omega_i, \times \mathscr{S}_i, \times \mu_i)$, where $\Omega_i = [0, 1]$, $\mathscr{S}_i$ is the class of Borel sets in Ω_i, and μ_i is the Lebesgue measure on $\mathscr{S}_i$. Again we let X_i, Y_i be the coordinate r.v. on $\times R_i$ and $\times \Omega_i$, respectively, so that $X_i(x) = x_i$, where $x = (x_i, i \in I)$ and $Y_i(y) = y$. If $r_i = F_i^{-1}Y$, $F_{r_i}(x_i) = F_i(x_i)$ by following the procedure in (1).[1]

EXERCISES

1. An experiment consists of a sequence of n repeated trials; in each trial, k possible, mutually exclusive, independent events $E_1, \ldots, E_k$ can occur.

(a) *Bernoulli Scheme.* Let $S_{ij} = 1$ if the event E_j occurs on the ith trial and $S_{ij} = 0$ otherwise; suppose that $P(s_{ij} = 1) = p_j$. Construct an appropriate probability space so that the s_{ij} are well defined.

(b) Let $r_i = (s_{i1}, s_{i2}, \ldots, s_{ik})$. Show that the random vectors r_i are independent.

(c) Let $s = \sum_{i=1}^{n} r_i = (s_1, s_2, \ldots, s_k)$, where $s_i = \sum_{j=1}^{n} s_{ij}$ for $i = 1, \ldots, k$. What does the random vector represent?

(d) Find the probability generating function of r_j. Deduce the generating function of s.

(e) Find $P(s_1 = n_1, \ldots, s_k = n_k)$.

2. *Poisson Scheme.* Rework Exercise 1 under the hypothesis that $P(s_{ij} = 1) = p_j(i)$ —that is, the probability of the event E_j occurring on the ith trial depends on the trial number. In fact, it is not necessary for the events to be the same in each trial.

3. (a) *Laplace–Lexis Scheme.* Again we consider n independent trials. Suppose that on each trial an event A occurs or does not occur. Let $r_i = 1$ if the event occurs and $r_i = 0$ otherwise, and let $s = \sum_{i=1}^{n} r_i$. Let $P(r_i = 1) = p$ be a r.v. that is fixed before the experiment begins. In other words, p has a probability distribution F and we randomly choose[2] a value of p, which remains fixed throughout the n trials. Construct a probability space on which the r.v.'s r_i are defined. Show that $P(s = k) = \binom{n}{k} E(p^k(1 - p)^{n-k})$.

(b) *Coolidge Scheme.* Now suppose that the event A occurs with different probabilities in each trial—that is, $P(r_i = 1) = p_i$. Each p_i is determined beforehand and the random vector $(p_1, p_2, \ldots, p_n)$ has the d.f. $F(x_1, \ldots, x_n)$. Rework part (a) under the new assumptions.

4. Construct an experiment that consists of drawing balls from urns of different compositions to illustrate Exercises 1–3.

[1]Frequently we must randomly choose a value of a r.v. with a given d.f. $F(x)$. In the jargon of the statistician, we take a sample from the given d.f. To do this we divide the interval [0, 1] into a large number of parts, say n, and choose an integer between 1 and n at random—for instance, by using a table of random numbers. What we are actually doing is approximating a uniform distribution by its discrete analog. Let y be the coordinate of the midpoint of the chosen interval. Then $F^{-1}(y) = x$ is a random value of the r.v.; that is, $P(x < r \leq x + dx) \doteq dF$. This follows from (1).

[2]See footnote 1.

11.6 RADEMACHER FUNCTIONS

To illustrate the notion of independence and some topics that will be covered in the remainder of the book, we introduce the *Rademacher functions*, which were first studied by H. Rademacher [7].

DEFINITION 1

Let $X = [0, 1]$ be the unit interval with Lebesgue measure. For every integer n, we define the Rademacher function r_n on X by setting $r_n(x) = 1$ or -1 depending on whether the integer i for which $(i - 1)/2^n \leq x < i/2^n$ is odd or even.

In other words, the function $r_n = 1, -1, 1, -1, \ldots, -1$ on the intervals $[0, 1/2^n)$, $[1/2^n, 2/2^n)$, $[2/2^n, 3/2^n), \ldots, [(2^n - 1)/2^n, 1)$ respectively. The Rademacher functions are also connected with the digits x_i in the binary expansion of x,

(1) $$x = \frac{x_1(x)}{2} + \frac{x_2(x)}{2^2} + \frac{x_3(x)}{2^3} + \cdots \qquad (0 \leq x \leq 1)$$

where each $x_i = 0$ or 1, depending on x. To insure uniqueness, we adopt the convention of writing terminating expansions in the form in which all the digits from a certain point on equal zero. Then

(2) $$r_n(x) = 1 - 2x_n(x) \qquad (n = 1, 2, 3, \ldots)$$

and, by using (1),

(3) $$1 - 2x = \sum_{n=1}^{\infty} \frac{r_n(x)}{2^n}.$$

There is an analytic formula for the r_n—that is,

(4) $$r_n(x) = sg(\sin 2\pi 2^{n-1}x).$$

We can also interpret the Rademacher functions as r.v.'s on the probability space $(\Omega, \mathscr{S}, P)$, where $\Omega = X$, $\mathscr{S}$ is the class of Borel sets contained in X, and P is the Lebesgue measure on $\mathscr{S}$. More concretely, the events $(r_n = 1)$ and $(r_n = -1)$ can be interpreted to mean that we get heads and tails respectively on the nth toss of a fair coin. Hence the sequence $r_n(x)$ represents the possible outcomes when a fair coin is tossed an infinite number of times. We have

(5) $$P(r_n = 1) = P(r_n = -1) = \frac{1}{2},$$

since $r_n = 1$ on 2^{n-1} intervals of lengths $1/2^n$.

THEOREM 1

The r.v.'s r_n, where $n = 1, 2, 3, \ldots$, are independent.

Proof

Let $x_i = \pm 1$ for $i = 1, 2, \ldots, k$; then

$$P(r_{n_1} = x_1, r_{n_2} = x_2, \ldots, r_{n_k} = x_k) = \frac{1}{2^k} = \prod_{i=1}^{k} P(r_{n_i} = x_i),$$

where the extreme right-hand side follows from (5). The extreme left-hand side follows from the fact that if $n_2 > n_1$, we can decompose the unit interval into 2^{n_2} intervals of lengths $1/2^{n_2}$, as in the statement following Definition 1, in two stages. First we subdivide the interval $[0, 1]$ into 2^{n_1} intervals of lengths $1/2^{n_1}$ in the standard manner so that $r_{n_1} = 1$ on half of these and $r_{n_1} = -1$ on the remainder. Then we subdivide each of these intervals into $2^{n_2 - n_1}$ parts of lengths $1/2^{n_2}$. Note that $r_{n_2} = 1$ on half of these and $r_{n_2} = -1$ on the other half. Hence $r_{n_2} = x_2$ on half of the intervals on which $(r_{n_1} = x_1)$. Since the total length of the intervals on which $(r_{n_1} = x_1)$ is $\frac{1}{2}$, it follows that $P(r_{n_1} = x_1, r_{n_2} = x_2) = \frac{1}{4}$. A repetition of this process yields the desired equality.

The next three results are immediate consequences of the definition of the Rademacher functions.

THEOREM 2

$$Er_n = 0 \quad \text{and} \quad \text{Var } r_n = 1.$$

THEOREM 3

The r_n have the same d.f.—namely,

$$F_n(x) = \begin{cases} 0 & (x < -1) \\ \frac{1}{2} & (-1 \leq x < 1) \\ 1 & (x \geq 1) \end{cases}$$

THEOREM 4

The r_n have the same c.f. $\chi_n = \cos \lambda$.

This last result enables us to find the probability of getting k heads among the first n tosses of a fair coin without the usual recourse to combinational methods. Under our interpretation, this problem is reduced to finding the measure of the set of x's such that exactly k of the numbers $r_1(x), r_2(x), \ldots, r_n(x) = 1$. This is equivalent to the condition that

$$s_n = r_1(x) + r_2(x) + \cdots + r_n(x) = 2k - n$$

and so

$$P(s_n = 2k - n) = \lim_{v\to\infty} \frac{1}{2v} \int_{-v}^{v} e^{-i\lambda(2k-n)} \cos^n \lambda \, d\lambda$$

by Theorem 10.5.4, since the c.f. of s_n is $\cos^n \lambda$. The last integral can be reduced to

$$P(s_n = 2k - n) = \frac{1}{2\pi} \int_0^{2\pi} (\cos (2k - n)\lambda) \cos^n \lambda \, d\lambda$$

by setting $v = 2m\pi$, dividing the range $-2m\pi$ to $2m\pi$ into intervals of lengths 2π, and using the fact that the trigonometric functions are periodic to simplify the resulting integrals. We leave the evaluation of the last integral to the reader. A more elementary method of deriving this result is given in Exercise 1.

Note that $\{r_n\}$ is an orthonormal system in L^2; that is,

$$\int_0^1 r_n(x)\overline{r_m(x)}\, dx = \delta_{mn},$$

where $\delta_{mn} = 1$ if $m = n$ and $\delta_{mn} = 0$ if $m \neq n$. This system is not complete, since the function $r_0(k) \equiv 1$ is orthogonal[3] to all r_n. The orthonormality enables us to prove the following theorem, commonly called the *weak law of large numbers*. We will discuss this result and generalize it in the next chapter.

THEOREM 5

For any $\epsilon > 0$,

(6) $$\lim_n P\left(\left|\frac{r_1 + \cdots + r_n}{n}\right| > \epsilon\right) = 0.$$

Proof

By the method that was used in proving Chebyshev's inequality,

$$\int_0^1 (r_1 + r_2 + \cdots + r_n)^2 \, dx \geq \int_{|r_1+\cdots+r_n|>\epsilon n} (r_1 + r_2 + \cdots + r_n)^2 \, dx$$
$$> \epsilon^2 n^2 P(|r_1 + \cdots r_n| > \epsilon n).$$

[3] $\int_0^1 r_n r_0 \, dx = 0$ if $n \neq 0$.

From the orthonormality property, it follows that

$$\int_0^1 (r_1 + \cdots + r_n)^2\, dx = n,$$

which, when substituted in the last inequality, yields (6).

This last result has a simple physical interpretation. Suppose that a fair coin is tossed, and that we win one dollar each time heads comes up and lose one dollar each time tails comes up. Then the gain (or loss) after n tosses is simply $r_1(x) + \cdots + r_n(x)$ and so the average gain is $[r_1(x) + \cdots + r_n(x)]/n$. The weak law of large numbers states that the probability of our average gain or loss being large is very small for very large n.

The weak law of large numbers states that $(r_1 + \cdots + r_n)/n \to 0(P)$. This law can be strengthened so that the average gain $\to 0$ (a.s.). It is then called the *strong law of large numbers*. This was discovered by E. Borel. As we shall see in the next chapter, this law has been generalized.

THEOREM 6

$$P\left(\lim_{n\to\infty} \frac{r_1 + \cdots + r_n}{n} = 0\right) = 1.$$

Proof

Let $s_n(x) = \{[r_1(x) + \cdots + r_n(x)]/n\}^4$. By using the orthonormality property,

$$\int_0^1 s_n(x)\, dx = \frac{1}{n^3} + \frac{4!}{2!2!}\binom{n}{2}\frac{1}{n^4};$$

therefore, $\sum_{n=1}^{\infty} \int_0^1 s_n(x)\, dx < \infty$. Since the s_n are positive, it follows that $\sum_{n=1}^{\infty} s_n(x)$ converges a.e. by Corollary 2 of Theorem 7.5.1. Consequently, the nth term of this series approaches zero a.e., which yields the result.

By using equation (2), we see that Theorem 6 implies that

$$\lim_{n\to\infty} \frac{x_1(x) + x_2(x) + \cdots + x_n(x)}{n} = \frac{1}{2} \text{ (a.s.)}.$$

This means that almost every number x has the same number of zeros and ones in its binary expansion. Of course, this result can be generalized for an arbitrary base $b > 1$ (see Exercise 3). It can be shown that each allowable digit in x, $0 \leq x \leq 1$ appears with the frequency $1/b$ for almost every x.

Furthermore, every number is normal.[4] Although nearly every number is normal, it is quite difficult to exhibit such a number and prove that it is normal. A simple example is

$$0.12345678910111213141516\cdots,$$

which is written in decimal notation and where after the decimal point all the positive integers appear in succession. The proof that this number is normal is not trivial [8]. Further applications of probability theory to number theory appears in [9].

In addition to generalizing the laws of large numbers, a large part of the remainder of the book will be concerned with generalizing the so-called *central limit theorem* for Rademacher functions.

THEOREM 7

$$\lim_{n\to\infty} P\left(\frac{r_1 + \cdots + r_n}{\sqrt{n}} \leq x\right) = \frac{1}{\sqrt{2\pi}} \int_{-\infty}^{x} e^{-t^2/2}\, dt.$$

The proof is a direct consequence of Theorem 10.6.5 applied to the c.f. of $(r_1 + \cdots + r_n)/\sqrt{n}$, since $\lim_{n\to\infty} (\cos \lambda/\sqrt{n})^n = e^{-\lambda/2}$ (c.f., Example 10.6.2).

The following theorem, discovered by Khintchine, is known as the *law of the iterated logarithm.* Even for Rademacher functions, a simple proof is not known (to the author). Hence we will postpone the proof of this theorem until we reach Section 12.7.

THEOREM 8

$$P\left(\overline{\lim_{n\to\infty}} \frac{r_1 + \cdots + r_n}{\sqrt{2n \log \log n}} = 1\right) = 1.$$

In 1922, H. Steinhaus posed the problem of determining the probability that the series

$$\sum_{1}^{\infty} \pm a_k \qquad (a_k \text{ is real})$$

converges if the signs of each term are chosen independently and each has a probability of $\frac{1}{2}$. The same problem was also posed by N. Wiener. However,

[4]A number x is said to be *normal* if the number of times $N(n)$ that any k digits selected in advance will appear among the first n digits in the decimal expansion of x is $N(n) = 10^{-k}n + o(n)$. We have shown that this equation holds for an arbitrary x when $k = 1$; it is also true for every k.

this problem had already been solved by H. Rademacher. To answer this question, we will use the Riesz–Fischer theorem. This theorem states that if $\sum |a_k|^2 < \infty$, where the a_k are complex numbers, and if $\{\phi_i\}$ is an orthonormal sequence of complex-valued functions on a set A (i.e., $\int_A \phi_i \phi_j \, dx = \delta_{ij}$), then there exists a function $f \in L^2$ such that

$$\lim_{n\to\infty} \int_A \left| f(x) - \sum_{k=1}^{n} a_k \phi_k(x) \right|^2 dx = 0.$$

We note that the above problem is equivalent to finding the measure of the set of x's for which $\sum_1^\infty a_k r_k(x)$ converges.

THEOREM 9

Let a_n be a sequence of complex numbers; then $\sum_{n=1}^\infty a_n r_n$ converges a.e. if

$$\sum_{n=1}^{\infty} |a_n|^2 < \infty.$$

Proof

Since the Rademacher functions are orthonormal, there exists a function f such that

$$\int_0^1 |f(x)|^2 \, dx < \infty$$

and

$$\lim_{n\to\infty} \int_0^1 \left| f(x) - \sum_{k=1}^{\infty} a_k r_k(x) \right|^2 dx = 0 \tag{6}$$

by the Riesz–Fischer theorem. Then for any x, y, with $0 \le x < y \le 1$,

$$\int_x^y \left| f(t) - \sum_{k=1}^{n} a_k r_k(t) \right| dt \le (y - x)^{1/2} \left(\int_0^1 \left| f(t) - \sum_1^n a_k r_k(t) \right|^2 dt \right)^{1/2}$$

by Schwarz's inequality. Hence by (6)

$$\int_x^y f(t) \, dt = \sum_1^\infty a_k \int_x^y r_k(t) \, dt. \tag{7}$$

Let x_0 be a point such that $f(x_0)$ is defined. We can assume that x_0 is not a dyadic rational since these points form a set of measure zero. Suppose that

$$x_m = \frac{k_m}{2^m} < x_0 < \frac{k_m + 1}{2^m} = y_m.$$

Note that

$$\int_{x_m}^{y_m} r_k(t)\,dt = 0 \qquad (k > m)$$

and

$$\int_{x_m}^{y_m} r_k(t)\,dt = (y_m - x_m)r_k(x_0). \qquad (k \leq m)$$

Therefore, setting $y = y_m$ and $x = x_m$ in (7) yields

$$\text{(8)} \qquad \frac{1}{y_m - x_m}\int_{x_m}^{y_m} f(t)\,dt = \sum_{1}^{m} a_k r_k(x_0).$$

Now let $m \to \infty$ and choose x_m, y_m so that $\lim_{m\to\infty} x_m = \lim_{m\to\infty} y_m = x_0$. Therefore, from (8) and the fundamental theorem of calculus,[5] we obtain, by letting $m \to \infty$,

$$f(x_0) = \sum_{1}^{\infty} a_k r_k(x_0)$$

and so our series converges.

Later we will see that this result remains valid for arbitrary r.v.'s, a fact discovered by A. Kolmogorov.

In order to discover what happens to the series $\sum_{k=1}^{\infty} a_k r_k$ when $\sum_{k=1}^{\infty} |a_k|^2 = \infty$, we require the following lemma discovered by Burstin. The simple proof was derived by Hartman and Kirshner [10].

LEMMA 1

If $f(t)$ is a Lebesgue-integrable function on $[a, b]$ that has arbitrarily small periods, then $f(t)$ is constant a.e.

[5]If $\int_0^1 f(t)\,dt < \infty$, then

$$\lim_{m\to\infty} \frac{1}{y_m - x_m}\int_{x_m}^{y_m} f(t)\,dt = f(x_0) \quad \text{(a.e.)}$$

where $y_m \leq x_0 \leq x_m$ and $x_m \to y_m$. See E. Titchmarsh, *Theory of Functions.* (London: Oxford University Press, 1939), p. 362.

Proof

Let p_n be the periods of $f(t)$; i.e., $f(t + p_n) = f(t)$, such that $\lim_{n\to\infty} p_n = 0$. Then

$$(9) \qquad I = \int_a^b f(t)\,dt = \sum_{k=0}^{l_n} \int_{a+kp_n}^{a+(k+1)p_n} f(t)\,dt + \int_{a+(l_n+1)p_n}^{b} f(t)\,dt,$$

where

$$a + (l_n + 1)p_n \leq b < a + (l_n + 2)p_n.$$

From this inequality, it follows that $l_n p_n \to b - a$ and $a + (l_n + 1)p_n \to b$.

Thus the integral at the extreme right of equation (9) is equal to ϵ_n, where $\epsilon_n \to 0$ as $n \to \infty$. Due to the periodicity of $f(t)$, all of the summands appearing in (9) are equal, as can easily be seen by making a change of variables; therefore,

$$(10) \qquad I = [(l_n + 1)p_n] \frac{1}{p_n} \int_{a+k_n p_n}^{a+(k_n+1)p_n} f(t)\,dt + \epsilon_n$$

for some fixed k_n, where $0 \leq k_n \leq l_n$. Let x_0 bé such that $a + k_n p_n < x_0 < a + (k_n + 1)p_n$ and

$$\lim_{n\to\infty} \frac{1}{p_n} \int_{a+k_n p_n}^{a+(k_n+1)p_n} f(t)\,dt = f(x_0).$$

Almost every x_0 has this property, by the fundamental theorem of calculus. Therefore, letting $n \to \infty$ in (10), $f(x_0) = I/(b - a)$.

As a consequence of this lemma we can prove the zero–one law for Rademacher functions. We will see that this result remains valid for arbitrary independent r.v.'s. To prove this theorem, we extend the domain of the definition of the $r_k(x)$ by defining $r_k(x + 1) = r_k(x)$. For these new periodic functions, we define $r_k(t) = r_1(2^{k-1}t)$.

LEMMA 2

The set of convergence of

$$(11) \qquad \sum_{k=1}^{\infty} a_k r_k(x)$$

must be either of measure zero or measure one.

Proof

If x is a point of convergence of the series, then so is $x + 2^{-n}$ for $n = 0, 1, 2, 3, \ldots$. This statement is a consequence of the periodicity of the

r_k, since $r_k(x + 2^{-n}) = r_k(x)$ for $k \geq 2^n$; therefore, only a finite number of terms of (11) are changed. This does not affect convergence. Let f be the indicator function of the set of convergence. Then f has arbitrarily small periods; hence by Lemma 1 it must be constant a.e. over $[0,1]$. By definition of f, the constant must be either zero or one.

The zero–one law enables us to investigate what happens if

$$\sum_1^\infty |a_k|^2 = \infty.$$

THEOREM 10

Let a_n be a sequence of complex numbers and let $\sum_1^\infty |a_n|^2 = \infty$. Then the series $\sum_1^\infty a_k r_k(t)$ diverges with a probability of one.

Proof

Suppose that $|a_n| \to 0$ (otherwise the theorem is trivial) and that the given series converges on a set of positive measure. Therefore, by Lemma 2,

$$\lim_{n\to\infty} \sum_1^\infty a_k r_k(x) = f(x) \quad \text{(a.e.)},$$

where $f(x)$ is a measurable function. Hence for every real $\alpha \neq 0$,

$$\lim_{n\to\infty} \exp\left(i\alpha \sum_1^n a_k r_k(x)\right) = e^{i\alpha f(x)} \quad \text{(a.e.)},$$

and so

$$\lim_{n\to\infty} \int_0^1 \exp\left(i\alpha \sum_1^n a_k r_k(x)\right) dx = \int_0^1 e^{i\alpha f(x)}\, dx \tag{12}$$

by the Lebesgue dominated convergence theorem. By using the independence of the $r_k(x)$, the integral on the left-hand side of (12) can easily be evaluated; it yields

$$\lim_{n\to\infty} \prod_{k=1}^n \cos \alpha a_k.$$

The facts that $a_k \to 0$ and $\sum_1^\infty |a_k|^2 = \infty$ imply that this last limit is zero if $\alpha \neq 0$ and so

$$\int_0^1 e^{i\alpha f(x)}\, dx = 0 \tag{13}$$

from (12) for every real $\alpha \neq 0$. Replace α by α_n in (13), where each $\alpha_n \neq 0$ and $\lim_{n\to\infty} \alpha_n = 0$. Taking limits as $n \to \infty$, it follows from the dominated convergence theorem that the limit is 1; therefore, $1 = 0$, which is a contradiction.

EXERCISES

1. (a) Show that

$$\int_0^1 \exp\left[i \sum_1^n a_k r_k(x)\right] dx = \prod_{k=1}^n \cos a_k = \prod_{k=1}^n \int_0^1 e^{ia_k r_k(x)}\, dx.$$

(b) Verify that the indicator function $g(t)$ of the event $(s_n = 2k - n)$, where $s_n = r_1 + r_2 + \cdots + r_n$, is

$$g(t) = \frac{1}{2\pi} \int_0^{2\pi} e^{ix(s_n(t)-(2k-n))}\, dx.$$

[*Hint:* $(1/2\pi) \int_0^2 e^{imx}\, dx = \delta_{m0}$.]

(c) Finally, show that $P(s_n = 2k - n) = 1/2^n \binom{n}{k}$.

2. Prove that if $i_1 < i_2 < \cdots < i_n$, then

$$\int_0^1 r_{i_1}(x) r_{i_2}(x) \cdots r_{i_n}(x)\, dx = 0.$$

3. Let x $(0 \leq x \leq 1)$ be expressed as a decimal fraction using the radix (base) b; i.e.,

$$x = 0.x_1(x)x_2(x)x_3(x) \cdots = \frac{x_1(x)}{b} + \frac{x_2(x)}{b^2} + \frac{x_3(x)}{b^3} + \cdots,$$

where each x_i is one of the digits $0, 1, \ldots, b - 1$. Let $F_n(t)$ denote the number of times the digit k $(0 \leq k \leq b - 1)$ appears among $x_1 \cdots x_n$. Show that $\lim_{n\to\infty} [F_n(x)/n] = 1/b$ (a.e.).

4. (a) Let $X = X_p(x)$ for $0 < p < 1$ be defined as follows:

$$X(x) = \frac{x}{p} \qquad (0 \leq x \leq p)$$

$$= \frac{x - p}{1 - p} \qquad (p < x \leq 1)$$

Next define the functions $x_i^{(p)}(x) = x_i(x)$, which are generalizations of the binary digits for $p = \frac{1}{2}$, by

$$x_1(x) = f(x) \qquad x_2(x) = f(X(x)) \qquad x_3(x) = f(X(X(x))),$$

where $f(x) = 0$ for $0 \leq x \leq p$ and $f(x) = 1$ for $p < x \leq 1$. Show that the functions x_i are independent.

(b) Prove that

$$\mu(x_1 + \cdots + x_n = k) = \binom{n}{k} p^k(1-p)^{n-k} \qquad (0 \leq k \leq n)$$

where μ is the Lebesgue measure on [0, 1].

(c) Use the x_i to construct a model for independent tosses of a coin, where the probability for heads is p.

REFERENCES

1. Steinhaus, H., and M. Kac, "Sur les fonctions indépendantes (v)," *Studia Math.*, **6** (1936).
2. Kolmogorov, A., Grundbegriffe der Wahrscheinlichkeitsrechnung," *Ergebn. der Mathematik*, II (1933).
3. Hartman, S., "Sur deux notions de fonctions indépendantes," *Coll. Math.*, **1** (1948).
4. Doob, J. L., "On a problem of Marczewski," *Coll. Math.*, **1** (1948).
5. Jessen, B., "On two notions of independent functions," *Coll. Math.*, **1** (1948).
6. Halmos, P. R., *Measure Theory*. Princeton, N. J.: Van Nostrand Co., Inc., 1950.
7. Rademacher, H., "Einige Satze über Reihen von allgemeinen orthogonalfunktionen," *Math. Ann.*, **87** (1922).
8. Champernowne, D. G., "The construction of decimals normal in the scale of ten," *London Math. Soc.*, **8** (1933).
9. Kubilius, J., *Probabilistic Methods in the Theory of Numbers*, Amer. Math. Soc., Providence, R. I., 1964. (Russian translation.)
10. Hartman, P., and R. Kirshner, "The structure of monotone functions," *Amer. Jour. Math.*, **59** (1937).

CHAPTER 12

SERIES OF INDEPENDENT RANDOM VARIABLES

12.1 ZERO-ONE LAW

One of the most striking properties of a sequence of independent r.v.'s is that it converges or diverges a.s. This is an immediate consequence of the zero–one law developed by Kolmogorov [1]. In order to state this theorem precisely, we will introduce the following definition. For brevity, we will suppose that all events and r.v.'s are defined with respect to a fixed probability space $(\Omega, \mathscr{S}, P)$ unless otherwise stated.

DEFINITION 1

Let $\{\mathscr{S}_n\}$ be a given sequence of σ-subfields of $\mathscr{S}$ and let $\mathscr{C}_n$ be the compound σ-field with components $\mathscr{S}_i$ for $i \geq n$. The events of the σ-field $\mathscr{S}_\infty = \bigcap_{n=1}^{\infty} \mathscr{C}_n$ are called *asymptotic*[1] events.

Let $\mathscr{S}(r_1, \ldots, r_n)$ denote the σ-subfield of $\mathscr{S}$ induced in Ω by $r_1, \ldots, r_n$

[1]Also known as *ultimate* or *tail* events.

(n is finite or infinite)—that is, the smallest σ-field with respect to which all of the indicated r.v.'s are measurable. In particular, if $\mathscr{S}_n = \mathscr{S}(r_n)$ for r.v.'s r_n, then

$$\mathscr{S}_\infty = \bigcap_{n=1}^{\infty} \mathscr{S}(r_n, r_{n+1}, \ldots). \tag{1}$$

THEOREM 1

Let $\{r_n\}$ be a sequence of independent real-valued r.v.'s and let $A \in \mathscr{S}_\infty$, where $\mathscr{S}_\infty$ is given by (1), be an ultimate event. Then $PA = 0$ or $PA = 1$.

Proof

By Definition 11.3.1, which defines independence, and by Theorem 11.2.1, $\mathscr{S}(r_1, \ldots, r_m)$ and $\mathscr{C}_n = \mathscr{S}(r_n, r_{n+1}, \ldots)$ are independent if $n > m$. Since $\mathscr{S}_\infty \subset \mathscr{C}_n$, it follows that $\mathscr{S}_\infty$ and $\mathscr{S}(r_1, \ldots, r_m)$ are independent. Accordingly, $\mathscr{S}(r_i)$ for $i \leq m$ and $\mathscr{S}_\infty$ are independent; because m is arbitrary, this implies that $\mathscr{C}_1$ and $\mathscr{S}_\infty$ are independent. Finally, since $\mathscr{S}_\infty \subset \mathscr{C}_1$, $\mathscr{S}_\infty$ and $\mathscr{S}_\infty$ are independent; in other words,

$$P(A \cap A) = (PA)(PA)$$

for every $A \in \mathscr{S}_\infty$, which is equivalent to $PA = 0$ or $PA = 1$.

COROLLARY 1

If r is a r.v. measurable on $(\Omega, \mathscr{S}_\infty)$, then there is a constant c such that $P(r = c) = 1$.

Proof

First note that there must be a real number a such that $P(a - 1 < r \leq a) > 0$. Since r is measurable, it follows by Theorem 1 that $P(a - 1 < r \leq a) = 1$. Hence either $P(a - 1 < r \leq a - \frac{1}{2}) > 0$ or $P(a - \frac{1}{2} < r \leq a) > 0$. Suppose that the latter probability is positive; then $P(a - \frac{1}{2} < r \leq a) = 1$. A repetition of this procedure yields the result.

COROLLARY 2

For every sequence $\{r_n\}$ of independent r.v.'s, the sequences $\{r_n\}$, $\{\sum_{i=1}^{n} r_i\}$, and $\{a_n \sum_{i=1}^{n} r_i\}$, where the real numbers $a_n \to 0$ as $n \to \infty$, converge or diverge a.s. Moreover, these limits, if they exist, are degenerate r.v.'s.

Proof

Corollary 2 follows from the fact that the set of points of convergence are asymptotic events. In fact $A = (\lim r_n \text{ exists})$, $B = (\sum r_n \text{ converges})$, and $C = (\lim a_n \sum_1^n r_i \text{ exists})$ belong to $\mathscr{S}(r_n, r_{n+1}, \ldots)$ for every n because $A = (\lim_{i\to\infty} r_{n+i} \text{ exists})$, $B = (\sum_{i=n}^{\infty} r_i \text{ converges})$, and $C = (\lim_{i\to\infty} a_{n+i} \sum_{k=n}^{n+i} r_k)$. Therefore, these events have a probability of one or zero, by the zero–one law. The last part of Corollary 2 follows from Corollary 1.

If the r_i are not independent, then an $\mathscr{S}_\infty$-measurable r.v. is not necessarily a constant a.s. Suppose that $r_1 = r_2 = \cdots$ (a.s.); then $r = r_1$ satisfies the conditions of Corollary 1 and does not have to be a constant.

In studying the convergence of a sequence of real numbers, a finite number of terms can be neglected. This same principle holds for sequences of r.v.'s, since the set of points of convergence is an asymptotic event. Accordingly, two sequences of r.v.'s can be equivalent with respect to convergence without being identical.

DEFINITION 2

The sequences r_n and s_n are said to be *asymptotically equivalent* if $P(\limsup (r_n \neq s_n)) = 0$.

If $\omega \notin \limsup (r_n \neq s_n)$, then ω does not belong to an infinite number of the sets $(r_n \neq s_n)$; therefore, there exists a number $N = N(\omega)$ such that for all $n \geq N$, $r_n(\omega) = s_n(\omega)$. Hence the definition reduces to the statement that for almost all $\omega \in \Omega$, there exists a finite number $N(\omega)$ such that for all $n \geq N(\omega)$, the two sequences $r_n(\omega)$ and $s_n(\omega)$ are the same.

LEMMA 1

If the series $\sum P(r_n \neq s_n)$ converges, then the sequences $\{r_n\}$ and $\{s_n\}$ are asymptotically equivalent.

Proof

The proof is a consequence of the following inequality:

$$P(\limsup (r_n \neq s_n)) = \lim_n P \bigcup_{k=n}^{\infty} (r_k \neq s_k) \leq \lim_n \sum_{k=n}^{\infty} P(r_k \neq s_k).$$

THEOREM 2

Let $\{r_n\}$ and $\{s_n\}$ be asymptotically equivalent. Then $\{r_n\}$, $\{\sum_{k=1}^{n} r_k\}$, and $\{a_n \sum_{k=1}^{n} r_n\}$, where the real number $a_n \to 0$, converge a.s. if and only if the corresponding sequences with r_n replaced by s_n converge a.s. and to the same limits.

Proof

Suppose that r_n converges a.s. Let A be the set of points for which the sequence s_n does not converge to the same limit. Then for every N,

$$A \subset (s_n \neq r_n, \text{ some } n \geq N) \subset \bigcup_{n=N}^{\infty} (r_n \neq s_n);$$

hence

$$P(A) \leq P \bigcup_{n=N}^{\infty} (r_n \neq s_n).$$

Since N is arbitrary, it follows from Definition 2 that $P(A) = 0$. The remaining parts of the theorem are proved in a similar fashion.

The following example illustrates the use of the zero–one law.

EXAMPLE 1

Let E_i be a sequence of independent events. What is the probability that infinitely many of the given events will occur?

To phrase the problem mathematically, let A be the event that infinitely many E_i will occur; then A is the set of points

(2) $$A = \bigcap_{n=1}^{\infty} \bigcup_{m=n}^{\infty} E_m.$$

The problem is to evaluate PA. In order to apply the zero–one law, we define the independent r.v.'s r_i to be the indicator functions of the events E_i. Then $A \in \mathscr{S}_\infty$, since $A = (\overline{\lim}\, r_i = 1)$; therefore, by the zero–one law, $PA = 0$ or $PA = 1$. In particular, if $E_n \to E$, then $PE = 0$ or $PE = 1$.

The following criterion, known as the Borel–Cantelli lemma, distinguishes the two possibilities.

THEOREM 3

Let E_n be a sequence of events, and let A be the event that infinitely many E_n's occur.

1. If $\sum_{n=1}^{\infty} PE_n < \infty$, then $PA = 0$.

2. If the events E_i are independent and $\sum_{n=1}^{\infty} PE_n = \infty$, then $PA = 1$.

Proof

1. Observing from (2) that $A \subset \bigcup_{i=n}^{\infty} E_i$ for every n, we have

$$0 \leq PA \leq \sum_{i=n}^{\infty} PE \to 0$$

as $n \to \infty$, which implies that $PA = 0$.
2. Instead of proving that $PA = 1$, we will show equivalently that $PA' = 0$. Again, from (2), $A' = \lim_{n\to\infty} A_n$, where $A_n = \bigcap_{m=n}^{\infty} E'_m$. Now by the assumption of independence,

$$0 \leq PA_n = P\left(\bigcap_{m=n}^{\infty} E'_m\right) \leq P\left(\bigcap_{m=n}^{N} E'_m\right) = \prod_{m=n}^{N} PE'_m$$

$$= \prod_{m=n}^{N} (1 - PE_m) \leq \exp\left(-\sum_{m=n}^{N} PE_m\right),$$

since $(1 - x) \leq e^{-x}$ for $x \in [0, 1]$. However, the quantity on the right approaches zero as $N \to \infty$, since $\sum_{n=1}^{\infty} PE_n = \infty$. Hence $PA' = P(\lim A_n) = \lim PA_n = 0$.

COROLLARY

If the r.v.'s r_n are independent and $r_n \to 0$ (a.s.), then $\sum_{n=1}^{\infty} P(|r_n| \geq c)$ converges for every real number $c > 0$.

Proof

Since the r.v.'s r_n are independent, then so are the events $E_n = (|r_n| \geq c)$. Now $r_n \to 0$ (a.s.) implies that infinitely many of the events E_n cannot occur with positive probability; that is, $P(\limsup E_n) = 0$. Hence $\sum PE_n < \infty$; otherwise, part 2 of the theorem would give a contradiction.

Variations in the statement of Theorems 1 and 3 appear in the Exercises.

EXAMPLE 2

What is the probability that in repeatedly tossing a coin we will obtain the pattern *ttt* infinitely often?

Let E_i be the event that we obtain tails on trials i, $i + 1$, and $i + 2$. The events $E_1, E_4, E_7, \ldots$ are independent. Since $PE_i = (\frac{1}{2})^3$, the series $PE_1 + PE_4 + PE_7 + \cdots$ diverges; thus by Theorem 3, the pattern ttt occurs infinitely often with a probability 1.

EXERCISES

1. (a) Show that $\sup A_n$, $\inf A_n$, $\limsup A_n$, and $\liminf A_n$ are asymptotic events for $\mathscr{S}_\infty$ generated by the I_{A_n}, the indicator of A_n.

(b) The event $[(1/b_n)\sum_{k=1}^{n} r_k$ converges], where $b_n \to \infty$, is an ultimate event.

2. Prove that the events $A = (\sum a_n r_n$ converges) and $B = (\sum a_n r_n$ is bounded) are ultimate events. Show that if the r_n are independent r.v.'s, then $PA = 0$ or 1; show the same thing for PB.

3. *Zero-One Law.* Consider a product probability space $(\times\Omega_i, \times\mathscr{S}_i, \times\mu_i)$ and let $\mu = \times_{i=1}^{\infty} \mu_i$. Let A be an event with the property that if $x = (x_1, x_2, \ldots) \in A$ and if $y = (y_1, y_2, \ldots)$ satisfies $x_n = y_n$ for $n \geq N$, then $y \in A$ (for arbitrary N). Show that $\mu(A) = 0$ or 1.

4. Let r_n be an independent sequence of r.v.'s and let $\phi(x_1, x_2, \ldots) = \phi(x)$ be a Borel function of x, where each x_i is real, such that $\phi(x) = \phi(y)$ if $x_n = y_n$ for all but a finite number of n's. Prove that $\phi(x)$ is constant a.s. in $(\times\Omega_i, \times\mathscr{S}_i, \times\mu_i)$.

5. *Borel-Cantelli.* Let r_n be a sequence of independent r.v.'s that only take the values 0 and 1, and suppose that $P(r_n = 1) = p_n$. Prove that

(a) $\sum_{n=1}^{\infty} p_n < \infty$ if and only if $P(\sum r_n < \infty) = 1$.

(b) $\sum_{n=1}^{\infty} p_n = \infty$ if and only if $P(\sum r_n = \infty) = 1$.

6. Let E_n be a sequence of independent events and let I_n be the indicator function of E_n. The number $N = \sum_{n=1}^{\infty} E(I_n)$ represents the expected number of E_n events that occur. Let $p = P(\cup E_j)$ be the probability that at least one E_j event will occur. Show that

$$P \leq N \leq p + 0(p^2).$$

[*Hint*: $p \leq N = \sum_j p_j$, where $p_j = PE_j$. Write $p_j = -\log(1 - p_j) + 0(p_j^2)$.]

12.2 INEQUALITIES

In order to study convergence of sums of independent r.v.'s we will require certain inequalities, which are presented in this section. The first theorem is Chebyshev's inequality. This is just a generalization of Theorem 2.3.3, which was proved for simple r.v.'s. Since the proof is the same for arbitrary r.v.'s, we will omit the details.

THEOREM 1

If r is a r.v. such that Er^2 exists, then for every $\epsilon > 0$,

$$P(|r| \geq \epsilon) \leq \epsilon^{-2} E(r^2). \tag{1}$$

COROLLARY

$$P(|r - \mu| \geq \epsilon) \leq \frac{1}{\epsilon^2} E(r - \mu)^2.$$

Now for a sequence r_i of independent r.v.'s,

$$P(|r_1 + \cdots + r_n| \geq \epsilon) \leq \epsilon^{-2}\left(\sum_{i=1}^{n} Er_i^2\right)$$

[by applying inequality (1) to their sum and using the fact that $E(\sum_{i=1}^{n} r_i)^2 = \sum_{i=1}^{n} Er_i^2$]. However, by taking greater advantage of the property of independence, stronger inequalities can be proved. The following inequality was derived by Hajek and Rényi [2].

THEOREM 2

Let r_n be a sequence of independent r.v.'s such that $Er_n = 0$ and $\operatorname{Var} r_n = \sigma_n^2$. If a_k is a decreasing sequence of positive constants, then for any positive integers m and n, with $m < n$ and arbitrary $\epsilon > 0$,

$$P(\max_{m \leq k \leq n} a_k|r_1 + \cdots + r_k| \geq \epsilon) \leq \frac{1}{\epsilon^2}\left(a_m^2 \sum_{k=1}^{m} \sigma_k^2 + \sum_{k=m+1}^{n} a_k^2 \sigma_k^2\right). \tag{2}$$

Proof

Let $r = \sum_{k=m}^{n-1} s_k^2(a_k^2 - a_{k+1}^2) + a_n^2 s_n^2$, where $s_k = \sum_{j=1}^{k} r_j$. Then Er reduces to the expression in parentheses appearing on the right-hand side of (2). Now we integrate over part of the space, as in the proof of Chebyshev's inequality, obtaining

$$Er \geq \int_A r\, dP, \tag{3}$$

where $A = (\max_{m \leq k \leq n} a_k|s_k| \geq \epsilon)$. The event A occurs if and only if $a_k|s_k| \geq \epsilon$ for at least one value of k. However, this last event is equivalent to the event $\sum_{k=m}^{n} A_k$, where

$$A_k = (a_j|s_j| < \epsilon \quad (m \leq j < k); \qquad a_k|s_k| \geq \epsilon).$$

Hence, denoting the indicator function of A_k by I_k,

$$\int_A r\,dP = \sum_{i=m}^{n} \int_{A_i} r\,dP = \sum_{i=m}^{n} \int rI_i\,dP. \tag{4}$$

To simplify the last integral, we first evaluate for $k \geq i$:

$$E(s_k^2 I_i) = E\left[\left(s_i^2 + 2\sum_{j=i+1}^{k} s_i r_j + \sum_{j=i+1}^{k} r_j^2 + 2\sum_{k\geq j>h>i} r_j r_h\right) I_i\right]$$

$$\geq E\left[\left(s_i^2 + 2\sum_{j=i+1}^{k} s_i r_j + 2\sum_{k\geq j>h>i} r_j r_h\right) I_i\right].$$

Since s_i and I_i depend only on $r_1, \ldots, r_i$,

$$E(s_i r_j I_i) = E(s_i I_i)E(r_j) = 0$$

for $j > i$ by the assumed independence of the r.v.'s r_n. Similarly, for $j > h > i$, $E(r_j r_h I_i) = 0$ and so

$$E(s_k^2 I_i) \geq E(s_i^2 I_i) \geq \frac{\epsilon^2}{a_i^2} PA_i,$$

since the event A_i has occurred. Using this last estimate and the definition of r,

$$E(rI_i) \geq \sum_{k=1}^{n-1} \left[\frac{\epsilon^2}{a_i^2}(a_k^2 - a_{k+1}^2) + \frac{\epsilon^2}{a_i^2} a_n^2\right] PA_i \tag{5}$$

$$\geq \left[\sum_{k=i}^{n-1} (a_k^2 - a_{k+1}^2) + a_n^2\right] \frac{\epsilon^2}{a_i^2} PA_i = \epsilon^2 PA.$$

If we insert the estimate (5) into (4), (3) reduces to

$$Er \geq \epsilon^2 \sum_{i=m}^{n} PA_i = \epsilon^2 PA,$$

which is just (2) in a different notation.

COROLLARY

If the independent r.v.'s r_n are integrable, then for every $\epsilon > 0$,

$$P(\max_{1\leq k\leq n} |s_k - Es_k| \geq \epsilon) \leq \frac{1}{\epsilon^2} \sum_{k=1}^{n} \sigma^2 r_k, \tag{6}$$

where $s_k = r_1 + r_2 + \cdots + r_k$.

Proof

If one of the variances is infinite, then the inequality (6) is trivial. On the other hand, if the variances are finite, inequality (6) follows immediately from (2) by choosing $a_i = 1$ for $i = 1, \ldots, n$ and replacing r_k by $r_k - Er_k$.

Inequality (6) is known as Kolmogorov's inequality. The next theorem gives an estimate in the opposite direction for the expression appearing on the left-hand side of inequality (6).

THEOREM 3

If the independent r.v.'s r_k are integrable and $|r_k| \leq c < \infty$, then

$$1 - \frac{(\epsilon + 2c)^2}{\sum_{k=1}^{n} \sigma^2 r_k} \leq P(\max_{1 \leq k \leq n} |s_n - Es_k| \geq \epsilon) \tag{7}$$

for every $\epsilon > 0$, where $s_k = \sum_{i=1}^{k} r_i$.

Proof

Since $|r_k| \leq c$, $|Er_k| \leq c$ and so $|r_k - Er_k| \leq 2c$, which implies that the variances $\sigma^2 r_k$ are finite. Moreover, the hypotheses of the theorem are still valid for the r.v.'s $\bar{r}_k = r_k - Er_k$ and $E\bar{r}_k = 0$. To save writing, we drop the bar over the r_k. However, note that $|r_k| \leq 2c$.

Let $A_0 = \Omega$ and

$$A_k = (|s_1| < \epsilon, |s_2| < \epsilon, \ldots, |s_k| < \epsilon) = (\max_{j \leq k} |s_j| < \epsilon)$$

$$B_k = A_{k-1} - A_k = (|s_1| < \epsilon, |s_2| < \epsilon, \ldots, |s_{k-1}| < \epsilon, |s_k| \geq \epsilon);$$

then the events B_k are disjoint, $A_k \supset A_{k+1}$, and

$$A_n' = \sum_{k=1}^{n} B_k. \tag{8}$$

Since $s_{k-1} + r_k = s_k$,

$$s_{k-1} I_{k-1} + r_k I_{k-1} = s_k I_{k-1}, \tag{9}$$

where I_k is the indicator function of the event A_k. Letting J_k denote the indicator function of B_k, $J_k = I_{k-1} - I_k$ and so

$$s_k I_{k-1} = s_k I_k + s_k J_k. \tag{10}$$

From (9) and (10), it follows that

$$s_{k-1} I_{k-1} + r_k I_{k-1} = s_k I_k + s_k J_k. \tag{11}$$

Squaring both sides of (11) and taking expectations yields

$$E(s_{k-1} I_{k-1})^2 + \sigma^2 r_k P A_{k-1} = E(s_k I_k)^2 + E(s_k J_k)^2, \tag{12}$$

since $I_k J_k = 0$ and $E(s_{k-1} I_{k-1} r_k I_{k-1}) = E(s_{k-1} I_{k-1}) E r_k = 0$. Recalling that $|r_k| \leq 2c$, it follows that

(13) $$|s_k J_k| \leq |s_{k-1} J_k| + |r_k J_k| \leq (\epsilon + 2c) J_k.$$

Since $A_{k-1} \supset A_n$, $PA_{k-1} \geq PA_n$. Using this last estimate for the left-hand side of (12) and the estimate in (13) for the right-hand side,

$$E(s_{k-1} I_{k-1})^2 + \sigma^2 r_k PA_n \leq E(s_k I_k)^2 + (\epsilon + 2c)^2 PB_k.$$

Summing over $k = 1, 2, \ldots, n$ leads to

(14) $$\left(\sum_{k=1}^{n} \sigma^2 r_k\right) PA_n \leq E(s_n I_n)^2 + (\epsilon + 2c)^2 \sum_{k=1}^{n} PB_k \leq (\epsilon + 2c)^2$$

upon using (8) and the fact that $E(s_n I_n)^2 \leq \epsilon^2 PA_n$. A slight simplification of (14) yields inequality (7).

The following inequality was derived by Ottaviani.

THEOREM 4

Let r_i $(i = 1, \ldots, n)$ be a set of independent r.v.'s; suppose that

$$P(|r_{k+1} + \cdots + r_n| > \epsilon) < \frac{1}{2} \qquad (k = 0, 1, \ldots, n-1)$$

for some $\epsilon > 0$. Then

$$P(\max_{1 \leq k \leq n} |r_1 + \cdots + r_k| > 2\epsilon) \leq 2P(|r_1 + \cdots + r_n| > \epsilon).$$

Proof

As in the previous theorems,

$$P(\max_{1 \leq k \leq n} |s_k| > 2\epsilon) = \sum_{k=1}^{n} P(|s_1| \leq 2\epsilon, \ldots, |s_{k-1}| \leq 2\epsilon, |s_k| > 2\epsilon);$$

hence

$$P(\max_{1 \leq k \leq n} |s_k| > 2\epsilon) \leq 2 \sum_{k=1}^{n} P(|s_1| \leq 2\epsilon, \ldots, |s_{k-1}| \leq 2\epsilon, |s_k| > 2\epsilon) P(|s_n - s_k| < \epsilon),$$

since by assumption, $P(|s_n - s_k| > \epsilon) < \frac{1}{2}$. Since the r_i are independent, the last inequality becomes

$$\begin{aligned} P(\max_{1 \leq k \leq n} |s_k| > 2\epsilon) &\leq 2 \sum_{k=1}^{n} P(|s_1| \leq 2\epsilon, \ldots, |s_{k-1}| \leq 2\epsilon, |s_k| > 2\epsilon, |s_n - s_k| \leq \epsilon) \\ &\leq 2 \sum_{k=1}^{n} P(|s_1| \leq 2\epsilon, \ldots, |s_{k-1}| \leq 2\epsilon, |s_k| > 2\epsilon, |s_n| > \epsilon) \leq 2P(|s_n| > \epsilon). \end{aligned}$$

EXERCISES

1. Prove Theorem 1.

2. (a) Employ the same method of proof as in the proof of Chebyshev's inequality to show that if $u(x) \geq 0$ and $u(r) \geq t$ for $r \in A$, then

$$P(r \in A) \leq t^{-1} E(u(r)),$$

provided that the expectation on the right-hand side exists.
(b) Suppose that $Er = 0$ and $u = (x + a)^2$; then use part (a) to show that

$$P(r \geq t) \leq \frac{\sigma^2 + a^2}{(t + a)^2}$$

for $a > 0$ and $t > 0$. Minimize the right-hand side by a proper choice of a and thus prove that

$$P(r \geq t) \leq \frac{\sigma^2}{\sigma^2 + t^2} \cdot$$

3. (a) Verify the following inequality if $0 \leq u(r) < 1$ for $r \in A$ and $u(r) \leq 0$ otherwise: $E(u(r)) \leq P(r \in A)$.
(b) Suppose that $r \geq 0$, $Er = 1$, and $Er^2 = a$; then for $0 < t < 1$,

$$P(r \geq t) \geq \frac{(1 - t)^2}{(1 - t)^2 + a - 1} \cdot$$

[*Hint:* The polynomial $u(x) = (x - t)(r + 2l - x)l^{-2}$ is positive if $(t < x < t + 2l)$ and $u(x) \leq 1$. Hence

$$P(t < r < t + 2l) \geq Eu(r) = [2l(1 - t) - (t - 1) - (1 - t)^2]l^{-2};$$

now choose l so that the right-hand side is maximized.]

12.3 CONVERGENCE

We recall that a sequence of independent r.v.'s either converges a.s. or diverges a.s. The following theorem, formulated by Khintchine and Kolmogorov [3] offers a criterion for convergence.

THEOREM 1

Let r_i be independent r.v.'s with variances $\sigma_i^2 < \infty$. If $\sum_{i=1}^{\infty} \sigma_i^2 = \sigma^2 < \infty$, then $\sum_1^{\infty} (r_i - Er_i)$ converges a.s.

Proof

From Kolmogorov's inequality (2.6) for $s_m - s_{m+j}$,

$$P(\max_{1 \leq j \leq k} |s_m - s_{m+j} - E(s_m - s_{m+j})| \geq \epsilon) \leq \frac{1}{\epsilon^2} \sum_{j=m+1}^{m+k} \sigma_j^2 \leq \frac{1}{\epsilon^2} \sum_{j=m+1}^{\infty} \sigma_j^2.$$

When $k \to \infty$, this becomes

$$P(\sup_{n>m} |s_m - Es_m - (s_n - Es_n)| \geq \epsilon) \leq \frac{1}{\epsilon^2} \sum_{j=m+1}^{\infty} \sigma_j^2, \tag{1}$$

provided that ϵ is a point of continuity of the d.f. of $\sup_{n>m} |s_m - Es_m - (s_n - Es_n)|$. Since the points of discontinuity form a set of measure zero, (1) is valid for every $\epsilon > 0$, as can be seen by using a continuity argument. Since $\lim_{m\to\infty} \sum_{j=m+1}^{\infty} \sigma_j^2 = 0$, it follows from (1), by applying Corollary 2 of Theorem 7.3.3, that $s_n - Es_n$ converges a.s.

COROLLARY

If, in addition to the hypotheses of the theorem, $\sum_{i=1}^{\infty} Er_i$ converges, then $\sum_{i=1}^{\infty} r_i$ converges a.s.

THEOREM 2

If the r_i are independent r.v.'s that are uniformly bounded, then $\sum_{i=1}^{\infty} (r_i - Er_i)$ converges a.s. if and only if $\sum_{i=1}^{\infty} \sigma_i^2$ converges.

Proof

By Theorem 1, the convergence of $\sum_{i=1}^{\infty} \sigma_i^2$ implies the a.s. convergence of $\sum_{i=1}^{\infty} (r_i - Er_i)$. To prove the "only if" part, we will show that if $\sum_{i=1}^{\infty} \sigma_i^2 = \infty$, then $\sum_{i=1}^{\infty} (r_i - Er_i)$ diverges a.s. Following the proof leading to (1) but using inequality 2.7 instead gives

$$P(\sup_{n>m} |s_m - Es_m - (s_n - Es_n)| \geq \epsilon) \geq 1 - \frac{(\epsilon + 2c)^2}{\sum_{j=m+1}^{\infty} \sigma_j^2}$$

and, since $\sum_{j=m+1}^{\infty} \sigma_j^2 = \infty$ for arbitrarily large m, $s_n - Es_n$ diverges a.s.

COROLLARY

If the r_i are independent r.v.'s that are uniformly bounded, then $\sum_{i=1}^{\infty} r_i$ converges a.s. if and only if $\sum_{i=1}^{\infty} \sigma_i^2$ and $\sum_{i=1}^{\infty} Er_i$ converge.

Proof

We have already proved that if $\sum_{i=1}^{\infty} \sigma_i^2$ and $\sum_{i=1}^{\infty} Er_i$ converge, then $\sum_{i=1}^{\infty} r_i$ converges a.s. To prove the converse result, let r_n and x_n be identically distributed r.v.'s for $n = 1, 2, 3, \ldots$, where $r_1, x_1, r_2, x_2, r_3, \ldots$ is a sequence of independent r.v.'s. Let $y_n = r_n - x_n$; then

$$Ey_n = Er_n - Ex_n = 0 \qquad \sigma^2 y_n = \sigma^2 r_n + \sigma^2 x_n \equiv 2\sigma_n^2$$

and $|y_n| \leq 2c$ if $|r_n| \leq c$. If $\sum r_n$ converges a.s., so does $\sum x_n$ and $\sum r_n - \sum x_n = \sum y_n$. Accordingly, the above theorem implies that $\sum \sigma^2 y_n = 2 \sum \sigma_n^2$ converges. Another application of the theorem shows that $\sum (r_n - Er_n)$ converges a.s.; hence $\sum Er_n = \sum r_n - \sum (r_n - Er_n)$ converges.

The following generalization of Theorems 11.6.9 and 11.6.10 was derived by Steinhaus.

THEOREM 3

Let r_i be a sequence of independent r.v.'s such that $Er_i = 0$, $\sigma^2 r_i = 1$, and $P(1/c \leq |r_i| \leq c) = 1$ for some $c > 0$ $(i = 1, 2, 3, \ldots)$. Then $\sum_{i=1}^{\infty} a_i r_i$ converges a.s. if and only if $\sum_{i=1}^{\infty} a_i^2$ converges.

Proof

Since $y_i = a_i r_i$ are independent r.v.'s and $\sigma^2 y_i = a_i^2$, the a.s. convergence of $\sum_{i=1}^{\infty} a_i r_i$ follows from the fact that $\sum_{i=1}^{\infty} a_i^2 < \infty$ according to Theorem 1. We will prove the converse result by showing that $\sum_{i=1}^{\infty} a_i^2 = \infty$ implies that $\sum_{i=1}^{\infty} a_i r_i$ diverges a.s. If $|a_i| \leq K$ for all i, then $|y_i| \leq cK$ (a.s.) and the a.s. divergence of $\sum_{i=1}^{\infty} a_i r_i$ follows from the corollary to Theorem 2. On the other hand, there might be an infinite subsequence a_{n_i} such that $a_{n_i} \to \infty$; in this case, $\sum a_n r_n$ does not converge because $|r_i| \geq 1/c$ and so $a_i r_i$ does not approach zero as $i \to \infty$.

The next convergence criterion, known as Kolmogorov's three-series theorem (see [1] and [4]), is based upon the fact that if r_i is eventually a.s. uniformly bounded and if we set $y_i = r_i$ when r_i is bounded and $y_i = 0$ otherwise, then $\sum y_i$ and $\sum r_i$ converge or diverge together. This process of truncation is used quite frequently in probability theory; we will state it formally as a definition for ease in reference.

DEFINITION 1

The r.v. r is *truncated* at $a > 0$ when we replace r by $r^a = r$ if $|r| < a$ and by $r^a = 0$ if $|r| \geq a$.

Note that even though all moments of r do not exist, the moments of r^a always exist. Furthermore, we can always choose a sufficiently large, in order to make $P(r \neq r^a) = P(|r| \geq a)$ arbitrarily small. From this last fact, it follows that

$$P \bigcup_{j=1}^{\infty} (r_j \neq r_j^{a_j}) \leq \sum_{j=1}^{\infty} P(|r_j| \geq a_j) < \epsilon$$

for an arbitrary $\epsilon > 0$ when a_j is chosen so that $P(|r_j| \geq a_j) \leq \epsilon/2^j$. Hence if we are interested in asymptotic properties—for example, the convergence of the series $\sum_{i=1}^{\infty} r_i$—then an arbitrary sequence of r.v.'s r_i can be replaced with impunity by a bounded sequence $r_i^{c_i}$ of r.v.'s, using Theorem 1.2 and Lemma 1.1.

THEOREM 4

The series $\sum_{i=1}^{\infty} r_i$ of independent summands converges a.s. if and only if for a fixed $a > 0$ the following three series converge:

1. $\sum_{i=1}^{\infty} P(|r_i| \geq a)$.
2. $\sum_{i=1}^{\infty} Er_i^a$.
3. $\sum_{i=1}^{\infty} \sigma^2 r_i^a$.

Proof

Conditions 2 and 3 imply that $\sum r_i^a$ converges a.s. by the corollary to Theorem 1. Theorem 1.2 and Lemma 1.1, along with condition 1, imply that $\sum r_i$ converges a.s.

Now suppose that $\sum r_i$ converges a.s. Then $r_i \to 0$ with a probability 1 and so $\sum_{i=1}^{\infty} P(|r_i| \geq a)$ converges by the corollary to Theorem 1.3. Therefore, Lemma 1.1 and Theorem 1.2 imply that $\sum_{i=1}^{\infty} r_i^a$ converges a.s. and, by the corollary to Theorem 2, the series 2 and 3 converge.

For independent r.v.'s, nearly all modes of convergence are equivalent, as the following two theorems show. The first theorem was derived by Lévy.

THEOREM 5

For a series $\sum_{i=1}^{\infty} r_i$ with independent summands, the following are all equivalent:

1. Convergence a.s.
2. Convergence in probability.
3. Convergence in distribution.[2]

Proof

We recall that part $1 \to 2 \to 3$ (see Theorems 7.3.4 and 8.3.3). To show that part 2 implies part 1, we require the following generalization of Theorem 2.4: If $P\left(|\sum_{i=m}^{\infty} r_i| > \epsilon\right) < \frac{1}{2}$, then

$$P\left(\sup_{n>m} |s_n - s_m| > 2\epsilon\right) \leq 2P\left(|\sum_{i=m}^{\infty} r_i| > \epsilon\right),$$

which is proved in the same manner as (1). The result follows immediately from this last inequality by Corollary 2 of Theorem 7.3.3. To complete the proof, we need only show that part 3 implies part 2. Let F_i and χ_i be the d.f. and c.f., respectively, of the r.v. r_i. Since the r.v.'s r_i are independent, it follows that the d.f. and c.f. of $s_n = r_1 + \cdots + r_n$ are $F_1 * F_2 * \cdots * F_n = G_n$ and $\phi_n = \chi_1 \cdots \chi_n$ respectively. By hypothesis, $G_n \to G(c)$ and therefore $\phi_n \to \phi$, the c.f. of G on R. The Cauchy criterion for convergence of infinite products shows that $\phi_{mn} = \chi_m \chi_{m+1} \cdots \chi_n \to 1$ as $m, n \to \infty$. Therefore, $s_n - s_m \to 0$ in distribution by Lévy's continuity theorem 10.6.5 and so, according to Theorem 8.3.5, $s_n - s_m \to 0(P)$, which implies that $\sum_{i}^{\infty} r_i$ converges in probability.

COROLLARY

A series $\sum r_i$ of independent r.v.'s converges a.s. if and only if $\prod_{k=1}^{n} \chi_n \to \chi$ and χ is continuous at the origin or $\chi \neq 0$ on a set of positive Lebesgue measure (see [6]).

THEOREM 6

If the independent r.v.'s r_i are uniformly bounded and $Er_i = 0$ for $i = 1, 2, 3, \ldots$, then convergence a.s., convergence in probability, convergence in quadratic mean,[3] and convergence in distribution are all equivalent.

[2]Also called "convergence in laws."
[3]Also called "convergence in mean of order 2."

Proof

If we show that convergence in quadratic mean (convergence in q.m.) is equivalent to convergence a.s., the last theorem would yield the result. The series $\sum r_i$ converges in q.m. if and only if

$$E(s_m - s_n)^2 = \sum_{m+1}^{n} \sigma^2 r_i \to 0$$

as $m, n \to \infty$. Therefore, $\sum_{i=1}^{\infty} \sigma^2 r_i$ converges and so, by Theorem 2, $\sum_{i=1}^{\infty} r_i$ converges a.s.; the converse is also true.

Although the proofs of the theorems in this section were given for real-valued r.v.'s, they can easily be extended to complex-valued r.v.'s.

EXERCISES

1. Let r_i be independent r.v.'s with variances σ_i^2. Show that $\sum_i r_i = r$, where the convergence is in q.m., if and only if the series $\sum \sigma_i$ and $\sum Er_i$ converge. (*Hint:* Apply the Cauchy convergence criterion.)

2. Prove that $Er = \sum_{i=1}^{\infty} Er_i$ and $Er^2 - (Er)^2 = \sigma^2 = \sum_{i=1}^{\infty} \sigma_i^2$. [*Hint:* Use the following properties of mean convergence: $r_i \to r$ implies that $Er_i \to Er$ and $Er_i^2 \to Er^2$.]

3. Show that if for every $\epsilon > 0$

$$P(\sup_{n>m} |r_n - r_m| \geq \epsilon) \leq \frac{1}{\epsilon^2} \sum_{m+1}^{\infty} \sigma_i^2,$$

then the inferior and superior limits of the r_i are a.s. finite and

$$P(\overline{\lim}\, r_i - \underline{\lim}\, r_i \geq 2\epsilon) \leq \frac{1}{\epsilon^2} \sum_{m+1}^{\infty} \sigma_i^2.$$

4. Prove the following generalization of the three-series theorem. Let r_i be independent r.v.'s and let $\{a_n\}$, $\{b_n\}$ be sequences of numbers whose inferior limits are positive and whose superior limits are bounded. Define $r'_n = r_n$ for $-a_n \leq r_n \leq b_n$ and $r'_n = 0$ otherwise. Then $\sum_{i=1}^{\infty} r_i$ converges a.s. if and only if the series $\sum_1^{\infty} P(r_i \neq r'_i)$, $\sum_1^{\infty} Er'_i$, $\sum_1^{\infty} \sigma^2 r'_i$ converge.

5. Give a direct proof of the fact that convergence in probability of the series $\sum r_i$ with independent summands r_i implies the convergence of the products $\prod_{i=1}^{\infty} \chi_i$ as $n \to \infty$ uniformly in t on every compact subset of R. [*Hint:* Use the fact that $|e^{itr_n} - e^{itr}| > \epsilon$ implies that $|r_n - r|\,|t| > \epsilon$.]

6. In the corollary to Theorem 5, what happens if $\prod_{i=1}^{n} \chi_i \to 0$?

7. Show that the corollary of Theorem 2 remains true if we only require that $P(\sum r_i \text{ converges}) > 0$.

8. We say that r is *centered* at c if we replace r by $r - c$.

(a) An r.v. r is centered at its expectation if $Er = 0$.
(b) Show that $\sigma^2(r - c) = \sigma^2 r$.

9. (a) Let $\bar{r} = r$ for $|r| < a$ and $\bar{r} = a$ for $|r| \geq a$; show that there exists a centering constant $-c$ such that $E(\overline{r + c}) = 0$.
(b) Prove that either c is unique or else there is a whole interval I of such constants on which the d.f. F of the r.v. r is a constant.

12.4 SYMMETRIZATION

In the last section we proved the corollary to Theorem 3.2 by replacing r_n by $r_n - x_n$, where r_n and x_n are independent, identically distributed r.v.'s. This technique is used quite frequently in probability theory and is called *symmetrization*. The reason for this name appears in the following definition.

DEFINITION 1

A r.v. r is said to be *symmetric* if for every x,

(1) $$P(r \leq x) = P(r \geq -x).$$

The *symmetrization procedure* consists of assigning to a family of r.v.'s $r = \{r_i, i \in I\}$ the *symmetrized family* ${}^s r = \{r_i - r'_i, i \in I\}$, where the family $r' = \{r'_i, i \in I\}$ is independent of r and has the same distribution.

The d.f. and c.f. of a symmetric r.v. are also said to be symmetric. Equation (1) is equivalent to $F(-x - 0) = 1 - F(x)$, or $F(a) - F(b) = F(-b) - F(-b)$, for every pair of continuity points $a < b$ of F, or $\chi = \overline{\chi}$ is real. If r has a density function f, then r is symmetric if and only if $f(x) = f(-x)$; that is, f is symmetric about the y-axis.

EXAMPLE 1

1. (*Triangular Density*). The r.v. r whose density function is $f(x) = 1 - |x|$ for $|x| \leq 1$ and $f(x) = 0$ otherwise is symmetric.
2. The discrete r.v. r such that $P(r = 1) = P(r = -1) = P(r = -2) = P(r = 2) = \frac{1}{4}$ is symmetric.
3. Symmetrization of the r.v. that is distributed uniformly on [0, 1] leads to the r.v. that has the triangular density described above.

A useful feature of the symmetrization method is that it often enables us to reduce problems to simpler ones involving symmetric r.v.'s whose c.f.'s are real valued and nonnegative.

THEOREM 1

Let r be a r.v. with a d.f. F and a c.f. χ. Then the symmetrized r.v. ${}^s r$ has a d.f. $F^s = F * (1 - F(-x))$ at a continuity point and a c.f. $\chi^s = |\chi|^2$.

Proof

Since ${}^s r = r + (-r)$ and the r.v.'s r and $-r'$ are independent, we have $F^s = F_r * F_{-r}$ and $\chi^s = \chi_r \chi_{-r}$. Substituting $F_{-r} = P(-r \leq x) = P(r \geq -x) = 1 - P(r < -x) = 1 - F(-x - 0)$ and $\chi_{-r} = \overline{\chi}_r$ [see equation (2) of Section 10.2] in the last expressions yields the desired results (see Exercise 2).

There are many useful inequalities connecting ${}^s r$ and r. The next theorem is an example of such an inequality which involves the notion of a median.

DEFINITION 2

The *median* of a r.v. r is a number m (or mr) such that

(2) $$P(r \geq m) \geq \frac{1}{2} \qquad P(r \leq m) \geq \frac{1}{2}.$$

LEMMA 1

A r.v. r either has a unique finite median or has all points of a closed interval of R as medians.

Proof

Inequalities (2) are equivalent to

(3) $$F(m) \geq \frac{1}{2} \geq F(m - 0).$$

The line $y = \frac{1}{2}$ intersects the curve in no points, a single point, or a segment parallel to the x-axis. In the first two situations we have a unique median, while in the last we have a closed interval of medians.

The following theorem bounds the truncated variance for the symmetrized r.v. To avoid clumsy notation, we introduce the function

$$g(x) = \begin{cases} x^2 & (|x| \leq 1) \\ 1 & (|x| > 1) \end{cases}.$$

Then the truncated variance of a r.v. r with $Er = 0$ "truncated" at 1 is simply $Eg(r)$.[4]

THEOREM 2

Let m be the median of a r.v.; then

$$Eg({}^s r) \geq \frac{1}{2} Eg(r - m).$$

Proof

We can assume that $m = 0$ without loss of generality; otherwise we can consider the r.v. $r - m$. Let $N = (x \geq 0, y \leq 0) + (x < 0, y \geq 0)$ be the set of points in R_2 such that x and y are of opposite sign. For such points, $g(x - y) \geq g(x)$; therefore,

$$\begin{aligned} Eg({}^s r) &= \iint g(x - y)\, dF(x)\, dF(y) \\ &= \iint_N g(x - y)\, dF(x)\, dF(y) \\ &\geq \int_{y \leq 0} dF(y) \int_{x \geq 0} g(y)\, dF(x) + \int_{y \leq 0} dF(y) \int_{x < 0} g(x)\, dF(x) \\ &\geq \frac{1}{2} Eg(r), \end{aligned}$$

since the median $m = 0$.

THEOREM 3

Suppose that r_i $(i = 1, 2, \ldots, n)$ are independent r.v.'s whose medians are m_i. Then for every ϵ,

$$\frac{1}{2} P(\max_{i \leq n} (r_i - m_i) \geq \epsilon) \leq P(\max_{i \leq n} {}^s r_i \geq \epsilon). \tag{4}$$

Proof

The r.v.'s r'_i that appear in ${}^s r_i = r_i - r'_i$ have the same d.f.'s as the r_i; therefore, $mr'_i = m_i$. Let

$$A_i = (r_i - m_i \geq \epsilon) \qquad B_i = (r'_i - m_i \leq 0) \qquad C_i = ({}^s r_i \geq \epsilon). \tag{5}$$

By writing a union as a disjoint sum,

[4]Here r truncated at 1 is the r.v. $s = r$ for $|r| \leq 1$ and $s = 1$ for $|r| > 1$.

$$P(\bigcup_{i=1}^{n} A_i B_i) = P(A_1 B_1 + A_2 B_2 (A_1 B_1)' + A_3 B_3 (A_2 B_2)'(A_1 B_1)' + \cdots)$$
$$\geq P(A_1 B_1) + P(A_2 A_1' B_2) + P(A_3 A_2' A_1' B_3) + \cdots.$$

Because of the independence of the families $\{r_i\}$ and $\{r_i'\}$, the last line reduces to

$$PA_1 PB_1 + P(A_2 A_1')PB_2 + P(A_3 A_2' A_1')PB_3 + \cdots$$
$$\geq \frac{1}{2}(PA_1 + P(A_2 A_1') + P(A_3 A_2' A_1') + \cdots) = \frac{1}{2} P(\bigcup_{i=1}^{n} A_i)$$

by the fact that $PB_i \geq \frac{1}{2}$ from (2). Hence,

$$\frac{1}{2} P\left(\bigcup_{i=1}^{n} A_i\right) \leq P\left(\bigcup_{i=1}^{n} A_i B_i\right) \leq P \bigcup_{i=1}^{n} C_i, \tag{6}$$

where the left-hand inequality just summarizes the above calculations while the right-hand inequality is a consequence of $A_i B_i \subset C_i$. A reinterpretation of (6) according to (5) shows that it is just equivalent to (4).

THEOREM 4

Suppose that r_i $(i = 1, 2, \ldots, n)$ are independent r.v.'s whose medians are m_i. Then for every ϵ and every a_i,

$$\frac{1}{2} P(\max_{i \leq n} |r_i - m_i| \geq \epsilon) \leq P(\max_{i \leq n} |{}^s r_i| \geq \epsilon) \tag{7}$$
$$\leq 2P(\max_{i \leq n} |r_i - a_i| \geq \frac{\epsilon}{2}).$$

Proof

Replacing r_i by $-r_i$ in (4) leads to

$$\frac{1}{2} P(\max_{i \leq n} (r_i - m_i) \leq -\epsilon) \leq P(\max_{i \leq n} {}^s r_i \leq -\epsilon)$$

which, when added to (4), yields the left-hand inequality in (7).

To prove the right-hand inequality, we note that the occurrence of the event $(\max_{i \leq n} |{}^s r_i| \geq \epsilon)$ implies that $(\max_{i \leq n} |r_i - a_i| + |r_i' - a_i| \geq \epsilon)$ holds. The occurrence of this last event implies the occurrence of $(\max_{i \leq n} |r_i - a_i| + \max_{i \leq n} |r_i' - a_i| \geq \epsilon)$, which in turn implies that the event $(\max_{i \leq n} |r_i - a_i| \geq \epsilon/2) \cup (\max_{i \leq n} |r_i' - a_i| \geq \epsilon/2)$ holds. Hence

$$P(\max_{i \leq n} |{}^s r_i| \geq \epsilon) \leq P[(\max |r_i - a_i| \geq \epsilon/2) \cup (\max |r_i' - a_i| \geq \epsilon/2)]$$
$$\leq 2P(\max |r_i - a_i| \geq \epsilon/2),$$

since the r.v.'s $\max |r_i - a_i|$ and $\max |r_i' - a_i|$ have the same d.f.

COROLLARY

If $\{r_i\}$ is a sequence of independent r.v.'s, then $r_i - a_i \to 0$ (a.s.) if and only if ${}^s r_i \to 0$ (a.s.) and $a_i - m_i \to 0$.

Proof

Inequality (7) leads to

$$\frac{1}{2} P(\sup_{n \le k} |r_k - m_k| \ge \epsilon) \le P(\sup_{n \le k} |{}^s r_k| \ge \epsilon) \le 2P(\sup_{n \le k} |r_k - a_k| \ge \frac{\epsilon}{2}).$$

Suppose that $r_i - a_i \to 0$ (a.s.) as $i \to \infty$. Taking limits as $n \to \infty$ in the above inequality, it follows from Corollary 2 of Theorem 7.3.3 that $\lim_{n\to\infty} P(\sup_{n \le k} |{}^s r_k| \ge \epsilon) = 0$ and $\lim_{n\to\infty} P(\sup_{n \le k} |r_k - m_k| \ge \epsilon) = 0$. By another application of Corollary 2 of Theorem 7.3.3, ${}^s r_k \to 0$ (a.s.) and $r_k - m_k \to 0$ (a.s.); therefore, $a_i - m_i \to 0$.

Conversely, if ${}^s r_i \to 0$ (a.s.), it follows in a similar manner that $r_i - m_i \to 0$ (a.s.) and, since $m_i - a_i \to 0$, $r_i - a_i \to 0$ (a.s.).

Further symmetrization inequalities appear in the Exercises.

EXERCISES

1. (a) Let c be finite; show that $m(cr) = cm(r)$.
(b) Prove that if r is integrable, then

$$|mr - Er| \le \sqrt{2\sigma^2 r}.$$

[*Hint:* Use Chebyshev's inequality to show that $P(|r - Er| \ge \sqrt{2\sigma^2 r}) \le \frac{1}{2}$.]

2. Show that if h is integrable and F is a d.f., then $\int_{-\infty}^{\infty} h\, dF(x) = \int_{-\infty}^{\infty} h\, dF(x-)$.

[*Hint:* The Borel field is generated by the class of open intervals.]

3. Let ${}^s r = r - r'$; show that if the origin is the median for r, then

$$P(|{}^s r| > t) \ge \frac{1}{2} P(|r| > t).$$

[*Hint:* ${}^s r > t$ whenever $r > t$ and $r' \le 0$.]

4. Prove the following inequalities, proposed by P. Lévy, for independent r.v.'s r_i and arbitrary ϵ:

(a) $P(\max_{i \le n} (s_i - m(s_i - s_n)) \ge \epsilon) \le 2P(s_n \ge \epsilon)$.

(b) $P(\max_{i \le n} |s_i - m(s_i - s_n)| \ge \epsilon) \le 2P(|s_n| \ge \epsilon)$. [*Hint:* Set $s_0 = 0$, $s_k^* = \max_{i \le k} (s_i - m(s_i - s_n))$, $A_k = (s_{k-1}^* < \epsilon,\ \ s_k - m(s_k - s_n) \ge \epsilon)$, and $B_k = (s_n - s_k -$

$m(s_n - s_k) \geq 0)$. Prove part (a) by showing that $P(s_n \geq \epsilon) \geq \sum_{i=1}^{n} PA_i PB_i \geq \frac{1}{2} \sum_{i=1}^{n} PA_i = \frac{1}{2} P(s_n^* \geq \epsilon)$.]

5. Show that if r_i are independent r.v.'s that have a symmetric distribution, then $s_n = r_1 + r_2 + \cdots + r_n$ has a symmetric distribution and

$$P(|s_n| > t) \geq \frac{1}{2} P(\max |r_i| > t).$$

[*Hint:* Let m be the first r_i, which is the largest in absolute value, and let $v = s_n - m$. The four combinations $(\pm m, \pm v)$ have the same distribution. Deduce that

$$P(s_n > t) = P(m + v > t) \geq P(m > t, v \geq 0) \geq \frac{1}{2} P(m > t).]$$

12.5 WEAK LAW OF LARGE NUMBERS

A sequence $\{r_k\}$ of r.v.'s is said to satisfy the *weak law of large numbers* if the sequence $(1/n) \sum_{k=1}^{n} (r_k - Er_k)$ converges to zero in probability. In particular, suppose that $Er_k = \mu$, a constant, and the r_k are independent observations or measurements of some random phenomenon. Then the above law has the interesting interpretation that the ordinary arithmetic mean of n observations converges in probability to the expectation.

The weak law of large numbers (see the corollary to Theorem 3.6.3) was first discovered by J. Bernoulli. We refer the reader to Section 3.6 for some examples illustrating this result.

Suppose that a player A receives one dollar if he gets a head and loses one dollar if he gets a tail. Let s_n be the player's accumulated gain (or loss) after n tosses. If $s_n > 0$ or $s_n = 0$ but $s_{n-1} > 0$, we will say that the player leads at time n. Let us interpret the fact that A leads by using the law of large numbers. This law states that $P(|s_n/n| < \epsilon) \to 1$ as $n \to \infty$ or, for a large value of n, $P(0 \leq s_n < n\epsilon) + P(-n\epsilon < s_n \leq 0) \doteqdot 1$. By symmetry we have $P(0 \leq s_n < n\epsilon) = P(-n\epsilon < s_n \leq 0) \doteqdot \frac{1}{2}$. Since n is fixed and $\epsilon > 0$ is arbitrary, this last statement means that in a large number of different coin-tossing games, the frequency of those in which A leads is close to $\frac{1}{2}$ at time n, for sufficiently large n. The law of large numbers does not assert anything about the fluctuations of the lead in a fixed game. In fact, in a single coin-tossing game it is much more likely that A will lead (or will not lead) throughout the 10^6 trials than that A will lead in approximately $\frac{1}{2}(10)^6$ trials. Further details about fluctuations in coin tossing appear in Chapter 3 of [7].

In 1837, Poisson discovered the following generalization of Bernoulli's result.

THEOREM 1

If the probability of occurrence of an event A in a sequence of independent trials is p_k for the kth trial, then for any $\epsilon > 0$,

$$\lim_{n\to\infty} P\left(\left|\frac{s_n}{n} - \frac{p_1 + p_2 + \cdots + p_n}{n}\right| < \epsilon\right) = 1,$$

where s_n is a r.v. that denotes the number of times the event A occurs in the first n trials.

Both Theorem 1 and Theorem 2 are particular cases of Chebyshev's theorem, which appears below.

THEOREM 2

If r_i form a sequence of pairwise independent r.v.'s possessing finite variances that are uniformly bounded—that is, Var $r_i \leq c$—then for any $\epsilon > 0$,

$$\lim_{n\to\infty} P\left(\left|\frac{1}{n}\sum_{i=1}^{n} r_i - \frac{1}{n}\sum_{i=1}^{n} Er_i\right| < \epsilon\right) = 1.$$

At the beginning of the twentieth century A. A. Markov, realizing that the crux of the proof of Chebyshev's theorem was the application of Chebyshev's inequality coupled with the fact that the variance approached zero, discovered the generalization given in Theorem 3.

THEOREM 3

If a sequence of r.v.'s r_i is such that $(1/n^2)\,\mathrm{Var}\,(\sum_{i=1}^{n} r_i) \to 0$ as $n \to \infty$, then for any $\epsilon > 0$,

$$\lim_{n\to\infty} P\left(\left|\frac{1}{n}\sum_{i=1}^{n} r_i - \frac{1}{n}\sum_{i=1}^{n} Er_i\right| < \epsilon\right) = 1.$$

Note that the r.v.'s r_i do not have to be independent. However, if the r_i are pairwise independent, then Markov's condition reduces to $(1/n^2) \sum_{i=1}^{n} \mathrm{Var}\, r_i \to 0$. From this, it is clear that Chebyshev's theorem is a special case of Markov's theorem.

We leave the direct proofs of these theorems as exercises for the reader. We will deduce them from the more general result stated below.

For a long time, little progress was made in the study of the weak law of large numbers. It was not until 1926 that A. Kolmogorov [4] and [5] succeeded in finding necessary and sufficient conditions for a sequence of independent r.v.'s to obey the law of large numbers. He studied a double sequence of r.v.'s

$$
\begin{array}{cccc}
r_{11} & r_{12} & \cdots & r_{1k_1} \\
r_{21} & r_{22} & \cdots & r_{2k_2} \\
\cdots & \cdots & \cdots & \cdots \\
r_{n1} & r_{n2} & \cdots & r_{nk_n}
\end{array} \tag{1}
$$

Such an array will be denoted by $||r_{nk}||$. The r.v.'s in each row are independent; however, in general, the r.v.'s in different rows are dependent.

DEFINITION 1

The sequence $\{s_n\}$ of r.v.'s is called *stable* if and only if there exists a sequence of constants $\{d_n\}$ such that $s_n - d_n \to 0\,(P)$ as $n \to \infty$.

Note that if $s_n - d_n \to 0\,(P)$, then the fact that $c_n - d_n \to 0$ as $n \to \infty$ implies that $s_n - c_n \to 0\,(P)$ as $n \to \infty$; the converse is also true.

It follows from equation 4.7, by letting $n = 1$, that

$$\frac{1}{2}\,P(|s_i - m_i| \geq \epsilon) \leq 2P(|s_i - d_i| \geq \frac{\epsilon}{2}),$$

where m_i is the median of s_i. Thus if the sequence $\{a_n\}$ is stable, the constants d_n can be taken to be the medians m_n.

DEFINITION 2

The double sequence of r.v.'s $||r_{nk}||$ obeys the weak law of large numbers if and only if the sequence of sums

$$s_n = r_{n1} + r_{n2} + \cdots + r_{nk_n}$$

is stable.

The next definition is required for the statement of Kolmogorov's result that appears in Theorems 4 and 5.

DEFINITION 3

Two double sequences of r.v.'s $||r_{nk}||$ and $||\bar{r}_{nk}||$ are said to be *equivalent* if and only if $k_n = \bar{k}_n$ and $P(s_n \neq \bar{s}_n) \to 0$ as $n \to \infty$.

THEOREM 4

If there exists a system $||\bar{r}_{nk}||$, equivalent to $||r_{nk}||$, for which

$$\sum_{k=1}^{k_n} \text{Var}(\bar{r}_{nk}) \to \infty, \qquad (n \to \infty) \tag{2}$$

then $||r_{nk}||$ obeys the weak law of large numbers.

Proof

For every $\epsilon > 0$,

$$P(|\bar{s}_n - E\bar{s}_n| \geq \epsilon) \leq \epsilon^{-2}\left(\sum_{k=1}^{k_n} \text{Var}(\bar{r}_{nk})\right)$$

by Chebyshev's inequality (Theorem 2.1). It follows from relation (2) that $\bar{s}_n - E\bar{s}_n \to 0(P)$. By the hypothesis of equivalence, $s_n - \bar{s}_n \to 0\ (P)$ which, when added to the last result, yields $s_n - E\bar{s}_n \to 0\ (P)$.

The following lemma gives a criterion for the existence of such an equivalent system in terms of the d.f.'s F_{nk} of the r.v.'s r_{nk}.

LEMMA 1

Let m_{nk} be the median of the r.v. r_{nk}. If

$$\sum_{k=1}^{k_n} \int_{|x|>1} dF_{nk}(x + m_{nk}) \to 0 \tag{3}$$

as $n \to \infty$, then $||\bar{r}_{nk}||$ defined by

$$||\bar{r}_{nk}|| = \begin{cases} r_{nk} & (|r_{nk} - m_{nk}| \leq 1) \\ m_{nk} & (|r_{nk} - m_{nk}| > 1) \end{cases}$$

is equivalent to $||r_{nk}||$.

Proof

If $s_n \neq \bar{s}_n$, then $r_{nk} \neq \bar{r}_{nk}$ for some value of k, which implies that $|r_{nk} - m_{nk}| > 1$. Then

$$P(s_n \neq \bar{s}_n) \leq P\bigcup_{k=1}^{k_n} (|r_{nk} - m_{nk}| > 1)) \leq \sum_{k=1}^{k_n} P(|r_{nk} - m_{nk}| > 1).$$

To evaluate this last sum, we will find the d.f. of $t_{nk} = r_{nk} - m_{nk}$. Now $P(r_{nk} \leq x) = P(r_{nk} \leq x + m_{nk}) = F_{nk}(x + m_{nk})$. Therefore,

$$\sum_{k=1}^{k_n} P(|r_{nk} - m_{nk}| > 1) = \sum_{k=1}^{k_n} \int_{|x|>1} dF_{nk}(x + m_{nk}),$$

which proves the equivalence of the two systems by virtue of relation (3).

It is useful to calculate the stability constants $d_n = E\bar{s}_n$ that appear in Theorem 4 for the system $||\bar{r}_{nk}||$ described in Lemma 1. Then

$$d_n = E\left(\sum_{k=1}^{k_n} \bar{r}_{nk}\right) = \sum_{k=1}^{k_n} \left(m_{nk} + \int_{|x|\le 1} x\, dF_{nk}(x + m_{nk})\right). \tag{4}$$

Condition (2) can also be translated in terms of d.f.'s of the r_{nk} by finding Var $\bar{r}_{nk}$. It is not difficult to see by using relation (3) that an equivalent condition is

$$\sum_{k=1}^{k_n} \int_{|x|\le 1} x^2\, dF_{nk}(x + m_{nk}) \to 0. \qquad (n \to \infty) \tag{5}$$

The converse of Theorem 4 is also true. We will use c.f.'s to simplify Kolmogorov's original proof. Let χ_{nk} be the c.f. of the r.v. r_{nk}.

THEOREM 5

If $||r_{nk}||$ obeys the weak law of large numbers, then there exists an equivalent system $||\bar{r}_{nk}||$ (described in Lemma 1) for which relation (2) holds.

Proof

Since the weak law of large numbers holds, there exist constants d_n such that $s_n - d_n \to 0(P)$ as $n \to \infty$. By Theorem 8.3.5 the d.f. of the sum $s_n - d_n$ converges to a degenerate d.f. (with a jump at the origin); in terms of c.f.'s, $e^{-id_n\lambda} \prod_{k=1}^{k_n} \chi_{nk}(\lambda) \to 1$ uniformly in any finite interval as $n \to \infty$ (see Theorem 10.6.3). Therefore, as $n \to \infty$,

$$\prod_{k=1}^{k_n} |\chi_{nk}(\lambda)| \to 1 \tag{6}$$

uniformly in any finite interval, say $|\lambda| \le 1$.

Setting $a = 1 - |\chi_{nk}(\lambda)|$ in the inequality

$$\log(1 - a) \ge -a \qquad (0 \le a < 1)$$

and taking logarithms of relation (6) we find that as $n \to \infty$,

$$\sum_{k=1}^{k_n} (1 - |\chi_{nk}(\lambda)|) \to 0. \qquad (|\lambda| \le 1) \tag{7}$$

We will first prove the result under the assumption that all the summands r_{nk} are symmetric.

Let $\chi(\lambda)$ be the c.f. of the symmetric r.v. r; then $mr = 0$ and

$$\chi(\lambda) = \int_{-\infty}^{\infty} \cos \lambda x\, dF(x).$$

Hence

$$\frac{1}{2}\int_{-1}^{1}(1-\chi(\lambda))\,d\lambda=\int\left(1-\frac{\sin x}{x}\right)dF(x).$$

From the inequalities $1-(\sin x)/x\geq\frac{1}{10}$ for $|x|>1$ and $1-(\sin x)/x\geq x^2/8$ for $|x|\leq 1$, it follows that

$$\int_{-1}^{1}(1-\chi(\lambda))\,d\lambda\geq\frac{1}{4}\int_{|x|\leq 1}x^2\,dF+\frac{1}{5}\int_{|x|>1}dF. \tag{8}$$

To apply the above estimates to the sum in relation (7), we must be able to remove the absolute value signs. Now if $0\leq a_i\leq 1$, then $1-\prod_{i=1}^{n}a_i\geq 1-a_i$ which, along with relation (6), enables us to deduce that as $n\to\infty$,

$$\sup_{1\leq k\leq k_n}(1-|\chi_{nk}(\lambda)|)\to 0. \qquad (|\lambda|\leq 1)$$

Hence for sufficiently large n, $|\chi_{nk}(\lambda)|=\chi_{nk}(\lambda)$, since the c.f.'s are real (see Theorem 10.2.4). Therefore,

$$\int_{-1}^{1}\sum_{k=1}^{k_n}(1-\chi_{nk}(\lambda))\,d\lambda\geq\frac{1}{4}\sum_{k=1}^{k_n}\int_{|x|\leq 1}x^2\,dF_{nk}+\frac{1}{5}\sum_{k=1}^{k_n}\int_{|x|>1}dF_{nk}\geq 0.$$

From this and from relation (7) it is not difficult to see that conditions (3) and (5) hold, which proves our result for the symmetric summands r_{nk}.

In the general case, we consider the symmetrized r.v.'s ${}^sr_{nk}$ with corresponding d.f.'s F_{nk}^s. By our previous result for symmetric r.v.'s,

$$\sum_{k=1}^{k_n}\int_{|x|\leq 1}x^2\,dF_{nk}^s\to 0$$

$$\sum_{k=1}^{k_n}\int_{|x|>1}dF_{nk}^s\to 0$$

as $n\to\infty$. These last two conditions can be stated simply by

$$\sum_{k=1}^{k_n}Eg({}^sr_{nk})\to 0$$

as $n\to\infty$, where g is the function appearing in Theorem 4.2. The general result now follows from Theorem 4.2.

Theorems 4 and 5 can be restated in terms of d.f.'s with the aid of Lemma 1.

THEOREM 6

In order for $\|r_{nk}\|$ to obey the weak law of large numbers, it is necessary and sufficient that as $n\to\infty$,

$$\sum_{k=1}^{k_n} \int_{|x|>1} dF_{nk}(x + m_{nk}) \to 0$$

$$\sum_{k=1}^{k_n} \int_{|x|\le 1} x^2\, dF_{nk}(x + m_{nk}) \to 0.$$

The function g appearing in Theorem 4.2 satisfies the inequality

$$\frac{x^2}{1+x^2} \le g(x) \le 2\frac{x^2}{1+x^2}.$$

Hence the conditions of Theorem 6 are equivalent to the condition that as $n \to \infty$,

$$\sum_{k=1}^{k_n} E\left(\frac{(r_{nk} - m_{nk})^2}{1 + (r_{nk} - m_{nk})^2}\right) \to 0. \tag{9'}$$

The above versions of the weak law of large numbers require the existence of variances or second moments for the r.v.'s. This condition was weakened by A. A. Markov [10], who obtained the following result.

THEOREM 7

Let r_i be a sequence of independent r.v.'s with a common d.f. Then the weak law of large numbers holds if for any $\delta > 0$, $E|r_i|^{1+\delta}$ all exist and are bounded.

This result was generalized by A. Khintchine [9], who proved that the existence of the expectation is sufficient. Khintchine's result is a corollary of the following theorem by Kolmogorov [4].

THEOREM 8

Let r_i be independent r.v.'s with the same d.f. F—that is, $P(r \le x) = F(x)$. Then $s_n = (r_1 + \cdots + r_n)/n$ is stable if and only if

$$nP(|r| > n) \to 0 \tag{9}$$

as $n \to \infty$.

Proof

Assume that (9) holds or $n \int_{|x|>n} dF(x) \to 0$ as $n \to \infty$. It is not difficult to show that this last condition is equivalent to

$$h(n) = n \int_{|x|>n} dF(x + a) \to 0 \qquad (n \to \infty) \tag{10}$$

for any real number a. We will prove our result by applying Theorem 6 with $r_{nk} = r_n/n$ and $m_{nk} = m/n$, where $m = mr_n$. The first condition becomes

$$n \int_{|x|>n} dF(x + m) \to 0 \qquad (n \to \infty)$$

and is automatically satisfied since it is merely relation (10). The second condition is

$$\frac{1}{n} \int_{|x|\leq n} x^2 \, dF(x + m) \to 0. \qquad (n \to \infty) \tag{11}$$

To verify that this condition holds, let us apply the Cauchy criterion for convergence to relation (10). If $n > b$ and n is sufficiently large, then

$$|h(n) - h(b)| \geq n \int_{b<|x|\leq n} dF(x + m). \tag{12}$$

Hence the right-hand side can be made arbitrarily small by choosing b and n sufficiently large. Now

$$\frac{1}{n} \int_{|x|\leq n} x^2 \, dF(x + m) \leq \frac{1}{n} \int_{|x|\leq b} x^2 \, dF(x + m) + n \int_{b<|x|\leq n} dF(x + m).$$

The first term in the last sum converges to zero for any fixed b. The second term can be made small by choosing b properly. Hence condition (11) is satisfied; this proves the sufficiency of relation (9).

The necessity of (9) follows immediately. If the s_n are stable, then (10) holds by Theorem 6.

COROLLARY

Let $\{r_i\}$ be a sequence of independent r.v.'s with a common d.f. If $Er_i = \mu$ exists and is finite, then $\{r_i\}$ satisfies the weak law of large numbers.

Proof

If Er_i exists and is finite, then $\left(\int_{-\infty}^{-n} + \int_n^{\infty} x \, dF(x)\right) \to 0$ as $n \to \infty$. However, these last integrals are larger than $n \int_{|x|>n} dF$ and so condition (9) is satisfied. To complete the proof, we need only calculate the stability constants d_n. From (4),

$$d_n = m + \int_{|x|\leq n} x \, dF(x + m).$$

However, the constants

$$c_n = \int_{|x| \leq n} x\, dF(x)$$

have the property that $c_n - d_n \to 0$ as $n \to \infty$. Thus c_n are suitable stability constants by the remark following Definition 1. This completes the proof, since $\lim c_n = \mu$.

Incidentally, we have shown that $s_n - c_n \to 0(P)$ as $n \to \infty$ if and only if condition 9 holds.

Example 1.1 shows that Theorem 8 can hold even if the expectation does not exist.

EXAMPLE 1

1. Let $\{r_i\}$ be a sequence of independent r.v.'s with a common density function $f(x) = c/(x^2 \log x)$, where c is chosen so that $\int_2^\infty c/(x^2 \log x)\, dx = 1$. Then Er_i is not finite. However, condition 9 is valid, since

$$nP(|x| > n) = \int_n^\infty \frac{c}{(x^2 \log x)}\, dx \leq \frac{cn}{\log n} \int_n^\infty \frac{1}{x^2}\, dx = \frac{c}{\log n}$$

2. Theorem 8 does not apply to the independent r.v.'s r_i with a common density function $f(x) = 1/x^2$ for $1 \leq x < \infty$.
3. (*Petersburg Paradox*).[5] A and B play the game of heads or tails according to the following rule: If B wins the first $n - 1$ games and A wins the nth, then B pays A 2^n dollars and the game ends. How much should A pay B to play the game, assuming that the game is to be fair?

A's gain for such a game is a r.v. r that assumes the values $2, 2^2, 2^3, \ldots$ with corresponding probabilities $2^{-1}, 2^{-2}, 2^{-3}, \ldots$; therefore, $Er = \infty$. A should be willing to pay an arbitrarily large amount to play this game.

After playing n such games, A's accumulated gain is

$$s_n = r_1 + r_2 + \cdots + r_n, \tag{13}$$

where the r_i are independent r.v.'s with the same d.f. as r. Condition 9 is not satisfied and so s_n/n is not stable.

To obtain an analog of the law of large numbers, let us consider for a moment a game for which the gain is a r.v. r with a finite expectation μ. The accumulated gain s_n after n games is still given by (13). This time, however,

[5]This problem was posed by Daniel Bernoulli (1700–1782).

$s_n/n\mu \to 1(P)$ as $n \to \infty$, according to the weak law of large numbers. In this situation, it would be fair to pay an entrance fee of μ to play a single game, since the ratio of the accumulated gain to the accumlated entrance fees approaches 1. If Er does not exist, it seems reasonable to try to replace $n\mu$ by c_n, where $c_n \to \infty$ as $n \to \infty$. The following result derived by Feller [8] shows that this is indeed possible,

THEOREM 9

Let r_i be independent r.v.'s with a common d.f. F such that $F(0) = 0$. Let $s_n = r_1 + r_2 + \cdots + r_n$. Then there exist constants c_n such that[6]

$$c_n^{-1}s_n - 1 \to 0\,(P) \tag{14}$$

as $n \to \infty$ if and only if

$$\phi(t) = \frac{1}{t(1 - F(t))}\int_0^t x\,dF(x) \to \infty \tag{15}$$

as $t \to \infty$.

Proof

Assume that (14) holds. From the first condition of Theorem 6, for $r_{nk} = c_n^{-1}r_k$ and $k_n = n$,

$$n(1 - F(c_n)) \to 0 \tag{16}$$

as $n \to \infty$. Since the stability constants are 1, equation (4) implies[7] that $n/c_n \to 0$ and so

$$c_n^{-1}n\int_0^{c_n} x\,dF(x) \to 1 \tag{17}$$

as $n \to \infty$. Dividing relation (17) by relation (16), it follows that $\phi(c_n) \to \infty$ as $n \to \infty$. Assuming that $c_{n+1}/c_n \to 1$ as $n \to \infty$, it is not difficult to prove that $\phi(t) \to \infty$ as $t \to \infty$. To verify this last assumption, observe that $r_{n+1}/c_n \to 0(P)$ and so $s_{n+1}/c_n \to 1(P)$ or

$$(s_{n+1}/c_{n+1})(c_{n+1}/c_n) \to 1(P). \tag{18}$$

Since $s_{n+1}/c_{n+1} \to 1(P)$, it is obvious from (18) that $u = \overline{\lim}\,(c_{n+1}/c_n) < \infty$ and $l = \underline{\lim}\,(c_{n+1}/c_n) > 0$. In fact, choosing a sequence $\{n_k\}$ such that $\lim\,(c_{n_k+1}/c_{n_k}) = u$, it follows from (18) that $u = 1$. Similarly, $l = 1$ and so $\lim\,(c_{n+1}/c_n) = 1$. This completes the proof of necessity.

Now assume (15). If for some real number x, $F(x) = 1$, then the expectation μ is finite. Hence (14) follows trivially by Khintchine's theorem with

[6]Relation (14) implies that $c_n \to \infty$ as $n \to \infty$ (see Exercise 7).

[7]Otherwise, by Theorem 8, equation (15) holds trivially.

$c_n = n\mu$. On the other hand, if $F(x) \neq 1$ for every real x, it is possible to find a sequence $\{t_n\}$ such that $t_n \to \infty$ as $n \to \infty$ and, for any $\epsilon > 0$,

$$\lim_{n\to\infty} n(1 - F(t_n)) = \epsilon. \tag{19}$$

To prove this, choose t_n so that

$$1 - F(t_n^-) \geq \epsilon/n \geq 1 - F(t_n). \tag{20}$$

Since $\sum_{n=1}^{\infty} (F(t_n) - F(t_n^-)) \leq 1$, $\lim_{n\to\infty} F(t_n) = \lim_{n\to\infty} F(t_n^-)$ and so (20) implies (19). It is not difficult to verify that (19) and (15) imply that

$$\lim_{n\to\infty} \frac{n}{t_n} \int_0^{t_n} x\,dF = \infty. \tag{21}$$

Furthermore, by choosing a sequence $\{\epsilon_n\}$ tending to zero from above, it is possible to find a sequence $\{t_n\}$, with $t_n \to \infty$ as $n \to \infty$, such that (21) and

$$n(1 - F(t_n)) \to 0 \qquad (n \to \infty) \tag{22}$$

hold simultaneously. We define

$$c_n = n \int_0^{t_n} x\,dF; \tag{23}$$

then $c_n/t_n \to \infty$ from (21).

To show that $d_n = 1$ are suitable stability constants, it is only necessary to verify, according to (4), that $c_n^{-1} n \int_0^{c_n} x\,dF \to 1$ as $n \to \infty$. By definition of the c_n in (23), it is only necessary to show that $(n/c_n) \int_{t_n}^{c_n} x\,dF \to 0$ as $n \to \infty$. This is a consequence of the inequality

$$\frac{n}{c_n} \int_{t_n}^{c_n} x\,dF \leq n(F(c_n) - F(t_n)) \leq n(1 - F(t_n))$$

upon using relation (22). Thus the first condition of Theorem 6 is valid. The second condition of Theorem 6 follows from the estimate

$$\frac{n}{c_n^2} \int_0^{c_n} x^2\,dF \leq \frac{nt_n}{c_n^2} \int_0^{t_n} x\,dF + \frac{n}{c_n^2} \int_{t_n}^{c_n} x^2\,dF \leq \frac{t_n}{c_n} + n(1 - F(t_n)).$$

Hence Theorem 6 implies that relation (14) is valid.

The above limit law is valid for the r.v.'s described in parts 2 and 3 of Example 1. In the Petersburg paradox, we can set $c_n = n \log n$, where $\log n$ is a logarithm to the base 2. A further generalization of the above result is given in Feller's paper (see Exercise 6).

Finally, we obtain a condition that is necessary and sufficient for the law of large numbers to hold for a sequence of arbitrary r.v.'s (not necessarily independent r.v.'s). This result appears in [11].

THEOREM 10

Let r_i be a sequence of r.v.'s and let $s_n = (r_1 + \cdots + r_n)n^{-1}$. Then, for any $\epsilon > 0$,

$$\lim_{n\to\infty} P(|s_n - Es_n| < \epsilon) = 1 \tag{24}$$

if and only if

$$E\left(\frac{(s_n - Es_n)^2}{1 + (s_n - Es_n)^2}\right) \to 0 \tag{25}$$

as $n \to \infty$.

Proof

Suppose that (25) holds. If F_n is the d.f. of the r.v. $s_n - Es_n$, then

$$P(|s_n - Es_n| \geq \epsilon) = \int_{|x|\geq\epsilon} dF_n \leq \frac{1+\epsilon^2}{\epsilon^2} \int_{|x|\geq\epsilon} \frac{x^2}{1+x^2}\, dF_n$$
$$\leq \frac{1+\epsilon^2}{\epsilon^2} \int \frac{x^2}{1+x^2}\, dF_n = \frac{1+\epsilon^2}{\epsilon^2} E\left(\frac{(s_n - Es_n)^2}{1+(s_n - Es_n)^2}\right).$$

This inequality shows that (24) holds.

Now assume (24). It is easy to verify that

$$P(|s_n - Es_n| \geq \epsilon) = \int_{|x|\geq\epsilon} dF_n \geq \int \frac{x^2}{1+x^2}\, dF_n - \int_{|x|<\epsilon} \frac{x^2}{1+x^2}\, dF_n$$
$$\geq \int \frac{x^2}{1+x^2}\, dF_n - \epsilon^2 = E\left(\frac{(s_n - Es_n)^2}{1+(s_n - Es_n)^2}\right) - \epsilon^2.$$

This implies that relation (25) holds.

Theorems 1–3 follow from the above result, since

$$E\left(\frac{(s_n - Es_n)^2}{1+(s_n - Es_n)^2}\right) \leq E(s_n - Es_n)^2 = \frac{1}{n^2} \operatorname{Var} \sum_{k=1}^{n} r_k.$$

EXERCISES

1. Prove Theorems 1–3 directly.

2. Suppose that the identical independent r.v.'s r_i assume the values $2^{n-2\log n}$ for $n = 1, 2, 3, \ldots$ with probabilities $1/2^n$. Show that the law of large numbers holds but that the r_i do not satisfy the condition for Markov's theorem.

3. Prove that the law of large numbers holds for the sequence of independent r.v.'s r_n for which

$$P(r_n = n^2) = P(r_n = -n) = \frac{1}{2}$$

if and only if $\lambda < \frac{1}{2}$.

4. Does the law of large numbers hold for the following sequences of independent r.v.'s r_k $(k = 1, 2, 3, \ldots)$ with the distribution indicated below?

(a) $P(r_k = 2^{n-2\log n - 2\log\log n}) = \dfrac{1}{2^n} \qquad (n = 1, 2, 3, \ldots)$

(b) $P(r_k = \pm 2^k) = \dfrac{1}{2}$.

(c) $P(r_k = \pm 2^k) = 2^{-(2k+1)} \qquad P(r_k = 0) = 1 - 2^{-2k}$.

(d) $P(r_k = n) = \dfrac{c}{n^2 \log^2 n} \qquad \left(n \geq 2;\ c^{-1} = \displaystyle\sum_{n=2}^{\infty} \frac{1}{n^2 \log^2 n}\right)$.

(e) $P(r_k = \pm k) = \dfrac{1}{2}\, k^{-1/2} \qquad P(r_k = 0) = 1 - k^{-1/2}$.

5. (*Bernstein's Theorem*). Let $\{r_i\}$ be a sequence of r.v.'s for which $\operatorname{Var} r_n \leq C$ and $\rho_{ij} \to 0$ as $|i - j| \to \infty$ (ρ_{ij} is the correlation coefficient of r_i and r_j). Show that the law of large numbers holds for the given sequence.

6. (*Feller*). Let $\{r_i\}$ be a sequence of independent r.v.'s. Prove the following result. For a given sequence $B_n > 0$, there exist constants A_n such that for every $\epsilon > 0$,

$$P\left(\left|\frac{r_1 + \cdots + r_n}{B_n} - A_n\right| > \epsilon\right) \to 0 \qquad (n \to \infty)$$

if and only if

$$\sum_{k=1}^{n} \int \frac{x^2}{B_n^2 + x^2}\, dF_k(x + m_k) \to 0, \qquad (n \to \infty)$$

where m_k is a median of the r.v. r_k. Suitable constants A_n are given by

$$A_n = \frac{1}{B_n} \sum_{k=1}^{n} \left[m_k + \int_{|x| < \delta B_n} x\, dF_k(x + m_k)\right],$$

where δ is an arbitrary positive number. [*Hint:* Use the second form of Theorem 6.]

7. Show that if relation (14) is valid, then $c_n \to \infty$. [*Hint:* Use the inequality in Exercise 4.5.]

12.6 STRONG LAW OF LARGE NUMBERS

A sequence $\{r_k\}$ of r.v.'s is said to satisfy the *strong law of large numbers* if the sequence $(1/n)\sum_{k=1}^{n} (r_k - Er_k)$ converges to zero a.s. The law of large numbers was first discovered in the following form by E. Borel in 1909.

THEOREM 1

Consider a sequence of n independent trials in each of which an event A occurs with a probability p or does not occur. Let s_n be the number of occurrences of the event A; then $s_n/n \to p$ (a.s.).

We leave the direct proof of this result as an exercise for the reader; we will deduce it as a particular case of Theorem 3.

A similar result was obtained by F. P. Cantelli [12] in 1917 and a slightly more general theorem was found by G. Polya [13] in 1921. In 1913, Hausdorff had discussed certain special cases of the above result.

In 1930, Kolmogorov [14] discovered the following version of the law of large numbers.

THEOREM 2

If the sequence r_i of independent r.v.'s satisfies the condition

$$\sum_{n=1}^{\infty} \frac{\operatorname{Var} r_n}{n^2} < \infty,$$

then it obeys the strong law of large numbers.

Proof

There is no loss in generality in assuming that $Er_i = 0$. In the Hajek–Rényi inequality (2.2), set $a_k = 1/k$; then

$$P\left(\max_{m\le k\le n} \frac{1}{k}\left|\sum_{i=1}^{k} r_i\right| \ge \epsilon\right) \le \frac{1}{\epsilon^2}\left(\frac{1}{m^2}\sum_{k=1}^{m}\sigma_k^2 + \sum_{k=m+1}^{n}\frac{1}{k^2}\sigma_k^2\right).$$

Letting $n \to \infty$ yields

$$P\left(\sup_{m\le k} \frac{1}{k}\left|\sum_{i=1}^{k} r_i\right| \ge \epsilon\right) \le \frac{1}{\epsilon^2}\left(\frac{1}{m^2}\sum_{k=1}^{m}\sigma_k^2 + \sum_{k=m+1}^{\infty}\frac{1}{k^2}\sigma_k^2\right).$$

Since $\sum_{k=1}^{\infty} \sigma_k^2/k^2 < \infty$ by hypothesis, it follows from this last inequality that

$$\lim_{m\to\infty} P\left(\sup_{m\le k} \frac{1}{k}\left|\sum_{i=1}^{k} r_i\right| \ge \epsilon\right) = 0.$$

[Corollary 2 of Theorem 7.3.3 implies that $(1/k)\sum_{i=1}^{k} r_i \to 0$ (a.s.)]

This theorem immediately yields the following result.

COROLLARY

If $|\operatorname{Var} r_k| \le C$ for $k = 1, 2, 3, \ldots,$ then the sequence of independent r.v.'s r_i obeys the strong law of large numbers.

Kolmogorov [14] used Theorem 2 to prove the sufficiency part of the following more general theorem. An outline of this proof appears in Exercise 3. We will use another method.

THEOREM 3

The existence of the expectation is a necessary and sufficient condition for the strong law of large numbers, under the hypothesis that the r.v.'s are independent and identically distributed.

Proof

Let $\bar{r}_n = r_n$ for $|r_n| < n$ and let $\bar{r}_n = 0$ for $|r_n| \geq n$; then the sequences $\{r_n\}$ and $\{\bar{r}_n\}$ are asymptotically equivalent (see the remarks following Definition 3.1). By Theorem 1.2 with $a_n = 1/n$, it follows that the sequence $\{r_n\}$ obeys the strong law of large numbers if and only if the sequence $\{\bar{r}_n\}$ obeys this law. Since $Er < \infty$, then

$$\sum_{n=1}^{\infty} \frac{\operatorname{Var} \bar{r}_n}{n^2} = \sum_{n=1}^{\infty} \frac{1}{n^2} \int_{-n}^{n} x^2 \, dF < \infty$$

by Theorem 9.5.4. Therefore, by Theorem 2, $\{\bar{r}_n\}$ obeys the strong law of large numbers. This completes the sufficiency part of the proof; the necessity is obvious.

The converse result in Theorem 3 can be generalized. If, for some constant, $(1/n) \sum_{j=1}^{n} r_j \to c$ (a.s.) as $n \to \infty$, then

$$\frac{r_n}{n} = \frac{1}{n} \sum_{j=1}^{n} r_j - \frac{n-1}{n} \frac{1}{n-1} \sum_{j=1}^{n-1} r_j \to 0 \text{ (a.s.)}.$$

Hence if $A_n = (|r_n| \geq n)$, the probability that infinitely many events A_n will occur is zero. The independence of the r.v.'s r_i implies the independence of the events A_i; therefore, from the Borel–Cantelli lemma (Theorem 1.3), we conclude that $\sum_{n=1}^{\infty} PA_n < \infty$. However, the convergence of this last series implies that Er_i is finite by Theorem 9.5.2. The sufficiency part of Theorem 3 shows that $c = Er_i$.

Two principles play a fundamental role in the theory of probability and its applications. The first is the law of large numbers. Suppose that only the weak law of large numbers held for identically distributed r.v.'s r_i with finite expectations. Then with a probability arbitrarily close to 1 there would exist an infinite number of values of n such that the arithmetic mean $a_n(\omega) = [r_1(\omega) + \cdots + r_n(\omega)]/n$ would not be close to the expectation. Under these circumstances, it seems doubtful that the arithmetic mean could be taken as an approximation to the measured quantity. The second principle is the property of randomness; that is, the arithmetic mean clusters about the same value as before if the results of some trial are not counted at all. However, the criterion of acceptance or rejection must depend on past results and not on future results. The experimenter is not allowed to be clairvoyant. The mathematical statement of this principle was given in Section 3.3. Theorem 3.3.1 remains valid for arbitrary factor spaces. These two principles were used by von Mises as the basis of his frequency theory of probability [15].

An unsolved problem is to find the conditions that are necessary and sufficient in order for the law of large numbers to hold for independent r.v.'s r_i. A partial solution has been obtained by Prochorov [16] using Theorem 4 below with $b_n = n$. A generalization of Prochorov's result and its proof appears in Reference [17]. To state this theorem concisely, we introduce the following terminology. Let $s_n = r_1 + r_2 + \cdots + r_n$, $T_k = (s_{n_k} - s_{n_{k-1}})/b_{n_k}$, and $t_k^2 = \sigma^2 T_k$.

THEOREM 4

Let r_i be a sequence of independent r.v.'s. Suppose that b_n is a sequence of constants such that $b_n \uparrow \infty$ as $n \to \infty$ and $0 < \delta \leq b_{n_{k+1}}/b_{n_k} \leq c < \infty$. Finally, assume that $(|r_n|/b_n) \log(\log b_n) \to 0$ as $n \to \infty$. Then $s_n/b_n \to 0$ (a.s.) if and only if for every $\epsilon > 0$, the series $\sum_{k=1}^{\infty} \exp(-\epsilon^2/t_k^2)$ converges.

Most mathematicians agree that this criterion is unsatisfactory, since it involves nonoverlapping sums. Although there is no formal definition of a satisfactory solution, presumably mathematicians will agree on a satisfactory set of conditions if and when they are found. For instance, a criterion involving individual summands would probably be satisfactory.

EXERCISES

1. Prove Borel's theorem directly. [*Hint:* Show that $\sum_{k=1}^{n} P(|r_{k^2} - p| \geq 1/m)$ converges for every m. This implies that $r_{k^2} \to p$ (a.s.). Also show that $|r_n - r_{k^2}| \leq 4/k$.]

2. The series $\sum r_n$ is *essentially convergent* if, for some constants a_n, $\sum (r_n - a_n)$ converges a.s.; otherwise it is *essentially divergent*.

(a) Show that if the series is essentially divergent, then $\sum (r_n - a_n)$ diverges a.s. for *any* constants a_n.

(b) Show that the series $\sum r_n$ is essentially convergent if and only if the symmetrized series converges a.s. [*Hint:* For the sufficiency, apply Theorem 4.4 to show that $\sum P(|r_n - mr_n| \geq \epsilon) < \infty$. Then show that $\sum \sigma^2(r_n - mr_n)^\epsilon < \infty$ and apply Theorem 3.4 to prove that the series $\sum (r_n - mr_n - E(r_n - mr_n)^\epsilon)$ converges a.s.]

3. Prove that a series of independent r.v.'s $\sum r_n$ is essentially convergent if $\lim \prod_{k=1}^{n} |\chi_k| \neq 0$ on a set of positive Lebesgue measure and is essentially divergent if $\lim \prod_{k=1}^{n} |\chi_k| = 0$ (a.e.). [*Hint:* Use Exercise 4(b) and the corollary to Theorem 3.5.]

12.7 LAW OF THE ITERATED LOGARITHM

The study of the law of the iterated logarithm seems to have originated from the following problem in number theory. Let $x \in (0, 1)$ and let $x = 0.\,x_1x_2x_3 \cdots$ be its binary expansion. Let $r_k(x) = 1$ if $x_k = 1$ and let $r_k(x) = -1$ if $x_k = 0$. Then $s_n = r_1 + \cdots + r_n$ represents the excess of occurrences of the digit 1 over 0 in the first n places of x. The Borel–Cantelli version of the strong law of large numbers asserts that $s_n = o(n)$ (a.s.). The following refinements indicate the historic development of the problem. All of the assertions hold a.s.

1. Hausdorff, [18], 1913: $s_n = o(n^{1/2+\epsilon})$, $\epsilon > 0$.
2. Hardy and Littlewood, [19], 1914: $s_n = O((n \log_n)^{1/2})$.
3. Steinhaus, [20], 1922: $\overline{\lim}_{n\to\infty} s_n/(2n \log n)^{1/2} \leq 1$.
4. Khintchine, [21], 1923: $s_n = O((n \log \log n)^{1/2})$.
5. Khintchine, [22], 1924: $\overline{\lim}_{n\to\infty} s_n/(2n \log \log n)^{1/2} = 1$.

In 1927, Kolmogorov [23] discovered a generalization of these results. The connection with number theory is hardly visible in this theorem, which appears below. His proof is based upon the following lemmas.

LEMMA 1

Let r be a r.v. such that $Er = 0$ and $|r| \leq c < \infty$. If $x > 0$ and $cx \leq 1$, then

$$\exp\left(\frac{x^2\sigma^2}{2}(1 - cx)\right) < Ee^{xr} < \exp\left(\frac{x^2\sigma^2}{2}\left(1 + \frac{cx}{2}\right)\right). \tag{1}$$

Proof

Since $|Er^n| \leq c^n$ and

$$Ee^{rx} = 1Erx + Er^2\frac{x^2}{2!} + Er^3\frac{x^3}{3!} + \ldots, \tag{2}$$

it follows that

(3) $$Ee^{rx} < 1 + \frac{x^2\sigma^2}{2}\left(1 + \frac{cx}{3} + \frac{c^2x^2}{3\cdot 4} + \cdots\right) < 1 + \frac{x^2\sigma^2}{2}\left(1 + \frac{xc}{2}\right),$$

using the fact that $cx/3^{n-2} > (cx)^{n-2}/(3\cdot 4 \cdots n)$. Applying the inequality $e^t > 1 + t$ $(t > 0)$ to the extreme right-hand side in inequality (3) with $t = (x^2\sigma^2/2)(1 + cx/2)$ yields

$$Ee^{rx} < \exp\left(\frac{x^2\sigma^2}{2}\left(1 + \frac{cx}{2}\right)\right).$$

Replacing Er^n in (2) by $-c^{n-2}\sigma^2$ for $n \geq 3$ and performing a similar calculation to that used in deriving inequality (3) gives

(4) $$Ee^{rx} > 1 + \frac{\sigma^2x^2}{2}\left(1 - \frac{cx}{2}\right).$$

The estimate $1 + t > e^{t(1-t)}$ for $0 < t < 1$, when applied to the right-hand side of inequality (4), yields

(5) $$Ee^{rx} > \exp\left(\frac{\sigma^2x^2}{2}\left(1 - \frac{cx}{2}\right)\left(1 - \frac{x^2\sigma^2}{2}\left(1 - \frac{cx}{2}\right)\right)\right).$$

Finally, since

$$\frac{\sigma^2x^2}{2}\left(1 - \frac{cx}{2}\right)\left(1 - \frac{\sigma^2x^2}{2}\left(1 - \frac{cx}{2}\right)\right) > \frac{x^2\sigma^2}{2}(1 - xc)$$

by canceling $\sigma^2x^2/2$ and replacing the $\sigma^2x^2/2$ in the brackets by $c^2x^2/2$, and since the exponential function is an increasing function, it follows from inequality (5) that

$$Ee^{rx} > \exp\frac{x^2\sigma^2}{2}(1 - xc).$$

In the remainder of this section, $\{r_n\}$ will denote a sequence of independent r.v.'s such that $Er_n = 0$ and so $\sigma_n^2 = Er_n^2$. Furthermore, let $s = \sum_{i=1}^{n} r_i$ and $t^2 = \sigma^2 s = \sum_{i=1}^{n} \sigma_i^2$.

LEMMA 2

Let $c = \max_{k \leq n} |r_k/t| < \infty$ and $\epsilon > 0$; then

$$P\left(\frac{s}{t} > \epsilon\right) < \begin{cases} \exp\left(-\left(\frac{\epsilon^2}{2}\right)\left(1 - \frac{\epsilon c}{2}\right)\right) & (\epsilon c \leq 1) \\ \exp\left(-\frac{\epsilon}{4c}\right) & (\epsilon c \geq 1) \end{cases}$$

Proof

By integrating only over the set of points $(s/t > \epsilon)$, it is not difficult to show that for $x \geq 0$,

(6) $$P\left(\frac{s}{t} > \epsilon\right) \leq e^{-x\epsilon} Ee^{xs/t}.$$

We now estimate $E \exp(xs/t)$. Since

$$E \exp\left(\frac{xs}{t}\right) = \prod_{k=1}^{n} E \exp \frac{xr_k}{t},$$

we obtain, assuming that $cx \leq 1$ and $x \geq 0$,

(7) $$\exp\left(\frac{x^2}{2}(1 - cx)\right) < E \exp\left(x \frac{s}{t}\right) < \exp\left(\frac{x^2}{2}\left(1 + \frac{xc}{2}\right)\right)$$

by using inequality (1). The result follows from inequalities (6) and (7) when x is replaced by ϵ if $\epsilon c \leq 1$ and by $1/c$ if $\epsilon c \geq 1$.

LEMMA 3

Let $\rho > 0$; if $c = c(\rho) = \max_{k \leq n} |r_k/t|$ is sufficiently small and $\epsilon = \epsilon(\rho)$ is sufficiently large, then

(8) $$P\left(\frac{s}{t} > \epsilon\right) > \exp\left(-\left(\frac{\epsilon^2}{2}\right)(1 + \rho)\right).$$

Proof

Let $v = s/t$ and let $F(x) = P(v > x)$. Consider

(9) $$Ee^{\lambda v} = -\int e^{\lambda x}\, dF(x) = \lambda \int e^{\lambda x} F(x)\, dx, \qquad (\lambda > 0)$$

where the last integral is obtained by integrating by parts.[8] We decompose the range of integration $(-\infty, \infty)$ into the five intervals $I_1 = (-\infty, 0]$, $I_2 = (0, \lambda(1 - b)]$, $I_3 = (\lambda(1 - b), \lambda(1 + b)]$, $I_4 = (\lambda(1 + b), 8\lambda]$, and $I_5 = (8\lambda, \infty)$, where $0 < b < 1$. Let J_i for $i = 1, \ldots, 5$ be the value of the last integral appearing in (9) over I_i. By calculating the upper bounds of this integral over I_1, I_2, I_3, and I_5, we will show that

(10) $$J_1 + J_2 + J_4 + J_5 < \frac{1}{2} Ee^{v}$$

and hence it will follow from (9) that

[8]By inequality (7), the integral exists; moreover, $\lim_{x \to \infty} e^{\lambda x} F(x) = 0$ by Lemma 2 if $4\lambda < 1/c$.

(11) $$J_3 = \lambda \int\limits_{\lambda(1-b)}^{\lambda(1+b)} e^{\lambda x} F(x)\,dx > \frac{1}{2} Ee^{\lambda v}.$$

Finally, we will show that this last inequality implies inequality (8) if b is chosen properly.

Since $F(x) < 1$,

$$J_1 = \lambda \int\limits_{-\infty}^{0} e^{\lambda x} F(x)\,dx < \lambda \int\limits_{-\infty}^{0} e^{\lambda x}\,dx = 1.$$

If c is sufficiently small, $1/c > 8\lambda$ and so $1/c \in I_5$. Thus on I_5 we have, by Lemma 2,

$$F(x) < \exp\left(-\frac{x}{4c}\right) < \exp(-2\lambda x), \qquad x \geq \frac{1}{c},$$

$$F(x) < \exp\left(-\frac{x^2}{2}\left(1 - \frac{xc}{2}\right)\right) \leq \exp\left(-\frac{x^2}{4}\right) < \exp(-2\lambda x), \qquad x < \frac{1}{c}.$$

Therefore,

$$J_5 = \lambda \int\limits_{8\lambda}^{\infty} e^{\lambda x} F(x)\,dx < \lambda \int\limits_{8\lambda}^{\infty} e^{-\lambda x}\,dx < 1$$

for sufficiently small c; hence

(12) $$J_1 + J_5 < 2.$$

According to (7), we can take c small enough so that

(13) $$Ee^{\lambda v} > \exp\left(\frac{\lambda^2}{2}(1 - a)\right),$$

where $0 < a < 1$; a will be expressed in terms of b later on. If λ is sufficiently large, it follows from inequality (12) that

(14) $$J_1 + J_5 < \frac{1}{4} Ee^{\lambda v}.$$

Since $1/c \in J_5$, then $x < 1/c$ on the intervals I_2 and I_4; therefore, by Lemma 2,

$$e^{\lambda x} F(x) < \exp\left(\lambda x - \frac{x^2}{2}\left(1 - \frac{cx}{2}\right)\right) \leq \exp\left(\lambda x - \frac{x^2}{2}(1 - 4\lambda c)\right) \equiv e^{g(x)}.$$

The quadratic expression $g(x)$ attains its maximum at $x = \lambda/(1 - 4\lambda c)$, which lies in I_3, for sufficiently small c. Therefore, for $x \in I_2$,

$$g(x) \leq g(\lambda(1 - b)) = \frac{\lambda^2}{2}(1 - b)(1 + b + 4\lambda c - 4c\lambda b) < \frac{\lambda^2}{2}\left(1 - \frac{b^2}{2}\right)$$

for sufficiently small c.[9] Using this estimate,

[9]Actually, if $c < [b/(1 - b)]^2(1/8\lambda)$ for $0 < b < 1$, the inequality is valid.

$$J_2 = \lambda \int_0^{\lambda(1-b)} e^{\lambda x} F(x)\, dx < \lambda \int_0^{\lambda(1-b)} e^{g(x)}\, dx < \lambda^2 \exp\left(\frac{\lambda^2}{2}\left(1 - \frac{b^2}{2}\right)\right).$$

Similarly,

$$J_4 = \lambda \int_{\lambda(1+b)}^{8\lambda} e^{\lambda x} F(x)\, dx < \lambda \int_{\lambda(1+b)}^{8\lambda} e^{g(x)}\, dx < 8\lambda^2 \exp\left(\frac{\lambda^2}{2}\left(1 - \frac{b^2}{2}\right)\right).$$

Hence

$$J_2 + J_4 < 9\lambda^2 \exp\left(\frac{\lambda^2}{2}\left(1 - \frac{b^2}{2}\right)\right)$$

$$= 9\lambda^2 \exp\left(\frac{\lambda^2}{2}\left(a - \frac{b^2}{2}\right)\right) \exp\left(\frac{\lambda^2}{2}(1 - a)\right).$$

By choosing $a = b^2/4$, the argument in the first exponential function above becomes negative. Therefore, for sufficiently large λ,

$$\frac{1}{4} > 9\lambda^2 \exp\left(-\frac{\lambda^2}{2}\left(\frac{b^2}{4}\right)\right),$$

and it follows from inequality (13) that

$$J_2 + J_4 < \frac{1}{4} Ee^{\lambda v}. \tag{15}$$

The exact value of λ will be specified below. Now (14) and (15) imply (10) and so (11) is valid.

We increase the value of the integrand in (11) by replacing $F(x)$ by $F(\lambda(1 - b))$ and $e^{\lambda x}$ by $\exp(\lambda^2(1 + b))$, obtaining

$$2\lambda^2 b e^{\lambda^2(1+b)} F(\lambda(1 - b)) > J_3 > \frac{1}{2} Ee^{\lambda v} > \frac{1}{2} \exp\left(\frac{\lambda^2}{2}(1 - a)\right),$$

where the last inequality is a consequence of (13). Setting $\lambda = \epsilon/(1 - b)$ yields

$$F(\epsilon) > \frac{1}{4\lambda^2 b} \exp\left(\frac{\lambda^2}{2}(1 - a)\right) \exp(-\lambda^2(1 + b)).$$

The right-hand side of the above inequality reduces to

$$\frac{1}{4\lambda^2 b} \exp\left(\frac{\lambda^2}{2} a\right) \exp\left(-\frac{\lambda^2}{2}(1 + 2a + 2b)\right).$$

Since $(1/4\lambda^2 b) \exp((\lambda^2/2)a) \to \infty$ as $\lambda \to \infty$, it follows that for sufficiently large ϵ,

$$F(\epsilon) > \exp\left(-\frac{\lambda^2}{2}(1 + 2a + 2b)\right) = \exp\left(-\left(\frac{1 + 2b + b^2/2}{(1 - b)^2}\right)\left(\frac{\epsilon^2}{2}\right)\right)$$

upon replacing a and λ by their values. However, given $\rho > 0$, there exists a b for $0 < b < 1$ such that

$$\frac{1 + 2b + b^2/2}{(1 - b)^2} \leq 1 + \rho.$$

Hence, for sufficiently large $\epsilon = \epsilon(\rho)$ and sufficiently small $c = c(\rho)$,

$$F(\epsilon) > \exp\left(-\left(\frac{\epsilon^2}{2}\right)(1+\rho)\right).$$

LEMMA 4

If r is integrable, then $|Er - mr| \leq \sqrt{2\sigma^2 r}$.

Proof

By Chebyshev's inequality, $P(|r - Er| \geq \sqrt{2\sigma^2 r}) \leq \frac{1}{2}$; thus, by the definition of the median mr,

$$Er - \sqrt{2\sigma^2 r} \leq mr \leq Er + \sqrt{2\sigma^2 r}.$$

COROLLARY

Let r_i be independent r.v.'s let $Er_i = 0$, and let $Er_i^2 < \infty$, where $i = 1, 2, \ldots, n$; then

(16) $$P(\max_{k \leq n} s_k \geq \epsilon) \leq 2P(s_n \geq \epsilon - \sqrt{2\sigma^2 s_n}).$$

Proof

Lévy's inequality [see Exercise 4(a) of Section 12.4] yields

(17) $$P(\max_{k \leq n} (s_k - m(s_k - s_n)) \geq \epsilon) \leq 2P(s_n \geq \epsilon).$$

From the lemma,

$$|m(s_k - s_n)| \leq \sqrt{2\sigma^2(s_n - s_k)} \leq \sqrt{2\sigma^2 s_n}$$

and so $-\sqrt{2\sigma^2 s_n} \leq -m(s_k - s_n)$; hence

$$[(\max_{k \leq n} s_k - \sqrt{2\sigma^2 s_n} \geq \epsilon)] \subset [\max_{k \leq n} (s_k - m(s_k - s_n)) \geq \epsilon].$$

It follows from (17) that

$$P(\max_{k \leq n} (s_k \geq \epsilon + \sqrt{2\sigma^2 s_n})) \leq 2P(s_n \geq \epsilon);$$

replacing ϵ by $\epsilon - \sqrt{2\sigma^2 s_n}$ in the above inequality gives (16).

LEMMA 5

Let $\{s_n\}$ be an increasing sequence of real numbers such that $s_n \to \infty$ as $n \to \infty$. If

(18) $$\lim_{n \to \infty} \frac{s_{n+1}}{s_n} = 1,$$

then for every $a > 1$ there exists a sequence $n_k = n_k(a) \uparrow \infty$, as $k \to \infty$, such that $s_{n_k} \sim a^k$.

Proof

Since $s_n \uparrow$ and $s_n \to \infty$, there exists a large integer n_k such that $s_{n_k} \leq a^k < s_{n_k+1}$ or $1 \leq a^k/s_{n_k} < s_{n_k+1}/s_{n_k}$. It follows from (18) that $s_{n_k} \sim a^k$.

Finally we are prepared to prove the law of the iterated logarithm. To state the theorem concisely, we introduce the following notation. Instead of "lim sup A_n," we will write "A_n i.o.," which is read "A_n's occur infinitely often." Following P. Lévy, we will say that a numerical sequence a_n belongs to the *upper class* or the *lower class* of a sequence of r.v.'s s_n, according to whether $P(s_n > a_n \text{ i.o.}) = 0$ or 1.

THEOREM 1

Let $\{r_i\}$ be a sequence of independent r.v.'s such that $Er_i = 0$. Let $s_n = \sum_{i=1}^{n} r_i$ and let $t_n^2 = \sigma^2 s_n$. Suppose that $t_n^2 \to \infty$ and

(19) $$\frac{|r_n| l_n}{t_n} \to 0 \qquad (l_n = (2 \log \log t_n^2)^{1/2})$$

as $n \to \infty$. Then

(20) $$P(\limsup s_n/t_n l_n = 1) = 1.$$

Proof

1. To prove (20), we need only show that for sufficiently small $\epsilon > 0$, the sequence $(1 - \epsilon)t_n l_n$ belongs to the lower class of the sequence s_n and the sequence $(1 + \epsilon)t_n l_n$ belongs to the upper class. Of course, any smaller value of ϵ will also be satisfactory.

We begin by proving the latter statement. By using (19) and the definition of t_n, it can be shown that $t_{n+1}^2/t_n^2 \to 1$. Since $t_n \to \infty$, the hypotheses of Lemma 5 are satisfied. Choose $n_k = n_k(a)$ as described in Lemma 5; the value of a will be specified below. Let $m_{n_k} = \max_{n \leq n_k} s_n$. To prove that the sequence $(1 + \epsilon)t_n l_n$ belongs to the upper class of the sequence s_n, we need only show that the right-hand side of the self-evident inequality

(21) $$P(s_n > (1 + \epsilon)t_n l_n \text{ i.o.}) \leq P(m_{n_k} > (1 + \epsilon)t_{n_{k-1}} l_{n_{k-1}} \text{ i.o.}),$$

where $n_{k-1} \leq n < n_k$ is zero. Now

(22) $$(1 + \epsilon)t_{n_{k-1}} l_{n_{k-1}} \sim \frac{1 + \epsilon}{a} t_{n_k} l_{n_k}.$$

For a fixed positive $\epsilon_1 < \epsilon$, we select $a > 1$ so that $(1 + \epsilon)/a > 1 + \epsilon_1$. Then it follows from (22) that for sufficiently large k,

(23) $$P(m_{n_k} > (1+\epsilon)t_{n_{k-1}}l_{n_{k-1}} \text{ i.o.}) \le P(m_{n_k} > (1+\epsilon_1)t_{n_k}l_{n_k} \text{ i.o.}).$$

Our assertion follows from (21) if we can show that the right-hand side of inequality (23) is zero. However, by inequality (16), it is sufficient to prove that

$$2P\left(s_{n_k} > \left(1+\epsilon_1 - \frac{\sqrt{2}}{l_{n_k}}\right)t_{n_k}l_{n_k} \text{ i.o.}\right) = 0.$$

Since $1 + \epsilon_1 - \sqrt{2}/l_{n_k} > 1 + \epsilon_2$ for a positive $\epsilon_2 < \epsilon_1$ and sufficiently large k,

$$P\left(s_{n_k} > \left(1+\epsilon_1 - \frac{\sqrt{2}}{l_{n_k}}\right)t_{n_k}l_{n_k} \text{ i.o.}\right) \le P(s_{n_k} > (1+\epsilon_2)t_{n_k}l_{n_k} \text{ i.o.}),$$

and we only must show that the right-hand side of this last inequality is zero. By the Borel–Cantelli theorem (1.3) we only must prove that

(24) $$\sum_{k=1}^{\infty} P(s_{n_k} > (1+\epsilon_2)t_{n_k}l_{n_k}) < \infty.$$

An application of Lemma 2 with $c_k = \max_{j \le n_k} (|r_j|/t_{n_k})$ and $\epsilon_k = (1+\epsilon_2)l_{n_k}$ to the general term of the above series yields, since $\epsilon_k c_k \to 0$ by (19),

$$P(s_{n_k} > (1+\epsilon_2)t_{n_k}l_{n_k}) \le \exp\left(-\frac{1}{2}(1+\epsilon_2)^2 l_{n_k}^2\left(1 - \frac{\epsilon_k c_k}{2}\right)\right)$$

$$\le \exp\left(-\frac{1}{2}(1+\epsilon_2)^2 \log\log t_{n_k}^2\right) \sim (2k \log a)^{-(1+\epsilon_2)}.$$

Since $\sum (2k \log a)^{-(1+\epsilon_2)}$ is convergent, (24) is valid.

2. We now show that the sequence $(1-\epsilon)t_n l_n$ belongs to the lower class of the sequence s_n. To prove this assertion, we need only verify that it holds for a subsequence s_{n_k}. The $n_k = n_k(a)$ are chosen as in the first part of the proof; the constant a will be specified below. Here we cannot apply the Borel–Cantelli theorem since the sums s_{n_k} are not independent. For this reason we first consider $s_{n_k} - s_{n_{k-1}}$, which are nonoverlapping sums of independent r.v.'s and which therefore are independent. We will require the variances a_k and b_k, where

$$a_k^2 = t_{n_k}^2 - t_{n_{k-1}}^2 \sim t_{n_k}^2\left(1 - \frac{1}{a^2}\right)$$

$$b_k = (2 \log\log a_k^2)^{1/2} \sim (2 \log\log t_{n_k}^2)^{1/2} = l_{n_k}.$$

Choose an $\epsilon_1 > 0$ such that $1 > \epsilon > \epsilon_1$ and set

$$E_k = (s_{n_k} - s_{n_{k-1}} > (1-\epsilon_1)a_k b_k).$$

We will show that $P(E_k \text{ i.o.}) = 1$. By the Borel–Cantelli theorem, it suffices to prove that $\sum PE_k = \infty$. We estimate PE_k by Lemma 3, which is applicable because $\epsilon_k = (1 - \epsilon_1)b_k \to \infty$ and $c_k = \max_{n_{k-1}<n\leq n_k} (|r_n|/a_k) \to 0$ as $k \to \infty$. Therefore, setting $1 + \rho = 1/(1 - \epsilon_1)$, we have

$$PE_k > \exp\left(-\frac{1}{2}(1 - \epsilon_1)b_k^2\right) = \exp\left(-\frac{1}{2}(1 - \epsilon_1)2 \log\log a_k^2\right)$$
$$\sim \frac{1}{(2k \log a)^{1-\epsilon_1}}.$$

Since $\sum (2k \log a)^{\epsilon_1 - 1}$ diverges, $\sum PE_k = \infty$ and $P(E_k \text{ i.o.}) = 1$.

The last step in the proof is to show that the $s_{n_{k-1}}$ appearing in E_k can be neglected. Let $F_k = (|s_{n_{k-1}}| \leq 2t_{n_{k-1}}l_{n_{k-1}})$. Noting that the hypotheses of the theorem are unaltered if any r_n is replaced by $-r_n$, it follows that part 1 of the proof remains valid for $|s_{n_{k-1}}|$. Therefore, according to the end of part 1, $P(F_k' \text{ i.o.}) = 0$ and so $P(E_kF_k \text{ i.o.}) = 1$. Moreover, $s_{n_{k-1}}$ can be eliminated since the fact that $P(F_k' \text{ i.o.}) = 0$ implies that a.s. $s_{n_{k-1}} \geq -2t_{n_{k-1}}l_{n_{k-1}}$ i.o.; therefore,

$$(E_kF_k \text{ i.o.}) \subset (s_{n_k} > (1 - \epsilon_1)a_kb_k - 2t_{n_{k-1}}l_{n_{k-1}} \text{ i.o.}).$$

Finally, from the fact that

$$(1 - \epsilon_1)a_kb_k - 2t_{n_{k-1}}l_{n_{k-1}} \sim \left((1 - \epsilon_1)\left(1 - \frac{1}{a^2}\right)^{1/2} - \frac{2}{a}\right)t_{n_k}l_{n_k},$$

it follows by choosing a sufficiently large that

$$(1 - \epsilon_1)\left(1 - \frac{1}{a^2}\right)^{1/2} - \frac{2}{a} > 1 - \epsilon,$$

since $\epsilon > \epsilon_1$. Hence

$$(E_kF_k \text{ i.o.}) \subset (s_{n_k} > (1 - \epsilon)t_{n_k}l_{n_k} \text{ i.o.}),$$

which completes the proof.

COROLLARY

Under the hypotheses of the above theorem,

$$P\left(\liminf \frac{s_n}{t_nl_n} = -1\right) = 1.$$

The proof follows immediately from the observation that the hypotheses of the theorem are unaltered if every r_n is replaced by $-r_n$.

This result has been generalized by Lévy [24], Kolmogorov, and Erdös [25]. The most general result was obtained by Feller [26].

EXERCISES

1. (a) Let s_n be a r.v. that denotes the number of successes in n Bernoulli trials with a probability p for success, and set $s_n^* = (s_n - np)/(npq)^{1/2}$. Show that with a probability 1, we have

$$\limsup_{n\to\infty} \frac{s_n^*}{(2 \log \log n)^{1/2}} = 1.$$

(b) Deduce Theorem 11.6.8.

REFERENCES

1. Kolmogorov, A., "Grundbegriffe der Wahrscheinlichkeitsrechnung," *Erg. Mat.*, **2,** No. 3 (1933).
2. Hajek, J., and A. Rényi, "Generalization of an inequality of Kolmogorov," *Acta. Math. Acad. Sci. Hungar.*, **6** (1955).
3. Khintchine, A., and A. Kolmogorov, "Über Konvergenz von Reihen deren Glieder durch den Zufall bestimmt werden," *Rec. Math.* (*Mat Sbornik*), **32** (1924), 668–677.
4. Kolmogorov, A., "Über die Summen durch den Zufall bestimmter unabhängigen Grössen," *Math. Ann.*, **99** (1928), 309–319.
5. Kolmogorov, A., "Bemerkungken zu meiner Arbeit 'Uber die Summen Züfalliger Grossen,'" *Math. Ann.*, **102** (1930), 1184–1488.
6. Kawata, T., and M. Udagawa, "On infinite convolutions", *Kodai Math. Sem. Rep.*, No. 3 (1949), 15–22.
7. Feller, W., *An Introduction to Probability Theory and Its Applications*, Vol. I. New York: John Wiley & Sons, Inc., 1957.
8. Feller, W., "Über das Gesetz der grossen Zahlen," *Acta Univ. Szeged E* (1937), 191–201.
9. Khintchine, A., "Sur la loi forte des grands nombres," *C. R. Acad. Sci* (*Paris*), **189** (1929), 477–479.
10. Markov, A., *Calculus of Probability*. Leningrad, 1924.
11. Gnedenko, B. V., *Theory of Probability*. New York: Chelsea Publishing Co., 1962.
12. Cantelli, F. P., "Sulla probabilità come limite della frequenza," *Rend. Accad. Naz. dei Lincei* [5] **26** (1912), 39.
13. Polya, G., "Eine Ergänzung zu dem Bernoullischen Satz der Wahrscheinlichkeitsrechnung," *Nach. Akad. Wiss. Göttingen Math. Phys. Kl* (1921).

14. Kolmogorov, A., "Sur la loi forte des grands nombres," *C. R. Acad. Sci. (Paris)*, **191** (1930), 910ff.

15. Von Mises, R., *Mathematical Theory of Probability and Statistics*. New York: Academic Press Inc., 1964.

16. Prochorov, U., "On the strong law of large numbers," *Dokl. Akad. Nauk USSR*, **69** (1949) (in Russian).

17. Loève, M., *Probability Theory*. D. Van Nostrand Co., Inc., 1963.

18. Hausdorff, F., *Grundzüge der Mengenlehre*. Leipzig, 1913.

19. Hardy, G. H., and J. E. Littlewood, "Some problems of diophantine approximation," *Acta. Math.*, **37** (1914), 155–239.

20. Steinhaus, H., "Les probabilités dénombrables et leur rapport à la théorie de la mesure," *Fund. Math.*, **4** (1922), 286–310.

21. Khintchine, A., "Über dyadische Brüche," *Math. Zeit.*, **18** (1923), 109–116.

22. Khintchine, A., "Über einen Satz der Wahrscheinlichkeitsrechnung," *Fund. Math.*, **6** (1924), 9–20.

23. Kolmogorov, A., "Über das Gesetz der grossen Zahlen," *Math. Ann.*, **96** (1926), 156–168.

24. Lévy, P., "Sur un théorème de M. Khintchine," *Bull. Sci. Math.* (2), **55** (1931), 145–160.

25. Erdös, P., "On the law of the iterated logarithm," *Ann. Math.* (2), **43** (1942), 419–436.

26. Feller, W., "The general form of the so-called law of the iterated logarithm," *Trans. Am. Math. Soc.*, **54** (1943).

CHAPTER 13

LIMIT THEOREMS FOR SUMS OF INDEPENDENT RANDOM VARIABLES

13.1 CLASSICAL LIMIT THEOREMS

The DeMoivre–Laplace theorem (Theorem 4.3.1) served as the starting point for fundamental investigations in limit theorems. To see how these problems were motivated, it is convenient to rewrite this theorem in a different form. Let $r_i = 1$ or 0 according to whether an event A occurs or does not occur in n independent trials. Then $s_n = r_1 + r_2 + \cdots + r_n$ is the number of times the event A occurs in n trials. Set

$$a_i = Er_i \qquad b_i^2 = \text{Var}\, r_i = E(r_i - a_i)^2$$

and

$$A_n = a_1 + a_2 + \cdots + a_n$$
$$B_n^2 = b_1^2 + b_2^2 + \cdots + b_n^2.$$

In particular, for Bernoulli trials, we have $a_i = p$ and $b_i^2 = qp$. With this notation, the DeMoivre–Laplace theorem can be expressed in the following form: As $n \to \infty$,

$$(1) \qquad P\left(a < \frac{s_n - A_n}{B_n} \leq b\right) \to \frac{1}{\sqrt{2\pi}} \int_a^b e^{-x^2/2}\, dx.$$

Chebyshev was the first one to generalize this limit theorem. He considered independent r.v.'s r_i with finite expectations a_i and variances b_i, and asked what additional conditions would insure that relation (1) would hold. To solve this problem, Chebyshev created the method of moments. His solution [1] is based on a lemma that was proved later by Markov [2]). The following generalization was discovered by Lyapunov.

THEOREM 1

Let r_i be a sequence of independent r.v.'s. If a positive number δ can be found such that as $n \to \infty$,

$$\frac{1}{B_n^{2+\delta}} \sum_{k=1}^{n} E|r_k - a_k|^{2+\delta} \to 0,$$

then as $n \to \infty$,

$$P\left(\frac{1}{B_n} \sum_{k=1}^{n} (r_k - a_k) \leq x\right) \to \frac{1}{\sqrt{2\pi}} \int_{-\infty}^{x} e^{-t^2/2}\, dt.$$

In proving this result, Lyapunov [3] discovered the method of characteristic functions, which is now one of the chief methods used in studying limit theorems. He also studied the speed of convergence to the normal law. These last results were greatly improved upon by Berry [4] and Esseen [5]. We will not prove Lyapunov's theorem directly; instead, we will deduce it as a consequence of the following more general result by Lindeberg [6].

THEOREM 2

If a sequence r_i of independent r.v.'s satisfies the condition

$$\lim_{n\to\infty} \frac{1}{B_n^2} \sum_{k=1}^{n} \int_{|x-a_k|>\epsilon B_n} (x - a_k)^2\, dF_k(x) = 0 \tag{2}$$

for any $\epsilon > 0$, then as $n \to \infty$,

$$P\left(\frac{1}{B_n} \sum_{k=1}^{n} (r_k - a_k) \leq x\right) \to \frac{1}{\sqrt{2\pi}} \int_{-\infty}^{x} e^{-t^2/2}\, dt$$

uniformly in x.

Proof

We first normalize the r.v.'s; set

$$r_{nk} = \frac{r_k - a_k}{B_n} \qquad F_{nk}(x) = P(r_{nk} \leq x).$$

Since $Er_{nk} = 0$ and $\operatorname{Var} r_{nk} = b_k^2/B_n^2$, then

(3) $$\sum_{k=1}^{n} \operatorname{Var} r_{nk} = 1.$$

Condition (2) is transformed into

(4) $$\lim_{n\to\infty} \sum_{k=1}^{n} \int_{|x|>\epsilon} x^2\, dF_{nk}(x) = 0.$$

The c.f. ϕ_n of the sum $\sum_{k=1}^{n} r_{nk}$ is $\phi_n(\lambda) = \prod_{k=1}^{n} \chi_{nk}(\lambda)$, where χ_{nk} is the c.f. of the r_{nk}.

To prove the theorem, we need only show that $\lim_{n} \phi_n(\lambda) = e^{-\lambda^2/2}$ or $\lim_{n\to\infty} \log \phi_n(\lambda) = -\lambda^2/2$. Proceeding formally,

(5) $$\log \phi_n(\lambda) = \sum_{k=1}^{n} \log \chi_{nk}(\lambda) = \sum_{k=1}^{n} \log (1 + (\chi_{nk}(\lambda) - 1))$$
$$= \sum_{k=1}^{n} (\chi_{nk}(\lambda) - 1) + R_n,$$

where log is the principal value of the logarithm and

$$R_n = \sum_{k=1}^{n} \sum_{s=2}^{\infty} \frac{(-1)^{s-1}}{s} (\chi_{nk}(\lambda) - 1)^s.$$

To justify this expansion, we must show that $|\chi_{nk}(\lambda) - 1| < 1$, where $|\lambda| \le \lambda_0 (\lambda_0 > 0)$ for sufficiently large n and $1 \le k \le n$. This will be a consequence of the fact that the factors $\chi_{nk}(\lambda)$ tend to 1 as $n \to \infty$ for $1 \le k \le n$. Since $Er_{nk} = \int t\, dF_{nk}(t) = 0$,

$$\chi_{nk}(\lambda) - 1 = \int (e^{i\lambda t} - 1 - i\lambda t)\, dF_{nk}(t).$$

For any real α,

(6) $$|e^{i\alpha} - 1 - i\alpha| \le \alpha^2/2$$

by Lemma 10.4.1, and so

(7) $$|\chi_{nk}(\lambda) - 1| \le \frac{\lambda^2}{2} \int t^2\, dF_{nk}(t).$$

The last integral can be bounded in the following manner:

$$\int t^2\, dF_{nk}(t) = \int_{|t|\le\epsilon} t^2\, dF_{nk}(t) + \int_{|t|>\epsilon} t^2\, dF_{nk}(t) \le \epsilon^2 + \int_{|t|>\epsilon} t^2\, dF_{nk}(t),$$

for any $\epsilon > 0$. According to (4), the last term can be made smaller than ϵ^2 for sufficiently large values of n. Hence for all sufficiently large n and for

$|\lambda| \leq \lambda_0$, it follows from (7) that $|\chi_{nk}(\lambda) - 1| \leq \epsilon^2\lambda_0$ for $1 \leq k \leq n$. This implies that

(8) $$\lim_{n\to\infty} \max_{1\leq k\leq n} \chi_{nk}(\lambda) = 1$$

and that for all sufficiently large n, the inequality

(9) $$|\chi_{nk}(\lambda) - 1| < \tfrac{1}{2}$$

is valid for $|\lambda| \leq \lambda_0$. Therefore, expansion (5) is valid.

We will show that as $n \to \infty$, $R_n \to 0$ uniformly for λ in an arbitrary finite interval $|\lambda| \leq \lambda_0$. The remainder

$$|R_n| \leq \sum_{k=1}^{n} \sum_{s=2}^{\infty} \frac{1}{2}|\chi_{nk}(\lambda) - 1|^s = \frac{1}{2} \sum_{k=1}^{n} \frac{|\chi_{nk}(\lambda) - 1|^2}{1 - |\chi_{nk}(\lambda) - 1|}$$
$$\leq \sum_{k=1}^{n} |\chi_{nk}(\lambda) - 1|^2,$$

since by inequality (9), the infinite sum converges and $[1 - |\chi_{nk}(\lambda) - 1|]^{-1} \leq 2$. From

$$\sum_{k=1}^{n} |\chi_{nk}(\lambda) - 1| = \sum_{k=1}^{n} (e^{i\lambda t} - 1 - i\lambda t)\, dF_{nk}(t) \leq \frac{\lambda^2}{2} \sum_{k=1}^{n} t^2\, dF_{nk}(t) = \frac{\lambda^2}{2},$$

where the last equality is a consequence of (3), it follows that

$$|R_n| \leq \frac{\lambda^2}{2} \max_{1\leq k\leq n} |\chi_{nk}(\lambda) - 1|.$$

Thus, (8) implies that, as $n \to \infty$, $R_n \to 0$ uniformly for $|\lambda| \leq \lambda_0$.

To complete the proof, we see from (5) that we only must show

(10) $$\lim_{n\to\infty} \sum_{k=1}^{n} (\chi_{nk} - 1) = -\lambda^2/2$$

uniformly in every finite interval $|\lambda| \leq \lambda_0$. However,

$$\sum_{k=1}^{n} (\chi_{nk}(\lambda) - 1) = -\lambda^2/2 + \rho_n,$$

where

$$\rho_n = \lambda^2/2 + \sum_{k=1}^{n} \int (e^{it\lambda} - 1 - it\lambda)\, dF_{nk}(t),$$

since $\sum_{k=1}^{n} \int (i\lambda t)^2/2\, dF_{nk} = -\lambda^2/2$ by (3). The problem reduces to showing that $\lim_{n\to\infty} \rho_n = 0$ uniformly in every finite interval $|\lambda| \leq \lambda_0$. For any $\epsilon > 0$,

$$\rho_n = \sum_{k=1}^{n} \int_{|t|\le\epsilon} (e^{i\lambda t} - 1 - i\lambda t + \lambda^2 t^2/2)\, dF_{nk}(t) \tag{11}$$
$$+ \sum_{k=1}^{n} \int_{|t|>\epsilon} (e^{i\lambda t} - 1 - i\lambda t + \lambda^2 t^2/2)\, dF_{nk}(t).$$

By Lemma 10.4.1,

$$|e^{i\alpha} - 1 - i\alpha + \alpha^2/2| \le |\alpha|^3/6 \tag{12}$$

for any real α. Using (12) in the first integral in (11) and (6) in the second integral in (11) leads to

$$|\rho_n| \le \frac{|\lambda|^3}{6} \sum_{k=1}^{n} \int_{|t|\le e} |t|^3\, dF_{nk}(t) + \lambda^2 \sum_{k=1}^{n} \int_{|t|>\epsilon} t^2\, dF_{nk}(t)$$
$$\le \frac{|\lambda|^3}{6}\epsilon \sum_{k=1}^{n} \int_{|t|\le\epsilon} t^2\, dF_{nk}(t) + \lambda^2 \sum_{k=1}^{n} \int_{|t|>\epsilon} t^2\, dF_{nk}(t)$$
$$= \frac{|\lambda|^3}{6}\epsilon + \lambda^2\left(1 - \frac{|\lambda|}{6}\epsilon\right) \sum_{k=1}^{n} \int_{|t|>\epsilon} t^2\, dF_{nk}.$$

The second term can be made arbitrarily small by condition (4), while the first is small by the arbitrariness of ϵ for $|\lambda| \le \lambda_0$. Therefore, $\lim_{n\to\infty} \rho_n = 0$ uniformly in every finite interval $|\lambda| \le \lambda_0$.

COROLLARY

Suppose that the independent r.v.'s. r_i for $i = 1, 2, 3, \ldots$ are identically distributed and have finite nonzero variances. Then as $n \to \infty$,

$$P\left(\frac{1}{B_n} \sum_{k=1}^{n} (r_k - Er_k) \le x\right) \to \frac{1}{\sqrt{2\pi}} \int_{-\infty}^{x} e^{-x^2/2}\, dx$$

uniformly in x.

Proof

We need only verify that condition (4) is satisfied. Let $\operatorname{Var} r_i = b$ for $i = 1, \ldots, n$; then $B_n = b\sqrt{n}$. Setting $Er_i = a$, condition (4) becomes

$$(1/nb^2)n \int_{|x-a|>\epsilon B_n} (x-a)^2\, dF(x) = \int_{a+\epsilon b\sqrt{n}}^{\infty} + \int_{-\infty}^{a-\epsilon b\sqrt{n}} (x-a)^2\, dF(x),$$

which approaches zero since the variance is finite and $b \ne 0$.

Finally we prove Theorem 1, Lyapunov's theorem, by showing that condition (2) implies condition (4). This follows from the inequality

$$1/B_n^2 \sum_{k=1}^{n} \int_{|x-a_k|>\epsilon B_n} (x-a_k)^2\, dF_k \leq \frac{1}{B_n^2(\epsilon B_n)^\delta} \sum_{k=1}^{n} \int_{|x-a_k|>\epsilon B_n} |x-a_k|^{2+\delta}\, dF_k(x)$$

$$\leq \frac{1}{\epsilon\delta} \left(\sum_{k=1}^{n} \int |x-a_k|^{2+\delta}\, dF_k(x)\right) B_n^{-2-\delta}.$$

Condition (2) is called *Lindeberg's condition.* It is instructive to investigate its meaning; for this purpose let

$$E_k = (|r_k - a_k| > \epsilon B_n). \qquad (k = 1, 2, \ldots, n)$$

Since

$$P\left(\max_{1\leq k\leq n} |r_k - a_k| < \epsilon B_n\right) = p\left(\bigcup_{i=1}^{n} E_i\right) \leq \sum_{i=1}^{n} PE_i$$

and

$$PE_i = \int_{|x-a_i|>\epsilon B_n} cF_i(x) \leq \frac{1}{(\epsilon B_n)^2} \int_{|x-a_i|>\epsilon B_n} (x-a_i)^2\, dF_i(x),$$

it follows that

$$P\left(\max_{1\leq k\leq n} |r_k - a_k| > \epsilon B_n\right) \leq \frac{1}{\epsilon^2 B_n^2} \sum_{i=1}^{n} \int_{|x-a_i|>\epsilon B_n} (x-a_i)^2\, dF_i(x). \tag{13}$$

This last sum tends to zero as $n \to \infty$ by Lindeberg's condition. Hence if Lindeberg's condition is fulfilled, the terms $(r_k - a_k)/B_n \to 0(P)$ for all k. It is intuitively obvious that this last condition must be fulfilled if $1/B_n \sum_{k=1}^{n} (r_k - a_k)$ converges to the normal distribution law or any other law.

We have seen that Lindeberg's condition is sufficient for the d.f.'s of sums to converge to the normal law. In 1935, Feller [7] showed the necessity of this condition. Later we will deduce the necessity from a more general theorem.

EXERCISES

1. Determine whether Theorem 2 is valid for the following sequences of independent r.v.'s r_k,

(a) $P(r_k = \pm 2^k) = \frac{1}{2}$.
(b) $P(r_k = \pm k) = 1/2k^{1/2}$ $\qquad P(r_k = 0) = 1 - k^{-1/2}$.
(c) $P(r_k = \pm 2^k) = 2^{-(2k+1)}$ $\qquad P(r_k = 0) = 1 - 2^{-2k}$.

2. Find sufficient conditions on $\{x_k\}$ so that Theorem 2 will hold for the independent r.v.'s r_k if r_k assumes the values x_k, $-x_k$, and 0 with probabilities p_k, p_k, and $1 - 2p_k$.

3. The r.v.'s r_k, where r_k assumes the values $-k^\alpha$ and k^α each with probability $\frac{1}{2}$, are independent. Prove that for $\alpha > -\frac{1}{2}$, Lyapunov's theorem is applicable to r_k.

4. Show that

$$\lim_{n\to\infty} n^{n/2} 2^{n/2} \Gamma\left(\frac{n}{2}\right)^{-1} \int_0^{1+t\sqrt{2/n}} x^{n/2-1} e^{-nx/2}\, dx = \frac{1}{\sqrt{2\pi}} \int_{-\infty}^{t} e^{-x^2}\, dx.$$

(*Hint:* Apply Theorem 1 to the χ^2 distribution.)

5. Show that

$$\lim_{n\to\infty} e^{-n} \sum_{k=0}^{n} \frac{n^k}{k!} = \frac{1}{2}.$$

(*Hint:* Apply Theorem 1 to r.v.'s that have a Poisson distribution with a parameter 1.)

13.2 INFINITELY DIVISIBLE DISTRIBUTIONS

In Chapter 4 we saw that under certain conditions, sums of independent r.v.'s converged to the Poisson law. The question naturally arises of what laws besides the Poisson and Normal laws could be limits for sums of independent r.v.'s. It turns out that the class of limit laws for sums of independent r.v.'s coincides with the class of infinitely divisible laws defined below. This fact was discovered by Bawly [8] and Khintchine [9]. The infinitely divisible laws were discovered by B. de Finetti [10].

DEFINITION 1

A r.v. r is said to be *infinitely divisible* (i.d.) if, for every integer n, there exist (on some probability space) n independent and identically distributed r.v.'s r_{ni} such that $r = \sum_{i=1}^{n} r_{ni}$.

We know that, given any n independent r.v.'s, there exists some probability space on which they are defined. However, the following example[1] shows that the probability space on which the r_{ni} are defined is not, in general, the same as the probability space on which r is defined. This space may not be complex enough to support the r.v.'s r_{ni}.

[1]Proposed by J. L. Doob (see Appendix to Reference [24]).

EXAMPLE 1

Let $\Omega = \{\omega_o, \omega_1, \omega_2, \ldots\}$ and let r be a Poisson r.v. defined on Ω such that $r(\omega_k) = k$. Let $\mathcal{S}$, the σ-field of events, be the class of all subsets of Ω and let $P(r = k) = (e^{-1})(1/k!)$. Then the representation $r = r_1 + r_2$, where r_1 and r_2 are independent and have a common distribution, is not possible.

Since $Er = 1$, r_1 and r_2 must each have a Poisson distribution with an expectation of $\frac{1}{2}$. Hence $P(r_1 = 0) = e^{1/2} \approx 1.6487$. However, the event $(r_1 = 0)$ is a subset of Ω and so its probability is a sum of terms of the form $e^{-1}(1/k_1! + 1/k_2! + \cdots)$, where the k_i are nonnegative integers. In other words, $e^{1/2}$ can be written as the sum of certain terms of the sum

$$(1) \qquad e = 1 + \frac{1}{1!} + \frac{1}{2!} + \frac{1}{3!} + \cdots = 2.7183.$$

We will show that such a representation is impossible.

To obtain $e^{1/2}$, we must use at least one of the 1's appearing in the sum in equation (1). If not, the sum of the remaining terms is $0.7183 < e^{1/2}$. If $\frac{1}{2}!$ is not used, the sum of the remaining terms in (1) is $2.7183 - 1.5 = 0.2183$; therefore, 1 plus any remaining terms $\leq 1.2183 < e^{1/2}$. We now have 1.5. A similar argument shows that we must use $\frac{1}{3}!$ However, $1 + \frac{1}{2}! + \frac{1}{3}! \approx 1.6667 > e^{1/2}$. Hence $e^{1/2}$ cannot be written as a sum of certain terms of the sum for e and so $r \neq r_1 + r_2$ on $(\Omega, \mathcal{S}, P)$.

We will see below that a Poisson r.v. is i.d.

The c.f.'s of i.d. r.v.'s will be called *infinitely divisible* c.f.'s. The d.f.'s of i.d. r.v.'s will be called *infinitely divisible distribution* functions. In practice, to ascertain whether a d.f. or r.v. is i.d., the following lemma is frequently used.

LEMMA 1

A r.v. r is i.d. if and only if its c.f. $\mathcal{X}$ is, for every natural number n, the nth power of some c.f. $\mathcal{X}_n$, that is,

$$(2) \qquad \mathcal{X} = (\mathcal{X}_n)^n.$$

In solving (2) for $\mathcal{X}_n$, we obtain $\mathcal{X}_n = \mathcal{X}^{1/n}$. However, this does not determine $\mathcal{X}_n(t)$ in terms of the values of $\mathcal{X}(t)$, since the nth root has n values. However, if $\mathcal{X}(t)$ does not vanish in an interval containing $t = 0$, it is possible to determine $\mathcal{X}_n(t)$ uniquely from the fact that $\mathcal{X}_n(0) = 1$ and $\mathcal{X}_n(t)$ is continuous. For if $\mathcal{X} \neq 0$, then $\log \mathcal{X}$, the principal branch vanishing at $t = 0$, exists and is finite, and $\mathcal{X}_n = e^{(1/n)(\log \mathcal{X})}$. Unless otherwise stated, we will take the function defined by the preceding equality as the nth root of $\mathcal{X}$.

Theorem 1 will show that the c.f. χ of an i.d. law[2] never vanishes for real t. Therefore, χ_n and its corresponding d.f. F_n are uniquely determined by χ.

Some examples of i.d. laws follow.

EXAMPLE 2

1. A r.v. r distributed according to a Poisson law is i.d. Suppose that $P(r = a + kb) = \lambda^k e^{-\lambda}/k!$ for $\lambda > 0$ and $k = 0, 1, 2, \ldots$. Then the c.f. of r is $\chi(t) = \exp[iat + (e^{itb} - 1)]$; thus,

$$\chi^{1/n} = \exp[iat/n + (e^{itb} - 1)/n] \qquad (n = 1, 2, 3, \ldots)$$

is the c.f. of a r.v. that is also distributed according to a Poisson law.

This proves our assertion by Lemma 1.

2. A normally distributed r.v. r is i.d.

If $Er = \mu$ and $\operatorname{Var} r = \sigma^2$, then the c.f. of r is

$$\chi(t) = \exp(i\mu t - \sigma^2 t^2/2);$$

hence for every $n > 0$,

$$\chi_n(t) = \exp(i\mu t/n - \sigma^2 t^2/2n)$$

is the c.f. of a normal law.

3. A r.v. r distributed according to a Cauchy law

$$F(x) = \frac{1}{\pi}\left(\frac{\pi}{2} + \tan^{-1}\frac{x - b}{a}\right) \qquad (a > 0)$$

is i.d., since its c.f. is

$$\chi(t) = \exp(ibt - a|t|).$$

THEOREM 1

The c.f. of an i.d. law never vanishes.

Proof

For every n, $\chi = \chi_n^n$ (by Lemma 1), where χ_n is a c.f. Hence $|\chi_n|^2 = |\chi|^{2/n} \to \phi$ with $\phi(t) = 0$ if $\chi(t) = 0$ and $\phi(t) = 1$ if $\chi(t) \neq 0$, since $\chi(t)| \leq 1$. There exists a neighborhood of the origin where $|\chi(t)| > 0$, since $\chi(0) = 1$ and χ is continuous. Hence $\phi(t) = 1$ on the whole real line; consequently, $\chi \neq 0$.

The converse of this theorem is not necessarily true, as shown by the following example.

[2]All d.f.'s of the same type.

EXAMPLE 3

Let r be a r.v. that takes the values -1, 0, and 1 with probabilities $\frac{1}{8}$, $\frac{3}{4}$, and $\frac{1}{8}$ respectively. Its c.f. $\chi = (3 + \cos t)/4 > 0$. Suppose that $r = r_1 + r_2$, where r_1 and r_2 are independent, identically distributed r.v.'s. By examining the various possible values of $r_1 + r_2 = r$, it can be seen that each r_i $(i = 1, 2)$ can take only two values c_1 and c_2 $(c_1 < c_2)$, with $P(r_i = c_1) = p$ and $P(r_i = c_2) = q$, where $p + q = 1$. Since $c_1 < c_2$, we have $2c_1 = -1$, $c_1 + c_2 = 0$, and $2c_2 = 1$; therefore, $P(r = -1) = p^2$, $P(r = 0) = 2pq$, and $P(r = 1) = q^2$. Thus $p^2 = \frac{1}{8}$, $2pq = \frac{3}{4}$, and $q^2 = \frac{1}{8}$. However, these last three equations are inconsistent and so $r \neq r_1 + r_2$, which implies that r is not i.d.

THEOREM 2

The sum of a finite number of independent i.d. r.v.'s is i.d.

Proof

We will prove the result for two r.v.'s; the general result follows by induction. Let χ and ϕ be the c.f.'s of these r.v.'s. Since the r.v.'s are i.d., $\chi = (\chi_n)^n$ and $\phi = (\phi_n)^n$ for every natural number n, where χ_n and ϕ_n are c.f.'s. The c.f. of their sum is $\chi\phi = (\chi_n\phi_n)^n$ for every n, which proves the theorem.

The converse of this theorem is not true. Example 3.2 will show that there are r.v.'s that are not i.d. but whose sum is i.d.

THEOREM 3

If χ_n is an i.d. c.f. and $\chi_n \to \chi$, which is a c.f., then χ is i.d.

Proof

For every integer m, $|\chi_n|^{2/m} \to |\chi|^{2/m}$. By the continuity theorem (10.6.5), $|\chi|^{2/m}$ is a c.f. Hence $|\chi|^2$ is an i.d.c.f. and therefore $\chi \neq 0$ by Theorem 1. Since $\log \chi$ exists and is finite, we have $\chi_n^{1/m} = \exp[(1/m)(\log \chi_n] \exp [(1/m)(\log \chi)] = \chi^{1/m}$. By another application of Theorem 10.6.5 it follows that $\chi^{1/m}$ is a c.f., so that χ is i.d.

COROLLARY

If F_n is an i.d. distribution and $F_n \to F(c)$, then F is i.d.

THEOREM 4

If χ is an i.d.c.f., then for every $a > 0$, the function χ^a is also a c.f.

Proof

If $a = 1/n$, where n is a natural number, the result follows from the definition of an i.d.c.f. The assertion remains true for every rational number $a > 0$ by the theorem on the multiplication of c.f.'s. Finally, if c is an irrational number, the continuous function χ^c is the pointwise limit of the functions χ^{c_n}, where c_n are rational numbers and $c_n \to c$. The assertion follows from Theorem 10.6.5.

THEOREM 5

A c.f. is i.d. if and only if it is the product of a finite number of Poisson type c.f.'s or the limit of such products.

Proof

A product of a finite number of Poisson type c.f.'s is i.d. by Theorem 2. A c.f. which is the limit of such products is i.d. by Theorem 3.

We now prove the converse. Let χ be an i.d. c.f. Then $\log \chi$ exists and is finite and as $n \to \infty$,

$$n(\chi^{1/n} - 1) \to \log \chi. \tag{3}$$

Since χ is i.d., $\chi^{1/n}$ is a c.f., and there exists a d.f. F_n such that $(\chi(\lambda))^{1/n} = \int e^{i\lambda t}\, dF_n(t)$. Therefore,

$$n(\chi^{1/n}(\lambda) - 1) = \int n(e^{i\lambda t} - 1)\, dF_n(t). \tag{4}$$

We represent the integral appearing in (4) as the limit of the Stieltjes sum

$$\sum_{k=1}^{m} n(e^{i\lambda t_k} - 1)(F_n(t_k) - F_n(t_{k-1})) \to n(\chi^{1/n}(\lambda) - 1) \tag{5}$$

as $m \to \infty$.

Setting $c_k = c_k(n) = n(F_n(t_k) - F_n(t_{k-1}))$, the last sum can be written as $s_m = \sum_{k=1}^{m} c_k(e^{i\lambda t_k} - 1)$. From relations (3)–(5), it follows that there exists a sequence of integers m_i such that $\lim m_i = \infty$ and $s_{m_i} \to \log \chi$. This completes the proof, since each of the summands in s_{m_i} is the logarithm of a c.f. of the Poisson type.

EXAMPLE 4

Show that the function $\chi(t) = (1 - a)(1 - ae^{it})^{-1}$ for $0 < a < 1$ is an i.d.c.f.

The r.v. r, where $P(r = n) = (1 - a)a^n$ for $n = 0, 1, 2, 3, \ldots$, has the c.f. $\chi(t) = (1 - a)\sum_{n=0}^{\infty} a^n e^{int}$. Since

$$\log \chi(t) = \sum_{k=1}^{\infty} (e^{ikt} - 1)\frac{a^k}{k}$$

and since each summand is the logarithm of a c.f. of the Poisson type, the assertion follows by Theorem 5.

EXERCISES

1. Show that if $\chi = \chi_n^n$, where χ_n is a c.f. for $n = 1, 2, \ldots,$ then $\chi_n \to 1$ and $\chi \neq 0$.

2. Prove that the following distributions are i.d.:

(a) (*Pascal*). $P(r = k) = \dfrac{a^k}{(1 + a)^{k+1}}$. $\qquad (a > 0; k = 1, 2, 3, \ldots)$

(b) (*Polya*). $P(r = k) = \left(\dfrac{\alpha}{1 + \alpha\beta}\right)^k \dfrac{(1 + \beta)(1 + 2\beta) \cdots (1 + (k - 1)\beta)}{k!}$

for all $k > 0$, where $\alpha > 0$, $\beta > 0$, and

$$p_0 = P(r = 0) = (1 + \alpha\beta)^{-1/\beta}.$$

3. The r.v. r with the density function

$$f(x) = \begin{cases} 0 & (x \leq 0) \\ \dfrac{\beta^\alpha}{\Gamma(\alpha)} x^{\alpha-1} e^{-\beta x} & (x > 0) \end{cases}$$

where $\alpha > 0$ and $\beta > 0$ are constants, is i.d. Deduce that the χ^2 distribution and the Maxwell distribution are i.d.

4. Prove that for any positive constants α and β,

$$\chi(t) = (1 + t^2/\beta^2)^{-\alpha}$$

is an i.d.c.f. Deduce that the Laplace distribution is i.d.

5. Let $\zeta(z) = \zeta(x + iy)$ be the Riemann zeta function defined for $x > 1$ by means of the series

$$\zeta(z) = \sum_{n=1}^{\infty} n - n^{-2}$$

or by the Euler product

$$\zeta(z) = \prod_p (1 - p^{-z})^{-1}$$

extended over all prime numbers. Show that for every $x > 1$, the function $\chi(t) = \zeta(z)/\zeta(x)$ is an i.d.c.f. (*Hint:* Apply Theorem 5.)

13.3 CANONICAL REPRESENTATION OF INFINITELY DIVISIBLE LAWS

In 1932, Kolmogorov [11] discovered that a r.v. with a finite variance is i.d. if and only if its c.f. χ can be written in a certain form. This fact was generalized by Lévy [12]. A simplified version was obtained by Khintchine [13], who showed that $\log \chi$ has the following canonical form (see Theorem 4):

$$(1) \qquad \log \chi(t) = i\gamma t + \int \left(e^{itu} - 1 - \frac{itu}{1+u^2}\right)\left(\frac{1+u^2}{u^2}\right) dG(u),$$

where γ is a real number and G is a general d.f. ($G(-\infty) = 0, G(\infty) \leq 1$, and G is increasing and continuous from the right). The value of the integrand at $u = 0$, defined by continuity, is $-t^2/2$.

Throughout this section, the χ, γ, and G with the same indexes (if any) are related to each other by (1) and are to be interpreted as a c.f., a real constant, and a d.f., respectively.

THEOREM 1

A function χ satisfying (1) is an i.d.c.f.

Proof

The integrand in (1) is bounded and continuous in u (for fixed t) and so the integral is the limit of the Riemann–Stieltjes sums

$$\sum_{k=1}^{m} \left(e^{itu_k} - 1 - \frac{itu_k}{1+u_k^2}\right)\left(\frac{1+u_k^2}{u_k^2}\right)(G(u_k) - G(u_{k-1})). \qquad (u_k \neq 0)$$

Each term is the logarithm of the c.f. of a Poisson law; therefore, χ is a c.f. and is i.d. by Theorem 2.5.

It remains to be shown that if a c.f. χ is i.d., it has a unique representation of the form (1). In order to prove uniqueness, we will use the continuity theorem for c.f.'s. To do this we introduce for each function $\log \chi$, which is a c.f. ϕ (with the same indexes as χ, if any) defined by

$$(2) \qquad \phi(t) = \log \chi(t) - \int_0^1 \frac{\log \chi(t+h) + \log \chi(t-h)}{2}\, dh.$$

LEMMA 1

The function ϕ defined by (2) is a c.f. for every c.f. χ satisfying (1).

Proof

Replacing $\log \chi$ by its value given by (1) and interchanging the integrations, (2) reduces to

$$\phi(t) = \int_0^1 \left(\int e^{itu}(1 - \cos hu)\frac{1 + u^2}{u^2}\, dG(u) \right) dh \tag{3}$$

$$= \int e^{itu}\left(1 - \frac{\sin u}{u}\right)\frac{1 + u^2}{u^2}\, dG(u).$$

Define $F(u)$ by

$$F(u) = \int_{-\infty}^{u} \left(1 - \frac{\sin x}{x}\right)\left(\frac{1 + x^2}{x^2}\right) dG(x); \tag{4}$$

then $F(u)$ is a general d.f. The proof of this last statement follows. Since

$$0 < a \le \left(1 - \frac{\sin x}{x}\right)\left(\frac{1 + x^2}{x^2}\right) \le b < \infty,$$

where a and b are constants, then we see that F is increasing on R with $a \operatorname{Var} G \le \operatorname{Var} F \le b \operatorname{Var} G < \infty$.[3] Hence (3) can be rewritten as $\phi(t) = \int e^{itu}\, dF(u)$ and so ϕ is a c.f.

THEOREM 2

A c.f. χ satisfying (1) is uniquely determined by γ and G; the converse is also true.

Proof

Obviously γ and G determine χ by equation (1). Conversely, if χ is given, χ determines a function ϕ defined by (2). Now ϕ is a c.f. (by Lemma 1) and so by the inversion formula, ϕ determines its corresponding d.f. F. From (4), F determines G by

$$G(u) = \int_{-\infty}^{u} dF \Big/ \left(1 - \frac{\sin x}{x}\right)\left(\frac{1 + x^2}{x^2}\right). \tag{5}$$

Since G and $\log \chi$ determine γ, the proof is complete.

To show that the logarithm of every i.d.c.f. χ has the form of equation (1) we require the following convergence theorem for c.f.'s χ which satisfy (1) (see Reference [14]).

[3] $\operatorname{Var} G = G(\infty) - G(-\infty)$.

THEOREM 3

If $\gamma_n \to \gamma$ and $G_n \to G\,(c)$, then $\log \chi_n \to \log \chi$. Conversely, if $\log \chi_n \to g$ which is continuous at 0, then $\gamma_n \to \gamma$ and $G_n \to G\,(c)$, and

$$g = i\gamma t + \int \left(e^{itu} - 1 - \frac{itu}{1+u^2}\right)\left(\frac{1+u^2}{u^2}\right) dG(u).$$

Proof

The first statement is a direct consequence of the Helly–Bray theorem (9.3.2). Conversely, if $\chi_n \to e^g$ which is continuous at the origin, then the e^g is an i.d.c.f. and consequently $e^g \neq 0$. Hence g is finite and continuous on R and $\chi_n \to e^g$ uniformly on every finite interval. Then $\log \chi_n \to g$ uniformly on every finite interval and so from equation (2),

$$(6) \qquad \phi_n(t) \to g(t) - \int_0^1 \frac{g(t+h) + g(t-h)}{2}\, dh.$$

Therefore, the c.f.'s ϕ_n approach a function that is continuous on the whole real line R. At this point we would like to apply the continuity theorem (10.6.5) to deduce the existence of a d.f. F such that $F_n \to F(c)$, where F_n is the d.f. of ϕ_n. However, we recall that the proof of the continuity theorem requires Theorem 8.3.2, which in turn requires the uniform boundedness of the F_n. The uniform boundedness of the F_n follows from relation (6) since

$$\operatorname{Var} F_n = \phi_n(0) \to \int_0^1 \frac{g(h) + g(-h)}{2}\, dh$$

and the function g is finite. It follows from (5) and the Helly–Bray theorem (9.3.2) that $G_n \to G(c)$. Another application of this theorem yields

$$\begin{aligned} it\gamma_n &= \log \chi_n(t) - \int \left(e^{itu} - 1 - \frac{itu}{1+u^2}\right)\left(\frac{1+u^2}{u^2}\right) dG_n(u) \\ &\to g(t) - \int \left(e^{itu} - 1 - \frac{itu}{1+u^2}\right)\left(\frac{1+u^2}{u^2}\right) dG(u) = i\gamma t. \end{aligned}$$

THEOREM 4

A function χ is the c.f. of an i.d.r.v. if and only if its logarithm has the form

$$\log \chi(t) = i\gamma t + \int \left(e^{itu} - 1 - \frac{itu}{1+u^2}\right)\frac{1+u^2}{u^2}\, dG(u),$$

where γ is a real constant and G is a general d.f. This representation is unique.

Proof

By Theorem 1, χ is an i.d. c.f. Conversely, if χ is i.d., then for every n, $\chi = \chi_n^n$. Let F_n be the corresponding d.f.; then

$$\log \chi = \lim n(\chi^{1/n} - 1) = \lim n(\chi_n - 1) = \lim \int (e^{itu} - 1)n\, dF_n$$

$$= \lim \left(i\gamma_n t + \int \left(e^{iut} - 1 - \frac{iut}{1+u^2}\right) dG_n\right),$$

where

$$\gamma_n = \int \frac{nu}{1+u^2}\, dF_n \qquad dG_n = n\, dF_n.$$

It follows from Theorem 3 that $\gamma_n \to \gamma$ and $G_n \to G(c)$. The uniqueness of γ and G is a consequence of Theorem 2.

The above canonical representation of an i.d.c.f. will be called the *Lévy–Khintchine form.* For i.d.r.v.'s with a finite variance, we have the following result.

THEOREM 5

A function χ is the c.f. of an i.d.r.v. r with finite variance if and only if its logarithm has the form

$$\log \chi(t) = i\gamma t + \int (e^{itu} - 1 - itu)/u^2\, dK(u), \tag{6}$$

where γ is a constant[4] and $K(u)$ is a general d.f. The representation of $\log \chi$ by this formula is unique.

Proof

The result follows at once from Theorem 4 by setting

$$K(u) = \int_{-\infty}^{u} (1 + x^2)\, dG(x).$$

We will call (6) *Kolmogorov's formula.* Here $\gamma = Er$ and $K(\infty) = Er^2$ as follows from

$$\left[\frac{d}{dt} \log \chi(t)\right]_{t=0} = iEr = i\gamma$$

$$\left[\frac{d^2}{dt^2} \log \chi(t)\right]_{t=0} = -\, Er^2 = \int dK(u).$$

[4]This is not the same γ that appears in (1).

Lévy's formula for i.d.c.f.'s,

$$\log \chi(t) = i\gamma t - \sigma^2/2t^2 + \int_{-\infty}^{0} \left(e^{iut} - 1 - \frac{iut}{1+u^2}\right) dM(u) \tag{7}$$

$$+ \int_{0}^{\infty} \left(e^{iut} - 1 - \frac{iut}{1+u^2}\right) dN(u),$$

can be deduced from (1) by setting

$$M(u) = \int_{-\infty}^{u} \frac{1+x^2}{x^2}\, dG(x) \qquad (u < 0) \tag{8}$$

$$N(u) = -\int_{u}^{\infty} \frac{1+x^2}{x^2}\, dG(x) \qquad (u > 0)$$

$$\sigma^2 = G(0+) - G(0-).$$

The functions $M(u)$ and $N(u)$

1. Are respectively increasing in the intervals $(-\infty, 0)$, $(0, \infty)$.
2. Satisfy the relations

$$M(-\infty) = N(\infty) = 0.$$

3. Are continuous if and only if G is continuous.

Conversely, any two functions M and N satisfying numbers 1 and 2 and any constant $\sigma > 0$ uniquely determine the c.f. of some i.d.r.v., since Lévy's form (7) can be transformed into (1) by means of (8); the result follows from Theorem 4.

Finally, (1) can be transformed into the following form:

$$\log \chi(t) = i\gamma(\tau)t - \frac{\sigma^2}{2}t^2 + \int_{-\infty}^{-\tau} (e^{iut} - 1)\, dM(u) \tag{9}$$

$$+ \int_{\tau}^{\infty} (e^{iut} - 1)\, dN(u) + \int_{-\tau}^{0} (e^{iut} - 1 - iut)\, dM(u)$$

$$+ \int_{0}^{\tau} (e^{iut} - 1 - iut)\, dN(u),$$

where M and N are defined as in (8), τ is an arbitrary constant chosen so that τ and $-\tau$ are continuity points of M and N respectively, and

$$\gamma(\tau) = \gamma + \int_{|u|<\tau} u\, dG(u) - \int_{|u|\geq\tau} 1/u\, dG(u). \tag{10}$$

The following example illustrates the canonical representation for the normal and Poisson laws.

EXAMPLE 1

1. For the normal law

$$F(x) = \frac{1}{\sigma\sqrt{2\pi}} \int_{-\infty}^{x} e^{-(y-\mu)^2/2\sigma^2}\, dy,$$

we have $\gamma = \mu$ and

$$K(u) = G(u) = \begin{cases} \sigma^2 & (u > 0) \\ 0 & (u \leq 0) \end{cases}$$

in the Lévy–Khintchine (1) and Kolmogorov (6) forms. In Lévy's formula (7), $\gamma = \mu$, $M = 0$, $N = 0$, and $\sigma = \sigma$.

2. For the c.f.

$$\chi(t) = \exp\left(\lambda(e^{it} - 1)\right)$$

of a r.v. distributed according to the Poisson law,

$$\gamma = \lambda/2 \qquad G(u) = \begin{cases} 0 & (u \leq 1) \\ \lambda/2 & (u > 1) \end{cases}$$

in (1),

$$\gamma = \lambda \qquad K(u) = \begin{cases} 0 & (u \leq 1) \\ \lambda & (u > 1) \end{cases}$$

in (6), and $\gamma = \lambda/2$, $\sigma = 0$, $M(u) = 0$, $N(u) = -\lambda$ for $u \leq 1$, and $N(u) = 0$ for $u > 1$ in (7).

We will use Theorem 4 to show that a c.f. χ exists which is not i.d.; however, $|\chi|$ is i.d.

EXAMPLE 2

Let r be a r.v. such that

$$P(r = -1) = \left(\frac{1-\beta}{1+\alpha}\right)\alpha$$

$$P(r = n) = \left(\frac{1-\beta}{1+\alpha}\right)(1 + \alpha\beta)\beta^n, \qquad (n = 0, 1, 2, \ldots)$$

where $0 < \alpha \leq \beta < 1$. The c.f. of r is

$$\chi(t) = \left(\frac{1-\beta}{1+\alpha}\right)\left(\frac{1+\alpha e^{-it}}{1-\beta e^{it}}\right).$$

However, $\chi(t)$ is not an i.d.c.f. We have

$$\log \chi(t) = \sum_{n=1}^{\infty} \left[(-1)^{n-1} \frac{\alpha^n}{n}(e^{-int} - 1) + \frac{\beta^n}{n}(e^{int} - 1) \right]$$

$$= i\gamma t + \int \left(e^{itu} - 1 - \frac{itu}{1+u^2} \right) \frac{1+u^2}{u^2}\, dG(u),$$

where

$$\gamma = \sum_{n=1}^{\infty} \frac{\beta^n + (-1)^n \alpha^n}{1+n^2}$$

and $G(u)$ is a function of bounded variation, having jumps at the points $u = \pm 1, \pm 2, \ldots$ of magnitudes $n\beta^n/(n^2+1)$ for $u = n$ and $(-1)^{n-1} n\alpha^n/(n^2+1)$ for $u = -n$. Hence G is not increasing and so χ is not an i.d.c.f. by Theorem 4.

The function

$$\bar{\chi} = \left(\frac{1-\beta}{1+\alpha} \right) \left(\frac{1+\alpha e^{it}}{1-\beta e^{it}} \right)$$

is also a c.f. which is not i.d. However,

$$g(t) = \chi\bar{\chi} = |\chi|^2$$

is an i.d.c.f. since

$$\log g(t) = \sum_{n=1}^{\infty} \frac{1}{n}(\beta^n + (-1)^{n-1}\alpha^n)(e^{-int} - 1)$$

$$+ \sum_{n=1}^{\infty} \frac{1}{n}(\beta^n + (-1)^{n-1}\alpha^n)(e^{int} - 1)$$

$$= \int \left(e^{itu} - 1 - \frac{itu}{1+u^2} \right) \frac{1+u^2}{u^2}\, dG(u),$$

where G is an increasing function with jumps at the points $u = \pm 1, \pm 2, \pm 3, \ldots$. The jumps at the points n and $-n$ are the same; their magnitude is

$$\frac{n}{1+n^2}(\beta^n + (-1)^{n-1}\alpha^n)$$

for $n > 0$.

EXERCISES

1. Show that if χ is an i.d. c.f., then so is $|\chi|$.

2. Show that if the logarithm of a c.f. is representable in the Lévy-Khintchine form, where G is a function of bounded variation (not necessarily increasing), then such a representation is unique.

3. Let the r.v. r have the density function

$$f(x) = (2\alpha)^{-1} \exp(-|x - \mu|/\alpha)$$

(the Laplace distribution); then r is i.d. Find Kolmogorov's formula for the logarithm of the c.f. of r.

4. Find a suitable canonical form for the c.f.'s of the distributions of Exercise 2.2.

5. For the r.v. of Exercise 2.3, show that K and γ in Kolmogorov's formula are given by

$$K(x) = \alpha \int_0^x u e^{-\beta u}\, du \qquad \gamma = \alpha/\beta.$$

6. Prove that if the sum of two independent i.d.r.v.'s is distributed according to the Poisson (normal) law, then each term is distributed according to the Poisson (normal) law (cf., Section 10.8).

7. (*Khintchine*). Show that there exists an i.d.c.f. χ which can be decomposed into the product of an i.d.c.f. and two c.f.'s, neither of which is decomposable. [*Hint:* Expand $\psi(t) = \log(5 + 4\cos t)/9$ in a Fourier series $\psi(t) = \sum_{n=0}^{\infty} a_n \cos nt$. Let p_i be the nonnegative and $-n_i$ the negative numbers among the a_i. Prove that

$$\prod_{k=1}^{n} \exp(p_k \cos l_k t - 1) = \frac{5 + 4\cos t}{9} \prod_{k=1}^{n} \exp(n_k(\cos m_k t - 1)).$$

Finally write

$$\frac{5 + 4\cos t}{9} = \left(\frac{2 + e^{it}}{3}\right)\left(\frac{2 + e^{-it}}{3}\right).\Big]$$

13.4 CONDITIONS FOR CONVERGENCE

Theorem 3.3 states that a sequence $\{F_n\}$ of i.d. d.f.'s converges to an i.d.d.f. F if and only if $\lim_{n\to\infty} G_n(x) = G(x)\,(c)$ and $\lim_{n\to\infty} \gamma_n = \gamma$, where the functions G_n and G and the constants γ_n and γ are defined by the Lévy–Khintchine formula for F_n and F respectively. Here we will formulate some conditions for convergence of i.d.d.f.'s in terms of the functions and constants appearing in the canonical representation (9). The corresponding result for Kolmogorov's form is left as an exercise.

THEOREM 1

The i.d.d.f.'s $F_n \to F\,(c)$ as $n \to \infty$ if and only if

1. $M_n \to M\,(c) \qquad N_n \to N\,(c)$.

2. $\gamma_n(\tau) \to \gamma(\tau)$.

3. $\lim_{\epsilon\to 0} \overline{\lim_{n\to\infty}} \left(\int_{-\epsilon}^{0} u^2\, dM_n(u) + \sigma_n^2 + \int_0^{\epsilon} u^2\, dN_n(u) \right) = \lim_{\epsilon\to 0} \underline{\lim_{n\to\infty}} \left(\int_{-\epsilon}^{0} u^2\, dM_n(u) + \sigma_n^2 + \int_0^{\epsilon} u^2\, dN_n(u) \right) = \sigma^2,$

where the functions M_n, N_n and M, N and the constants σ_n, $\gamma_n(\tau)$ and σ, $\gamma(\tau)$ are defined by equations 3.8 and 3.10 for the d.f.'s F_n and F respectively (see Reference [14]).

Proof

Let $F_n \to F(c)$. Then by Theorem 3.3, $G_n \to G(c)$. It follows from the formula defining M, N, and γ and the Helly–Bray theorem (9.3.1 and 9.3.2) that conditions 1 and 2 are fulfilled.

To prove that condition 3 holds, let $\pm\epsilon$ be continuity points of the functions M_n, M, N_n, and N and set

$$I_n(\epsilon) = \int_{-\epsilon}^{0} \frac{u^2}{1+u^2}\, dM_n + \sigma_n^2 + \int_0^{\epsilon} \frac{u^2}{1+u^2}\, dN_n = G_n(\epsilon) - G_n(-\epsilon);$$

$I(\epsilon)$ is defined similarly, with the subscripts omitted. Since $\pm\epsilon$ are also continuity points of G_n and G, $G_n(\epsilon) \to G(\epsilon)$ and so

$$I_n(\epsilon) \to I(\epsilon). \qquad (n \to \infty) \tag{1}$$

From the inequalities

$$\frac{1}{1+\epsilon^2} \int_{-\epsilon}^{0} u^2\, dM_n \le \int_{-\epsilon}^{0} \frac{u^2}{1+u^2}\, dM_n \le \int_{-\epsilon}^{0} u^2\, dM_n$$

$$\frac{1}{1+\epsilon^2} \int_0^{\epsilon} u^2\, dN_n \le \int_0^{\epsilon} \frac{u^2}{1+u^2}\, dN_n \le \int_0^{\epsilon} u^2\, dN_n,$$

it follows that

$$\frac{1}{1+\epsilon^2} J_n(\epsilon) \le I_n(\epsilon) \le J_n(\epsilon), \tag{2}$$

where

$$J_n(\epsilon) = \int_{-\epsilon}^{0} u^2\, dM_n(u) + \sigma_n^2 + \int_0^{\epsilon} u^2\, dN_n(u).$$

From relations (1) and (2),

$$\overline{\lim_{n\to\infty}} \frac{1}{1+\epsilon^2} J_n(\epsilon) \le I(\epsilon) \le \underline{\lim_{n\to\infty}} J_n(\epsilon). \tag{3}$$

Finally, letting $\epsilon \to 0$, both sides of inequality (3) have the same limit—that is, $\lim_{\epsilon \to 0} I(\epsilon) = \sigma^2$. An analogous proof holds for $\underline{\lim}$. This completes the proof that condition 3 holds; thus the necessity of the conditions has been established.

To prove the sufficiency of the conditions, we will prove that they imply the conditions of Theorem 3.3.

Let $u < 0$ be a continuity point of M_n and M and consequently of G_n and G. Then by the Helly–Bray theorem,

$$G_n(u) = \int_{-\infty}^{u} \frac{t^2}{1+t^2}\, dM_n \to \int_{-\infty}^{u} \frac{t^2}{1+t^2}\, dM = G(u); \qquad (u < 0,\ n \to \infty) \tag{4}$$

therefore,

$$\lim_{\epsilon \to 0} \lim_{n \to \infty} G_n(-\epsilon) = G(-0). \tag{5}$$

For $u > 0$, a continuity point of G_n and G, the fact that $G_n \to G$ as $n \to \infty$ is harder to prove, since $G_n(u)$ is defined by

$$G_n(u) = G_n(\epsilon) - \int_{\epsilon}^{u} \frac{x^2}{1+x^2}\, dN_n(x) \qquad (u > 0,\ \epsilon > 0)$$

and $G_n(\epsilon)$ is not known explicitly. We will first show that

$$\lim_{\epsilon \to 0} \lim_{n \to \infty} G_n(\epsilon) = G(+0) = \lim_{\epsilon \to 0} \overline{\lim_{n \to \infty}}\, G_n(\epsilon). \tag{5'}$$

By (2),

$$G_n(-\epsilon) + \frac{1}{1+\epsilon^2} J_n(\epsilon) \le G_n(\epsilon) \le G_n(-\epsilon) + J_n(\epsilon).$$

Hence by (5) and condition 3 of the theorem,

$$\lim_{\epsilon \to 0} \overline{\lim_{n \to \infty}}\, G_n(\epsilon) = \lim_{\epsilon \to 0} \underline{\lim_{n \to \infty}}\, G_n(\epsilon) = G(-0) + \sigma^2 = G(0),$$

where the last equality follows from the definition of σ^2. Now choosing ϵ and u to be continuity points of N_n and N,

$$\int_{\epsilon}^{u} \frac{t^2}{1+t^2}\, dN_n \to \int_{\epsilon}^{u} \frac{t^2}{1+t^2}\, dN; \qquad (n \to \infty)$$

therefore,

$$\overline{\lim_{n \to \infty}}\, G_n(u) = \lim_{\epsilon \to 0} \overline{\lim_{n \to \infty}} \left(G_n(\epsilon) - \int_{\epsilon}^{u} \frac{t^2}{1+t^2}\, dN_n \right) \tag{6}$$

$$= \lim_{\epsilon \to 0} \underline{\lim_{n \to \infty}} \left(G_n(\epsilon) - \int_{\epsilon}^{u} \frac{t^2}{1+t^2}\, dN_n \right) = \underline{\lim_{n \to \infty}}\, G_n(u) = G(u). \qquad (u > 0)$$

Hence we have proved that $G_n(u) \to G(u)$ as $n \to \infty$ at all continuity points of G.

We still must consider $u = \infty$. Thus

$$G_n(\infty) = \int_{-\infty}^{-\epsilon} \frac{u^2}{1+u^2}\, dM_n + \int_{-\epsilon}^{0} \frac{u^2}{1+u^2}\, dM_n + \sigma_n^2$$

$$+ \int_{0}^{\epsilon} \frac{u^2}{1+u^2}\, dN_n + \int_{\epsilon}^{\infty} \frac{u^2}{1+u^2}\, dN_n.$$

Equation (4), the Helly–Bray theorem applied to $\int_\epsilon^\infty [u^2/(1 + u^2)]\, dN_n$ and condition 3 of the theorem [c.f., equation (3)] imply that as $n \to \infty$, $G_n(\infty) \to G(\infty)$.

Finally, $\gamma_n \to \gamma$ follows from formula 3.10 defining γ_n and γ, from condition 2 of the theorem, and the Helly–Bray theorem.

EXERCISE

1. Let F_n be i.d.d.f.'s with finite variances. Prove that $F_n \to F(c)$ if and only if

(a) $K_n \to K(c)$,
(b) $\gamma_n \to \gamma$,

as $n \to \infty$, where the functions K_n, K and the constants γ_n, γ are defined by Kolmogorov's formula (3.6). [*Hint:* Sufficiency follows from $\log \chi_n \to \log \chi$, using Helly's theorem. Conversely, variances are uniformly bounded and so $K_{n'} \to K$ for some subsequence $n' \to \infty$. Apply Helly's theorem and use uniqueness.]

13.5 INFINITESIMAL R.V.'S

As in Section 12.5, we will try to generalize the classic limit theorems by considering a double sequence r_{nk} of r.v.'s. Again the r.v.'s in each row are independent; however, the r.v.'s in different rows can be dependent. Let

$$s_n = r_{n1} + r_{n2} + \cdots + r_{nk_n}.$$

Are there constants a_n so that the d.f.'s of the sums $s_n - a_n$ converge to a limit? What are the properties of the limit d.f.?

The questions posed above are too general, since any d.f. F can appear as the limit of the d.f.'s of the sum $s_n - a_n$. For example, if we let the d.f. of r_{n1} be $F(x)$ while $r_{nk} = 0$ (a.s) for $k \neq 1$, then s_n has a d.f. $F(x)$. Hence some additional conditions must be imposed on the r_{nk}.

To discover the nature of these conditions, let us consider the classic situation discussed in Section 1. Here $r_{nk} = (r_k - a_n)B_n^{-1}$; we observed from equation (1.13) that $r_{nk} \to 0\,(P)$ as $n \to \infty$ for all k. This condition is emphasized in the following definition.

DEFINITION 1

The r.v.'s r_{nk} are called *infinitesimal*[5] if and only if for every $\epsilon > 0$,

$$\max_{1 \le k \le k_n} P(|r_{nk}| \ge \epsilon) = \max_k \int_{|x| \ge \epsilon} dF_{nk} \to 0\,, \tag{1}$$

as $n \to \infty$.

We will see later that the following less stringent restriction will be required.

DEFINITION 2

The variables r_{nk} are called *asymptotically constant* if and only if it is possible to find constants a_{nk} so that for every $\epsilon > 0$,

$$\max_{1 \le k \le k_n} P(|r_{nk} - a_{nk}| \ge \epsilon) = \max_k \int_{|x| \ge \epsilon} dF_{nk}(x + a_{nk}) \to 0 \tag{2}$$

as $n \to \infty$.

With the above restrictions, it turns out that the limit laws are i.d. (see Section 6).

LEMMA 1

If the r.v.'s r_{nk} are asymptotically constant, then it is possible to take $a_{nk} = m_{nk}$ in (2), where m_{nk} is a median of r_{nk}.

Proof

Let r_{nk} be asymptotically constant. It is sufficient to prove that as $n \to \infty$, $\max_{1 \le k \le k_n} |m_{nk} - a_{nk}| \to 0$. By the definition of "asymptotically constant," for every $\epsilon > 0$ there exists an $N = N(\epsilon)$ such that for $n \ge N$,

$$\max_{1 \le k \le k_n} P(|r_{nk} - a_{nk}| > \epsilon) < \frac{1}{2}$$

and so

$$\min_{1 \le k \le k_n} P(|r_{nk} - a_{nk}| \le \epsilon) > \frac{1}{2}.$$

[5]They also are called *uniformly asymptotically negligible* (u.a.n.).

If the probability of r_{nk} lying in some interval is greater than $\frac{1}{2}$, it follows that every median m_{nk} belongs to this interval (see Exercise 1); therefore,

$$\max_k |m_{nk} - a_{nk}| \leq \epsilon.$$

LEMMA 2

The r.v.'s r_{nk} are infinitesimal if and only if as $n \to \infty$,

$$\max_{1 \leq k \leq k_n} \int \frac{x^2}{1 + x^2} \, dF_{nk}(x) \to 0. \tag{3}$$

Proof

Suppose that the r.v.'s r_{nk} are infinitesimal and that $0 < \epsilon < \frac{1}{2}$. For sufficiently large n,

$$\max_{1 \leq k \leq k_n} \int \frac{x^2}{1 + x^2} \, dF_{nk} \leq \max_{1 \leq k \leq k_n} \left(\int_{|x| < \epsilon} x^2 \, dF_{nk} + \int_{|x| \geq \epsilon} dF_{nk} \right) \leq \epsilon^2 + \frac{\epsilon}{2} \leq \epsilon.$$

Conversely, by the inequalities

$$\max_k \int \frac{x^2}{1 + x^2} \, dF_{nk} \geq \max_k \int_{|x| \geq \epsilon} \frac{x^2}{1 + x^2} \, dF_{nk} \geq \frac{\epsilon^2}{1 + \epsilon^2} \max_k \int_{|x| \geq \epsilon} dF_{nk},$$

Equation (1) follows from relation (3).

COROLLARY

The r.v.'s r_{nk} are asymptotically constant if and only if

$$\max_{1 \leq k \leq k_n} \int \frac{x^2}{1 + x^2} \, dF_{nk}(x + m_{nk}) \to 0$$

as $n \to \infty$.

LEMMA 3

The r.v.'s r_{nk} are infinitesimal if and only if

$$\max_{1 \leq k \leq k_n} |\chi_{nk}(t) - 1| \to 0 \tag{4}$$

as $n \to \infty$.[6]

Proof

Assume that the r.v.'s r_{nk} are infinitesimal. For every $\epsilon > 0$, upon replacing χ_{nk} by its defining relation,

[6]Recall that convergence of c.f.'s is uniform on every finite interval.

$$\max_k |\chi_{nk}(t) - 1| \leq \max_k \int_{|x|<\epsilon} |e^{itu} - 1|\, dF_{nk} + 2 \max_k \int_{|x|\geq\epsilon} dF_{nk}$$

since $|e^{itu} - 1| \leq 2$. Upon using the inequality $|e^{itu} - 1| \leq |t|\epsilon$ in the first integral in the above inequality, the validity of relation (4) is established.

Now assume that relation (4) holds; i.e., $\chi_{nk} \to 1$ uniformly in k as $n \to \infty$. Then by Lévy's continuity theorem (10.6.5) and Theorem 8.3.5 it follows that $r_{nk} \to 0(P)$ uniformly in k.

For the general limit theorem of Section 6, we do not require the existence of the variances for the summands r_{nk}. The theoretical part of the proof is the same as for the case of finite variances. However, the computations are more difficult. The following theorems derive certain necessary estimates.

LEMMA 4

If the d.f.'s of the sums

(5) $$s_n = r_{n1} + \cdots + r_{nk_n} - a_n$$

of independent r.v.'s r_{nk} converge to a limit d.f. for suitably chosen constants a_n, then there exists a constant $c < \infty$ such that for all n,

(6) $$\sum_{k=1}^{k_n} \int \frac{u^2}{1+u^2}\, dF_{nk}(u+m_{nk}) < c.$$

Proof

Since the d.f.'s of the sums s_n converge to a limit, it follows that for every sequence of constants $\alpha_n \to 0$, $\alpha_n s_n \to 0\,(P)$ as $n \to \infty$. Moreover, for $0 < \alpha < 1$,

$$E\frac{\alpha^2 r^2}{1+\alpha^2 r^2} = \alpha^2 E\frac{r^2}{1+\alpha^2 r^2} \geq \alpha^2 E\frac{r^2}{1+r^2}.$$

Therefore, for sufficiently large n,

$$\sum_{k=1}^{k_n} E\frac{\alpha_n^2 \bar{r}_{nk}^2}{1+\alpha_n^2 \bar{r}_{nk}^2} \geq \alpha_n^2 \sum_{k=1}^{k_n} E\frac{\bar{r}_{nk}^2}{1+\bar{r}_{nk}^2},$$

where $\bar{r}_{nk} = r_{nk} - m_{nk}$. By equation 12.5.9 the left-hand side of the above inequality approaches zero as $n \to \infty$; consequently, as $n \to \infty$,

$$\alpha_n^2 \sum_{k=1}^{k_n} \int \frac{u^2}{1+u^2}\, dF_{nk}(u+m_{nk}) \to 0.$$

Since this relation holds for every sequence, it follows that inequality (6) is valid.

LEMMA 5

If the r.v.'s r_{nk} are infinitesimal, then

(7)
$$\max_{1\le k\le k_n} |m_{nk}| \to 0, \qquad (n\to\infty)$$

where m_{nk} is a median of r_{nk}.

Proof

The result follows from Lemma 1, since $a_{nk} = 0$.

LEMMA 6

If the r.v.'s r_{nk} are infinitesimal, then

(8)
$$\max_{1\le k\le k_n} |\alpha_{nk}| \to 0, \qquad (n\to\infty)$$

where

(9)
$$\alpha_{nk} = \int_{|x|<\tau} x\, dF_{nk}(x)$$

and τ is any positive constant.

Proof

$$|\alpha_{nk}| = \left|\int_{|x|<\tau} x\, dF_{nk}(x)\right| \le \int_{|x|\le\epsilon} |x|\, dF_{nk} + \int_{\epsilon\le|x|<\tau} |x|\, dF_{nk}$$
$$\le \epsilon + \tau P(|r_{nk}| \ge \epsilon).$$

Choosing $\epsilon > 0$ sufficiently small and n sufficiently large, $\sup_{1\le k\le k_n} |\alpha_{nk}|$ can be made arbitrarily small.

LEMMA 7

If, for suitably chosen constants a_n, the d.f.'s of the sums of independent infinitesimal r.v.'s in equation (5) converge to a limit d.f. as $n\to\infty$, then there exists a constant c such that

(10)
$$\sum_{k=1}^{k_n} \int \frac{x^2}{1+x^2}\, dF_{nk}(x+\alpha_{nk}) < c,$$

where α_{nk} is defined by (9).

Proof

(11)
$$\int \frac{x^2}{1+x^2}\, dF(x+\alpha_{nk}) = \int \frac{(x+m_{nk}-\alpha_{nk})^2}{1+(x+m_{nk}-\alpha_{nk})^2}\, dF_{nk}(x+m_{nk})$$
$$\le \int \frac{2x^2\, dF_{nk}(x+m_{nk})}{1+(x+m_{nk}-\alpha_{nk})^2} + 2(m_{nk}-\alpha_{nk})^2.$$

The last inequality follows from $(a + b)^2 \leq 2(a^2 + b^2)$. We only must show that inequality (10) is valid for sufficiently large n. From relation (7) and the fact that $|\alpha_{nk}| < \tau$, it follows that $|m_{nk} - \alpha_{nk}|$ is uniformly bounded. Therefore, there exists a constant M such that

$$\frac{Mx^2}{1 + x^2} \geq \frac{2x^2}{1 + (x + m_{nk} - \alpha_{nk})^2}$$

and so

$$(12) \quad \int \frac{2x^2}{1 + (x + m_{nk} - \alpha_{nk})^2} \, dF_{nk}(x + m_{nk}) \leq M \int \frac{x^2}{1 + x^2} \, dF_{nk}(x + m_{nk}).$$

Next we estimate $(m_{nk} - \alpha_{nk})^2$. For sufficiently large n, by using relation (7),

$$\begin{aligned}
(\alpha_{nk} - m_{nk})^2 &= \left(\int_{|x|<\tau} (x - m_{nk}) \, dF_{nk} - \int_{|x|\geq\tau} m_{nk} \, dF_{nk} \right)^2 \\
&\leq 2\left(\int_{|x+m_{nk}|<\tau} |x| \, dF_{nk}(x + m_{nk}) \right)^2 \\
&\quad + 2\left(\int_{|x+m_{nk}|\geq\tau} m_{nk} \, dF_{nk}(x + m_{nk}) \right)^2 \\
&\leq 2\left(\int_{|x|<2\tau} x \, dF_{nk}(x + m_{nk}) \right)^2 \\
&\quad + 2m_{nk}^2 \left(\int_{|x|>\tau/2} dF_{nk}(x + m_{nk}) \right)^2.
\end{aligned}$$

An application of Cauchy's inequality yields

$$\begin{aligned}
(13) \quad (\alpha_{nk} - m_{nk})^2 &\leq 2 \int_{|x|<2\tau} x^2 \, dF_{nk}(x + m_{nk}) + 2m_{nk}^2 \int_{|x|>\tau/2} dF_{nk}(x + m_{nk}) \\
&\leq 2\,(1 + 4\tau^2) \int_{|x|<2\tau} \frac{x^2}{1 + x^2} \, dF_{nk}(x + m_{nk}) \\
&\quad + 2m_{nk}^2 \left(\frac{1 + (\tau/2)^2}{(\tau/2)^2} \right) \int_{|x|>\tau/2} \frac{x^2}{1 + x^2} \, dF_{nk}(x + m_{nk}).
\end{aligned}$$

The result follows from the inequalities (11)–(13) by applying Lemmas 4 and 5.

LEMMA 8

If, for suitably chosen constants a_n, the d.f.'s of the sums of independent infinitesimal r.v.'s in equation (5) converge to a limit d.f. as $n \to \infty$, then

$$(14) \qquad \sum_{k=1}^{k_n} \int_{|x|\geq\tau} d\bar{F}_{nk} \leq \frac{1 + \tau^2}{\tau^2} c$$

$$(15) \qquad \sum_{k=1}^{k_n} \left| \int_{|x|<\tau} x \, d\bar{F}_{nk} \right| \leq 2\left(\frac{4+\tau^2}{\tau}\right)c$$

$$(16) \qquad \sum_{k=1}^{k_n} \int_{|x|<\tau} x^2 \, d\bar{F}_{nk} \leq (1+\tau^2)c,$$

where $\bar{F}_{nk}(x) = F_{nk}(x + \alpha_{nk})$ and α_{nk} is defined by (9).

Proof

By Lemma 7,

$$\sum_{k=1}^{k_n} \int_{|x|\geq\tau} d\bar{F}_{nk} \leq \frac{1+\tau^2}{\tau^2} \sum_{k=1}^{k_n} \int_{|x|\geq\tau} \frac{x^2}{1+x^2} \, d\bar{F}_{nk} \leq c\,\frac{1+\tau^2}{\tau^2}.$$

To prove (15), choose n sufficiently large so that $|\alpha_{nk}| < \tau/2$ [recall relation (8)] and consider

$$(17) \quad \left| \int_{|x|<\tau} x \, d\bar{F}_{nk} - \int_{|x+\alpha_{nk}|<\tau} x \, d\bar{F}_{nk} \right| \leq \int_{\tau/2\leq|x|\leq 3\tau/2} |x| \, d\bar{F}_{nk} \leq \frac{3\tau}{2} \int_{|x|\geq\tau/2} d\bar{F}_{nk}.$$

Since

$$\left| \int_{|x+\alpha_{nk}|<\tau} x \, d\bar{F}_{nk} \right| = \left| \int_{|x|<\tau} (x - \alpha_{nk}) \, dF_{nk} \right|$$

$$= \left|\alpha_{nk} \int_{|x|\geq\tau} dF_{nk}\right| \leq \tau/2 \int_{|x|\geq\tau/2} d\bar{F}_{nk},$$

it follows from (17) that

$$\left| \int_{|x|<\tau} x \, d\bar{F}_{nk} \right| \leq 2\tau \int_{|x|\geq\tau/2} d\bar{F}_{nk}.$$

Using (14) in the above inequality yields (15).

Equation (16) is an immediate consequence of Lemma 7 when the estimate $(1 + t^2)/(1 + x^2) \geq 1$ for $|x| < \tau$ is used.

EXERCISES

1. Let r be a r.v. and let m be its median. Prove that

(a) If $P(r \geq a) > \frac{1}{2}$, then $m \geq a$.
(b) If $P(r \leq b) > \frac{1}{2}$, then $m \leq b$.

Deduce that if $P(a \leq r \leq b) > \frac{1}{2}$, then $a \leq m \leq b$.

2. Show that if the r.v.'s r_{nk} are infinitesimal, then

$$\max_{k \le n} \int_{|x|<\tau} |x|^r \, dF_{nk} \to 0. \qquad (r > 0, 0 < \tau < \infty)$$

3. Let $\bar{\chi}_{nk}(t) = \int e^{iut} \, d\bar{F}_{nk}$. Show that if the r.v.'s r_{nk} are infinitesimal, then $\max_{k \le n} |\bar{\chi}_{nk} - 1| \to 0$ uniformly on every finite interval.

4. Prove the following inequalities:

(a) $c_1 \max_{|t| \le b} |\bar{\chi}_{nk}(t) - 1| \le \int \frac{x^2}{1+x^2} \, d\bar{F}_{nk}$. $\qquad (c_1 = c_1(\alpha_{nk}, b, \tau) > 0)$

(b) $\int \frac{x^2}{1+x^2} \, d\bar{F}_{nk} \le c_2 \int_0^b (1 - |\chi(t)|^2) \, dt,$

where $c_2 = c_2(m_{nk}, b, \tau)$ and $\tau > |m_{nk}|$.

13.6 THE GENERAL LIMIT THEOREM

The following theorem was proposed by B. V. Gnedenko [14], [15], and [16]. For its proof it is convenient to introduce the following notation:

$$\bar{F}_{nk}(x) = F_{nk}(x + \alpha_{nk})$$

$$\bar{\chi}_{nk}(t) = \int e^{itu} \, d\bar{F}_{nk}(u)$$

$$\beta_{nk} = \bar{\chi}_{nk} - 1.$$

Since $\chi_{nk} = \int e^{itu} \, dF_{nk} = e^{it\alpha_{nk}} \bar{\chi}_{nk}$, then

$$\sum \log \chi_{nk} = \log \bar{\chi}_{nk} - \sum it\alpha_{nk}.$$

THEOREM 1

The d.f.'s of the sequences of sums

(1) $$s_n = r_{n1} + r_{n2} + \cdots + r_{nk_n} - a_n$$

of independent infinitesimal r.v.'s r_{nk}, where a_n are constants, converge to a limit d.f. as $n \to \infty$ if and only if the sequence of i.d. laws whose c.f.'s χ_n have logarithms ψ_n given by

(2) $$\log \chi_n(t) \equiv \psi_n(t) = -ia_n t + \sum_{k=1}^{k_n} \left(it\alpha_{nk} + \int (e^{itx} - 1) \, dF_{nk}(x + \alpha_{nk}) \right)$$

converge to a limit law. Here

(3) $$\alpha_{nk} = \int_{|x|<\tau} x \, dF_{nk}(x),$$

and $\tau > 0$ is a constant. The limit d.f.'s for the two sequences coincide.

Proof

The d.f.'s of the r.v.'s s_n with c.f.'s $\phi_n = e^{-ita_n}\prod_{k=1}^{k_n} \chi_{nk}$ converge to a limit d.f. with a c.f. ϕ if and only if $\phi_n \to \phi$ uniformly on every finite interval as $n \to \infty$. To show that the i.d. laws converge to the same limit law as the s_n, we need only show that $|\chi_n - \phi_n| \to 0$ as $n \to \infty$ or, alternatively, that

$$(4) \qquad |\log \phi_n(t) - \psi_n(t)| = |-ita_n + \sum_{k=1}^{k_n} \log \chi_{nk}(t)$$

$$- (-ita_n + \sum_{k=1}^{k_n} (it\alpha_{nk} + \int (e^{itu} - 1)\, d\bar{F}_{nk})|$$

$$= |\sum_{k=1}^{k_n} \log \bar{\chi}_{nk} - \sum_{k=1}^{k_n} \beta_{nk}| \to 0$$

as $n \to \infty$ uniformly on every finite t-interval. Since the r.v.'s r_{nk} are infinitesimal,

$$(5) \qquad \sup_{1\le k\le k_n} |\beta_{nk}| \to 0 \qquad (n \to \infty)$$

(see Exercise 5.3) and so the expansion

$$\log \bar{\chi}_{nk} = \log (1 + \beta_{nk}) = \sum_{s=1}^{\infty} (-1)^{s+1} \beta_{nk}^s / s$$

is valid. Thus for sufficiently large n,

$$(6) \qquad \left|\sum_{k=1}^{k_n} \log \bar{\chi}_{nk} - \sum_{k=1}^{k_n} \beta_{nk}\right| \le \sum_{k=1}^{k_n} \sum_{s=2}^{\infty} \frac{1}{s} |\beta_{nk}|^s$$

$$\le \frac{1}{2} \sum_{k=1}^{k_n} \frac{|\beta_{nk}|^2}{1 - |\beta_{nk}|} \le \sup_k |\beta_{nk}| \sum_{k=1}^{k_n} |\beta_{nk}|.$$

Since

$$|\beta_{nk}| = \left| \int_{|x|<\tau} (e^{itx} - 1 - itx)\, d\bar{F}_{nk} + \int_{|x|\ge\tau} (e^{itx} - 1)\, d\bar{F}_{nk} + it \int_{|x|<\tau} x\, d\bar{F}_{nk} \right|$$

$$\le \frac{1}{2} |t|^2 \int_{|x|<\tau} x^2\, d\bar{F}_{nk} + 2 \int_{|x|\ge\tau} d\bar{F}_{nk} + |t| \int_{|x|<\tau} x\, d\bar{F}_{nk},$$

it follows from Lemma 5.8 that

$$(7) \qquad \sum_{k=1}^{k_n} |\beta_{nk}| \le \left[|t|^2 \left(\frac{1+\tau^2}{2}\right) + 2\left(\frac{1+\tau^2}{\tau^2}\right) + 2|t| \left(\frac{4+\tau^2}{\tau}\right) \right] c.$$

Thus, relations (5)–(7) imply that (4) is valid.

Now suppose that the i.d. laws whose c.f.'s have logarithms given by (4) converge to a limit. Again we need only verify that (4) is valid. The proof is exactly the same as before up to the point where we apply the estimates (14)–(16) of Lemma 5.8. We will show that these estimates are still valid. An examination of the proof of Lemma 5.8 shows that it depends on the validity of inequality (5.10). The validity of inequality (5.10) is established by writing (2) in the Lévy–Khintchine form. This leads to

$$G_n(u) = \sum_{k=1}^{k_n} \int_{-\infty}^{u} \frac{x^2}{1+x^2}\, dF_{nk}.$$

Since it is given that the i.d. laws converge to a limit, Theorem 3.3 yields

$$\int dG_n = \sum_{k=1}^{k_n} \int \frac{x^2}{1+x^2}\, d\bar{F}_{nk} \to \int dG, \qquad (n \to \infty)$$

where G is the d.f. defined by the Lévy–Khintchine formula. Therefore, inequality (5.10) is valid and once again it follows from relations (5)–(7) that (4) is valid.

An immediate consequence of Theorem 1 is the following fundamental result by A.Y. Khintchine [17].

THEOREM 2

F is the limit d.f. of the sums of infinitesimal independent r.v.'s in equation (1) if and only if F is i.d.

Proof

From Theorem 1, the limit d.f. F of the sums in (1) is simultaneously the limit of i.d. laws. Therefore, by the corollary to Theorem 2.3, F is i.d. The fact that every i.d. law is the limit law for sums of infinitesimal r.v.'s follows from Theorem 2.1 (in which we showed that $\mathcal{X}_n \to 1$) and Lemma 5.3.

If the r_{nk} for $n = 1, 2, 3, \ldots$ and $1 \leq k \leq k_n$ are asymptotically constant, then the r.v.'s $r_{nk} - m_{nk}$ are infinitesimal. Hence the class of limit laws for the sums of asymptotically constant r.v.'s in (1) coincides with the class of i.d. laws.

EXERCISES

1. Consider the double sequence r_{nk} of r.v.'s that are independent in each row, subject to the following conditions:

(a) $\sup_{1 \leq k \leq k_n} P(|r_{nk} - Er_{nk}| \geq \epsilon) \to 0.$ $\qquad (n \to \infty, \epsilon > 0)$

(b) The r_{nk} have finite variances and

$$\operatorname{Var}\left(\sum_{k=1}^{k_n} r_{nk}\right) = \sum_{k=1}^{k_n} \operatorname{Var}(r_{nk}) \leq c,$$

where c is a constant independent of n.

Give a direct proof of Theorem 1 for such r.v.'s. Here $\alpha_{nk} = Er_{nk}$ and $a_n = 0$.

2. Show that the limit laws for the r.v.'s described in Exercise 1 are i.d.

13.7 CONDITIONS FOR CONVERGENCE TO AN ARBITRARY I.D.D.F.

Theorem 6.1, although of theoretical importance, is not of much practical value. In this section we will obtain necessary and sufficient conditions for convergence to an i.d.d.f phrased in terms of the d.f.'s of the summands appearing in (1); we also will specify how to choose the a_n.

THEOREM 1

Let F be a given i.d.d.f. and let M and N be the functions and $\gamma(\tau)$ and σ^2 the constants determined by equation (3.9) for the logarithm of the c.f. of F. Then the d.f.'s of the sums

$$s_n = r_{n1} + r_{n2} + \cdots + r_{nk_n} - a_n$$

of the independent infinitesimal r.v.'s r_{nk} converge to the d.f. F if and only if

(a) At the continuity points of M and N,

$$\sum_{k=1}^{k_n} F_{nk}(u) \to M(u) \qquad (u < 0)$$

$$\sum_{k=1}^{k_n} (F_{nk}(u) - 1) \to N(u) \qquad (u > 0)$$

as $n \to \infty$.

(b)
$$\lim_{\epsilon \to 0} \overline{\lim_{n \to \infty}} \sum_{k=1}^{k_n} \left(\int_{|u|<\epsilon} u^2\, dF_{nk} - \left(\int_{|u|<\epsilon} u\, dF_{nk}\right)^2\right)$$

$$= \lim_{\epsilon \to 0} \underline{\lim_{n \to \infty}} \sum_{k=1}^{k_n} \left(\int_{|u|<\epsilon} u^2\, dF_{nk} - \left(\int_{|u|<\epsilon} u\, dF_{nk}\right)^2\right) = \sigma^2.$$

(c)
$$a_n = \sum_{k=1}^{k_n} \int_{|x|<\tau} x\, dF_{nk} - \gamma(\tau) + o(1),$$

where $\tau > 0$ and $-\tau$ are continuity points of M and N.

Proof

By Theorem 6.1, we need only investigate the conditions for convergence of the i.d. laws defined by equation (6.2). To write these laws in the form of equation (3.9) we should define

$$M_n(x) = \sum_{k=1}^{k_n} \int_{-\infty}^{x} d\bar{F}_{nk} \qquad (x < 0)$$

$$N_n(x) = -\sum_{k=1}^{k_n} \int_{x}^{\infty} d\bar{F}_{nk} \qquad (x > 0)$$

$$\sigma_n = 0$$

$$\gamma_n(\tau) = -a_n + \sum_{k=1}^{k_n} \left(\int_{|u|<\tau} u\, dF_{nk} + \int_{|u|<\tau} u\, d\bar{F}_{nk} \right),$$

where $\bar{F}_{nk}(u) = F_{nk}(u + \alpha_{nk})$ and α_{nk} is defined by equation (6.3). According to Theorem 4.1, necessary and sufficient conditions for the convergence (as $n \to \infty$) of the d.f.'s defined by equation (6.2) are:

(a′)
$$\sum_{k=1}^{k_n} \int_{-\infty}^{x} d\bar{F}_{nk} \to M(x) \qquad (x < 0)$$

$$-\sum_{k=1}^{k_n} \int_{x}^{\infty} d\bar{F}_{nk} \to N(x). \qquad (x > 0)$$

(b′)
$$\lim_{\epsilon\to 0} \lim_{n\to\infty} \sum_{k=1}^{k_n} \left(\int_{-\epsilon}^{0} u^2\, d\bar{F}_{nk} + \int_{0}^{\epsilon} u^2\, d\bar{F}_{nk} \right)$$

$$= \lim_{\epsilon\to 0} \underline{\lim}_{n\to\infty} \sum_{k=1}^{k_n} \left(\int_{-\epsilon}^{\epsilon} u^2\, d\bar{F}_{nk} \right)$$

$$= \lim_{\epsilon\to 0} \overline{\lim}_{n\to\infty} \sum_{k=1}^{k_n} \left(\int_{-\epsilon}^{\epsilon} u^2\, d\bar{F}_{nk} \right) = \sigma^2,$$

(c′)
$$-a_n + \sum_{k=1}^{k_n} \left(\int_{|u|<\tau} u\, dF_{nk} + \int_{|u|<\tau} u\, d\bar{F}_{nk} \right) \to \gamma(\tau),$$

where $\gamma(\tau)$ is defined by equation (3.10).

We will show that these last three conditions are equivalent to the conditions in the statement of the theorem.

Assume (a′) holds and let $l_n = \sup_k \alpha_{nk}$. We recall that for infinitesimal r.v.'s, $l_n \to 0$ as $n \to \infty$. Hence at a continuity point of M,

(1) $$I_n = \sum_{k=1}^{k_n} \int_{-\infty}^{u-l_n} d\bar{F}_{nk} \to M(u). \qquad (n \to \infty)$$

Now

(2) $$\sum_k \int_{-\infty}^{u} dF_{nk}(x) = \sum_k \int_{-\infty}^{u-\alpha_{nk}} dF_{nk}(x + \alpha_{nk}) \geq \sum_k \int_{-\infty}^{u-l_n} d\bar{F}_{nk} = I_n.$$

Similarly,

(3) $$J_n = \sum_{k=1}^{k_n} \int_{-\infty}^{u+l_n} d\bar{F}_{nk} \to M(u) \qquad (n \to \infty)$$

and

(4) $$J_n \geq \sum_k \int_{-\infty}^{u} dF_{nk}.$$

Equations (1)–(4) lead to

$$\sum_k \int_{-\infty}^{u} dF_{nk} = \sum_k F_{nk}(u) \to M(u). \qquad (n \to \infty)$$

It follows in exactly the same way that for u > 0,

$$-\sum_{k=1}^{k_n} \int_{u}^{\infty} dF_{nk} = \sum_{k=1}^{k_n} (F_{nk}(u) - 1) \to N(u).$$

We have proved that (a') implies (a). The converse result for the function M follows from the inequality

$$\sum_k \int_{-\infty}^{u-2l_n} dF_{nk} \leq I_n \leq \sum_k \int_{-\infty}^{u} dF_{nk},$$

and the corresponding result for the function N follows just as simply. Hence (a) implies (a') and so (a) and (a') are equivalent.

To establish the equivalence of (b) and (b') we require the following relations:

$$\lim_{\epsilon\to 0} \overline{\lim_{n\to\infty}} \sum_{k=1}^{k_n} \int_{|x|<\epsilon} x^2\, dF_{nk}(x + \alpha_{nk}) = \lim_{\epsilon\to 0} \overline{\lim_{n\to\infty}} \sum_k \int_{|x+\alpha_{nk}|<\epsilon} x^2\, dF_{nk}(x + \alpha_{nk})$$

$$= \lim_{\epsilon\to 0} \overline{\lim_{n\to\infty}} \sum_k \int_{|x|<\epsilon} (x - \alpha_{nk})^2\, dF_{nk}(x)$$

$$\lim_{\epsilon\to 0} \underline{\lim_{n\to\infty}} \sum_{k=1}^{k_n} \int_{|x|<\epsilon} x^2\, d\bar{F}_{nk} = \lim_{\epsilon\to 0} \underline{\lim_{n\to\infty}} \sum_k \int_{|x|<\epsilon} (x - \alpha_{nk})^2\, dF_{nk}(x).$$

Moreover,

$$\sum_{k=1}^{k_n} \int_{|x|<\epsilon} (x - \alpha_{nk})^2 \, dF_{nk}(x)$$

$$= \sum_k \left(\int_{|x|<\epsilon} x^2 \, dF_{nk} - 2\alpha_{nk} \int_{|x|<\epsilon} x \, dF_{nk} + \alpha_{nk}^2 - \alpha_{nk}^2 \int_{|x|\geq\epsilon} dF_{nk} \right)$$

$$= \sum_k \left(\int_{|x|<\epsilon} x^2 \, dF_{nk} - \left(\int_{|x|<\epsilon} x \, dF_{nk} \right)^2 \right) + R_n,$$

where

$$R_n = \sum_{k=1}^{k_n} \left[\left(\int_{\epsilon\leq|x|<\tau} x \, dF_{nk} \right)^2 - \alpha_{nk}^2 \int_{|x|\geq\epsilon} dF_{nk} \right].$$

From these last equations, it follows that conditions (*b*) and (*b*′) will be equivalent if we can show that $\lim_{n\to\infty} R_n = 0$. For this purpose, consider the following estimate:

$$|R_n| \leq \left(\tau^2 \sup_{1\leq k\leq k_n} \int_{|x|\geq\epsilon} dF_{nk} + l_n^2 \right) \sum_{k=1}^{k_n} \int_{|x|\geq\epsilon} dF_{nk}.$$

The first factor approaches zero since the r.v.'s are infinitesimal. We will show that the second factor is bounded. Since conditions (*a*) and (*a*′) are equivalent, it follows that for sufficiently large *n*,

$$\sum_{k=1}^{k_n} \int_{|x|>\epsilon} dF_{nk} \leq M(-\epsilon) - N(\epsilon) + 1.$$

Therefore, $\lim R_n = 0$ and so (*b*) and (*b*′) are equivalent.

Finally, assume that (*c*) holds. Then we must show that (*c*′) holds or, equivalently, that $\sum_k \int_{|u|<\tau} u \, d\bar{F}_{nk} \to 0$ as $n \to \infty$. This result follows from the inequality

$$\left| \sum_k \int_{|u|<\tau} u \, d\bar{F}_{nk} \right| \leq \left| \sum_k \int_{|u|<\tau} (u - \alpha_{nk}) \, dF_{nk} \right|$$

$$+ \left| \sum_k \left(\int_{|u-\alpha_{nk}|<\tau} (u - \alpha_{nk}) \, dF_{nk} - \int_{|u|<\tau} (u - \alpha_{nk}) \, dF_{nk} \right) \right|$$

$$\leq \sum_k \left[l_n \int_{|u|\geq\tau} dF_{nk} + (\tau + l_n) \left(\int_{\tau-l_n}^{\tau+l_n} dF_{nk} + \int_{-\tau-l_n}^{-\tau+l_n} dF_{nk} \right) \right]$$

by using condition (*a*), since $l_n \to 0$ as $n \to \infty$ and $\pm\tau$ are continuity points of *M* and *N*. Conversely, if (*c*′) holds, it follows that the constants a_n must be of the form indicated in (c).

A slight reformulation of the wording of the theorem yields the conditions that are necessary and sufficient for the d.f.'s of the sums s_n to converge to a limit d.f. Here we need only assume that there exist increasing functions M with $M(-\infty) = 0$ and N with $N(\infty) = 0$, defined in the intervals $(-\infty, 0)$ and $(0, \infty)$ respectively, and a constant $c \geq 0$ and a constant γ such that the conditions of the theorem are satisfied. We leave the details to the reader.

Exercise 1 formulates the above results for asymptotically constant summands.

EXERCISES

1. Prove the following theorem: The d.f.'s of the sums s_n of independent asymptotically constant r.v.'s converge to a limit d.f. if and only if there are d.f.'s M and N and a constant σ such that:

(a) At the continuity points of M and N, as $n \to \infty$,

$$\sum_{k=1}^{k_n} \int_{-\infty}^{u} dF_{nk}(x + m_{nk}) \to M(u) \qquad (u < 0)$$

$$\sum_{k=1}^{k_n} \int_{u}^{\infty} dF_{nk}(x + m_{nk}) \to -N(u). \qquad (u > 0)$$

(b) $$\lim_{\epsilon \to 0} \lim_{n \to \infty} \sum_{k=1}^{k_n} \left(\int_{|u|<\epsilon} u^2\, dF_{nk}(u + m_{nk}) - \left(\int_{|u|<\epsilon} u\, dF_{nk}(u + m_{nk}) \right)^2 \right) = \sigma^2.$$

(c) $$a_n = \sum_{k=1}^{k_n} \int_{|x|<\tau} x\, dF_{nk}(x + m_{nk}) + \sum_{k=1}^{k_r} m_{nk} - \gamma(\tau) + o(1),$$

where γ is any constant and $\pm\tau$ are continuity points of M and N. The logarithm of the c.f. of the limit law is given by Lévy's formula.

2. Prove the following theorem for r.v.'s subject to the conditions described in Exercise 6.1: The d.f.'s of the sequence of sums

$$s_n = r_{n1} + \cdots + r_{nk_n}$$

converge to a given d.f. F as $n \to \infty$; the variances of the sums converge to the variance of the limit distribution if and only if the following conditions are satisfied for $n \to \infty$:

(a) At the continuity points of K,

$$\sum_{k=1}^{k_n} \int_{-\infty}^{u} x^2\, d\bar{F}_{nk} \to K(u),$$

where $\bar{F}_{nk}(x) = F(x + Er_{nk})$,

(b) $\sum_{k=1}^{k_n} \int u^2 \, d\bar{F}_{nk} \to K(\infty)$,

(c) $\sum_{k=1}^{k_n} \int u \, dF_{nk} \to \gamma$,

where the function K and the constant γ are determined by Kolmogorov's formula.

3. (a) Show that if

$$\lim_{\epsilon \to 0} \overline{\lim_{n \to \infty}} \sum_{k=1}^{k_n} \left(\int_{|u|<\epsilon} u \, dF_{nk}(u + m_{nk}) \right)^2 = 0, \tag{1}$$

then the d.f.'s of the sums s_n of independent asymptotically constant r.v.'s converge to a limit if and only if there exists a d.f. G such that

$$G_n(u) = \sum_{k=1}^{k_n} \int_{-\infty}^{u} x^2/(1 + x^2) \, dF_{nk}(x + m_{nk}) \to G(u) \ (c)$$

as $n \to \infty$. Suitable constants a_n are given by

$$a_n = \sum_{k=1}^{k_n} \left(\int_{|u|>\tau} u \, dF_{nk}(u + m_{nk}) \right).$$

(b) Show that (1) holds whenever the limit law does not have a normal component or the r.v.'s r_{nk} are symmetrical with respect to the medians.

13.8 NORMAL, POISSON, AND UNITARY CONVERGENCE

We now apply the convergence criterion of the last section to particular d.f.'s F. A normal law

$$F(x) = \frac{1}{\sqrt{2\pi}\,\sigma} \int_{-\infty}^{x} e^{-(t-\mu)^2/2\sigma^2} \, dt \tag{1}$$

has $\log \chi(t) = i\mu t - \sigma^2 t^2/2$ and so for $u > 0$, $N(u) = M(-u) = 0$ and $\gamma(\tau) = \mu$ in equation 3.9.

THEOREM 1

The d.f. of the sums

$$s_n = r_{n1} + r_{n2} + \cdots + r_{nk_n} - a_n$$

of independent r.v.'s r_{nk} (with a_n defined as in (c) of Theorem 7.1) converges as $n \to \infty$ to the normal law (1); the summands r_{nk} are infinitesimal if and only if

1. $\sum_{k=1}^{k_n} \int_{|x|\geq\epsilon} dF_{nk} \to 0,$

2. $\sum_{k=1}^{k_n} \sigma_{nk}^2(\tau) \equiv \sum_{k=1}^{k_n}\left[\int_{|u|<\tau} u^2\,dF_{nk} - \left(\int_{|u|<\tau} u\,dF_{nk}\right)^2\right] \to \sigma^2$

for every $\epsilon > 0$ and $\tau > 0$ as $n \to \infty$.

Proof

Assume that the d.f. of s_n converges to the normal distribution and the r.v.'s r_{nk} are independent and infinitesimal. Since $M = N = 0$, the conditions in (a) of Theorem 7.1 immediately yield condition 1 of this theorem. Condition (b) of Theorem 7.1 states that

$$\lim_{\tau\to 0}\overline{\lim_{n\to\infty}}\sum_{k=1}^{k_n}\sigma_{nk}^2(\tau) = \lim_{\tau\to 0}\underline{\lim_{n\to\infty}}\sum_{k=1}^{k_n}\sigma_{nk}^2(\tau) = \sigma^2. \tag{2}$$

We will show that equation (2) is equivalent to condition 2 of the theorem because condition 1 holds. If $\epsilon < \tau$, then

$$\left|\sum_k \sigma_{nk}^2(\tau) - \sum_k \sigma_{nk}^2(\epsilon)\right|$$

$$= \sum_k\left[\int_{\epsilon\leq|u|<\tau} u^2\,dF_{nk} + \left(\int_{|u|<\epsilon} u\,dF_{nk}\right)^2 - \left(\int_{|u|<\tau} u\,dF_{nk}\right)^2\right].$$

Since $|x^2 - y^2| \leq |x - y|\,|x + y| \leq 2\max(x, y)|x - y|$, then

$$\left|\left(\int_{|u|<\epsilon} u\,dF_{nk}\right)^2 - \left(\int_{|u|<\tau} u\,dF_{nk}\right)^2\right|$$

$$\leq \left(2\int_{|u|<\tau} u\,dF_{nk}\right)\left|\int_{|u|<\tau} u\,dF_{nk} - \int_{|u|<\epsilon} u\,dF_{nk}\right|$$

$$\leq 2\tau\left|\int_{\epsilon\leq|u|<\tau} u\,dF_{nk}\right| \leq 2\tau^2\int_{\epsilon\leq|u|} dF_{nk}.$$

Hence

$$\left|\sum_k \sigma_{nk}^2(\tau) - \sum_k \sigma_{nk}^2(\epsilon)\right| \leq 3\tau^2\sum_k\int_{|u|\geq\epsilon} dF_{nk} \to 0$$

as $n \to \infty$ by condition 1 of the theorem. The same is true if $\epsilon > \tau$; it suffices to interchange ϵ and τ in the foregoing chain of inequalities. Therefore, $\lim \sigma_{nk}^2(\tau)$ is independent of τ and thus equation (2) is equivalent to condition 2 of the theorem.

Conversely, suppose that conditions 1 and 2 hold. Since

$$\sup_k P(|r_{nk}| \geq \epsilon) \leq \sum_k P(|r_{nk}| \geq \epsilon) \to 0,$$

then by condition 1, the r.v.'s r_{nk} are infinitesimal. Moreover, condition 1 implies condition (a) of Theorem 7.1, since $M = N = 0$. As in the preceding paragraph, condition 2 is equivalent to equation (2); accordingly, all the conditions of Theorem 7.1 are satisfied for F, the normal d.f. in equation (1).

COROLLARY 1

If the r_{nk} are independent r.v.'s, then the d.f.'s of the sums

$$s_n = r_{n1} + r_{n2} + \cdots + r_{nk_n}$$

converge as $n \to \infty$ to the normal law (1) and the summands r_{nk} are infinitesimal if and only if

1. $\displaystyle\sum_{k=1}^{k_n} \int_{|x|\geq\epsilon} dF_{nk} \to 0,$
2. $\displaystyle\sum_{k=1}^{k_n} \sigma_{nk}^2(\tau) \to \sigma^2,$
3. $\displaystyle\sum_{k=1}^{k_n} \int_{|u|<\tau} u\, dF_{nk} \to \mu,$

for every $\epsilon > 0$ and $\tau > 0$,[7] as $n \to \infty$.

The proof is a direct consquence of the theorem, since conditions 1 and 2 hold. Condition 3 is just a restatement of condition (c) of Theorem 7.1, since $a_n = 0$ in the corollary.

If the d.f.'s of the sums s_n converge to a limit d.f., then the limit d.f. is normal if and only if condition 1 is satisfied; therefore, we can deduce the following result [18].

COROLLARY 2

If the r.v.'s r_{nk} are independent and the d.f.'s of the sums

$$s_n = r_{n1} + r_{n2} + \cdots + r_{nk_n}$$

converge to a limit, then the limit d.f. is normal and the r.v.'s are infinitesimal if and only if $\sup_k |r_{nk}| \to 0\ (P)$.

Proof

Set $p_{nk} = P(|r_{nk}| \geq \epsilon)$; then

$$P(\sup_k |r_{nk}| \geq \epsilon) = 1 - \prod_k (1 - p_{nk})$$

because of the independence of the summands. By the inequality

[7]Note that one value of τ is enough to prove sufficiency.

$$1 - \exp\left(-\sum_k p_{nk}\right) \leq 1 - \prod_k (1 - p_{nk}) \leq \sum_k p_{nk},$$

it follows that the given condition is equivalent to condition 1 of Corollary 1.

The following example shows that infinitesimal r.v.'s need not satisfy the condition of corollary 2. In other words, the limit law (if it exists) for sums s_n of independent infinitesimal r.v.'s is i.d. but not necessarily normal.

EXAMPLE 1

Let $F_{nk}(u) = 0$ for $u < 0$, $F_{nk}(u) = 1 - 1/n$ for $0 \leq u < 1$, and $F_{nk}(u) = 1$ for $u \geq 1$, where $1 \leq k \leq n$. Then for $0 < \epsilon < 1$,

$$\sup_k P(|r_{nk}| > \epsilon) = 1/n \to 0. \qquad (n \to \infty)$$

However,

$$P(\sup_k |r_{nk}| > \epsilon) = 1 - \prod_k P(|r_{nk}| < \epsilon) = 1 - (1 - 1/n)^n \to 1 - e^{-1}.$$

$$(n \to \infty)$$

We will now specialize the above results to an ordinary sequence of independent r.v.'s to show that Lindeberg's condition is also a necessary condition in Theorem 1.2. The notation of Theorem 1.2 will be used.

THEOREM 2

Let $\{r_i\}$ be a sequence of independent r.v.'s. Then

$$P\left(\frac{1}{B_n}\sum_{k=1}^{n}(r_k - a_k) \leq x\right) \to \frac{1}{\sqrt{2\pi}}\int_{-\infty}^{x} e^{-t^2/2}\, dt$$

uniformly in x as $n \to \infty$ if and only if

$$\lim_{n\to\infty} \frac{1}{B_n^2}\sum_{k=1}^{n} \int_{|x-a_k|\geq\epsilon B_n} (x - a_n)^2\, dF_k = 0. \tag{3}$$

Proof

Normalizing the r.v.'s as in Theorem 1.2, equation (3) reduces to

$$\lim_{n\to\infty} \sum_{k=1}^{n} \int_{|u|\geq\epsilon} u^2\, dF_{nk} = 0. \tag{4}$$

For every n,

$$\sum_{k=1}^{n} \int u\, dF_{nk} = 0 \tag{5}$$

(6) $$\sum_{k=1}^{n} \int u^2 \, dF_{nk} = 1.$$

To prove our result, we need only show that conditions 1–3 of Corollary 1 of Theorem 1, with $\sigma = 1$ and $\mu = 0$, reduce to (4) for normalized r.v.'s r_{nk} satisfying (5) and (6).

From (4) and (6), it is obvious that for any $\tau > 0$,

(7) $$\lim_{n\to\infty} \sum_{k=1}^{n} \int_{|u|<\tau} u^2 \, dF_{nk} = 1.$$

The inequality

$$\int_{|u|\geq\epsilon} u^2 \, dF_{nk} \geq \epsilon \int_{|u|\geq\epsilon} |u| \, dF_{nk} \geq \epsilon \int_{|u|\geq\epsilon} u \, dF_{nk}$$

and (4) show that

$$\lim_{n\to\infty} \sum_{k=1}^{n} \int_{|u|\geq\epsilon} u \, dF_{nk} = 0,$$

and so by (5),

(8) $$\lim_{n\to\infty} \sum_{k=1}^{n} \int_{|u|<\tau} u \, dF_{nk} \to 0.$$

The proof of the theorem is completed, since (7) and (8) imply that conditions 2 and 3 of Corollary 1 of Theorem 1 are valid,

The Poisson d.f.

(9) $$P(x) = \sum_{k=0}^{x} e^{-\lambda} \lambda^k / k! \qquad (\lambda > 0)$$

has $\log \chi(t) = \lambda(e^{it} - 1)$. In equation (3.9) we should set $M = 0$, $N(t) = -\lambda$ for $0 < t < 1$, $N(t) = 0$ for $t \geq 1$, $\sigma = 0$, and $\gamma(\tau) = 0$ for $0 < \tau < 1$. Applying Theorem 7.1 and observing that the condition relative to $\sum_k \sigma_{nk}^2(\tau)$ reduces exactly as in the normal case, we obtain the following result (see References [15] and [19]).

THEOREM 3

The d.f.'s of the sums

$$s_n = r_{n1} + r_{n2} + \cdots + r_{nk_n}$$

of independent infinitesimal r.v.'s converge to the Poisson law in equation (9) as $n \to \infty$ if and only if for every ϵ $(0 < \epsilon < 1)$ and τ $(0 < \tau < 1)$,

1. $\sum_{k=1}^{k_n} \int_D dF_{nk} \to 0,$

2. $\sum_{k=1}^{k_n} \int_{|u-1|<\epsilon} dF_{nk} \to \lambda,$

3. $\sum_{k=1}^{k_n} \int_{|u|<\epsilon} u\, dF_{nk} \to 0,$

4. $\sum_{k=1}^{k_n} \int_{|u|<\epsilon} u^2\, dF_{nk} \to 0,$

where $D = (x : |x| \geq \epsilon, |x - 1| \geq \epsilon)$.

The unitary d.f.

$$F(x) = \begin{cases} 0 & (x < 0) \\ 1 & (x \geq 0) \end{cases} \tag{10}$$

can be considered as a degenerate normal, with $\sigma = 0$ and $\mu = 0$; therefore, the normal convergence criterion (Corollary 1 of Theorem 1) reduces to the following theorem.

THEOREM 4

If the r_{nk} are independent r.v.'s, then the d.f.s of the sums

$$s_n = r_{n1} + r_{n2} + \cdots + r_{nk_n}$$

converge as $n \to \infty$ to the unitary law in equation (10); the summands r_{nk} are infinitesimal if and only if

1. $\sum_{k=1}^{k_n} \int_{|u|\geq\epsilon} dF_{nk} \to 0,$

2. $\sum_{k=1}^{k_n} \left[\int_{|u|<\tau} u^2\, dF_{nk} - \left(\int_{|u|<\tau} u\, dF_{nk} \right)^2 \right] \to 0,$

3. $\sum_{k=1}^{k_n} \int_{|u|<\tau} u\, dF_{nk} \to 0,$

for every $\epsilon > 0$ and $\tau > 0$.

COROLLARY 1

If r_k are independent r.v.'s and the constants $b_n \uparrow \infty$, then the d.f.'s of s_n/b_n converge to the unitary d.f. as $n \to \infty$ if and only if

1. $\sum_{k=1}^{n} \int_{|u|\geq\epsilon b_n} dF_k \to 0,$

2. $\frac{1}{b_n^2}\sum_{k=1}^{n}\left[\int_{|u|<b_n} u^2\,dF_k - \left(\int_{|u|<b_n} u\,dF_k\right)^2\right] \to 0,$

3. $\frac{1}{b_n}\sum_{k=1}^{n}\int_{|u|<b_n} u\,dF_k \to 0,$

for any $\epsilon > 0$.

Proof

Set $r_{nk} = r_k/b_n$; then $F_{nk}(u) = F_k(b_n u)$. With this interpretation, conditions 1–3 of the corollary coincide with conditions 1–3 of the theorem if $\tau = 1$. Hence conditions 1–3 of the corollary imply that the d.f.'s of s_n/b_n converge to the unitary d.f.

Conversely, assume that the d.f.'s of s_n/b_n converge to the unitary d.f. or, equivalently, that $s_n/b_n \to 0\,(P)$ by Theorem 8.3.5. If we can show that this last condition implies that the r.v.'s r_{nk} are infinitesimal, then conditions 1–3 of the corollary will hold by the above theorem. Now the fact that $s_n/b_n \to 0\,(P)$ means that for sufficiently large $n \geq N(\epsilon, \delta)$, we have $P(|s_n/b_n| \geq \epsilon) \leq \delta$. Therefore, for $n \geq N$,

$$(11)\qquad P\left(\left|\frac{r_n}{b_n}\right| < 2\epsilon\right) = P\left(\left|\frac{s_n}{b_n} - \frac{b_{n-1}}{b_n}\frac{s_{n-1}}{b_{n-1}}\right| < 2\epsilon\right)$$

$$P\left(\left(\left|\frac{s_n}{b_n}\right| < \epsilon\right)\left(\left|\frac{b_{n-1}}{b_n}\frac{s_{n-1}}{b_{n-1}}\right| < \epsilon\right)\right) \geq P\left(\left(\left|\frac{s_n}{b_n}\right| < \epsilon\right)\left(\left|\frac{s_{n-1}}{b_{n-1}}\right| < \epsilon\right)\right),$$

where the last inequality is a consequence of $b_n \geq b_{n-1}$ and

$$\left(\left|\frac{s_{n-1}}{b_{n-1}}\right| < \left|\frac{b_n}{b_{n-1}}\right|\epsilon\right) \supset \left(\left|\frac{s_{n-1}}{b_{n-1}}\right| < \epsilon\right).$$

The fact that $P(AB) \geq 1 - PA' - PB'$ implies that

$$(12)\qquad P\left(\left(\left|\frac{s_n}{b_n}\right| < \epsilon\right)\left(\left|\frac{s_{n-1}}{b_{n-1}}\right| < \epsilon\right)\right) \geq 1 - 2\delta.$$

From equations (11) and (12), it follows that $r_n/b_n \to 0\,(P)$. Since $s_n/b_n \to 0\,(P)$, we see that $(s_{n-1}/b_{n-1}) = (s_n/b_n) - (r_n/b_n) \to 0\,(P)$. Applying the above reasoning to s_{n-1}/b_n, it follows that $r_{n-1}/b_n \to 0\,(P)$. Continuing this procedure, we can deduce that $r_k/b_n \to 0\,(P)$ for $1 \leq k \leq n$; in other words, the r.v.'s $r_{nk} = r_k/b_n$ are infinitesimal.

COROLLARY 2

If r_k are independent r.v.'s, then $s_n/n \to 0\,(P)$ if and only if

1. $\sum_{k=1}^{n}\int_{|x|\geq n} dF_k \to 0,$

2. $\frac{1}{n^2}\sum_{k=1}^{n}\left(\int_{|x|<n} x^2\,dF_k - \left(\int_{|x|<n} x\,dF_k\right)^2\right) \to 0,$

3. $\frac{1}{n}\sum_{k=1}^{n}\int_{|x|<n} x\,dF_k \to 0,$

as $n \to \infty$.

This result follows immediately from Corollary 1 by setting $\epsilon = 1$ and $b_n = n$.

A large number of theorems on the weak law of large numbers can be deduced by the above methods. The interested reader should consult the monograph of Gnedenko and Kolmogorov (Reference [24]) for further details.

EXERCISES

1. Suppose that the r.v.'s. r_{nk} satisfy the conditions stated in Exercise 6.1. Find the conditions that are necessary and sufficient for the d.f.'s of the sums $s_n = r_{n1} + \cdots + r_{nk_n}$ to converge to

(a) The Laplace d.f.
(b) The d.f. described in Exercise 2.2. (*Hint*: Use Exercise 7.2.)

2. In order that the d.f.'s of the sums of independent infinitesimal r.v.'s

$$s_n = r_{n1} + r_{n2} + \cdots + r_{nk_n} - a_n,$$

where a_n are defined as in (c) of Theorem 7.1, converge to the normal law ($\mu = 0$, $\sigma = 1$), it is necessary and sufficient that for every $\epsilon > 0$,

(a) $\sum_{k=1}^{k_n}\int_{|u|>\epsilon} dF_{nk}(u + \alpha_{nk}) \to 0,$

(b) $\sum_{k=1}^{k_n}\int_{|u|>\epsilon} u^2\,dF_{nk}(u + \alpha_{nk}) \to 1,$

as $n \to \infty$.

3. Deduce the following theorem by using Exercise 2. In order that the sums s_n of r.v.'s that are independent in each row obey the weak law of large numbers, it is necessary and sufficient that

$$\sum_{k=1}^{k_n}\int \frac{x^2}{1+x^2}\,dF_{nk}(x + m_{nk}) \to 0. \qquad (n \to \infty)$$

The constants a_n may be chosen according to the formula

$$a_n = \sum_{k=1}^{k_n}\left(m_{nk} + \int_{|u|<\tau} u\,dF_{nk}(u + m_{nk})\right),$$

where τ is a constant.

4. Deduce the following result from Theorem 1 (see Reference [7]): There exist constants a_n and $b_n > 0$ such that the d.f.'s of the sums

$$s_n = \frac{r_1 + r_2 + \cdots + r_n}{b_n} - a_n$$

of independent r.v.'s r_i converge to the normal law ($\mu = 0, \sigma = 1$) and the summands are infinitesimal if and only if there exists a sequence of constants c_n with $c_n \to \infty$ such that as $n \to \infty$,

(a) $\displaystyle\sum_{k=1}^{n} \int_{|u|>c_n} dF_k \to 0,$

(b) $\displaystyle\frac{1}{c_n^2} \sum_{k=1}^{n} \left[\int_{|u|<c_n} u^2\, dF_k - \left(\int_{|u|<c_n} u\, dF_k \right)^2 \right] \to \infty.$

13.9 NORMED SUMS

In 1936, A. Y. Khintchine formulated the problem of determining the class of d.f.'s that are limits of the normed sums

$$\sigma_n = s_n/b_n - a_n, \tag{1}$$

where a_n and $b_n > 0$ are suitably chosen real constants, and

$$s_n = r_1 + r_2 + \cdots + r_n,$$

the r_i being independent r.v.'s.

If we assume that the r.v.'s.

$$r_{nk} = r_k/b_n \qquad (1 \le k \le n; n = 1, 2, \ldots) \tag{2}$$

are infinitesimal, it follows that every limit distribution for the normed sums in (1) is i.d. As we shall see, the converse is not true. There are i.d.d.f.'s which cannot be the limiting distributions of normed sums for any choice of the constants $b_n > 0$ and a_n and any sequence r_i of independent r.v.'s.

DEFINITION 1

The d.f. F *belongs to the class* L if and only if we can find a sequence of independent r.v.'s r_i such that:

(a) For suitably chosen constants a_n and $b_n > 0$, the d.f.'s of the normed sums in equation (1) converge to F.

(b) The r.v.'s $r_{nk} = r_k/b_n$ for $1 \le k \le n$ are infinitesimal.

The degenerate laws belong to L. The first part of the proof of Theorem 8.4.1 shows that there always exist sequences a_n and $b_n > 0$ such that the limit laws of the sums in equation (1) are degenerate.

The class L was completely characterized by Lévy [20]. The proof of Lévy's theorem requires the following lemma (see Reference [18]).

LEMMA 1

If the limit d.f. of the sums in (1) of independent infinitesimal r.v.'s $r_{nk} = r_k/b_n$ is proper, then $b_{n+1}/b_n \to 1$ and $b_n \to \infty$ as $n \to \infty$.

Proof

Suppose that b_n is bounded for all n. By the Bolzano–Weierstrass theorem, there exist a subsequence b_{n_k} and a finite number b such that $b_{n_k} \to b$ as $n_k \to \infty$. Let t be any given number; then $t_k = tb_{n_k} \to tb$ as $k \to \infty$. By hypothesis, the r.v.'s r_{ns} are infinitesimal; hence their c.f.'s $\chi_{ns}(t) \to 1$ uniformly in s $(1 \leq s \leq n)$ as $n \to \infty$. Since $\chi_{ns}(t) = \chi_s(t/b_n)$, where χ_s is the c.f. of r_s, then (see Exercise 10.6.6)

$$\chi_s(t) = \chi_s(t_k/b_{n_k}) \to 1$$

uniformly in s $(1 \leq s \leq n_k)$; i.e., for every t, $\chi_s(t) = 1$ $(s = 1, 2, 3 \ldots)$. Let χ be the c.f. of F; by hypothesis,

$$e^{-ita_n} \prod_{s=1}^{n} \chi_s(t/b_n) \to \chi(t)$$

and $|\chi| \neq 1$. However, we have a contradiction, since the fact that $\prod_{s=1}^{n} \chi_s(t/b_n) = 1$ implies that $|\chi| = 1$.

Since $r_{n+1}/b_{n+1} \to 0\,(P)$, it follows by Theorem 8.3.4 that the d.f. of the sums

$$s_n/b_{n+1} - a_{n+1} \tag{3}$$

also converges to F. Denoting the d.f. of the sums in (1) by G_n, the d.f. of the sums in (3) is $G_n(b'_n u + a'_n)$, where $b'_n = b_{n+1}/b_n$ and $a'_n = (b_{n+1}/b_n)\, a_{n+1} - a_n$. By Theorem 8.4.2 it follows that $b_{n+1}/b_n \to 1$ as $n \to \infty$.

The class L is characterized by the following decomposability property.

DEFINITION 2

A c.f. χ and its corresponding d.f. F are *self-decomposable* if and only if for every $c \in (0, 1)$ there exists a c.f. χ_c such that

$$\chi(t) = \chi(ct)\chi_c(t) \tag{*}$$

for every t.

In terms of d.f.'s, (*) is equivalent to $F(u) = F(u/c) * F_c(u)$. A simple example of a self-decomposable c.f. is a degenerate c.f.

Although the restriction of c may seem strange, it is natural in the following sense. If (*) holds and χ is a nondegenerate c.f., then the fact that $c > 0$ implies that $c < 1$ (see Exercises 2 and 3).

To prove Lévy's theorem, we require the following lemmas.

LEMMA 2

If $\lim_{n\to\infty} b_{n+1}/b_n = 1$ and $b_n \to \infty$, then there exists a sequence of integers $m = m(n)$ such that as $n \to \infty$,

1. $m \to \infty$.
2. For any $c \in (0, 1)$, $b_m/b_n \to c$.

Proof

Since $b_n \to \infty$, there is some integer N such that $b_1/b_n \le c$ for all $n \ge N$. For fixed n, let $m = m(n) \le n$ be the largest integer such that

$$b_m/b_n \le c < b_{m+1}/d_n. \tag{4}$$

Such an integer m exists for $n \ge N$ since $b_1/b_n \le c < b_n/b_n$.

Now $m \to \infty$ as $n \to \infty$; this is true because if m is bounded for an infinite number of values of n, then taking limits as $n \to \infty$ in the right-hand inequality of (4) yields the contradictory statement $c \le 0$.

Again from the right-hand side of (4),

$$c < (b_{m+1}/b_m)(b_m/b_n). \tag{5}$$

Letting $n \to \infty$ and using the fact that $b_{m+1}/b_m \to 1$, it follows from equation (4) and inequality (5) that $b_m/b_n \to c$ as $n \to \infty$.

LEMMA 3

If χ is self-decomposable, then $\chi \ne 0$.

Proof

Suppose that $\chi(2a) = 0$ and $\chi(t) \ne 0$ for $0 \le t < 2a$; then $\chi_c(2a) = 0$. Setting $t = h = a$ in $|\chi_c(t + h) - \chi_c(t)| \le 2\,(1 - \mathscr{R}\chi_c(h))$ yields

$$|\chi_c(a)|^2 \le 2(1 - \mathscr{R}\chi_c(a)).$$

Letting $c \to 1$, $\chi_c(a) = \chi(a)/\chi(ca) \to 1$ and the last inequality reduces to $1 \le 0$, a contradiction.

THEOREM 1

A d.f. belongs to L if and only if it is self-decomposable.

Proof

We need only consider nondegenerate c.f.'s χ, since a degenerate law belongs to L.

Suppose that χ is self-decomposable; i.e., $\chi_c(u) = \chi(u)/\chi(cu)$. Replacing c by $(k - 1)/k$ and u by kt gives

$$\chi_{(k-1)/k}(kt) = \chi(kt)/\chi((k-1)t), \tag{6}$$

which is well defined because of Lemma 3.

For all n, let r_k ($k = 1, 2, \ldots, n$) be independent r.v.'s with c.f.'s $\psi_k(t) \equiv \chi_{(k-1)/k}(kt)$ defined by (6). Since $\max_{1 \le k \le n} \psi_k(t/n) \to 1$ as $n \to \infty$, the r.v.'s $r_{nk} = r_k/n$ are infinitesimal. The c.f. of s_n/n is given by $\prod_{k=1}^{n} \psi_k(t/n) = \chi(t)$. Therefore, it is obvious that the d.f. of s_n/n converges to the d.f. of χ; hence the d.f. of χ belongs to L.

Conversely, let the d.f. of χ belong to L. Then there exist normed sums $s_n/b_n - a_n$ with c.f. ϕ_n such that

$$\phi_n(t) = e^{-ita_n} \prod_k \chi_k(t/b_n) \to \chi(t), \tag{7}$$

where χ_k are the c.f.'s of the summands r_k. By Lemma 1, $b_n \to \infty$ and $b_{n+1}/b_n \to 1$ as $n \to \infty$. Choosing m and c as in Lemma 2, we can rewrite ϕ_n as follows:

$$\phi_n(t) = \left[e^{-itca_m} \prod_{k=1}^{m} \chi_k(b_m t/b_n b_m)\right]\left[e^{-it(a_n - ca_m)} \prod_{k=m+1}^{n} \chi_k(t/b_n)\right]. \tag{8}$$

From (7) and Lemma 2, it follows that the first bracket converges to $\chi(ct)$. Therefore, the c.f. defined by the second factor in (8) converges to the continuous function χ_c defined by $\chi_c(t) = \chi(t)/\chi(ct)$. The continuity theorem shows that χ_c is a c.f.

COROLLARY

A self-decomposable c.f. χ and its components χ_c are i.d.

Proof

Since χ is self-decomposable, it is i.d.

To prove that χ_c is i.d., let ψ_k be the c.f. of the r.v.'s r_k defined in the first part of the proof of the theorem. Choose $m = m(n) < n$ such that $m/n \to c$ as $n \to \infty$. Consider

$$\chi(t) = \prod_{k=1}^{m} \psi_k(mt/nm) \prod_{k=m+1}^{n} \psi_k(t/n).$$

The first product converges to $\chi(ct)$; therefore, the second one converges to $\chi_c(t)$. It follows that χ_c is the c.f. of the limit d.f. of the sums $\sum_{k=m+1}^{n} r_{nk}$, where the summands $r_{nk} = r_k/n$ are infinitesimal. Accordingly, this d.f. belongs to L and so χ_c is i.d.

The c.f. χ of a d.f. of the class L is i.d., thus the logarithm of its c.f. can be represented by Lévy's formula:

$$\text{(9)} \qquad \log \chi(t) = i\gamma t - \sigma^2 t^2/2 + \int_{-\infty}^{0} \left(e^{itu} - 1 - \frac{itu}{1+u^2}\right) dM + \int_{0}^{\infty} \left(e^{itu} - 1 - \frac{itu}{1+u^2}\right) dN.$$

The class of self-decomposable c.f.'s is smaller than the class of i.d.c.f.'s. The question naturally arises as to what special properties the functions M and N must have in order for χ to be self-decomposable. The answer, contained in Theorem 2 below, was discovered by Lévy [20]. To prove this result, we require the following lemma.

LEMMA 4

A d.f. F belongs to the class L if and only if the functions

$$\text{(10)} \qquad M(t) - M(t/c) \qquad N(t) - N(t/c)$$

are increasing. Here M and N are the functions appearing in (9) and $c \in (0, 1)$.

Proof

Let χ be the c.f. of F; then from (9),

$$\log \chi(ct) = i\gamma ct - \sigma^2 t^2 c^2/2 + \int_{-\infty}^{0} \left[e^{icut} - 1 - \frac{icut}{1+u^2}\right] dM + \int_{0}^{\infty} [\quad] \, dN$$

$$= i\gamma_1 t - \sigma^2 c^2 t^2/2 + \int_{-\infty}^{0} \left[e^{iut} - 1 - \frac{iut}{1+u^2}\right] dM(u/c) + \int_{0}^{\infty} [\quad] \, dN(u/c),$$

where

$$\gamma_1 = c\gamma + c \int_{-\infty}^{0} \left[\frac{u^3(1-c^2)}{(1+u^2)(1+c^2u^2)}\right] dM(u) + c \int_{0}^{\infty} [\quad] \, dN(u).$$

Hence,

$$\log \chi(t)/\chi(ct) = i\gamma_2(t) - \frac{\sigma^2(1-c^2)t^2}{2} + \int_{-\infty}^{0} \left[e^{itu} - 1 - \frac{itu}{1+u^2}\right] d(M(u) - M(u/c)) + \int_{0}^{\infty} [\quad] \, d(N(u) - N(u/c)).$$

Since $\chi_c(t)$ is i.d. and the above formula is its canonical representation, it follows that the functions (10) are increasing.

Conversely, if the functions (10) are increasing, then $\chi(t)/\chi(ct)$ will be the c.f. of an i.d. law.

THEOREM 2

A d.f. F belongs to the class L if and only if

(a) The functions M and N in (9) have right- and left-hand derivatives for every value of u.
(b) The functions $uM'(u)$ for $u < 0$ and $uN'(u)$ for $u > 0$ are decreasing. (Here M' and N' denote either right- or left-hand derivatives.)

Proof

Suppose that F belongs to L. By Lemma 4, the function $N(t) - N(t/c)$ is increasing, which is equivalent to

$$N(t_2/c) - N(t_1/c) \leq N(t_2) - N(t_1), \tag{11}$$

where $t_2 > t_1 > 0$. Let $b > a$, $h > 0$, and $c = e^{a-b}$. Setting $t_1 = e^a$ and $t_2 = e^{a+h}$ in (11) yields

$$N(e^{a+h}) - N(e^a) \geq N(e^{a+h}/c) - N(e^a/c) = N(e^{b+h}) - N(e^b).$$

If $N(e^t)$ is denoted by $f(t)$, the last inequality can be rewritten as

$$f(a+h) - f(a) \geq f(b+h) - f(b). \qquad (b > a) \tag{12}$$

Since f is increasing, its discontinuities can only be jumps. Suppose that $f(b+h) - f(b) = j > 0$. There can only be a finite number of jumps of a magnitude greater than or equal to j to the left of b. Hence for sufficiently small h, there are some a such that $f(a+h) - f(a) < j$. This is a contradiction [by equation (12)] and so f must be continuous at every point b.

Setting $b = a + h$ in (12), we obtain

$$f(a+h) \geq \frac{f(a) + f(a+2h)}{2}$$

and so the continuous function $f(t)$ is concave. Therefore, it has finite left- and right-hand derivatives at every point; the left-hand derivative is greater than or equal to the right-hand derivative, and both derivatives are decreasing.[8]

Since $f' = e^t N'(e^t) = uN'(u)$, where $u = e^t$, we see that the conditions of the theorem are necessary for the function N. A similar proof holds for M.

Now suppose that $uN'(u)$ is a decreasing function. Then for every $c \in (0, 1)$, $(u/c)N'(u/c) \leq uN'(u)$. Hence if $0 < t_1 < t_2$,

[8]See G. Hardy, J.E. Littlewood and G. Polya, Inequalities (New York: Cambridge University Press, 1934).

$$\int_{t_1}^{t_2} dN(u/c) = \int_{t_1}^{t_2} (1/c)N(u/c)\,du \leq \int_{t_1}^{t_2} N'(u)\,du = \int_{t_1}^{t_2} dN.$$

Thus inequality (11) is valid and so $N(t) - N(t/c)$ is an increasing function. A similar proof and result holds for M. Therefore, by Lemma 4, F belongs to L.

The normal distribution belongs to the class L, since $uM'(u) = uN'(u) = 0$.

Theorem 2 also explains why the Poisson laws remained isolated as long as only limit laws of normed sums were considered. The function N corresponding to a Poisson law P is discontinuous and so P does not belong to L.

Conditions necessary and sufficient for the existence and determination of norming sequences a_n and $b_n > 0$, where the corresponding normed sums of given independent r.v.'s approach a given d.f., appear in Exercise 6.

EXERCISES

1. Show that the class of limit distributions of the sums (1) of asymptotically constant summands $r_{nk} = r_k/b_n$ coincides with the class of limit distributions of the sums (1) of infinitesimal summands $\bar{r}_{nk} = (r_k - m_k)/B_n$.

2. If $c > 0$ and $\mathcal{X}_c$ is a degenerate c.f., then $\mathcal{X}(u) = \mathcal{X}_c(u)\mathcal{X}(cu)$ is a degenerate c.f. True or false?

3. Show that if a c.f. $\mathcal{X}$ has a corresponding number $c > 0$ and a nondegenerate c.f. $\mathcal{X}_c$ such that, for every t, $\mathcal{X}(t) = \mathcal{X}_c(t)\mathcal{X}(ct)$, then $c > 1$.

4. Prove the following theorem: In order for the d.f. F with finite variance to belong to the class L, it is necessary and sufficient for the function K in Kolmogorov's formula to have right- and left-hand derivatives at every point $u \neq 0$ and for the functions $K'(u)/u$, where K' denotes the right- or left-hand derivatives, to be decreasing for $u < 0$ and $u > 0$.

5. Verify that the d.f.'s of finite variance belonging to the class L are not exhausted by the normal and degenerate d.f.'s, by showing that the d.f. for which $K(u) = 0$ $(u < 0)$, $K(u) = u^2$ $(0 \leq u < 1)$, and $K(u) = 1$ $(u \geq 1)$ belongs to L.

6. Prove the following theorem:[9] There exist sequences $b_n > 0$ such that the d.f.'s of the sums (1) converge to a proper d.f. F for suitable a_n if and only if there exists an increasing function H with $H(-\infty) = 0$, and a finite variance $V = H(\infty)$; if we define $c_n > 0$ by

[9]See B.V. Gnedenko and A.V. Groshev, "On the convergence of distribution laws of normalized sums of independent random variables," *Rec. Math.* (*Mat Sbornik*), N.S. 6 (48), (1939), 521–541.

$$\frac{1}{2}\sum_{k=1}^{n}\int\frac{x^2}{c_n^2+x^2}\,d^sF_k(x)=V,$$

then

(a) For every u $(-\infty \leq u \leq \infty)$ except possibly $u=0$,

$$\sum_{k=1}^{n}\int_{-\infty}^{c_n u}\frac{x^2}{c_n^2+x^2}\,dF_k(x+\alpha_{nk})\to H(u), \qquad (n\to\infty)$$

where

$$\alpha_{nk}=\frac{1}{c_n}\int_{|x|>c_n} x\,dF_k.$$

(b) $\sup_{1\leq k\leq n}\int\frac{x^2}{c_n^2+x^2}\,dF_k\to 0.$ $\qquad (n\to\infty)$

13.10 STABLE LAWS

Lévy first investigated limit laws of the normed sums $s_n/b_n - a_n$ of independent and identically distributed summands r_k with an arbitrary common d.f. F and a c.f. χ. These limit laws L_1 are the family of laws whose c.f.'s Φ are such that

$$e^{-iua_n}\chi^n(t/b_n)\to\phi(t). \qquad (b_n\to\infty)$$

Since $b_n\to\infty$, it follows that the r.v.'s r_k/b_n are infinitesimal; therefore, $L_1\subset L$. The problem of characterizing L_1 naturally arises; the solution to this problem requires the introduction of a new concept.

DEFINITION 1

The d.f. F (and its c.f.) is called *stable* if to every $a_1>0$, $a_2>0$, b_1, and b_2 there correspond constants $a>0$ and b such that

(1) $$F(a_1x+b_1)*F(a_2x+b_2)=F(ax+b).$$

The normal and improper laws are stable.

Within one type, all laws are stable or none is stable. Therefore, we can say that a type is stable if it contains all the compositions of the laws belonging to it.

A stable law is frequently defined as in Definition 2.

DEFINITION 2

The c.f. χ (and its d.f.) is called *stable* if, for arbitrary $b>0$ and $b'>0$, there exist finite numbers a and $b''>0$ such that

(2) $$\chi(b''t) = e^{iua}\chi(bu)\chi(b'u).$$

Denoting the c.f. of F by χ,

$$\int e^{iut}\, dF(au + b) = \exp(-itb/a)\chi(t/a). \qquad (a > 0)$$

Applying this last result, (1) reduces to

(3) $$\chi(t/a) = \exp(it(b/a - b_1/a_1 - b_2/a_2))\chi(t/a_1)\chi(t/a_2).$$

It is not difficult to verify from this equation that Definitions 1 and 2 are equivalent.

The following theorem derived by Lévy [21] shows the importance of stable laws.

THEOREM 1

A d.f. belongs to L_1 if and only if it is stable.

Proof

We only must consider nondegenerate d.f.'s, since degenerate d.f.'s are stable.

Assume that the d.f.'s of the normed sums

$$\sigma_n = (r_1 + r_2 + \cdots + r_n)/B_n - A_n$$

converge to a proper d.f. F. By Lemmas 9.1 and 9.2 there exists an $m = m(n)$ such that $B_m/B_n \to a_2/a_1$ for $0 < a_2 < a_1 < \infty$ as $n \to \infty$. Let b_1 and b_2 be arbitrary real constants. Consider the identity

(4) $$a_1(\sigma_n - b_1) + \frac{B_m}{B_n}a_1\left(\frac{r_{n+1} + \cdots + r_{n+m}}{B_m} - A_m - b_2\right)$$
$$= a_1\left(\frac{r_1 + \cdots + r_{n+m}}{B_n}\right) - A,$$

where

$$A = a_1 A_n + a_1 b_1 + \frac{B_m}{B_n}a_1 A_m + \frac{B_m}{B_n}a_1 b_2.$$

Since the d.f.'s of the sums σ_n converge to F, it follows that the d.f.'s of the first summand in (4) converge to $F(a_1^{-1}x + b_1)$. The d.f. of $[(r_{n+1} + \cdots + r_{n+m})/B_m - A_m - b_2]$ is the same as the d.f. of $\sigma_m - b_2$ since the r.v.'s r_i are identically distributed. Therefore, the d.f. of the second summand in (4) converges to $F(a_2^{-1}x + b_2)$. Accordingly, the d.f.'s of the left-hand side of (4) must converge to the proper d.f.

$$F(a_1^{-1}x + b_1) * F(a_2^{-1}x + b_2).$$

Hence the right-hand side of (4) must converge to the same limit; by Theorem 8.4.1, it must be of the same type as F. In other words, we have proved that F satisfies (1) and therefore F is stable.

Conversely, assume that F is stable. Let r_i ($i = 1, 2, 3, \ldots$) be independent and identically distributed r.v.'s with a common d.f. F. Then by (1), their sum $r_1 + r_2 + \cdots + r_n$ is distributed according to the law $F(a_n x + b_n)$. Therefore, the r.v. $a_n(r_1 + \cdots + r_n) + b_n$ has the d.f. F.

Since the stable laws are i.d., we can use Lévy's canonical form to characterize the logarithms of their c.f.'s. The main results were derived by Khintchine and Lévy [22]. Their proof requires the following lemmas.

LEMMA 1

If N is an increasing function that is not a constant function for $x > 0$, then $N(ax) = N(bx)$ for all $x > 0$ if and only if $a = b$. Here a and b are two fixed positive real numbers.

Proof

If $a = b$, then $N(ax) = N(bx)$. The converse is more difficult to prove. If N is not a constant function, then there exists a point $x_1 \neq 0$ such that $N(x_1 - h) < N(x_1)$ for all $h > 0$. Assume that $b > a$; then if h is sufficiently small, $(x_1 - h)/a > x_1/b$. Choose a real number y such that $x_1/b < y < (x_1 - h)/a$. Then $ay < x_1 - h$ and $by > x_1$; thus $N(ay) < N(by)$, which is a contradiction. Therefore, $a \geq b$. However, a similar argument shows that $a < b$ is also impossible; accordingly, $a = b$.

LEMMA 2

A function χ is a stable c.f. if and only if it is a normal law or

$$\log \chi(t) = i\gamma t + c_1 \int_{-\infty}^{0} \left[e^{itu} - 1 - \frac{itu}{1 + u^2}\right] \frac{du}{|u|^{1+\beta}} + c_2 \int_{0}^{\infty} [\qquad] \frac{du}{u^{1+\beta}}, \tag{5}$$

with $c_1 \geq 0$, $c_2 \geq 0$, and $0 < \beta < 2$.

Proof

Taking logarithms of (3) yields

$$\log \chi(t/a) = \log \chi(t/a_1) + \log \chi(t/a_2) + i(b/a - b_1/a_1 - b_2/a_2)t. \tag{6}$$

Since χ is i.d., it can be represented by Lévy's formula, (9.9). Expressing both sides of (6) in Lévy's form and simplifying the result as in the first part of Lemma 9.4, we conclude, because of the uniqueness of the representation of an i.d. law, that

$$\sigma^2(1/a^2 - 1/a_1^2 - 1a_2^2) = 0 \tag{7}$$

$$M(au) = M(a_1 u) + M(a_2 u) \qquad (u < 0) \tag{8}$$

$$N(au) = N(a_1 u) + N(a_2 u). \qquad (u > 0) \tag{9}$$

From equations (7)–(9), we will find all continuous increasing solutions $M \neq 0$ and $N \neq 0$ that satisfy $M(-\infty) = N(\infty) = 0$. We will only consider the function N, since the argument for M is similar.

Setting $a_1 = a_2 = 1$ in (9) we conclude that $N(a(2)u) = 2N(u)$ for some number $a(2)$; similarly,

$$N(u) + N(u) + N(u) = N(a(2)u) + N(u) = N(a(3)u).$$

By induction it follows that

$$N(au) = nN(u), \tag{10}$$

where $a = a(n)$. If $N \neq 0$, it follows from (10) that the range of N is $(-\infty, 0]$ since N is continuous. Hence for every u there exists a u_1 such that $(1/n)N(u) = N(u_1)$.[10] Defining $a(1/n)$ by $a(1/n)u = u_1$,

$$N(u)/n = N(a(1/n)u). \tag{11}$$

To see that we can choose $a(1/n)$ independent of the point u, note that equation (11) holds if and only if

$$N(u) = nN(a(1/n)u) = N(a(n)a(1/n)u).$$

Therefore, we only must set $a(1/n) = 1/a(n)$. For an arbitrary rational number p/q, we define $a(p/q) = a(p)\,a(1/q)$; then

$$(p/q)N(u) = pN(a(1/q)u) = N(a(p)a(1/q)u) = N(a(p/q)u). \tag{12}$$

Let λ_i be a sequence of rational numbers such that $\lambda_i \uparrow \lambda$, an arbitrary positive real number. Since a is a strictly decreasing function on the set of rational numbers, we can define $a(\lambda-)$ by

$$a(\lambda-) = \lim_{\lambda_i \to \lambda} a(\lambda_i).$$

Similarly, we can define $a(\lambda+)$. From equation (12), it follows that

$$\lambda N(u) = N(a(\lambda+)u) = N(a(\lambda-)u).$$

By Lemma 1, it follows that $a(\lambda+) = a(\lambda-) = a(\lambda)$. Thus for every $\lambda > 0$,

$$\lambda N(u) = N(a(\lambda)u), \qquad (u > 0) \tag{13}$$

where $a(\lambda)$ is a strictly decreasing continuous function of λ.

In order to determine the function N, we need the result that either N is different from zero everywhere or $N \equiv 0$. We establish this by a proof by contradiction. Suppose that for some $w > 0$, $N(w) = 0$, and for some $v > 0$, $N(v) \neq 0$. Let u_0 be the infimum of those u for which $N(u) = 0$. We know that the function N is continuous; hence $N(u_0) = 0$ and so

[10]Geometrically, the function $a(\lambda)$, in particular $\lambda = 1/n$, such that (13) holds, is defined as follows: Find the point u_1 corresponding to the intersection of the line $y = \lambda N(u)$ and the curve $N(u)$, and set $u_1 = a(\lambda)u$. We then show that it is possible to choose $a(\lambda)$ independent of u. Since N is increasing, it is obvious that $a(\lambda)$ is a strictly decreasing function.

$$0 = 2N(u_0) = N(a(2)u_0).$$

Since a is decreasing and $a(1) = 1, a(2) < 1$; therefore, $a(2)u_0 < u_0$, which yields $N(a(2)u_0) < 0$. We have reached a contradiction.

Finally, we determine the form of the function N, supposing that $N(u) \neq 0$ and $a(\lambda)$ takes all values between zero and infinity as λ varies in the interval $(0, \infty)$. Differentiating (13) (since N has right- and left-hand derivatives for each u) yields

$$\lambda\, dN(u)/du = a\, dN(au)/d(au).$$

Another application of (13) reduces this last equation to

$$N'(u)/N(u) = [a\, dN(au)/d(au)]/N(au).$$

Setting $u = 1$ and $N'(1)/N(1) = -\beta$ $(\beta < 0)$[11] leads to

$$dN(a)/N(a) = -\beta\, da/a.$$

Hence

$$N(a) = -c_2 a^{-\beta}, \tag{14}$$

where c_2 is a positive constant. Since $\int_0^\epsilon u^2\, dN < \infty$ for $\epsilon > 0$, the integral

$$\int_0^1 u^2\, dN = c_2\beta \int_0^1 u^{1-\beta}\, du$$

must converge, so $0 < \beta < 2$.

Analogously,

$$M(u) = c_1/|u|^{\beta_1}, \tag{15}$$

where $c_1 > 0$ and $0 < \beta_1 < 2$.

Setting $a_1 = a_2 = 1$ in (8) and (9) and substituting the solutions (14) and (15) in the resulting equations gives

$$a^{-\beta} = a^{-\beta_1} = 2. \tag{16}$$

Hence $\beta = \beta_1$. From (7), with $a_1 = a_2 = 1$, it follows that $\sigma^2(1/a^2 - 2) = 0$. Since $\beta \neq 2$, $1/a^2 \neq 2$ or (16) is contradicted; therefore, $\sigma = 0$. On the other hand, if $\sigma \neq 0$, $M = 0$ and $N = 0$ so (16) does not hold.

Summarizing our results, we see that the logarithm of the c.f. of a stable law is either

$$\log \chi(t) = i\gamma t - \sigma^2/2t^2, \qquad (\sigma > 0) \tag{17}$$

which yields the normal law, or of the form (5).

Conversely, it is easily verified that (5) and (17) give stable laws.

[11]It follows from (13) and the continuity of $a(\lambda)$ that N is a strictly increasing function; hence $N' > 0$.

Lévy's theorem follows directly from Lemma 2 by evaluating the integrals appearing in (5).

THEOREM 1

A function χ is a stable c.f. if and only if either

$$\log \chi(t) = i\gamma t - c|t|^{\beta}\left(1 + i\alpha \frac{t}{|t|} \tan \frac{\pi}{2}\beta\right) \tag{18}$$

or

$$\log \chi(t) = i\gamma t - c|t|\left(1 + i\alpha \frac{t}{|t|} \frac{2}{\pi} \log |t|\right) \tag{19}$$

for real γ and

$$c \geq 0 \qquad |\alpha| \leq 1 \qquad \beta \in (0, 1) \cup (0, 2).$$[12]

Proof

We must consider three cases.

1. $0 < \beta < 1$. Since the integrals

$$\int_0^{\infty} \frac{u}{1 + u^2} \frac{du}{|u|^{1+\beta}} \qquad \int_0^{\infty} \frac{u}{1 + u^2} \frac{du}{u^{1+\beta}}$$

are finite, (5) can be written as

$$\log \chi(t) = i\gamma_1(t) + c_1 \int_{-\infty}^{0} (e^{itu} - 1) \frac{du}{|u|^{1+\beta}} + c_2 \int_0^{\infty} (e^{itu} - 1) \frac{du}{u^{1+\beta}}. \tag{20}$$

Let $t > 0$ and set $-ut = v$ in the first integral and $ut = v$ in the second integral appearing in (20). Then (20) reduces to

$$\log \chi(t) = i\gamma_1 t + t^{\beta}\left[c_1 \int_0^{\infty} (e^{-iv} - 1) \frac{dv}{v^{1+\beta}} + c_2 \int_0^{\infty} (e^{iv} - 1) \frac{dv}{v^{1+\beta}}\right]. \tag{21}$$

We will evaluate the integrals appearing in the bracket by using Cauchy's theorem. We integrate along the closed contour formed by the positive halves of the real and imaginary axes and an arc centered at the origin of radius r. Letting $r \to \infty$,

$$\int_0^{\infty} (e^{iv} - 1) \frac{dv}{v^{1+\beta}} = \int_0^{i\infty} (e^{iv} - 1) \frac{dv}{v^{1+\beta}} = i^{-\beta} \int_0^{\infty} (e^{-y} - 1) \frac{dy}{y^{1+\beta}} = e^{-i\pi\beta/2}\Gamma(\beta),$$

where

$$\Gamma(\beta) = \int_0^{\infty} (e^{-y} - 1) \frac{dy}{y^{1+\beta}} < 0.$$

[12] β is called the (*characteristic*) *exponent* of the stable law.

Since the first integral in (21) is the conjugate complex of the second integral, it follows that

$$\int_0^\infty (e^{-iv} - 1)\frac{dv}{v^{1+\beta}} = \exp(i\pi\beta/2)\Gamma(\beta).$$

Hence for $t > 0$,

$$\log \chi(t) = it\gamma_1 + t^\beta\Gamma(\beta)[(c_1 + c_2)\cos(\pi\beta/2) + i(c_1 - c_2)\sin(\pi\beta/2)]. \tag{22}$$

Set

$$c = -\Gamma(\beta)(c_1 + c_2)\cos\left(\frac{\pi}{2}\beta\right)$$

$$\alpha = (c_1 - c_2)/(c_1 + c_2).$$

Since $\Gamma(\beta) < 0$, $c_1 \geq 0$, $c_2 \geq 0$, and $0 < \beta < 1$, it follows that $c \geq 0$ and $|\alpha| \leq 1$. We find that (22) reduces to

$$\log \chi(t) = i\gamma_1 t - ct^\beta\left[1 + i\alpha\tan\frac{\pi}{2}\beta\right].$$

For $t < 0$,

$$\log \chi(t) = \log \overline{\chi(-t)} = -i\gamma_1(-t) - c(-t)^\beta\left[1 - i\alpha t\tan\frac{\pi}{2}\beta\right]$$

$$= i\gamma_1 t - c|t|^\beta\left[1 + i\alpha\frac{t}{|t|}\tan\frac{\pi}{2}\beta\right].$$

Therefore, for every t, (18) is valid.

2. $1 < \beta < 2$. From the identity

$$\int_0^\infty \frac{u}{1+u^2}\frac{du}{u^{1+\beta}} = \int_0^\infty \frac{u\,du}{u^{1+\beta}} - \int_0^\infty \frac{u^3}{1+u^2}\frac{du}{u^{1+\beta}}$$

and the fact that the last integral is convergent, it follows, by changing the constant γ, that (5) can be written as

$$\log \chi(t) = it\gamma_2 + c_1\int_{-\infty}^0 (e^{itu} - 1 - itu)\frac{du}{|u|^{1+\beta}} + c_2\int_0^\infty [\qquad]\frac{du}{u^{1+\beta}}.$$

Exactly as in the first case, we can reduce the above expression to

$$\log \chi(t) = it\gamma_2 + t^\beta\left[c_1\int_0^\infty (e^{-iv} - 1 + iv)\frac{dv}{v^{1+\beta}} + c_2\int_0^\infty (e^{iv} - 1 - iv)\frac{dv}{v^{1+\beta}}\right] \tag{23}$$

for $t > 0$. By Cauchy's theorem, using the same contour of integration as in the first case, the integrals in (23) are

$$\int_0^\infty (e^{iv} - 1 - iv)\frac{dv}{v^{1+\beta}} = \exp\left(-\frac{\pi}{2}i\beta\right)\Lambda(\beta)$$

$$\int_0^\infty (e^{-iv} - 1 + iv)\frac{dv}{v^{1+\beta}} = \exp\left(\frac{\pi}{2}i\beta\right)\Lambda(\beta),$$

where

$$\Lambda(\beta) = \int_0^\infty (e^{-y} - 1 + y)\frac{dy}{y^{1+\beta}} > 0.$$

Setting

$$c = -\Lambda(\beta)(c_1 + c_2)\cos\frac{\pi}{2}\beta \qquad (c > 0)$$

and defining α as in part 1, we find that (18) holds.
3. $\beta = 1$. Since

$$\int_0^\infty \frac{\cos tu - 1}{u^2}\,du = -(\pi/2)t,$$

we find that for $t > 0$,

$$\int_{+0}^\infty \left(e^{itu} - 1 - \frac{itu}{1+u^2}\right)\frac{du}{u^2} = \int_0^\infty \frac{\cos tu - 1}{u^2}\,du + i\int_{+0}^\infty \left(\sin tu - \frac{ut}{1+u^2}\right)\frac{du}{u^2}$$

$$= -\frac{\pi}{2}t + i\lim_{\epsilon\to 0+}\left[-t\int_\epsilon^{\epsilon t}\frac{\sin u}{u^2}\,du + t\int_\epsilon^\infty \left(\frac{\sin u}{u^2} - \frac{1}{u(1+u^2)}\right)du\right].$$

The limit of the first integral is log t and the second integral is finite. Since the first integral in (23) is the complex conjugate of the second integral, it follows that for $t > 0$,

$$\log \chi(t) = i\gamma' t - (c_1 + c_2)\frac{\pi}{2}t + i(c_1 - c_2)t\log t.$$

For $t < 0$,

$$\log \chi(t) = \log \overline{\chi(-t)} = i\gamma' t - (c_1 + c_2)\frac{\pi}{2}|t| - i(c_1 - c_2)t\log|t|.$$

Setting $c = (c_1 + c_2)\pi/2$ and $\alpha = (c_2 - c_1)/(c_1 + c_2)$ yields (19).

The explicit form of the density function for stable laws is known only for a few values of the constants in (18) and (19). For $\beta = 2$ we have the normal law, and for $\beta = 1$ and $\alpha = 0$ we have the Cauchy law. Lévy [23] has proved that the stable law for which $\beta = \frac{1}{2}$, $\alpha = 1$, $\gamma = 0$, and $c = 1$ has a probability density f, where

$$f(x) = \begin{cases} 0 & (x < 0) \\ \dfrac{1}{\sqrt{2\pi}} e^{-1/2x} x^{-3/2}. & (x > 0) \end{cases}$$

EXERCISES

1. Let $\mathcal{X}$, with or without indexes, denote a c.f. (nondegenerate). The *type* of $\mathcal{X}$ is the family of all c.f.'s defined by $\mathcal{X}(ct)$ for some $c > 0$. Prove the following statements:

(a) $\mathcal{X}$ is decomposable by every $\mathcal{X}^n$ ($n = 2, 3, \ldots$) if and only if $\mathcal{X}$ is degenerate.
(b) $\mathcal{X}$ is decomposable and every component belongs to its type if and only if $\mathcal{X}$ is normal.
(c) If for a positive $s \neq 1$, $\mathcal{X}^s$ belongs to the type of $\mathcal{X}$, then $\mathcal{X}$ is i.d.
(d) If there are two values s' and s'' of s and if $(\log s')/(\log s'')$ is irrational, then $\mathcal{X}$ is stable.

2. If $\mathcal{X}(t) = \mathcal{X}(ct)\mathcal{X}_c(t)$, then $\mathcal{X}$ is said to be *c-decomposable*, where $c \in (0, 1)$. Let L_c be the family of all c-decomposable laws, L_0 the family of all laws, and L_1 the family of self-decomposable laws. Prove the following statements:

(a) $L_0 \supset L_c \supset L_1$. If $(\log c)/(\log c')$ is rational, then $L_c = L_{c'}$. L_c is closed under compositions and passages to the limit.
(b) $\mathcal{X} \in L_0$ if and only if it is the limit of a sequence of c.f.'s of normed sums s_n/b_n of independent r.v.'s, with $b_n/b_{n+1} \to c$.
(c) $\mathcal{X} \in L_c$ if and only if it is the c.f. of

$$r(c) = \sum_{k=0}^{\infty} r_k c^k,$$

where the law of the series converges and the r_k are independent and identically distributed. Then the series converges a.s. and $\mathcal{X}_{rk} = \mathcal{X}_c$. If r_k is bounded, then $\mathcal{X}$ is not i.d.

3. A d.f. F is *attracted* to a d.f. G if G is the limit d.f. of suitably normed sums of independent r.v.'s with a common d.f. F. The totality of d.f.'s attracted to G is called the *domain of attraction* of the law G. Verify the following statements:

(a) Only stable laws have nonempty domains of attraction.
(b) The domains of attraction of two laws belonging to the same type coincide.
(c) F is attracted by a normal law if and only if as $t \to \infty$,

$$\frac{t \int_{|x| \geq t} dF}{\int_{|x| < t} x^2 \, dF} \to 0.$$

(d) F is attracted by a stable law with a characteristic exponent β $(0 < \beta < 2)$ if and only if as $x \to \infty$,

$$\frac{F(-x)}{1 - F(x)} \to c \qquad \frac{1 - F(x) + F(-x)}{1 + F(kx) + F(-kx)} \to k^{\beta}$$

for every $k > 0$.
(e) If F is attracted by a stable law with $\beta \leq 2$, then $\int |x|^r \, dF < \infty$ for $0 \leq r < \beta$.

4. Consider the following simplified version of a problem of finding the gravitational field resulting from a random distribution of stars. Assume that n points of mass $m > 0$ are distributed uniformly and independently in the interval $[-n, n]$ of the real line.

(a) If the gravitational constant is unity show that the force F_n exerted on a unit mass at the origin is

$$F_n = m \sum_{i=1}^{n} r_i^{-2} \, sg(r_i),$$

where r_i is the coordinate of the ith point.
(b) Let χ_n be the c.f. of F_n. Show that

$$\lim_{n \to \infty} \chi_n(\lambda) = \exp(-c|\lambda|^{1/2}) \qquad (c > 0)$$

and hence that the limiting c.f. is stable.

13.11 LIMIT THEOREMS FOR DENSITIES

The question naturally arises of whether or not there are limit theorems, similar to those in the preceding section, for density functions. We begin the investigation of this problem by first considering discrete r.v.'s—in particular, lattice r.v.'s.

Recall that a discrete r.v. r has a lattice distribution if there exist numbers a and $b > 0$ such that all possible values[13] of r are representable in the form $a + kb$, where k runs through integral values (though not necessarily all integral values). The number b is called the *span* of the distribution.

The Bernoulli and Poisson distributions are examples of lattice distributions.

DEFINITION 1

If b is a distribution span and it is impossible to represent all possible values of r in the form $a + kb_1$ for any a $(-\infty < a < \infty)$, and if $b_1 > b$, then b is a *maximal distribution span*.

An example will illustrate the distinction between the notions of the maximal distribution span and the distribution span.

[13]Those values assumed with a positive probability.

Let r be a r.v. that takes on odd integral values. All values of r can be written in the form kb, with $b = 1$. However, the span b is not maximal, since all possible values of r may also be expressed in the form $a + kb_1$ with $a = 1$ and $b_1 = 2$.

The following condition needed for the span to be maximal also shows that there is a unique maximal distribution span.

LEMMA 1

A span b will be maximal if and only if the greatest common divisor of the differences of every pair of possible values of the r.v. r divided by b is equal to 1.

The straightforward proof of this lemma is left as an exercise.

Another criterion for a distribution span to be maximal is a consequence of the following lemma.

LEMMA 2

A r.v. r has a lattice distribution if and only if for some $t_1 \neq 0$, $|\chi(t_1)| = 1$, where χ is the c.f. of r.

Proof

If $|\chi(t_1)| = 1$, then $\chi(t_1) = e^{ia}$ for some real a. Therefore, $\psi(t_1) = \chi(t_1)e^{-ia} = 1$ and the result follows by applying Theorems 10.2.3 and 10.2.5 to the c.f. ψ.

If r is a lattice r.v., then $r = s + a$, where s is a r.v. that assumes the values kb. By Theorem 10.2.5, $\chi_s(t_1) = 1$ for some $t_1 \neq 0$. Since $\chi_r = \chi_r e^{ia}$, we see that $|\chi_r(t_1)| = 1$.

COROLLARY 1

A distribution span b is maximal if and only if

$$|\chi(t)| \begin{cases} < 1 & (0 < t < 2\pi/b) \\ = 1 & (t = 2\pi/b) \end{cases}.$$

Proof

By Theorem 10.2.5, $\chi(2\pi/b) = 1$ if and only if b is a distribution span. Suppose that b is a maximal distribution span and $b_1 > b$. If $t = 2\pi/b_1 < 2\pi/b$, then $|\chi(t)| < 1$; otherwise, b_1 would be a distribution span, which is impossible, since $b_1 > b$. On the other hand, suppose that $|\chi(t)| < 1$ for $0 < |t| < 2\pi/b$. Let $b_1 > b$; then $t = 2\pi/b_1 < 2\pi/b$ and so $|\chi(t)| \neq 1$, which implies that b_1 is not a distribution span. Therefore, b is a maximal distribution span. If b is a distribution span, it is maximal if and only if $|\chi(t)| < 1$ for $0 < t < 2\pi/b$.

COROLLARY 2

If b is a maximal distribution span, then for any $\epsilon > 0$, we can find a $c > 0$ for $c = c(\epsilon)$ such that

$$|\chi(t)| \leq e^{-c}. \qquad (\epsilon \leq t \leq 2\pi/(b - \epsilon))$$

This inequality follows immediately from Corollary 1.

Actually, it is sufficient to consider the maximal span for each lattice distribution. In fact, we can consider the maximal span to be 1, since the transformation $\bar{r} = (r - a)/b$ reduces the span to 1 and the constant a to 0. Hence it is sufficient to consider lattice r.v.'s that take only integral values k, where the greatest common divisor of all the differences $k' - k''$ for which $P(r = k') > 0$ and $P(r = k'') > 0$, is equal to 1.

Suppose now that $r_1, r_2, \ldots, r_n, \ldots$ are independent lattice r.v.'s having the same d.f. F. The sum

$$s_n = r_1 + r_2 + \cdots + r_n$$

is a lattice r.v. that also takes only integral values. Set

$$P_n(k) = P(s_n = k)$$

and set

$$P_1(k) = P(r_1 = k) = p_k.$$

We introduce the notation

$$z_{nk} = (k - A_n)/B_n,$$

where $A_n = Es_n = nEr_1$ and $B_n^2 = \operatorname{Var} s_n = n \operatorname{Var} r_1$.

The following generalization of the De Moivre–Laplace local limit theorem was proposed by B. V. Gnedenko.

THEOREM 1

Let $\{r_i\}$ be a sequence of independent identically distributed r.v.'s, each of which takes only integral values and has a finite mean and variance. Then the relation

$$B_n P_n(k) - 1/\sqrt{2\pi} \exp(-z_{nk}^2/2) \to 0 \qquad (n \to \infty) \tag{1}$$

holds uniformly in k for $-\infty < k < \infty$ if and only if the greatest common divisor of the differences of all the values of r_n having positive probabilities is equal to 1.

Proof

The necessity of the conditions of the theorem is clear. If the greatest common divisor d of the difference of the possible values of r_n is different from 1, then the difference between two consecutive possible values of s_n for $n = 1, 2, 3, \ldots$ cannot be less than d. Therefore, relation (1) cannot hold for all values of k.

The proof of sufficiency is more difficult. We first note that the condition concerning the greatest common divisor is equivalent to the fact 1 is a maximal distribution span.

The c.f. of r_k for $k = 1, 2, 3, \ldots$ is given by

$$\chi(t) = \sum_{k=-\infty}^{\infty} p_k e^{itk}$$

and the c.f. of the sum s_n by

$$\chi^n(t) = \sum_{j=-\infty}^{\infty} P_n(j) e^{itj}.$$

Solving for $P_n(k)$ by multiplying both sides of the last equation by e^{-itk} and integrating from $-\pi$ to π yields

$$2\pi P_n(k) = \int_{-\pi}^{\pi} \chi^n(t) e^{-itk}\, dt.$$

Replacing k in the above integral by $B_n z + A_n$ gives

$$2\pi P_n(k) = \int_{-\pi}^{\pi} \phi^n(t) e^{-itzB_n}\, dt,$$

where

$$\phi(t) = \exp(-itA_n/n)\chi(t).$$

Setting $x = tB_n$ gives

$$2\pi B_n P_n(k) = \int_{-\pi B_n}^{\pi B_n} e^{-izx} \phi^n(x/B_n)\, dx.$$

Since

$$1/\sqrt{2\pi}\exp(-z^2/2) = (1/2\pi)\int \exp(-izx - x^2/2)\, dx,$$

the difference

$$R_n = 2\pi[B_n P_n(k) - (1/\sqrt{2\pi})\exp(-z^2/2)]$$

can be represented as the sum of the four integrals

$$R_n = J_1 + J_2 + J_3 + J_4,$$

where

$$J_1 = \int_{-A}^{A} e^{-izx}[\phi^n(x/B_n) - e^{-x^2/2}]\,dx$$

$$J_2 = \int_{|x|>A} \exp(-izx - x^2/2)\,dx$$

$$J_3 = \int_{\epsilon B_n \le |x| \le \pi B_n} e^{-izx}\phi^n(x/B_n)\,dx$$

$$J_4 = \int_{A \le |x| < \epsilon B_n} e^{-izx}\phi^n(x/B_n)\,dx.$$

In the above integrals, $A > 0$ and $\epsilon > 0$ are fixed numbers whose precise values will be specified later.

By the corollary to Theorem 1.2, $\phi^n(t/B_n) \to e^{-t^2/2}$ as $n \to \infty$ uniformly for t in any finite interval. It follows that $J_1 \to 0$ as $n \to \infty$, no matter what the value of the constant A.

The inequality

$$|J_2| \le \int_{|x|>A} e^{-x^2/2}\,dx \le \frac{2}{A}\int_{A}^{\infty} xe^{-x^2/2}\,dx = \frac{2}{A}\,e^{-A^2/2}$$

shows that we can make J_2 arbitrarily small by choosing A sufficiently large.

Since 1 is a maximal distribution span, we apply Corollary 2 of Lemma 2 to ϕ, thereby obtaining the following inequality:

$$J_3 \le \int_{\epsilon B_n \le |x| \le \pi B_n} |\phi(x/B_n)|^n\,dx \le e^{-nc}2B_n(\pi - \epsilon).$$

From this it is clear that $J_3 \to 0$ as $n \to \infty$.

To estimate J_4, we expand $\phi(t)$ in the neighborhood of $t = 0$. Since the variance exists, we obtain

$$\phi(t) = 1 - \sigma^2 t^2/2 + o(t^2),$$

where $\sigma^2 = B_n^2/n$ (see Exercise 10.4.5, part (b)). If $|t| \le \epsilon$ and ϵ is sufficiently small, then

$$|\phi(t)| < 1 - \sigma^2 t^2/4 < \exp(-\sigma^2 t^2/4).$$

Hence for $|x| \le \epsilon B_n$,

$$|\phi(x/B_n)|^n < \exp(-n\sigma^2 x^2/4B_n^2) = \exp(-x^2/4).$$

Therefore,

$$|J_4| \le 2\int_{A}^{\epsilon B_n} \exp(-x^2/4)\,dx < 2\int_{A}^{\infty} \exp(-x^2/4)\,dx.$$

By choosing A sufficiently large, we can make the integral J_4 arbitrarily small. This completes the proof of the theorem.

We now turn to continuous r.v.'s. The following example shows that the d.f.'s of sums of independent r.v.'s can converge to a limit while the density functions of the sums do not converge.

EXAMPLE 1

Let r be a r.v. with a density function f defined by

$$f(x) = \begin{cases} 0 & (x \geq 1/e) \\ \frac{1}{2}\,|x| \log^2 |x| & (x < 1/e) \end{cases}$$

If r_i are independent r.v.'s having f as their common density, and

$$\sigma^2 = \int_0^{1/e} x/(\log^2 x)\, dx,$$

then as $n \to \infty$,

$$P((r_1 + \cdots + r_n)/\sigma\sqrt{n} \leq x) \to (1/\sqrt{2\pi}) \int_{-\infty}^{x} \exp(-t^2/2)\, dt,$$

since the conditions of Theorem 8.2 are satisfied. However, the probability densities of the sums

$$(r_1 + \cdots + r_n)/\sigma\sqrt{n}$$

do not converge to the normal density as $n \to \infty$.

The probability density of the sum $r_1 + r_2$ is

$$f_2(x) = \int_{-e^{-1}}^{e^{-1}} f(t) f(x - t)\, dt.$$

Assuming that $0 < x < e^{-2}$,

$$f_2(x) \geq \int_{-x}^{x} f(t) f(x - t)\, dt.$$

Since the minimum of the function $f(x - t)$ in the interval $0 \leq t < x$ is attained at the point $t = 0$, then

$$f_2(x) \geq (2x \log^2 x)^{-1} \int_{-x}^{x} (2|t| \log^2 |t|)^{-1}\, dt = (2x|\log^3 x|)^{-1}.$$

Similarly, the probability density f_3 of the sum $r_1 + r_2 + r_3$ satisfies the inequality

$$f_3(x) > c_3/(|x \log^4 |x||)$$

in a neighborhood of $x = 0$, where $c_3 > 0$ is a constant. By induction, it can be shown that the density f_n of the sum $r_1 + \cdots + r_n$ satisfies the relation

$$f_n(x) > c_n/(|x \log^{n+1}|x||) \qquad (c_n > 0)$$

in a neighborhood of $x = 0$.

Since for every n, the function f_n is infinite for $x = 0$, f_n cannot converge to the normal density under any normalization.

Various conditions under which density functions of sums converge to the density of the limit distribution appear in the Exercises. For further details, the reader should consult the monograph by Gnedenko and Kolmogorov [24].

Another approach to limit theorems that is based on the theory of linear operators and semi-groups can be found in [25]–[27].

EXERCISES

Prove the following theorems.

1. Let r_i be independent r.v.'s with a common density f and finite expectation and variance. Let the r.v.

$$s_n = (n \operatorname{Var} r_1)^{1/2} \sum_{k=1}^{n} (r_k - E_{r_k})$$

have a density $f_n(x)$ for $n \geq n_0$. Then

$$f_n(x) - (1/\sqrt{2\pi}) \exp(-x^2/2) \to 0$$

uniformly in x $(-\infty < x < \infty)$ as $n \to \infty$ if and only if there exists a number n_1 such that the function $f_{n_1}(x)$ is bounded.

2. Let $\{r_i\}$ be independent r.v.'s with a common density function f. If

(a) the probability density f_m $(m \geq 1)$ of the sum $r_1 + \cdots + r_m$ is integrable in the sth power $(1 < s \leq 2)$;

(b) the function $F(x) = \int_{-\infty}^{x} f(t)\, dt$ belongs to the domain of the stable law $G(x)$ with a characteristic exponent $\beta < 2$,

then the relation

$$B_n f_n(B_n x + A_n) - g(x) \to 0 \qquad (n \to \infty)$$

holds uniformly with respect to x in the interval $(-\infty < x < \infty)$, where $g(x) = G'(x)$ and the constants A_n and B_n are such that

$$P(((r_1 + \cdots + r_n) - A_n)/B_n \leq x) \to F(x). \qquad (n \to \infty)$$

REFERENCES

1. Chebyshev, P. L., "Sur deux théorèmes relatifs aux probabilités," *Acta Math*, **14** (1890), 305–315.
2. Markov, A., "The law of large numbers and the method of least squares," *Izv. Fiz.-Mat., Obshchestva Kazan. Univ.* (**2**), 8 (1899), 110–128.
3. Lyapunov, A. M., "Sur une proposition de la théorie des probabilités," *Bull. de l'Acad. Imp. des Sci. St. Petersbourg* (**5**), 13, No. 4 (1900), 359–386.
4. Berry, A. C., "The accuracy of the Gaussian approximation to the sum of independent variates," *Trans. Amer. Math. Soc.* **49** (1941), 122–136.
5. Esseen, C. G., "Fourier analysis of distribution functions. A mathematical study of the Laplace Gaussian Law," *Acta. Math.*, **77** (1945), 1–125.
6. Lindeberg, Y. W., "Eine neue Herleitung des Exponentialgesetzes in der Wahrscheinlichkeitsrechnung," *Math. Z.*, **15** (1922), 211–225.
7. Feller, W., "Über Zentralengrunzwertsatz der Wahrscheinlichkeitsrechnung," *Math. Zeit.*, **40** (1935), 521–559; **42** (1937), 301–312.
8. Bawly, G. M., "Über einige Verallgemeinerungen der Grenzwertsätze der Wahrscheinlichkeitsrechnung," *Rec. Math.* (*Mat. Sbornik*), N. S. 1 (43)(1963).
9. Khintchine, A. Y., "Zur Theorie der unbeschränkt teilbaren Verteilungsgesetze," *Rec. Math.* (*Mat. Sbornik*), N:S: 2 (44) (1937).
10. De Finetti, B., "Sulle funzioni a incremento aleatorio," *Rend. Accad. Naz. Lincei*, **10** (1929).
11. Kolmogorov, A., "Sulla forma generale di un processo stocastico omogeneo," *Atti. Acad. Naz. Lincei Rend. A. Sci. Fis. Matt. Nat.* (6) 15 (1932).
12. Lévy, P., "Sur les intégrales dont les éléments sont des variables aléatoires Independantes," *Annali R Scuola Norm. Sup. Pisa*, **3** (1934); **4** (1935).
13. Khintchine, A. Y., "Déduction nouvelle d'une formule de M. Paul Lévy," *Bull. Univ. d'Etat Moskow, Ser. Int. Sect. Al*, No. 1 (1937).
14. Gnedenko, B. V., "On the theory of limit theorems for sums of independent random variables," *Izvestiya Akad. Nauk SSSR. Ser. Math.*, 643–647 (1939), 181–232, 643–647.
15. Gnedenko, B. V., "Über die Konvergenz der Verteilungsgesetz von Summen voneinander unabhängigen Summanden," *C. R.* (*Doklady*) *Acad. Sci. U.R.S.S.* N. S., **18** (1938).
16. Gnedenko, B. V., "Limit theorems for sums of independent random variables," *Uspekhi Mat. Nauk*, **10** (1944). Translation 45, A. M. S., New York.
17. Khintchine, A. Y., "Zur Theorie der unbeschränkt teilbaren Verteilungsgesetze," *Rec. Mat.* (*Mat. Sbornik*) N. S., **2** (44) (1937).
18. Khintchine, A. Y., *Limit theorems for sums of independent random variables.* Moscow, Leningrad: Gonti, 1938.

19. Marcinkiewicz, J., "Sur les functions indépendantes II," *Fund. Mat.*, **30** (1938), 349–364.

20. Lévy, P., *Théorie de l'Addition des Variables Aléatoires.* Paris: Gauthier-Villars, 1937.

21. Lévy, P., *Calcul des Probabilités.* Paris: Gauthier-Villars, 1925.

22. Khintchine, A. Y., and P. Lévy, "Sur les lois stables," *C. R. Acad Sci.* (*Paris*), **202** (1936), 374–376.

23. Lévy, P., "Sur certains processus stochastiques homogènes," *Composit. Mathematica*, **17** (1940), 283–339.

24. Gnedenko, B. V., and A. N. Kolmogorov, *Distributions for Sums of Independent Random Variables.* Reading, Mass: Addison-Wesley, 1954.

25. Feller, W., *An Introduction to Probability Theory and its Applications*, Vol. 2, John Wiley and Sons, New York, 1966.

26. Hunt, G. A., "Semigroups of measures on Lie groups," *Trans. Amer. Math. Soc.*, **81** (1956) 264–293.

27. Trotter, F., "An elementary proof of the Central Limit Theorem," *Archiv. d. Math.*, **10** (1959), 226–234.

INDEX

D

E

F

G

H

I

J

K

L

N

O

P

R

S

T

U

V

W

Z